Mathematical Macroevolution in Diatom Research

Scrivener Publishing
100 Cummings Center, Suite 541J
Beverly, MA 01915-6106

Diatoms: Biology and Applications

**Series Editors: Richard Gordon (dickgordoncan@protonmail.com) and
Joseph Seckbach (Joseph.Seckbach@mail.huji.ac.il)**

Scope: The diatoms are a single-cell algal group, with each cell surrounded by a silica shell. The shells have beautiful attractive shapes with multiscalar structure at 8 orders of magnitude, and have several uses. 20% of the oxygen we breathe is produced by diatom photosynthesis, and they feed most of the aquatic food chain in freshwaters and the oceans. Diatoms serve as sources of biofuel and electrical solar energy production and are impacting on nanotechnology and photonics. They are important ecological and paleoclimate indicators. Some of them are extremophiles, living at high temperatures or in ice, at extremes of pH, at high or low light levels, and surviving desiccation. There are about 100,000 species and as many papers written about them since their discovery over three hundred years ago. The literature on diatoms is currently doubling every ten years, with 50,000 papers during the last decade (2006-2016). In this context, it is timely to review the progress to date, highlight cutting-edge discoveries, and discuss exciting future perspectives. To fulfill this objective, this new Diatom Series is being launched under the leadership of two experts in diatoms and related disciplines. The aim is to provide a comprehensive and reliable source of information on diatom biology and applications and enhance interdisciplinary collaborations required to advance knowledge and applications of diatoms.

Publishers at Scrivener
Martin Scrivener (martin@scrivenerpublishing.com)
Phillip Carmical (pcarmical@scrivenerpublishing.com)

Mathematical Macroevolution in Diatom Research

Janice L. Pappas

Department of Mathematics, University of Michigan, Ann Arbor, MI, USA

Scrivener
Publishing

This edition first published 2023 by John Wiley & Sons, Inc., 111 River Street, Hoboken, NJ 07030, USA and Scrivener Publishing LLC, 100 Cummings Center, Suite 541J, Beverly, MA 01915, USA
© 2023 Scrivener Publishing LLC
For more information about Scrivener publications please visit www.scrivenerpublishing.com.

Wiley Global Headquarters
111 River Street, Hoboken, NJ 07030, USA

For details of our global editorial offices, customer services, and more information about Wiley prod-ucts visit us at www.wiley.com.

Limit of Liability/Disclaimer of Warranty

Library of Congress Cataloging-in-Publication Data

ISBN 978-1-119-74985-1

Cover images: Tiling of 3D diatom surface models and partial diatom arrangement from the frontispiece by Janice L. Pappas
Frontispiece: An artful arrangement of 3D diatom surface models by Janice L. Pappas in the spirit of the art form of diatom arranging from the Victorian era, Klaus Kemp and Dr. Stephen S. Nagy
Cover design by Russell Richardson

Set in size of 11pt and Minion Pro by Manila Typesetting Company, Makati, Philippines

Printed in the USA

To Emmy, Jamie, Olaf, and a host of flying miracles, at once, lively and peaceful.

$$e^{i\pi} + 1 = 0$$

Leonhard Euler

"Writing is the geometry of the soul."

Plato

*"In all affairs it's a healthy thing now and then to hang a question mark
on the things you have long taken for granted."*

Bertrand Russell

*"One day the absurdity of the almost universal human belief in the slavery of other animals
will be palpable. We shall then have discovered
our souls and become worthier of sharing this planet with them."*

Martin Luther King, Jr.

Contents

List of Figures

Chapter 5: Cenozoic Diatom Origination, Extinction, Diversity

(girdle and valve). Row 7: *Tabularia fasciculata*. Row 8: *Cocconeis grovei*; *Surirella undulata*. *Actinoptychus* SEM from James M. Ehrman Mount Allison University, with permisson. *Arachnoidiscus ehrenbergii* SEM from Mary Ann Tiffany, with permission. *Cylindrotheca* sp. SEMs (Rows 3 and 4) from Shinya Sato, Fukui Prefectural University (SEMS taken at Royal Botanic Garden Edinburgh) with permission. *Entomoneis alata* (Row 4) and *E. grandis* nom. nud. SEMs are from Keigo Osada, The Nippon Dental University School of Life Dentistry at Niigata with permission. Light micrographs of *Campylodiscus* through *Tabularia* are from [11.9.57.].

Chapter 12: Diatom Phenotype Evolvability

List of Tables

Chapter 4: Diatom Adaptive Radiation

Chapter 5: Cenozoic Diatom Origination, Extinction, Diversity

Preface

Diatoms are ubiquitous siliceous shelled unicellular algae that are found in almost all environments, either free-living or attached to various substrates. They form the base of marine food webs and are one of the most important components of the carbon and silica cycles, primary productivity, and carbon sequestration. On the one hand, diatoms serve an important role as bioindicators of environmental and climate conditions, while on the other hand, they may be the source of toxic blooms or invasives forming dense mats and altering habitats. Humans cannot live without diatoms as they are important in nanoengineering in the health sciences, biofuel production, forensics analysis, and diatomaceous earth applications and manufacturing.

In diatom studies, the application of mathematics is useful in describing the variety of biological issues to be addressed. Mathematics has enabled us to extended our ability to delve into complicated biological problems in our quest to develop hypotheses, obtain insight into theory, formulate questions, or find solutions. Modeling, simulation, and other means of compiling our understanding and representation of processes, patterns and experimental or observational findings is expressed in our desire to use and apply mathematics. Computational complexities and the bases for how we devise more systematic, generalized or recurrent themes in biology are dependent upon mathematics. In diatom research, all of the previous sentiments are applicable, and at this juncture in such studies, there is no more urgent need for the application of mathematics.

Mathematical macroevolution is the study of macroevolution using mathematics more widely unlike those studies in macroevolution which have only a statistical framework. By mathematical macroevolution, I mean that geometric, algebraic or combinatorial bases or analyses are used. This approach does not preclude using statistics, but rather, the emphasis is on finding mathematical relations among variables or parameters using algebraic or combinatorial statements and applying statistical methods as well. Mathematics from such an approach opens up the opportunities to study various aspects of the characteristics, processes, and tempo and mode of biological existence that affect and describe diatoms.

Each chapter in this book was written as a complete original research study using the familiar framework given in scientific journals. Having said this, each chapter is also an introduction to the application of geometric, algebraic or combinatorial methods so that much time is spent on developing the mathematical methodology as well as obtaining specific biological results. Toward the end of each chapter, the "Discussion" and "Summary and Future Research" sections include information on how the developed methods may be used in additional research along the same lines or in application to other kinds of research. The multitude of possibilities awaits all of us in diatom research.

"*Mathematical Macroevolution in Diatom Research*" is the next volume in the series Diatoms: Biology & Applications by series editors: Richard Gordon & Joseph Seckbach:

Volume 1: *Diatoms: Fundamentals and Applications*
Volume 2: *Diatom Gliding Motility*
Volume 3: *Diatom Morphogenesis*
Volume 4: *Diatom Microscopy*
Volume 5: *The Mathematical Biology of Diatoms*

Acknowledgments

My heartfelt thanks go to all of those individuals who helped me to complete this book. No person can complete such a project without the contributions of others. For the contribution of light micrographs and scanning electron micrographs, I give many thanks to Mary Ann Tiffany (Chapters 2, 11), Amy Leventer (Chapter 4), Isabel Dove (Chapter 4), Gastón Almandoz (Chapter 4), James Ehrman (Chapter 11), Shinya Sato (Chapter 11), and Keigo Osada (Chapter 11). I also want to thank Sara Spaulding for permission to use information and images with specific attributions of individual contributors from the "Diatoms of North America" website (Chapters 1, 3), permission to use a digital image of *Licmophora* by Pauli Snoeijs from the "Nordic Microalgae and Aquatic Protozoa" website (Chapter 1), Mark Edlund for permission to use a digital image of *Surirella*, (Chapter 1), permission to use a digital image of *Grammatophora* by Evelyn Gaiser from the STRI Data Portal website, and Marina Montresor for permission to use information and drawings from the "Diatom New Taxon File at the Academy of Natural Sciences (ANSP)" (Chapter 1). Although I did not explicitly include the aforementioned images and drawings in Chapter 1, I did use them in analyses in obtaining shape outlines nonetheless. You will find acknowledgements of the fine work of these contributors throughout the book. What would we do without digital images and drawings of diatoms for research? Not much. What would we do without the diatom databases and compilations of data? Again, not much. Many thanks for your contributions.

My mentor, Professor Eugene Stoermer, told me many years ago that I should write a "how to" book for diatom researchers. At the time, I rolled my eyes and scoffed at the notion. Although he is not alive to see this book, I am sure if he were here, he would be laughing and saying, "See, I told you so." Because of his apparent clairvoyance and along with Bobbie (Barbara) Stoermer, I want to express my gratitude for many years of excellent conversation and support.

Additionally, my gratitude goes to the series editors, Richard Gordon and Joseph Seckbach, for the invitation and opportunity to write the volume, the typesetters Manila Typesetting Company, Philippines, who tirelessly perfect the art of painstaking tedium satisfying the author and publisher alike, and Martin Scrivener for his expertise and support in facilitating the publishing of this volume.

Finally, I would like to give my personal, special thanks to those individuals for their support, inspiration and dedication in helping me to complete this project. My family and close friends have been a source of inspiration and encouragement that is immeasurable during the three years that is took for me to complete this project. Rohn Federbush inspired me with her creativity as a published writer, and Professor Paul Federbush inspired me with his mathematical mastery at the cluster expansion level. My dear, longtime friend Cathy Zawacki, who bought me a laptop when I could least afford it and with her ongoing support, made it possible for me to do the necessary work for the book. My dear husband Mark Brahce, who kept me in good spirits, health and delectable meals with his love and caring, was undoubtably my bedrock for success.

Janice L. Pappas
University of Michigan
February, 2023

Prologue – Introductory Remarks

Mathematics applied to any field should be an expectation after many years of empirical, experimental and descriptive research. The maturation of any discipline from the natural to social to political to economic sciences inevitably veers toward mathematical applications to resolve ever more intricate and complex problems. Interdisciplinary and multidisciplinary studies are becoming increasingly the norm in that mathematical applications must be instituted via technological means to obtain results. Such studies in themselves exemplify a level of complexity that is difficult to comprehend and deal with by individual researchers with backgrounds in a single discipline. Such disciplines as computer science, pattern recognition, computer vision, image processing and analysis, and complex systems necessarily induce a more encompassing approach. With this in mind, this book was written.

Diatom research is a good example of a relatively small number of individuals studying and covering a large subject area that have to navigate all the ramifications of the multitude of subdisciplines of the biological sciences. I identify the biological sciences to be those disciplines that cover both extant, extirpated and extinct forms of life. Despite diatoms being unicells, they are one of the more complex organisms on Earth, and studying them requires being mindful of all biological scientific disciplines and applications therein. Many studies in diatom research are focused on systematics and taxonomy or applications in aquatic ecology and limnology. While many questions remain concerning these topics with respect to diatoms, other areas of the biological sciences are no less important. In fact, while many studies of diatoms are concerned with species to population level research, macroevolutionary studies are of great importance in developing an understanding of the place of diatoms within a broader picture of life on Earth. With the complexities of the life of organisms such as diatoms, and their life in the geological past, diatom research is poised to go in directions that has heretofore not been considered very widely.

In more recent times, diatoms have been recognized for their importance in biogeochemical cycles as well as food webs, and as such, could be called one of the "directors of our future." As more and more studies are published on the important role of diatoms, there is an even larger need to recognize that diatoms must be seen as the basis of studies involving complex systems that demand the application of mathematical and technological methodologies. In this book, diatoms provide the main interest in macroevolutionary studies ranging from morphological evolution to adaptive radiation to origination-extinction-diversity events to biogeography to complexity, symmetry and evolvability. The mathematics and technological techniques applied in these instances are necessary in order to elucidate the intricacies of such topics in diatom research.

Throughout the biological sciences, macroevolution has many definitions. In neontology, and paleontology, macroevolution entails organismal processes across multiple classification levels but not including population studies of species-level research. Macroevolution includes studies on such topics as origination and extinction events, tempo and mode of evolution, bauplans, morphospace analysis, and evolutionary mechanisms and processes. In this book, macroevolution is the study of evolution at the level of the diatom genus or above in classification systems and includes many of the topics of concern listed above and applied to either Recent or fossil diatom taxa or a combination of both. In some cases, diatom species are identified, but they are used in a macroevolutionary sense, representing larger classification levels such as the genus, clade or phylum.

This book is arranged in sections covering four categories of macroevolutionary studies. The first section is about morphological analysis, while the second section deals with analysis of larger scale systems of processes both temporally and spatially. Characteristics of macroevolutionary processes covers section three, and characteristics of morphological macroevolution encompass the fourth section. The overall view of the book is a tour of the topics in macroevolution that should be of interest to those in diatom research.

Each chapter is a complete original research study and should be approached as such. The mathematical notation used is consistent only within each chapter. One should not assume that a given variable or parameter from one chapter is identified as such throughout other chapters of the book. Each chapter's content varies with regard to the mathematics presented. In some cases, there is very little in the way of explicit mathematical statements in a given chapter, while in other cases, much of a particular chapter is elucidated mostly with equations. For one approach to using this book, one could use only the concepts elucidated in each chapter to get an idea of how a given topic was studied to gain some understanding of the results. For those who are interested in and familiar with the mathematics, one could approach each chapter as being directly explicated by the mathematics to understand the results. Each subsequent equation follows from the previous one so that the reader can follow the train of thought in explication and evaluation. No matter how one approaches the content of each chapter, the concepts should be readily extracted in order to see how the results were obtained.

Topics in diatom macroevolutionary studies were covered mathematically in various ways, including the following examples. Morphological analysis was presented in terms of a statistical distribution (Chapter 1) or multivariable analyses (Chapters 8, 10 and 11). Differential, partial differential, delay, or stochastic differential equations figured prominently in studies of adaptive radiation (Chapter 4), origination-extinction-diversity (Chapter 5), food web dynamics (Chapter 6), life cycle dynamics (Chapter 8), and evolvability (Chapter 12). Specialized areas used include probability theory in adaptative radiation (Chapter 4), combinatorics and optimization in diatom biogeography (Chapter 6), networks in a morphospace analysis (Chapter 8), ergodic and information theory in complexity (Chapter 9), and knot theory in symmetry (Chapter 10). Robotics, computer vision, or machine learning were used in surface analyses, (Chapters 2 and 3) as such applications were used to start off the book, while dynamical systems, group theory, and differential geometry (Chapters 10, 11 and 12) were the bases of studies that finished the book. As you can see, these are but a few ways to view the chapters of this book and glean information that may be used in areas of diatom research.

At the end of each chapter, the references provide a wide range of topics covered. In some cases, original research from the distant past is cited, while in other cases, the most recent publications from the literature are presented. You will notice that many fields of research are represented. References used from disciplines other than those typically found in the biological sciences cover topics mentioned above as well as those in medical statistics, computational physics and finance. The reference lists for each chapter provide a useful bibliographic introduction to the multidisciplinary nature of conducting mathematical studies in diatom macroevolution.

While this is a mathematically-based book, there was no attempt to provide theorems, lemmas, conjectures, proofs or any other formal mathematical discourse to the text. Instead, it is presumed that anyone with sufficient mathematical background could produce sufficient formal mathematical elaboration on what is presented in each chapter. As a result, each chapter provides a direct progression of mathematical expressions to provide the reader with a step-by-step walk from start to end, arriving at the actual equations used in obtaining numerical results. For the most part, computer coding of the expressions is developed and used but not explicitly stated because such coding could be recovered by anyone with sufficient background in computer science. The exception is Chapter 1 where the complete computer program is supplied to calculate Legendre coefficients from Legendre polynomials and the original and reconstructed outlines of diatoms.

And, especially, the intent of this book is to provide a more analytical way of thinking about topics in the complexities of diatom research. This book should provide sufficient explication for the reader so that enough information is obtained to develop a satisfactory understanding of the text. No matter what is one's background, it is hoped that this book will be used by anyone engaged in diatom research.

Part I

MORPHOLOGICAL MEASUREMENT IN MACROEVOLUTIONARY DISTRIBUTION ANALYSIS

1

Diatom Bauplan, As Modified 2D Valve Face Shapes of a 3D Capped Cylinder and Valve Shape Distribution

Abstract

Development and life cycle are understandable via the role of the bauplan in macroevolution studies. The diatom bauplan is indicative of the unique geometric shapes that exemplify the phylum Bacillariophyta. Valve shapes are geometrically quantifiable using shape analysis via orthogonal polynomial regression, and the coefficients extracted are useful as shape descriptors. Using a set of 41 different diatom valve shapes, outline shape analysis was conducted using Legendre polynomials to calculate Legendre shape coefficients which have been shown to be biologically interpretable. To determine the representative valve shape outline, the minimum number of Legendre coefficients was used to identify each shape. Along with basic geometric shapes, Legendre shape bins were constructed. Then, 418 diatom valve shapes were tabulated via selection of taxa from the Catalogue of Diatom Names and verified by the ANSP Diatom New Taxon File website. Taxa were binned with regard to the Legendre coefficient number that represented that taxon's shape. Diatom valve shapes were found to be hypergeometrically distributed and representative of the probability of occurrence as recurrent forms with respect to the shape bins. Valve shape distribution may be useful in evolution and micropaleontologic studies.

Keywords: Bauplan, valve face, shape analysis, Legendre polynomials, orthogonal coefficients, hypergeometric distribution, probability of occurrence, recurrent forms

1.1 Introduction

An important component in understanding the role of development in macroevolution is that of the bauplan or body plan of an organism. A bauplan encompasses the anatomical features or morphological characteristics commons to a phylum of organisms that occurs at some point during their life cycle [1.7.83.]. A phylum is a taxonomic level of classification according to sharing common anatomical features or morphological characteristics and is a monophyletic group of organisms [1.7.83.]. Commonly, the concept of a bauplan is applied to multicellular organisms, and usually this is animals; however, bauplan has been studied in plants as well (e.g., [1.7.13., 1.7.22.]).

Unfortunately, unicells are understudied with respect to the concept of a bauplan (e.g., [1.7.46.]). As pigmented protists, eukaryotic unicells such as diatoms have animal and plant features. The "anatomical" features and morphological characteristics of diatoms make them unique. No other unicell looks like a diatom, especially when considering their vegetative cell stage which is most often seen. Sometimes, diatoms are considered to be a phylum in their own right as Bacillariophyta [1.7.15., 1.7.48.] in taxonomic classification schemes [1.7.67.]. They are also considered to be a monophyletic group (e.g., [1.7.87.]), and because diatoms constitute an important and informative group with respect to evolution, they qualify for study of their bauplan.

Comparison to the geometry of other unicellular forms is helpful in finding the kinds of shapes that are not found in diatom valves. As one example, *Scenedesmus* forms have pairs to multiple sickle-shaped cells that are attached together. Another example, including *Staurastrum*, *Cosmarium* and other desmids, have very different lobed and spiked geometric shapes compared to diatoms. Radiolarians and silicoflagellates are other unicells that have rippled, conical shapes and open-woven, spiked forms, respectively, that cannot be mistaken for diatom forms.

The diatom vegetative cell with amorphous silica valves (or frustules) is striking in its glassy, rigid and ornamented appearance. Diatom frustule shape does not have multiple sharp edges or multiple convexities on its edges or over the surface. The lack of convexities with respect to valve shape is not necessarily the case concerning ornamentation on the frustule surface as this may occur as sharp edges but on a nanoscale. Morphological characters such as spines or other protrusions or pores with circular to hexagonal edges are present even in the range of the nanoscale. External valve shape may more importantly reflect life history and ecological niche more than purely species differentiation.

Constraints that dictate diatom valve shape are at work including the formation of new cells within the mother cell during vegetative reproduction and size diminution. Restrictions on valve shape are a function of genetic factors [1.7.47.]. Besides reproductive requirements, additional constraints for recurrent forms include frustule function, habitat and substrate shape, propensity for colonial formation and buoyancy requirements, or predation resistance.

Interlocking spatulate spines or cell adhesion of colonial forms may work in concert with valve shape in maintaining the colony to achieve buoyancy and prevent sinking (e.g., [1.7.61., 1.7.68.]). This has repercussions with respect to possible predation avoidance [1.7.30.]. Another constraint to consider is substrate attachment and valve shape versus substrate shape. Epilithic, epipsammic, epizoic, and epiphytic diatom valve shape may reflect their conformability to the substrates as the life habits of such diatoms, and stalk attachment may be related to diatom valve shape as well (e.g., [1.7.64.]). While considering size, colonial formation and spines, predation resistance as a function of valve shape may be evident in those diatom taxa that are scraped off of substrate by specific predators such as gastropods (e.g., [1.7.66.]) and polychaetes [1.7.80.] or crushed by herbivorous copepods that have silica coated mandibles (e.g., [1.7.77.]).

Diatoms with their great morphological diversity provide a unique opportunity to study the geometry of their morphology at the level of the phylum. The diatom bauplan consists of valves exhibiting the concavity of a Petri dish or pill box (Figure 1.1) as they fit together to form the diatom vegetative cell. This design presents a considerable constraint on the evolutionary pathway of the diatom as a unicell to any potential for multicellular differentiation [1.7.64.]. The geometry of the diatom bauplan is most simply put as a capped cylinder where changes in the sides of the cylinder are reflected in shape changes parallel to the edge of the two-dimensional (2D) capped end (Figure 1.2). Changes in 2D valve shape may be used to represent three-dimensional (3D) frustule changes geometrically.

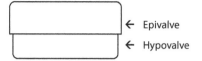

Figure 1.1 Simple drawing of a diatom cell with labeled valves in a Petri dish or pill box configuration.

Figure 1.2 Examples of diatom cells as capped and modified capped cylinders. From left to right, circular cell such as *Cyclotella*, modified square cell such as *Perissonoë*, and modified linear oval cell such as *Didymosphenia geminata*.

Because of their geometrically intricate, siliceous frustules, diatoms exhibit a variety of patterns and ornamentation [1.7.64.]. As a result of nanoscale silica spheres being deposited [1.7.24.–1.7.26.] with an arrangement in closely packed perforated layers, diatom frustule structure may be described as a sort of scaffolding configuration providing strength yet lightweight quality to the cell [1.7.53.]. However, it is because of their siliceous frustules that diatom geometry may be analyzed to study their bauplan configurations. Valve shape diversity constraints may be embodied in the kinds of geometric configurations of diatom forms found throughout geologic time. That is, diatom valve shape geometry may be used to study recurrent forms [1.7.57.] and which forms, if any, are prevalent throughout geologic time.

Diatom valve shapes change in subtle ways that depict combinations of multiple basic geometries. To analyze these forms, valve shape may be quantified. There are various means to accomplish this, but to include the geometric characteristics of shapes that can be binned and further analyzed, valve outline will be quantified using orthogonal polynomial regression. In this regard, the method using Legendre polynomials [1.7.55., 1.7.57.] has proven to be useful in obtaining geometric description as well as binning characteristics of valve outline shape, and interpreting the biological meaning of valve shape geometry is useful in binning recurrent forms as well.

1.1.1   Analytical Valve Shape Geometry

The cap of a cylinder is a circle. Geometrically, a generalized circle may be expressed in Cartesian coordinates as the quadratic equation

$$ax^2 + by^2 + cx + dy + e = 0 \qquad (1.1)$$

where a, b, c, d, e are coefficients and x and y are variables. The unit circle is $x^2 + y^2 < 1$ with boundary $x^2 + y^2 < 1$ [1.7.81.]. In polar coordinates, $x = r \cos \theta$ and $y = r \sin \theta$, and a circle is represented by radius, $r = \sqrt{x^2 + y^2}$, and angle, $\theta = \tan^{-1}\left(\dfrac{y}{x} \right)$ (Figure 1.3). To modify

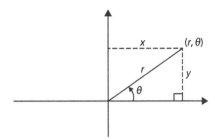

Figure 1.3 Diagram of Cartesian coordinates x, y and their relation to polar coordinates r, θ.

a circle, therefore we need to consider polynomials, that when changed, change the circle into the shapes of diatom valves.

We look for particular polynomials to achieve our goal. Such polynomials are harmonic, and additionally they are homogeneous [1.7.81.]. That is, we seek polynomials that are expansions of periodic functions $x = r \cos \theta$ and $y = r \sin \theta$ with respect to $x^2 + y^2$ that is homogeneous in degree two. Laplace's equation in the unit circle qualifies and is

$$\nabla^2 = \frac{\partial^2 u}{\partial r^2} + \frac{1}{r}\frac{\partial u}{\partial r} + \frac{1}{r^2}\frac{\partial^2 u}{\partial \theta^2} = 0 \tag{1.2}$$

as an elliptic second-order partial differential equation. A solution of $u(r, \theta)$ that is harmonic and satisfies Laplace's equation is

$$\sum_{1}^{\infty} r^n (a_n \cos n\theta + b_n \sin n\theta)(n(n-1) + n - n^2) \tag{1.3}$$

where a_n and b_n are nth coefficients. According to Sturm-Liouville theory, when a polynomial function is equal to zero at a or b or both and that function is positive for $a < \cos \theta < b$ [1.7.33., 1.7.81.], a case is defined having singular points that are not defined as boundaries, but the function remains continuous and is given as

$$\frac{d}{dx}\left[(1-(\cos\theta)^2)\frac{dy}{dx}\right] + \lambda y = 0 \tag{1.4}$$

where $-1 \leq \cos \theta \leq 1$ [1.7.33.]. Solution of this case is the Legendre polynomials, $P_n(\cos \theta)$, and $\lambda_n = n(n + 1)$. Laplace's equation in terms of polar coordinates may be solved via separation of variables, and the solutions contain Legendre polynomials in even or odd degrees that are expressed as $\int_0^\pi P_n(\cos\theta)^2 \sin\theta d\theta$ [1.7.81.]. The zeroth Legendre polynomial is $P_0(x) = 1$ and represents a circle.

More generally, by separation of variables,

$$x = r^m \sin m\theta \cos m\varphi \tag{1.5}$$

$$y = r^m \sin m\theta \sin m\varphi \tag{1.6}$$

are polynomial solutions in x and y of Laplace's equation, and

$$r^n P_n^m(\cos\theta)\cos m\varphi \tag{1.7}$$

$$r^n P_n^m(\cos\theta)\sin m\varphi \tag{1.8}$$

are homogeneous polynomials in terms of the Legendre polynomials, P_n^m [1.7.81.]. Changing the shape of a circle via the Legendre polynomials is thus achieved.

For additional successive Legendre coefficients, the nth degree Legendre polynomial given using Rodrigues formula is

$$P_n^m(\cos\theta) = \frac{1}{2^n n!}(1-(\cos\theta)^2)^{m/2}\frac{d^{m+n}}{d(\cos\theta)^{m+n}}((\cos\theta)^2 - 1)^n \tag{1.9}$$

where $n \geq m$ and $P_n^m(\cos\theta)$ is degree n in $\cos\theta$ and $\sin\theta$ [1.7.81.]. These associated Legendre polynomials are eigenfunctions and occur as even or odd with respect to whether $n - m$ is even or odd [1.7.81.]. If n is not an integer, but finite points, then the polynomials are Legendre functions of the first kind [1.7.82.]. Legendre functions of the second kind are

$$Q_n^m(\cos\theta) = (-1)^m(1-(\cos\theta)^2)^{m/2}\frac{d^m}{d(\cos\theta)^m}Q_n(\cos\theta) \tag{1.10}$$

where

$$Q_n(\cos\theta) = \frac{1}{2}P_n(\cos\theta)\ln\frac{1+(\cos\theta)}{1-(\cos\theta)} - \sum_{m=1}^{n}\frac{1}{m}P_{m-1}(\cos\theta)P_{n-m}(\cos\theta) \tag{1.11}$$

[1.7.1.], and this function provides solutions to the Legendre polynomials at singular points $\cos\theta = \pm 1$ [1.7.81.].

Legendre functions of the first and second kind are solutions to the Legendre differential equation which is expressed as

$$\frac{d^2 y}{d\theta^2} + \frac{\cos\theta}{\sin\theta}\frac{dy}{d\theta} + n(n+1)y = 0 \tag{1.12}$$

and with n as an integer, solutions to the Legendre differential equation are the Legendre polynomials [1.7.1.]. This second order ordinary differential equation is related to the hypergeometric differential equation

$$t(1-t)\frac{d^2 y}{dt^2}+(1-2t)\frac{dy}{dt}+n(n+1)y=0 \tag{1.13}$$

where $t=\dfrac{1-\cos\theta}{2}$ for singular points +1 and ∞, and transformation of t takes the form of $1 \to 0$, $-1 \to 1$ and $\infty \to \infty$ for the singular points to satisfy the equation [1.7.23., 1.7.82.]. Legendre function

$$P_n(\cos\theta)=F\left(-n,n+1;1;\frac{1-\cos\theta}{2}\right) \tag{1.14}$$

and

$$F(\alpha,\beta;\gamma;t)=\frac{n!}{(\alpha+1)_n}P_n^{(\alpha,\beta)}(1-2\cos\theta) \tag{1.15}$$

[1.7.1.] where $\alpha = -n$, $\beta = n+1$, and $\gamma = 1$ is the hypergeometric differential equation with respect to t in terms of the Legendre polynomial, P_n [1.7.23.].

Hypergeometric functions occur as a result of expansion of the Gauss hypergeometric series so that

$$F(\alpha,\beta;\gamma;t)=\sum_{n=0}^{\infty}\frac{\alpha_n\beta_n}{\gamma_n}\frac{t^n}{n!} \tag{1.16}$$

for $|t| < 1$, and

$$F(-m,\beta;\gamma;t)=\sum_{n=0}^{m}(-1)^n\binom{m}{n}\frac{\beta_n}{\gamma_n}t^n \tag{1.17}$$

for α, β as non-positive integers [1.7.52.]. For the Legendre polynomials, $P_n(x)$,

$$F(-m, \beta; \gamma; t) = F(-n, n+1; 1; x) = P_n(1-2x) \tag{1.18}$$

[1.7.1.] and $x = \cos\theta$. From this, coefficients may be obtained to characterize the successive additions in a hypergeometric series, and they are related to Legendre coefficients via Legendre polynomials.

1.1.2   Valve Shape Constructs of Diatom Genera

Starting off, the geometry of some of the valve face shapes follows recognizable basic geometries (Table 1.1) as found in the Stuart R. Stidolph Diatom Atlas [1.7.74.] and Schmidt's Atlas [1.7.65.]. Measurement of more complicated valve shape outlines using Legendre

Table 1.1 Basic geometries and diatom examples given for each geometric valve shape. For identifications and more information, see [1.7.65.] and [1.7.74.].

Geometric shape	Diatom valve face shapes - examples
⬭	*Aulacoseira granulata* *Cyclotella* *Stephanodiscus*
◻ ⬜ ✧	*Amphitetras subrotundata* *Biddulphia subjuncta* *Triceratium favus* fo. *quadrata* *Triceratium pentacrinus* fo. *quadrata* *Triceratium scitulum* *Trigonium arcticum* fo. *quadrata*
△	*Triceratium favus* *Triceratium reticulum* *Trigonium alternans* *Trigonium arcticum* var. *kerguelense*
⬠ ☆ ⬠	*Triceratium pentacrinus* *Trigonium arcticum* var. *pentagonalis* *Trigonium formosum* var. *pentagonalis* *Trigonium graeffeanum* *Biddulphia biddulphiana*
✡ ⬡	*Stictodiscus parallelus* fo. *hexagona* *Trigonium arcticum* var. *sexangulatum* *Hydrosera whampoensis*
⬡	*Triceratium septangulatum*
⯃	*Aulacodiscus recedens*
⬡	*Triceratium campechianum*

polynomials results in shape coefficients which may be ordinated in a morphospace using principal components analysis (PCA) to find species group delineations (e.g., [1.7.55., 1.7.57., 1.7.75.]). Unlike other shape outline techniques, Legendre polynomials are useful in interpreting biological meaning in valve shape constructs (e.g., [1.7.57.]). This is a result of the way in which such polynomials are used to reconstruct steps in the modification of a capped cylinder, and more specifically, modifications in the boundary of a circle.

Background on Legendre polynomials and coefficients used in shape analysis may be found in Stoermer and Ladewski [1.7.75.] and Pappas and Stoermer [1.7.55.]. Derivation of the method in terms of separation of variables, Sturm-Liouville Theory, as well as orthogonality and least-squares approximation may be found in Pappas and Stoermer [1.7.55.], and relation to the Fourier transform may be found in Pappas *et al.* [1.7.57.].

Legendre polynomials are a result of expansion of the Legendre-Fourier series [1.7.57.]. Legendre polynomial curve fitting involves using the Frobenius method [1.7.82.] to obtain the width function, *W*, as

$$W(x) = \sum_{n=0}^{N} b_n P_n(x) \tag{1.19}$$

where the *n*th to *N* polynomials, *P*, and coefficients, *b*, are expanded in a summation in a power series (Table 1.2). Using the width function, expansion of a diatom valve shape is perpendicular to the apical axis and is measured with successive additions of Legendre polynomials where a linear combination is used to produce the final valve shape (Table 1.3).

A number of diatom taxa in publications have been analyzed using Legendre polynomials (e.g., [1.7.34., 1.7.55., 1.7.63., 1.7.73., 1.7.75., 1.7.76., 1.7.79.]), and this information will be compiled with results from this study to produce a generalized picture of the contribution of each Legendre coefficient to reconstructing diatom valve shape. Despite the varieties of whole valve shape, different numbered coefficients define similar valve shape characteristics across taxa. Because of this, Legendre coefficients as quantified shape descriptors will be used with clustering techniques to determine bins for valve shape groups. Recurrent forms will be categorized for further analyses.

1.2   Methods: A Test of Recurrent Diatom Valve Shapes

Forty-one diatom taxa were chosen to represent a variety of valve shapes to be used in analyses. In most cases, full scientific names are given of the taxon used, and in other cases, only the generic name is given. Taxa represented are *Actinella punctata, Amphicampa, Asterionella formosa, Bleakeleya notata, Cylindrotheca, Cymatoneis sulcata, Cymatopleura, Cymbellonitzschia, Delphineis surirella, Didymosphenia geminata, Encyonema leibleinii, Epithemia, Eunotia serra, Fragilariforma constricta, Gomphonema acuminatum, Gomphonema affine, Grammatophora undulata, Gyrosigma, Hannaea, Hantzschia, Hippodonta capitata, Licmophora gracilis, Meridion, Navicula exigua* var. *capitata, Nitzschia sigma, Nitzschia sigma* var. *tabellaria, Oestrupia, Perissonoë crucifera, Peronia fibula, Plagiodiscus, Pseudostaurosira trainorii, Rhaphoneis carolinica, Sceptroneis, Semiorbis rotundus, Staurosirella leptostauron, Surirella peisonis, Tabellaria fenestrata, Tabellaria fenestrata* var. *geniculata, Tetracyclus emarginatus, Tryblionella levidensis,* and *Ulnaria capitata.* For complete authority information on each taxon, see Table 1.4. Choice of taxa is commensurate with representation of a variety of shapes that are common and amenable to 2D shape modeling. Shape analysis of a small group of taxa is used to produce an initial schema to facilitate the binning of a larger pool of diatom valve shapes. The computer program I wrote in 2002 is used for calculating Legendre coefficients and reconstruction of diatom outlines and is listed in the Appendix.

Table 1.2 A diagrammatic and descriptive representation of the width function, $W(x) = \sum_{n=0}^{N} b_n P_n(x)$. Example given: *Cymbella cistula*.

Diagrammatic width function	Steps in using the width function
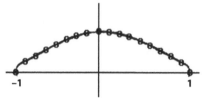	Find equidistant x-coordinates between -1 and 1 for the top half of the diatom valve outline.
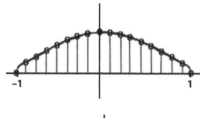	Perpendiculars from the coordinates indicate the width found by the width function for the top half.
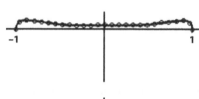	Flip the bottom half on the horizontal, and find equidistant x-coordinates between -1 and 1 for the bottom half.
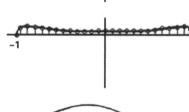	Perpendiculars from the coordinates indicate the width found by the width function for the bottom half.
	Flip the bottom half on the horizontal, and the whole outline is completed.

Shape analysis was conducted by obtaining outlines from digital light micrographs in: Krammer and Lange-Bertalot [1.7.41.]; Pappas and Stoermer [1.7.54.]; Jahn and Kusber [1.7.31.]; Diatoms of North America website contributors Spaulding and Edlund [1.7.70.– 1.7.72.], Beals [1.7.8.], Potapova [1.7.58.], Morales [1.7.50.], Alexson [1.7.2.], Burge and Edlund [1.7.11.], Burge *et al.* [1.7.12.], Bahls and Kimmich [1.7.4.], and Bryłka and Lee [1.7.10.]; Snoeijs, P. [1.7.69.], and STRI Data Portal website contributor Evelyn Gaiser [1.7.20.]. Drawings were the source of outlines obtained from ANSP Diatom New Taxon File [1.7.59.] and Schmidt's Atlas [1.7.65.]. Information on each taxon image or drawing used is given in Table 1.4. Legendre polynomials were calculated to quantify the entire valve face shape, using 50 x, y-equidistant coordinates chosen on each half of a diatom digital

Table 1.3 Partial valve reconstructions for *Cymbella cistula* in terms of number of Legendre coefficients used.

Valve shape reconstruction	Number of Legendre coefficients used
	Addition of the 6th coefficient
	Addition of the 5th coefficient
	Addition of the 4th coefficient
	Addition of the 3rd coefficient
	Addition of the 2nd coefficient
	Curve reconstruction using the 0th and 1st coefficients

image. Orthogonal polynomial regression was used via Legendre polynomials to get the best-fit valve outline shape. Legendre coefficients were calculated and used to represent quantified shape descriptors for each taxon and in additional analyses to determine valve shape bins.

1.2.1 Legendre Polynomials, Hypergeometric Distribution, and Probabilities of Valve Shapes

Because each successive Legendre coefficient represents clustered valve shapes, a distribution of Legendre coefficients may be determined from the 0th to n^{th} coefficients. As successive coefficients are added to the previous valve shape, probabilities of occurrence may be established as successive valve shapes are created.

Table 1.4 List of 41 diatom taxa used in Legendre shape analysis and source of light micrograph or drawing used. Authority information on taxa was acquired from [1.7.39.].

Taxon	Source of light micrograph or drawing outline
Actinella punctata F.W. Lewis, 1864	[1.7.8.]
Amphicampa (C.G. Ehrenberg) J. Ralfs in A. Pritchard, 1861	[1.7.72.]
Asterionella formosa A.H. Hassall, 1850	[1.7.54.]
Bleakeleya notata F.E. Round, 1990	[1.7.59.], *Asterionella notata* (Grun.) Grunow var. *recticostata* Körner n. var. 1970
Cylindrotheca L. Rabenhorst, 1859	[1.7.70.]
Cymatoneis sulcata (Greville) P.T. Cleve, 1894	[1.7.65.], Notebook 53, Table 212, no. 41 (*N.* (*Cymatoneis*) *sulcata* Grev.)
Cymatopleura W. Smith, 1851	[1.7.54.]
Cymbellonitzschia F. Hustedt in A. Schmidt *et al.*, 1924	[1.7.54.]
Delphineis surirella (Ehrenberg) G.W. Andrews, 1981	[1.7.59.]
Didymosphenia M. Schmidt in A. Schmidt *et al.*, 1899	[1.7.54.]
Encyonema F.T. Kützing, 1833	[1.7.2.]
Epithemia F.T. Kützing, 1844	[1.7.54.]
Eunotia serra C.G. Ehrenberg, 1837	[1.7.54.]
Fragilariforma constricta (Ehrenberg) D.M. Williams and Round, 1988	[1.7.12.]
Gomphonema acuminatum Ehrenberg, 1832	[1.7.36.]
Gomphonema affine F.T. Kützing, 1844	[1.7.41.]
Grammatophora C.G. Ehrenberg, 1840	[1.7.20.]
Grunowia tabellaria Grunow (Rabenhorst), 1864	[1.7.37.]
Gyrosigma A.H. Hassall, 1845	[1.7.54.]
Hannaea R. Patrick in Patrick and C.W. Reimer, 1966	[1.7.54.]
Hantzschia A. Grunow, 1877	[1.7.54.]

(Continued)

Table 1.4 List of 41 diatom taxa used in Legendre shape analysis and source of light micrograph or drawing used. Authority information on taxa was acquired from [1.7.39.]. (*Continued*)

Taxon	Source of light micrograph or drawing outline
Hippodonta capitata (Ehrenberg) Lange-Bertalot, Metzeltin and A. Witkowski, 1996	[1.7.58.]
Licmophora gracilis (Ehrenberg) Grunow, 1867	[1.7.69.]
Meridion C.A. Agardh, 1824	[1.7.54.]
Nitzschia sigma (Kützing) W. Smith, 1853	[1.7.38.]
Oestrupia H. Heiden *ex* F. Hustedt, 1935	[1.7.54.]
Perissonoë crucifera (Kitton in Pritchard) Desikachary, Gowthaman, Hema, Prasad & Prema, 1987	[1.7.59.]
Peronia fibula (Brébisson *ex* Kützing) R. Ross, 1956	[1.7.4.]
Navicula exigua var. *capitata* R. Patrick, 1945	[1.7.54.], *Placoneis exigua* var. *capitata*
Plagiodiscus A. Grunow & T. Eulenstein in A. Grunow, 1867	[1.7.65.], Hustedt (1914), Table 309, no. 9, *Surirella reniformis* Grun.
Pseudostaurosira trainorii E.A. Morales, 2001	[1.7.50.]
Rhaphoneis carolinica Andrews in Abbott and Andrews, 1879	[1.7.59.]
Sceptroneis praecaducea Hajós and Stradner in Hajós 1974	[1.7.59.]
Semiorbis rotundus G. Reid and D.M. Williams, 2010	[1.7.12.]
Staurosirella leptostauron (Ehrenberg) D.M. Williams and Round, 1987	[1.7.54.]
Surirella peisonis Pantocsek, 1902	[1.7.14.]
Tabellaria fenestrata (Lyngbye) F.T. Kützing, 1844	[1.7.54.]
Tabellaria fenestrata var. *geniculata* A. Cleve, 1899	[1.7.54.]
Tetracyclus emarginatus (Ehrenberg) W. Smith, 1856	[1.7.71.]
Tryblionella levidensis W. Smith, 1856	[1.7.54.]
Ulnaria capitata (Ehrenberg) P. Compère, 2001	[1.7.10.]

The Legendre-Gaussian quadrature rule [1.7.82.] states that

$$\int_{-1}^{1} f(x)dx \approx \sum_{i=1}^{n} w_i f(x_i) \tag{1.20}$$

where a function, f, represents the associated Legendre polynomials of degree $2n - 1$, and the weight function, w, is 1 [1.7.82.] and is a summation in a power series [1.7.79.]. The hypergeometric function as a hypergeometric series is also a summation in a power series as it is with the width function of Legendre polynomials. Valve shapes per Legendre coefficient follow a hypergeometric distribution so that the valve shape distribution as a probability distribution is hypergeometric. From this, a cumulative distribution function may be devised via a hypergeometric distribution and is given as

$$Prob(T \leq k) = 1 - \frac{\binom{n}{k+1}\binom{N-n}{M-k-1}}{\binom{N}{M}} F(\alpha, \beta; \gamma; t) \tag{1.21}$$

so that

$$Prob(X \leq k) = 1 - \frac{\binom{n}{k+1}\binom{N-n}{M-k-1}}{\binom{N}{M}} P_n(1-2x) \tag{1.22}$$

where the nth-Legendre polynomial is related to the probability of a particular valve shape represented by the nth-Legendre coefficient. In this way, valve shape bins may be used as a means to probabilistically determine the incidence of recurrent forms.

1.2.2 Multivariate Hypergeometric Distribution of Diatom Valve Shapes as Recurrent Forms

To determine the incidence of recurrent forms probabilistically that are based on Legendre polynomials of n-shapes, a multivariate hypergeometric schema is used as exemplified by the Urn Problem [1.7.9.]. From shape bin results, there are initially K_i valve groups with i-shapes. We randomly select j shape groups depleting the total number of shape groups, $N = \sum_{i=1}^{j} K_i$, to be sample $n - j$. The number of groups representing shape(s) in the sample is k_1, k_2, ..., k_{shape}, and the distribution of the sample is multivariate hypergeometric without replacement [1.7.60.].

The probability of successfully selecting a given shape bin is

$$Prob(x=k) = \frac{[Number\ of\ k\ successes][Number\ of\ N-k\ failures]}{Total\ number\ of\ shape\ bins\ to\ select} = \frac{\binom{n}{k}\binom{m}{N-k}}{\binom{m+n}{N}}$$

(1.23)

where k-shapes are available from N total shapes. For k-shapes, $N - k$ shapes are chosen without replacement. The probability that one of each shape bin is chosen is

$$P[(1|N-k_1),(1|N-k_2),\ldots,(1|N-k_{shape})] = \frac{\binom{N-k_1}{1}\binom{N-k_2}{1}\cdots\binom{N-k_{shape}}{1}}{\left(\dfrac{\sum_{i=1}^{k}(N_i-k_{shape})}{\sum_{i=1}^{N-k_{shape}}1_i}\right)}.$$

(1.24)

The probability of the kth success is $Prob(x=k) = \dfrac{n}{m+n}$, and the expectation that a

given shape bin will be selected is $\displaystyle\sum_{k=0}^{N} k\frac{\binom{n}{k}\binom{m}{N-k}}{\binom{m+n}{N}} = \frac{nN}{m+n}$ [1.7.82.].

A multivariate hypergeometric distribution is determined from Legendre shape coefficients from 41 taxa chosen for analysis. Then, genera listed in Fourtanier and Kociolek's [1.7.18.] "Catalogue of the Diatom Genera," updated as "Catalogue of Diatom Names" [1.7.19.]—hereafter called, Catalogue—are used as a source of input data for construction of a probability of occurrence distribution of diatom valve shapes. To qualify as input data, shapes in each genus listed in the Catalogue are verified by checking representative taxa as digital images, drawings and/or adequate description as given in the ANSP Diatom New Taxon File [1.7.59.] website. In some cases, synonymy with a valid taxon name is used in order to ascertain valve shape as stated in the description of a given genus. A genus is counted only once, though, discounting synonymies. When valve shapes per genus are not uniform, majority shape was estimated. When a majority shape could not be established, the designated type is used as representative of the genus. In the case of a genus such as *Triceratium* or *Trigonium*, having representatives of many basic geometric shapes, all valve shapes are counted.

Table 1.5 Valve shapes matched with minimum number of Legendre coefficient used in outline reconstruction. Shape bins and descriptors are devised with respect to Legendre coefficient number. Geometric and diatom shape terminology from [1.7.7.] is used for descriptions.

Legendre coefficient number	Valve shape description	Valve shape examples
40th	Reniform; cardioid	
22nd	Crenulate	
20th	Quadrate or square with multiple undulations.	
18th	Elongation; ends developed.	
16th	Triundulate	
14th	Clavate; hastate; cruciform; sigmoid; trapezoid; rhombic.	
12th	Expansion of the mid-valve region – defines shape dorsally more than ventrally; auricular.	
10th	Spatulate	
8th	Expansion of the mid-valve region – defines shape ventrally in a global sense; biundulate; arcuate; lunate; crescentic.	
6th	Expansion of the mid-valve region (e.g., gibbousness); lanceolate; undulate; bilobate; panduriform.	

(Continued)

Table 1.5 Valve shapes matched with minimum number of Legendre coefficient used in outline reconstruction. Shape bins and descriptors are devised with respect to Legendre coefficient number. Geometric and diatom shape terminology from [1.7.7.] is used for descriptions. (*Continued*)

Legendre coefficient number	Valve shape description	Valve shape examples
4th	Semi-lanceolate, or semi-circular – defines shape dorsally in a global sense; may have rostrate or capitate ends.	
2nd	Circular; ovoid; elliptical. Elongation in shape horizontally from end to end; greater expansion of the mid-valve region relative to expansion of the valve ends.	
0th	Square; rectangular	

Each valve shape is assigned to a shape bin based on number of Legendre coefficients given in Table 1.5. Only even coefficients are used because they represent shape, while odd coefficients represent symmetry (e.g., [1.7.55., 1.7.57., 1.7.63., 1.7.75.]). Because only Legendre coefficient number is used, taxa do not fall into the typical characterization of "centric" or "pennate," and uniform categories of shapes do not exist. Shape bins are used as the basis for further analysis of diatom valve shapes and recurrent forms.

Taxa from the Catalogue are used via assignment of each genus to a valve shape bin and determination of a distribution of valve shapes across genera. A hypergeometric distribution is devised based on the total number of taxa extracted from the Catalogue and assignment to each valve shape bin. A histogram is constructed where each bar represents the hypergeometric distribution of each valve shape bin, and from this, a probability distribution is devised based on the number of shapes per Legendre coefficient category. This distribution indicates the probability of encountering a given diatom valve shape and is an indicator of the probability of recurrent forms.

1.3   Results

Two to 40 Legendre coefficients were calculated for each taxon shape (Figure 1.4 and Table 1.5). The minimum number of coefficients used to recover the taxon shape defined each shape bin. A number of taxon valve outlines were reconstructed with 14 Legendre coefficients. Exceptions to this were necessary for taxa where one half of the valve was markedly

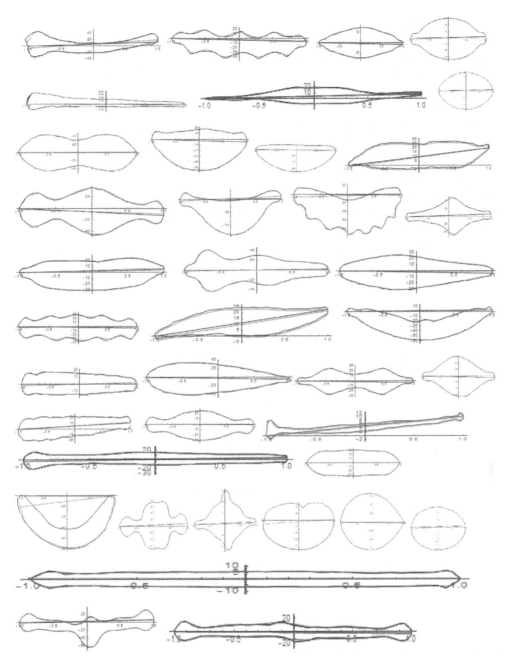

Figure 1.4 Reconstructed outlines plotted on top of valve outlines for 41 diatom taxa analyzed using Legendre polynomials.

different from the other half. The number of coefficients for these taxa were as follows: 14 coefficients were used to reconstruct the less wavy half valve shape of *Eunotia serra* in contrast to the wavier half valve shape which required 22 coefficients; *Hannaea* required 22 and 14 coefficients to reconstruct the valve outline; for *Plagiodiscus*, 40 coefficients were needed on the valve half with a sharp cut midway in the outline in contrast to 14 for the smooth concave half. Some taxa required more than 14 coefficients to reconstruct their outline, including: 20 coefficients were necessary for *Asterionella formosa*, *Perissonoë*, *Tabellaria fenestrata* var. *geniculata*, and *Ulnaria capitata*; 22 coefficients were necessary for *Meridion*; 30 coefficients were necessary to reconstruct the valve outline of *Tetracyclus emarginatus*. Reconstructed outlines superimposed on the taxon original outlines are given in Figure 1.4.

Thirteen shape bins were defined by the minimum Legendre coefficient number needed to reconstruct the valve shape from 41 taxa used in analysis (Table 1.5). From the CAS Catalogue [1.7.18.-1.7.19.] and verification using the ANSP Diatom New Taxon File website [1.7.59.], 418 taxa were assigned to a shape bin. The number of taxa per shape bin was tabulated and given in Table 1.6. This data was used to devise a multivariate hypergeometric distribution as a histogram (Figure 1.5). A probability density function (PDF) was constructed and depicted as an overlay of the histogram (Figure 1.5).

The probabilities of occurrence for each shape bin were calculated from the multivariate hypergeometric distribution (Table 1.6). These probabilities defined occurrences of diatom

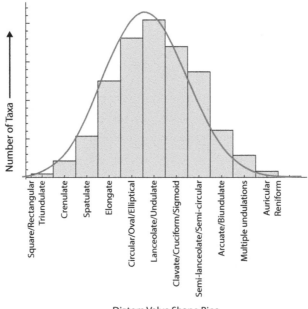

Figure 1.5 Histogram of the multivariate hypergeometric distribution of shape bins filled with assignment of 418 taxa from the Catalogue of the Diatom Genera. Overlay of the probability density function is depicted by the blue line curve.

valve shapes and are an indication of recurrent forms. The lowest probability of occurrence is defined for diatom valve shapes that are square to rectangular (0th Legendre coefficient) or cardioid (40th Legendre coefficient). Among the lower probabilities of occurrence are valve shapes defined by Legendre coefficients 12, 16 and 20 representing auricular, triundulate, and multiple undulated squarish forms, respectively. For Legendre coefficients 8 and 22, slightly higher probabilities were found for arched or biundulate to crenulate valve shapes, while probability occurrence for Legendre coefficient 10 was marginally higher for spatulate valve shapes. The highest probabilities were found for elongated valves with developed ends, semi-circular, and circular to oval or elliptical shapes, represented by the 18th, 4th and 2nd Legendre coefficients, respectively. The highest probability of occurrence is representative of diatom shapes defined by Legendre coefficient six, representing elongated, lanceolate, undulate, and panduriform valve shapes (Tables 1.5 and 1.6).

Table 1.6 Legendre coefficient number representing valve shape bins, tabulated number of taxon shapes, and probability of shape occurrence per shape bin.

Legendre coefficient number	Tabulated number of taxa	Probability
0	2	0.000894978
2	68	0.653543
4	46	0.429059
6	183	0.994368
8	9	0.0284528
10	12	0.0494644
12	3	0.0026376
14	47	0.440556
16	3	0.0026376
18	31	0.24964
20	4	0.00518226
22	8	0.0225259
40	2	0.000894978

1.4   Discussion

1.4.1   Valve Shape Probability Distribution

Diatom valve shape distribution was shown to be based on a hypergeometric function via Legendre polynomials. Orthogonal coefficients such as Legendre coefficients to quantify shape is useful not only in determinations of species shape groups, but also in devising a way to understand the interrelation among all probable diatom forms and their degree of occurrence. The highest number of recurrent forms are the lanceolate and ovoid valve shapes, and the probability of occurrence of other forms decreases as one approaches the tails of the probability distribution. Obtaining a probability distribution via Legendre coefficients is efficacious in quantifying biological attributes. More generally, orthogonal coefficients may be obtained in other ways, and even though they may not embody biological meaning explicitly, they may be used to quantify valve shape and devise similar distributions of taxa.

1.4.2   Hypergeometric Functions and Other Shape Outline Methods

Legendre polynomials and coefficients were used to quantify valve shape outline not only because of their biologically interpretable meaning, but also because they are related to the hypergeometric function and distribution. Taking a wider view, the Legendre polynomials belong to the class of Jacobi polynomials that includes, Chebyshev, Gegenbauer (ultraspherical), Hermite, and Laguerre polynomials [1.7.1.]. Chebyshev, Gegenbauer, and the generalized Jacobi polynomials are related to hypergeometric functions and given as

$$F\left(-n,n;\frac{1}{2};x\right) = T_n(1-2x) \tag{1.25}$$

$$F\left(-n,n+2\alpha;\alpha+\frac{1}{2};x\right) = \frac{n!}{(2\alpha)_n}C_n^{(\alpha)}(1-2x) \tag{1.26}$$

$$F(-n,\alpha+1+\beta+n;\alpha+1;x) = \frac{n!}{(\alpha+1)_n}P_n^{(\alpha,\beta)}(1-2x) \tag{1.27}$$

respectively [1.7.1.].

Laguerre polynomials expressed as

$$M(-n,\alpha+1,x) = \frac{n!}{(\alpha+1)_n}L_n^{(\alpha)}(x) \tag{1.28}$$

$$U(-n,\alpha+1,x) = (-1)^n n! L_n^{(\alpha)}(x) \tag{1.29}$$

correspond to confluent hypergeometric functions of the first and second kind, respectively [1.7.82.]. Hermite polynomials are related to the confluent hypergeometric functions of the first and second kind, respectively, as

$$M\left(-n, \frac{1}{2}, \frac{1}{2}x^2\right) = \frac{n!}{(2n)!}\left(-\frac{1}{2}\right)^{-n} H_{2n}(x) \tag{1.30}$$

$$M\left(-n, \frac{1}{2}, \frac{1}{2}x^2\right) = \frac{n!}{(2n+1)!}\left(-\frac{1}{2}\right)^{-n}\frac{1}{x} H_{2n+1}(x) \tag{1.31}$$

$$U\left(\frac{1}{2} - \frac{1}{2}n, \frac{3}{2}, x^2\right) = 2^{-n} H_n \frac{(x)}{x} \tag{1.32}$$

and Hermite polynomials are a special case of Laguerre polynomials [1.7.1., 1.7.82.]. Relation among Legendre, Chebyshev, Gegenbauer, Laguerre, Hermite, and the generalized Jacobi polynomials may be devised with respect to different weight functions [1.7.81.]. All of these polynomials could be used to devise shape group bins and probability distributions of shapes.

Bessel functions are related to Legendre functions, and as a result are related to Legendre polynomials. For fixed values of μ and x, and $x > 0$,

$$J_\mu(x) = \lim_{\mu\to\infty}\left[v^\mu P_v^{-\mu}\left(\cos\frac{x}{v}\right)\right] \tag{1.33}$$

where J_μ is a Bessel function of the first kind for a Legendre polynomial of the first kind, P_v, and

$$-\frac{1}{2}\pi Y_\mu(x) = \lim_{\mu\to\infty}\left[v^\mu Q_v^{-\mu}\left(\cos\frac{x}{v}\right)\right] \tag{1.34}$$

Y_μ is a Bessel function of the second kind (also known as a Weber's function) for a Legendre polynomial of the second kind, Q_v [1.7.1.]. Bessel functions are related to confluent hypergeometric functions of the first kind as

$$M\left(v + \frac{1}{2}, 2v+1, 2iz\right) = \Gamma(1+v)e^{iz}\left(\frac{1}{2}z\right)^{-v} J_v(z) \tag{1.35}$$

$$M\left(-v + \frac{1}{2}, -2v+1, 2iz\right) = \Gamma(1+v)e^{iz}\left(\frac{1}{2}z\right)^{v}[\cos(v\pi)J_v(z) - \sin(v\pi)Y_v(z)] \tag{1.36}$$

where Γ is a gamma function, and a number of modified Bessel functions in various forms are related to confluent hypergeometric functions of the second kind [1.7.1.]. Bessel functions have been used to devise models of diatom valves (e.g., [1.7.21.]).

Expansion of the Legendre-Fourier series is an expression of the terms of the width function, thereby hinting at the relation between Fourier and hypergeometric functions. From Sturm-Liouville theory, general Fourier expansions are eigenfunctions of trigonometric functions, having the property of orthogonality [1.7.79.]. The continuous Fourier transform may be expressed as

$$\Im[f] = \int_{-\infty}^{+\infty} f(x)e^{i\omega x}dx \tag{1.37}$$

including both real and imaginary parts with respect to a power series [1.7.79.]. For diatom valve shape analysis, the discrete Fourier transform is used and expressed variously as summation equations (e.g., [1.7.35., 1.7.51., 1.7.56., 1.7.57., 1.7.84.]). For a discrete Fourier transform,

$$\Im[f] = \sum_{x=-\infty}^{+\infty} e^{-i\omega x} f(x) \tag{1.38}$$

representation as a hypergeometric function, F, is

$$\Im[f] = \sum_{x=-\infty}^{+\infty} e^{-i\omega x} F\left(\frac{\alpha+1}{2}; \frac{\alpha+3}{2}, \frac{1}{2}; -\frac{\pi^2 x^2}{4}\right) \tag{1.39}$$

so that in terms of trigonometric functions

$$\Im[f] = 2\sum_{x=1}^{+\infty} F\left(\frac{\alpha+1}{2}; \frac{\alpha+3}{2}, \frac{1}{2}; -\frac{\pi^2 x^2}{4}\right)\cos(kx) + 1 \tag{1.40}$$

and

$$\Im[f] = -2i\sum_{x=1}^{+\infty} x\left[F\left(\frac{\alpha+1}{2}; \frac{\alpha+3}{2}, \frac{1}{2}; -\frac{\pi^2 x^2}{4}\right)\right]\sin(kx) \tag{1.41}$$

[1.7.78.]. Indeed, the Fourier transform may represent a hypergeometric function (e.g., [1.7.3., 1.7.78.]).

All of the polynomials and transforms have the orthogonality property (1.7.81.; 1.7.1.), and could be used in diatom valve shape analysis. Associating valve shapes with orthogonal coefficients could be used to develop schema with regard to probability distributions of occurrence and recurrent forms. In this way, diatom valve shapes are amenable to study with regard to the diatom bauplan.

1.4.3   Application: Valve Shape Changes and Diversity during the Cenozoic

Diatom valve shape distribution may be related to diatom diversity over geologic time. The diatom fossil record from the Cretaceous to the Recent is generally complete and has been documented in many studies (e.g., [1.7.6., 1.7.16., 1.7.42.]). Diatoms are thought to have first appeared during the Jurassic, diversifying in the Cretaceous (e.g., [1.7.28., 1.7.43.]), and decreasing in diversity during the Eocene-Oligocene transition (e.g., [1.7.6., 1.7.42., 1.7.53.]). However, the Paleocene may rival diversification at the late Eocene where many Cretaceous diatom species actually survived [1.7.62.]. Reconstructing diatom diversification throughout the Cenozoic is still a work in progress (e.g., [1.7.42., 1.7.62.]).

The record of diatom diversification during the Cenozoic requires looking at particular taxa to determine if there is any relation to the change in valve shapes over geologic time. During the Cretaceous, many centric diatoms were present, representing the circular/oval/elliptic valve shape bin as well as lanceolate/undulate forms such as *Hemiaulus*. As the Paleocene proceeds, taxa including triangular forms such as *Trinacria* and *Triceratium* and clavate forms such as *Sceptroneis* appear in greater numbers as Cretaceous forms gradually decline [1.7.27., 1.7.62.]. During the Eocene-Oligocene transition, multi-undulate taxa are present such as *Cerataulus*, biundulate taxa are present such as *Biddulphia*, semi-circular/semi-lanceolate taxa such as *Amphora* and *Euodia* occur, and *Donkina*, *Huttonia*, *Navicula*, *Nitzschia*, and *Glyphodesmis* are present as lanceolate forms [1.7.85.]. Triangular, square and *n*-gon forms of *Triceratium*, *Trinacria*, and circular/oval/elliptic forms continue during the Eocene as well.

During the Oligocene-Miocene transition, occurrence of circular, oval, and elliptical forms continues as do triangular forms. The presence of elongated and semi-lanceolate forms such as *Synedra* and *Rhaphoneis*, respectively, as well as sigmoid and lanceolate forms such as *Pleurosigma* and *Clavicula*, respectively, are present during this time [1.7.5.].

The modern diatom flora may have originated in the late Miocene [1.7.42.], although changes from marine to freshwater circular forms occurred during the Middle Miocene Climatic Optimum [1.7.29.], and the presence of lanceolate, crenulate and auricular forms represented by *Pinnularia*, *Eunotia* and *Epithemia*, respectively, occurred during this time as well [1.7.45.]. Valve shapes represented by higher numbered Legendre coefficients are indicative of Miocene to Recent diatom taxa.

The brief discussion of change in diatom valve shapes during the Cenozoic is but one example that may be explored further with more detailed studies. Is there a difference in valve shape distribution between the Cretaceous and the Cenozoic? Is valve shape distribution the same throughout the Cenozoic? What is the relation between valve shape distribution and first and last appearance of diatom taxa? What, if any, is the relation between appearance of freshwater versus marine taxa throughout the Cenozoic and valve shape distribution? What are novelties, if any, in diatom shape that occurred during the Cenozoic, and what are the probabilities of occurrence? Such questions and many others may be addressed concerning diatom valve shape and recurrent forms throughout geologic time.

1.4.4   Diatom Valve Shape Distribution: Other Potential Studies

That the plethora of diatom valve shapes follows a multivariate hypergeometric distribution enables viewing shape as a probabilistic measure. Because of this, additional studies may be explored with regard to evolutionary studies. Valve shape distribution could be used in conjunction with models of evolution to project possible diatom valve shapes. Some of the models may include those using Monte Carlo simulation (e.g., [1.7.17.]) or the Metropolis algorithm with the Markov chain Monte Carlo method (e.g., [1.7.49.]). Knowing that diatom valve shapes are hypergeometrically distributed may be useful in devising evolutionary studies such as speciation (e.g., [1.7.32.]), the pairing of predictions of protein functions with cell formation (e.g., [1.7.40., 1.7.49.]), evolutionary pathways of multiple taxa over time (e.g., [1.7.44.]), or analysis of genotype and phenotype using the Urn model [1.7.9.]. Valve shape distribution may be of interest in various evolutionary studies.

1.5 Summary and Future Research

Usually, shape outline analysis is used to discern species groups. By using Legendre polynomials and coefficients to reconstruct diatom valve shape, the coefficients are used in devising a probability distribution of valve shape occurrences from a hypergeometric function. This quantitative comparison of diatom taxa at the genus level may provide a new avenue of study with regard to topics in evolutionary and micropaleontology studies. Probabilistic approaches merging one aspect of morphology, i.e., shape, and evolutionary pathways over geologic time or at the level of the cell may induce new ways of finding and obtaining predictive outcomes in evolutionary and micropaleontologic studies.

1.6 Appendix

The following is the computer program as verbatim text:

(* This Mathematica [1.7.86.] nb is a computer program written by Janice L. Pappas in 2002 and is used for calculating Legendre coefficients for least-squares fitting of high order orthogonal polynomials to an outline curve and plotting reconstructed curves. Follow all the instructions listed throughout. The example given for computation and plotting is Staurosirella leptostauron. Once values are inserted for variables, j, k, jj, and kk, and the selected coordinates are copied and pasted as given by the instructions, the nb is saved; then, SelectAll, then, Shift, Enter to evaluate the entire program and obtain results. *)

(* Import the image. You may need to select FilePath... from the Insert menu to place in between the brackets for Import. Be sure the complete file path is between double quote marks within the brackets. *)

tetra=Import["lepto.jpg"]

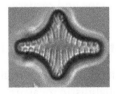

Show[tetra]

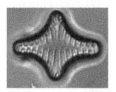

(* Next, convert the image to graphics and superimpose a grid on the graphics. The number of vertical and horizontal grid lines may need to be adjusted, depending on the graphics. The color of the grid lines may need to be adjusted, depending on the color of the graphics. The AspectRatio may need to be adjusted, depending on the length and width of the graphics. *)

```
gem=ImageData[tetra];
gem=gem/.Graphics->List;
Short[gem,15];
ex=Dimensions[gem];
xm=Extract[ex,{1}];
my=Extract[ex,{2}];
ldp=ListDensityPlot[gem[[All,All]],Mesh->False,AspectRatio->0.7]
```

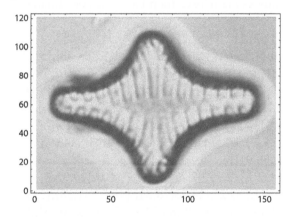

```
lines=Table[Line[{{i,0},{i,xm}}],{i,0,my,20}];
firstgrid=Show[{ldp,Graphics[{GrayLevel[1],lines}]}]
```

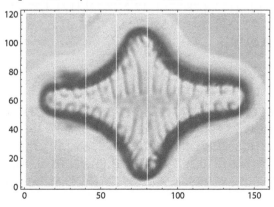

nel=Table[Line[{{0,n},{my,n}}],{n,0,xm,20}];
twogrd=Show[{firstgrid,Graphics[{GrayLevel[1],nel}]}]

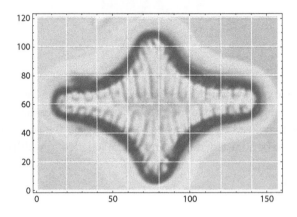

(* For each diatom outline half, choose 50 equi-distant coordinates using the Drawing Tools palette (Ctl-D), selecting the Get Coordinates tool (dashed line cross). Select 50 coordinates for the TOP half of the outline ONLY at this time. *)

(* Place first half coordinates next by selecting coordinates as stated above then select Shift-Ctl-C, which is CopyAs Plain Text, then Paste the coordinates on a new line. *)

{{10.38,60.43},{10.86,64.62},{12.77,67.77},{16.1,70.91},{20.87,71.43},{26.11,71.43},{31.36,
71.96},{37.08,74.05},{40.41,75.62},{44.23,78.24},{48.52,81.38},{51.85,83.48},{55.19,85.58},
{57.57,88.19},{59.48,90.81},{60.91,94.48},{63.77,98.67},{65.2,102.3},{66.63,105},{68.54,107.6},
{71.4,109.1},{73.78,110.2},{76.64,110.2},{79.98,109.1},{81.88,108.6},{84.27,106.5},{86.17,103.4},
{88.56,101.3},{90.46,98.67},{92.85,93.96},{95.23,89.77},{98.09,87.15},{101,84.53},{103.8,81.91},
{107.1,80.86},{110.5,79.81},{113.8,78.77},{117.2,77.72},{121.4,77.72},{125.3,77.19},{129.1,
77.19},{133.4,76.67},{137.7,74.57},{141,73.53},{142.9,71.43},{143.9,69.86},{144.3,67.77},{144.8,
64.62},{144.8,62},{144.8,59.91}}
{{10.38,60.43},{10.86,64.62},{12.77,67.77},{16.1,70.91},{20.87,71.43},{26.11,71.43},{31.36,
71.96},{37.08,74.05},{40.41,75.62},{44.23,78.24},{48.52,81.38},{51.85,83.48},{55.19,85.58},
{57.57,88.19},{59.48,90.81},{60.91,94.48},{63.77,98.67},{65.2,102.3},{66.63,105},{68.54,107.6},
{71.4,109.1},{73.78,110.2},{76.64,110.2},{79.98,109.1},{81.88,108.6},{84.27,106.5},{86.17,103.4},
{88.56,101.3},{90.46,98.67},{92.85,93.96},{95.23,89.77},{98.09,87.15},{101,84.53},{103.8,81.91},
{107.1,80.86},{110.5,79.81},{113.8,78.77},{117.2,77.72},{121.4,77.72},{125.3,77.19},{129.1,
77.19},{133.4,76.67},{137.7,74.57},{141,73.53},{142.9,71.43},{143.9,69.86},{144.3,67.77},{144.8,
64.62},{144.8,62},{144.8,59.91}}

coords=%

{{10.38,60.43},{10.86,64.62},{12.77,67.77},{16.1,70.91},{20.87,71.43},{26.11,71.43},{31.36,
71.96},{37.08,74.05},{40.41,75.62},{44.23,78.24},{48.52,81.38},{51.85,83.48},{55.19,85.58},
{57.57,88.19},{59.48,90.81},{60.91,94.48},{63.77,98.67},{65.2,102.3},{66.63,105},{68.54,107.6},
{71.4,109.1},{73.78,110.2},{76.64,110.2},{79.98,109.1},{81.88,108.6},{84.27,106.5},{86.17,103.4},
{88.56,101.3},{90.46,98.67},{92.85,93.96},{95.23,89.77},{98.09,87.15},{101,84.53},{103.8,81.91},

{107.1,80.86},{110.5,79.81},{113.8,78.77},{117.2,77.72},{121.4,77.72},{125.3,77.19},{129.1, 77.19},{133.4,76.67},{137.7,74.57},{141,73.53},{142.9,71.43},{143.9,69.86},{144.3,67.77},{144.8, 64.62},{144.8,62},{144.8,59.91}}

(* For each diatom outline half, choose 50 equi-distant coordinates using the Drawing Tools palette (Ctl-D), selecting the Get Coordinates tool (dashed line cross). Select 50 coordinates for the BOTTOM half of the outline ONLY at this time. *)

(* Place second half coordinates next by selecting coordinates as stated above then select Shift-Ctl-C, which is CopyAs Plain Text, then Paste the coordinates on a new line. *)

{{145.8,59.91},{144.8,56.76},{143.4,54.67},{141.5,52.05},{139.6,51},{136.2,49.43},{132.9,48.91}, {127.6,47.86},{124.3,47.34},{121,46.29},{118.6,45.24},{115.3,44.19},{111.9,42.62},{109.5, 41.57},{105.2,39.48},{103.3,37.38},{100.5,35.29},{98.09,32.67},{95.71,28.48},{93.8,24.81}, {92.85,21.67},{90.94,17.48},{89.51,13.29},{86.65,9.097},{82.84,7.002},{78.55,5.954},{75.21, 5.954},{70.92,9.097},{68.54,13.29},{68.06,17.48},{66.63,19.57},{65.2,23.24},{63.77,26.38}, {61.39,28.48},{58.05,31.62},{56.14,34.24},{53.76,35.81},{51.38,38.43},{49.47,40.53},{47.56, 41.57},{45.18,43.15},{41.84,44.19},{38.51,45.24},{34.22,46.29},{29.93,47.34},{24.68,47.86}, {19.92,49.43},{15.63,51},{12.77,53.62},{10.38,58.86}}
{{145.8,59.91},{144.8,56.76},{143.4,54.67},{141.5,52.05},{139.6,51},{136.2,49.43},{132.9,48.91}, {127.6,47.86},{124.3,47.34},{121,46.29},{118.6,45.24},{115.3,44.19},{111.9,42.62},{109.5, 41.57},{105.2,39.48},{103.3,37.38},{100.5,35.29},{98.09,32.67},{95.71,28.48},{93.8,24.81}, {92.85,21.67},{90.94,17.48},{89.51,13.29},{86.65,9.097},{82.84,7.002},{78.55,5.954},{75.21, 5.954},{70.92,9.097},{68.54,13.29},{68.06,17.48},{66.63,19.57},{65.2,23.24},{63.77,26.38}, {61.39,28.48},{58.05,31.62},{56.14,34.24},{53.76,35.81},{51.38,38.43},{49.47,40.53},{47.56, 41.57},{45.18,43.15},{41.84,44.19},{38.51,45.24},{34.22,46.29},{29.93,47.34},{24.68,47.86}, {19.92,49.43},{15.63,51},{12.77,53.62},{10.38,58.86}}

seccoords=%
{{145.8,59.91},{144.8,56.76},{143.4,54.67},{141.5,52.05},{139.6,51},{136.2,49.43},{132.9,48.91}, {127.6,47.86},{124.3,47.34},{121,46.29},{118.6,45.24},{115.3,44.19},{111.9,42.62},{109.5, 41.57},{105.2,39.48},{103.3,37.38},{100.5,35.29},{98.09,32.67},{95.71,28.48},{93.8,24.81}, {92.85,21.67},{90.94,17.48},{89.51,13.29},{86.65,9.097},{82.84,7.002},{78.55,5.954},{75.21, 5.954},{70.92,9.097},{68.54,13.29},{68.06,17.48},{66.63,19.57},{65.2,23.24},{63.77,26.38}, {61.39,28.48},{58.05,31.62},{56.14,34.24},{53.76,35.81},{51.38,38.43},{49.47,40.53},{47.56, 41.57},{45.18,43.15},{41.84,44.19},{38.51,45.24},{34.22,46.29},{29.93,47.34},{24.68,47.86}, {19.92,49.43},{15.63,51},{12.77,53.62},{10.38,58.86}}

(* First or top half of evaluation begins here. *)
topxy=Table[coords];
origreconst=ListPlot[topxy,Joined->True,PlotRange->All,AspectRatio->0.25`]

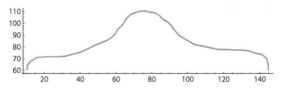

(* Set the values for the maximum number of Legendre coefficients calculated for the first or top half of the diatom. Set j = maximum number - 1, and set k = maximum number. The maximum number must be an even number. *)

```
j=13;
k=14;

xto=MatrixForm[ArrayFlatten[topxy]];
xlist=Transpose[topxy]//MatrixForm;
usex=xlist[[All,1]];
yuse=xlist[[All,2]];
middx=(usex[[1]]);
xmid=middx[[24]];
middy=(yuse[[1]]);
midpoint=xmid-xmid;
renewx=middx-xmid;
firsthalfx=Take[renewx,23];
todiv=First[firsthalfx*-1];
xnor=Divide[firsthalfx,todiv];
secondhalfx=Take[renewx,-26];
botodiv=Last[secondhalfx];
nory=Divide[secondhalfx,botodiv];
AppendTo[xnor,midpoint];
AppendTo[xnor,nory];
tophalfx=Flatten[xnor]//Table;
ydiv=First[middy];
ytopy=Subtract[middy,ydiv]//Table;
L=Table[LegendreP[n,tophalfx], {n,0,j}];
T=Transpose[L];
minor=L.T//Table;
invminor=Inverse[minor];
Lyproduct=L.ytopy;
b=invminor.Lyproduct//Table;
listbmean=Mean[b];
termscol=b*L;
cumtermscol=Accumulate[termscol];
reconstructedy=listbmean-cumtermscol;
onecon=Divide[3,5];
twocon=Divide[2,3];
numx=Plus[L[[4]],twocon*L[[2]]];
reconstrx=onecon*Divide[numx,L[[3]]];
xr=Table[reconstrx];
yr2=Table[reconstructedy];
posyr2=yr2[[k]]*-1;
xyorig=Join[{tophalfx,ytopy}];
```

xytrans=Transpose[xyorig];
lxy=ListPlot[xytrans,Joined->True,AspectRatio->0.25`]

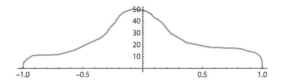

posapprows=Join[{xr,posyr2}];
TApos=Transpose[posapprows];
poslTA=ListPlot[TApos,Joined->True,AspectRatio->0.25`]

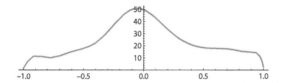

(* Plot of first or top half curve with reconstructed version superimposed. *)
tophalfdiatom=ListPlot[{xytrans,TApos},Joined->True,AspectRatio->0.4`]

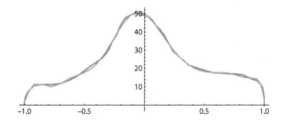

(* Second or bottom half of evaluation begins here. *)
uunderxybu=Table[seccoords];
underxyb=Reverse[uunderxybu];
origreconst=ListPlot[underxyb,Joined->True,AspectRatio->0.25`]

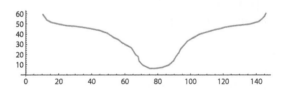

(* Set the values for the maximum number of Legendre coefficients calculated for the second or bottom half of the diatom. Set jj = maximum number - 1, and set kk = maximum number. The maximum number must be an even number. *)

jj=13;
kk=14;

```
bxtob=MatrixForm[ArrayFlatten[underxyb]];
bxlistb=Transpose[underxyb]//MatrixForm;
busexb=bxlistb[[All,1]];
byunderb=bxlistb[[All,2]];
bylb=(busexb[[1]]);
buyb=(byunderb[[1]]);
bxmidb=bylb[[24]];
bmidpointb=bxmidb-bxmidb;
brenewedxb=bylb-bxmidb;
bfirsthalfxb=Take[brenewedxb,24];
bunderdivb=First[bfirsthalfxb*-1];
undxnorb=Divide[bfirsthalfxb,bunderdivb];
bsecondhalfxb=Take[brenewedxb,-25];
divby=Last[bsecondhalfxb];
norybund=Divide[bsecondhalfxb,divby];
AppendTo[undxnorb,bmidpointb];
AppendTo[undxnorb,norybund];
underhalfxb=Flatten[undxnorb]//Table;
ydivideb=First[buyb];
yunderyb=Subtract[buyb,ydivideb]//Table;
Lb=Table[LegendreP[n,underhalfxb], {n,0,jj}];
Tb=Transpose[Lb];
minorproductb=Lb.Tb//Table;
underinvminorb=Inverse[minorproductb];
Lyproductunderb=Lb.yunderyb;
bc=underinvminorb.Lyproductunderb//Table;
bcmean=Mean[bc];
scolb=bc*Lb;
cumscolb=Accumulate[scolb];
recondyb=bcmean-cumscolb;
onecon=Divide[3,5];
twocon=Divide[2,3];
undernumxb=Plus[Lb[[4]],twocon*Lb[[2]]];
xreconxb=onecon*Divide[undernumxb,Lb[[3]]];
xrb=Table[xreconxb];
yr2underb=Table[recondyb];
posunderyr2b=yr2underb[[kk]]*-1;
xyunderorigb=Join[{underhalfxb,yunderyb}];
xytransbottomb=Transpose[xyunderorigb];
lxyunder=ListPlot[xytransbottomb,Joined->True,AspectRatio->0.25`]
```

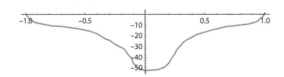

possb=Join[{xrb,posunderyr2b}];
TAposbottomb=Transpose[possb];
lposTAb=ListPlot[TAposbottomb,Joined->True,AspectRatio->0.25`]

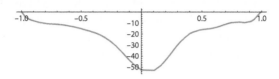

(* Plot of second or bottom half curve with reconstructed version superimposed. *)
bottomhalfdiatom=ListPlot[{xytransbottomb,TAposbottomb},AspectRatio->0.25`,
Joined->True]

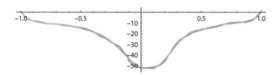

(* Original outline and superimposed outline for entire diatom. *)
originalxcoords=Append[tophalfx,underhalfxb];
wholex=Flatten[originalxcoords];
ycoordsoriginal=Append[ytopy,yunderyb];
ywhole=Flatten[ycoordsoriginal];
allorigcoords=Join[{wholex,ywhole}];
pairsoforiginalcoords=Transpose[allorigcoords]//Table;
reconstructedxcoords=Append[xr,xrb];
allreconx=Flatten[reconstructedxcoords];
ycoordsreconstructed=Append[yr2[[k]],yr2underb[[kk]]];
reconallycoords=Flatten[ycoordsreconstructed];
ys=reconallycoords*-1;
reconstructedall=Join[{allreconx,ys}];
reconstructallxys=Transpose[reconstructedall]//Table;
wholediatomsuperimposedoutlines=ListPlot[{pairsoforiginalcoords,reconstructallxys},
AspectRatio->0.55`,PlotStyle->{GrayLevel[0],Dashing[{Dot,Dashed}]},Joined->True]

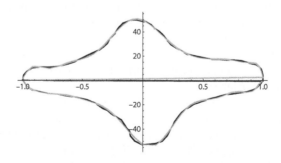

allcoeffstouse=Join[{b,bc}];
   (* Legendre coefficients for entire reconstructed outline. *)

orthogonalcoeffs=Flatten[allcoeffstouse]//Table
{23.7362,1.56283,-23.5096,2.8106,16.5219,-5.26927,-14.8049,4.21422,6.00771,-2.07892,
-6.4108,1.95733,0.406697,-2.66145,-22.0644,-0.961804,27.5578,3.32322,-18.7058,
-4.46331,17.3043,4.816,-8.56224,-4.04752,7.06964,4.68384,-2.51316,-3.23698}
   (* Whole outline plotted from all Legendre coefficients calculated. *)
wholediatomsuperimposedoutlines=ListPlot[{TApos,TAposbottomb},AspectRatio->
0.55`,PlotStyle->{GrayLevel[0],Dashing[{Dot,Dashed}]}],Joined->True,Axes->False]

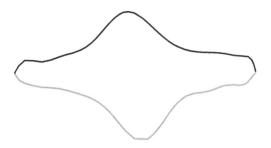

1.7   References

1.7.1.   Abramowitz, M. and Stegun, I.A. (eds.) (1972) *Handbook of Mathematical Functions with Formulas, Graphs, and Mathematical Tables*, 9th edition, Dover Publications, Inc., New York, USA.

1.7.2.   Alexson, E. (2014). *Encyonema leibleinii*. In Diatoms of North America. Retrieved March 23, 2020, from https://diatoms.org/species/encyonema_leibleinii

1.7.3.   Al-Lail, M.H. and Qadir, A. (2015) Fourier transform representation of the generalized hypergeometric functions with applications to the confluent and Gauss hypergeometric functions. *Applied Mathematics and Computation* 263, 392-397.

1.7.4.   Bahls, L., Kimmich, R. (2017) *Peronia fibula*. In Diatoms of North America. Retrieved March 23, 2020, from https://diatoms.org/species/peronia_fibula

1.7.5.   Baldauf, J.G. and Barron, J.A. (1987) *Oligocene marine diatoms recovered in dredge samples from the Navarin Basin Province, Bering Sea*, U.S. Geological Survey Bulletin 1765, U.S. Geological Survey, Department of the Interior, United States Government Printing Office, Washington, D.C.

1.7.6.   Baldauf, J.G. and Barron, J.A. (1990) Evolution of biosiliceous sedimentation patterns – Eocene through quaternary: paleoceanographic response to polar cooling. In: Bleil, U., Thiede, J. (eds.), *Geological History of the Polar Oceans: Arctic Versus Antarctic*. Kluwer Academic Publishers, The Netherlands: 575-607.

1.7.7.   Barber, H.G. and Haworth, E.Y. (1981) *A Guide to the Morphology of the Diatom Frustule: With a Key to the British Freshwater Genera*. Freshwater Biological Association Scientific Publication No. 44, Ambleside, Cumbria, U.K.

1.7.8.   Beals, J. (2011) *Actinella punctata*. In Diatoms of North America. Retrieved March 23, 2020, from https://diatoms.org/species/actinella_punctata

1.7.9.   Benaïm, M., Schreiber, S.J., Tarrès, P. (2004) Generalized urn models of evolutionary processes. *The Annals of Applied Probability* 14(3), 1455-1478.

1.7.10.   Bryłka, K., Lee, S. (2019) *Ulnaria capitata*. In Diatoms of North America. Retrieved March 23, 2020, from https://diatoms.org/species/ulnaria-capitata

1.7.11. Burge, D., Edlund, M. (2016) *Semiorbis rotundus*. In Diatoms of North America. Retrieved March 23, 2020, from https://diatoms.org/species/semiorbis_rotundus

1.7.12. Burge, D., Edlund, M., Brant, L. (2016) *Fragilariforma constricta*. In Diatoms of North America. Retrieved March 23, 2020, from https://diatoms.org/species/fragilariforma_constricta

1.7.13. Cronk, Q.C.B. (2001) Plant evolution and development in a post-genomic context. *Nature Reviews Genetics* 2, 607-619.

1.7.14. Edlund, M.E., Stoermer, E.F., Jamsran, Ts., Soninkhishig, N., Williams, R.M., *Welcome to the Mongolian Diatom Home Page, International Parternship for Research and Training in Mongolia—The diatom (Bacillariophyta) flora of ancient Lake Hovsgol*, http://www.umich.edu/~mongolia/index.html, accessed on March 25, 2020.

1.7.15. Engler, A. and Gilg, E. (1919) *Syllabus der Pflanzenfamilien: eine Übersicht über das gesamte Pflanzensystem mit besonderer Berücksichtigung der Medizinal- und Nutzpflanzen, nebst einer Übersicht über die Florenreiche und Florengebiete der Erde zum Gebrauch bei Vorlesungen und Studien über spezielle und medizinisch-pharmazeutische Botanik, 8th edition*, Gebrüder Borntraeger Verlag, Berlin, Germany.

1.7.16. Fenner, J. (1985) Late Cretaceous to Oligocene planktic diatoms. In: Bolli, H.M., Saunders, J., Perch-Nielsen, K. (eds.), *Plankton Stratigraphy*. Cambridge University Press, Cambridge, UK: 713-762.

1.7.17. Fog, A. (2008) Sampling methods for Wallenius' and Fisher's noncentral hypergeometric distributions. *Communications in Statistics—Simulation and Computation* 37, 241-257.

1.7.18. Fourtanier, E. and Kociolek, J.P. (1999) Catalogue of the diatom genera. *Diatom Research* 14(1), 1-9.

1.7.19. Fourtanier, E. and Kociolek, J.P. (2011) Catalogue of Diatom Names, California Academy of Sciences, http://research.calacademy.org/research/diatoms/names/index.asp

1.7.20. Gaiser, E.E., Smithsonian Tropical Research Institute-Digital File Manager, STRI Data Portal (https://stricollections.org/portal/imagelib/index.php) *Grammatophora undulata*, accessed on March 23, 2020, and https://biogeodb.stri.si.edu/bioinformatics/dfmfiles/files/c/23739/23739.jpg, accessed on May 10, 2022.

1.7.21. Gordon, R. and Tiffany M.A. (2011) Possible buckling phenomena in diatom morphogenesis. In: *The Diatom World*. J. Seckbach and J.P. Kociolek, (eds.) Springer, Dordrecht, The Netherlands: 245-272.

1.7.22. Graham, L.E., Cook, M.E., Busse, J.S. (2000) The origin of plants: body plan changes contributing to a major evolutionary radiation. *Proceedings of the National Academy of Sciences* 97(9), 4535-4540.

1.7.23. Hall, L.M. (1995) Special Functions, Chapter 4, https://docplayer.net/124258672-Special-functions-leon-m-hall-professor-of-mathematics-university-of-missouri-rolla-copyright-c-1995-by-leon-m-hall-all-rights-reserved.html, accessed on-line 28 February 2020.

1.7.24. Hamm, C.E. (2005) The evolution of advanced mechanical defenses and potential technological applications of diatom shells. *Journal of Nanoscience and Nanotechnology* 5, 108-119.

1.7.25. Hamm, C.E. (2007) Interactions between biomineralization and function of diatom frustules. In: Bäuerlein, E. (ed.), *Handbook of Biomineralization: Biological Aspects and Structure Formation*. Wiley-VCH Verlag GmbH, Weinheim, Germany: 83-94.

1.7.26. Hamm, C.E., Merkel, R., Springer, O., Jurkojc, P., Maier, C., Prechtel, K., Smetacek, V. (2003) Architecture and material properties of diatom shells provide effective mechanical protection. *Nature* 421, 841-843.

1.7.27. Harwood, D.M. (1988) Upper Cretaceous and lower Paleocene diatom and silicoflagellate biostratigraphy of Seymour Island, eastern Antarctic Peninsula. *Geol. Soc. Am. Mem.* 169, 55-130.

1.7.28. Harwood, D.M. and Nikolaev, V.A. (1995) Cretaceous diatoms: morphology, taxonomy, biostratigraphy. *Short Courses in Paleontology* 8, 81-106.

1.7.29. Hayashi, T., Krebs, W.N., Saito-Kato, M., Tanimura, Y. (2018) The turnover of continental planktonic diatoms near the middle/late Miocene boundary and their Cenozoic evolution. *PLoS ONE* 13(6): e0198003.

1.7.30. Irigoien, X., Flynn, K.J., Harris, R.P. (2005) Phytoplankton blooms: a 'loophole' in micro-zooplankton grazing impact? *Journal of Plankton Research* 27(4), 313-321.

1.7.31. Jahn, R. and Kusber, W.-H. (2004) Algae of the Ehrenberg collection – 1. Typification of 32 names of diatom taxa described by C. G. Ehrenberg. *Willdenowia* 34, 557-595.

1.7.32. Kalinka, A.T. (2014) The probability of drawing intersections: extending the hypergeometric distribution. arXiv:1305.0717v5 [math.PR]

1.7.33. Kaplan, W. (2003) *Advanced Calculus, 5th edition*, Addison-Wesley, Reading, Massachusetts, USA.

1.7.34. Kingston, J.C. and Pappas, J.L. (2009) Quantitative shape analysis as a diagnostic and prescriptive tool in determining *Fragilariforma* (Bacillariophyta) taxon status. *Nova Hedwigia Beiheft* 135, 103-119.

1.7.35. Kloster, M., Kauer, G., Beszteri, B. (2014) SHERPA: an image segmentation and outline feature extraction tool for diatoms and other objects. *BMC Bioinformatics* 15:218.

1.7.36. Kociolek, P. (2011a) *Gomphonema acuminatum*. In Diatoms of North America. Retrieved March 23, 2020, from https://diatoms.org/species/gomphonema_acuminatum

1.7.37. Kociolek, P. (2011b) *Grunowia tabellaria*. In Diatoms of North America. Retrieved March 25, 2020, from https://diatoms.org/species/grunowia_tabellaria

1.7.38. Kociolek, P. (2011c) *Nitzschia sigma*. In Diatoms of North America. Retrieved March 25, 2020, from https://diatoms.org/species/nitzschia_sigma1

1.7.39. Kociolek, J.P., Balasubramanian, K., Blanco, S., Coste, M., Ector, L., Liu, Y., Kulikovskiy, M., Lundholm, N., Ludwig, T., Potapova, M., Rimet, F., Sabbe, K., Sala, S., Sar, E., Taylor, J., Van de Vijver, B., Wetzel, C.E., Williams, D.M., Witkowski, A., Witkowski, J. (2020) DiatomBase. Accessed at: http://www.diatombase.org on 2020-03-25.

1.7.40. Kotaru, A.R., Shameer, K., Sundaramurthy, P., Joshi, R.C. (2013) An improved hypergeometric probability method for identification of functionally linked proteins using phylogenetic profiles. *Bioinformation* 9(7), 368-374.

1.7.41. Krammer, K. and Lange-Bertalot, H. (1991) *Bacillariophyceae. 4. Teil: Achnanthaceae. Kritische Ergänzungen zu* Navicula *(Lineolatae) und* Gomphonema, *Süßwasserflora von Mitteleuropa, band 2/4*, Ettl, H., Gerloff, J., Heynig, H., Mollenhauer, D. (eds.), Gustav Fischer Verlag, Stuttgart Jena, Germany.

1.7.42. Lazarus, D., Barron, J., Renaudie, J., Diver, P., Türke, A. (2014) Cenozoic planktonic marine diatom diversity and correlation to climate change. *PLoS ONE* 9: e84857.

1.7.43. Lewitus, E., Bittner, L., Malviya, S., Bowler, C., Morlon, H. (2018) Clade-specific diversification dynamics of marine diatoms since the Jurassic. *Nature Ecology & Evolution* 2, 175-1723.

1.7.44. Li, Y., Calvo, S.E., Gutman, R., Liu, J.S., Mootha, V.K. (2014) Expansion of biological pathways based on evolutionary inference. *Cell* 158, 213-225.

1.7.45. Loughney, K.M., Hren, M.T., Smith, S.Y., Pappas, J.L. (2019) Vegetation and habitat change in southern California through the Middle Miocene Climatic Optimum: paleoenvironmental records from the Barstow Formation, Mojave Desert, USA. *GSA Bulletin* Doi: 10.1130/B35061.1.

1.7.46. Lyons, D.C., Marindale, M.Q., Srivastava, M. (2012) The cell's view of animal body-plan evolution. *Integrative and Comparative Biology* 54(4), 658-666.

1.7.47. Mann, D.G. (1999) The species concept in diatoms. *Phycologia* 38, 437-495.

1.7.48. Mann, D.G., Crawford, R.M., Round, F.E. (2017) Bacillariophyta. In: Archibald, J., Simpson, A., Slamovitis, C. (eds.), *Handbook of the Protists*, Springer, Cham.

1.7.49. Mihalek, I., Reš, I., Lichtarge, O. (2006) A structure and evolution-guided Monte Carlo sequence selection strategy for multiple alignment-based analysis of proteins. *Bioinformatics* 22(2), 149-156.

1.7.50. Morales, E. (2013) *Pseudostaurosira trainorii*. In Diatoms of North America. Retrieved March 23, 2020, from https://diatoms.org/species/pseudostaurosira_trainorii.

1.7.51. Mou, D. and Stoermer E.F. (1990). Separating *Tabellaria* (Bacillariophyceae) shape groups based on Fourier descriptors. *Journal of Phycology* 28, 386-395.

1.7.52. Olver, F.W. and Lozier, D.W. (2010) *NIST Handbook of Mathematical Functions*, Cambridge University Press, New York, New York, USA.

1.7.53. Pappas, J.L. (2016) Multivariate complexity analysis of 3D surface form and function of centric diatoms at the Eocene-Oligocene transition. *Marine Micropaleontology* 122, 67-86.

1.7.54. Pappas, J.L. and Stoermer, E.F. (1995) Great Lakes Diatoms, (http://umich.edu/~phytolab/GreatLakesDiatomHomePage/top.html) accessed on February-March, 2020.

1.7.55. Pappas, J.L. and Stoermer, E.F. (2003) Legendre shape descriptors and shape group determination of specimens in the *Cymbella cistula* species complex. *Phycologia* 42(1), 90-97.

1.7.56. Pappas, J.L., Fowler, G.W., Stoermer, E.F. (2001) Calculating shape descriptors from Fourier analysis: shape analysis of *Asterionella* (Heterokontophyta, Bacillariophyceae). *Phycologia* 40(5), 440-456.

1.7.57. Pappas, J.L., Kociolek, J.P., Stoermer, E.F. (2014) Quantitative morphometric methods in diatom research. In: Nina Strelnikova Festschrift. J.P. Kociolek, M. Kulivoskiy, J. Witkowski, and D.M. Harwood, D.M. (eds.), *Nova Hedwigia, Beihefte* 143, 281-306.

1.7.58. Potapova, M. (2011) *Hippodonta capitata*. In Diatoms of North America. Retrieved March 23, 2020, from https://diatoms.org/species/hippodonta_capitata

1.7.59. Potapova, M.G., Minerovic, A.D., Veselá, J., Smith, C.R. (eds.) (2020) Diatom New Taxon File at the Academy of Natural Sciences (DNTF-ANS), Philadelphia. Retrieved on February-March, 2020, from http://dh.ansp.org/dntf, symbiont.ansp.org.

1.7.60. Puza, B. and Bonfrer, A. (2018) A series of two-urn biased sampling problems. *Communications in Statistics – Theory and Methods* 47(1), 80-91.

1.7.61. Raven, J.A. and Waite, A.M. (2004) The evolution of silicification in diatoms: inescapable sinking and sinking as an escape? *New Phytologist* 162, 45-61.

1.7.62. Renaudie, J., Drews, E.-L., Böhne, S. (2018) The Paleocene record of marine diatoms in deep-sea sediments. *Foss. Rec.* 21, 183-205.

1.7.63. Rhode, K.M, Pappas, J.L., Stoermer, E.F. (2001) Quantitative analysis of shape variation in type and modern populations in *Meridion* ag. (Bacillariophyta). *Journal of Phycology* 7, 175-183.

1.7.64. Round, F.E., Crawford, R.M. and Mann, D.G. (1990) *The Diatoms, Biology & Morphology of the Genera*. Cambridge University Press, Cambridge, U.K.

1.7.65. Schmidt, A. (1874-1959). Atlas der Diatomaceen-Kunde, von Adolf Schmidt, continued by Martin Schmidt, Friedrich Fricke, Heinrich Heiden, Otto Muller, Friedrich Hustedt. Reprint 1984, Koeltz Scientific Books, Konigstein, 480 plates.

1.7.66. Sitnikova, T.Y., Pomazkina, G.V., Sherbakova, T.A., Maximova, N.V., Khanaev, I.V., Bukin, Y.S. (2014) Patterns of diatom treatment in two coexisting species of filter-feeding freshwater gastropods. *Knowledge and Management of Aquatic Ecosystems* 413, 08. DOI: 10.1051/kmae/2014003

1.7.67. Šlapeta, J., Moreira, D., López-García, P. (2005) The extent of protist diversity: insights from the molecular ecology of freshwater eukaryotes. *Proc. Royal Society B* 272, 2073-2081.

1.7.68. Smetacek, W.S. (1985) Role of sinking in diatom life-history cycles: ecological, evolutionary and geological significance. *Mar. Biol.* 84, 239-251.

1.7.69. Snoeijs, P., http://media.nordicmicroalgae.org/original/Licmophora%20gracilis_1194440290. jpg, accessed on May 10, 2022.

1.7.70. Spaulding, S., Edlund, M. (2008a) *Cylindrotheca*. In Diatoms of North America. Retrieved March 23, 2020, from https://diatoms.org/genera/cylindrotheca

1.7.71. Spaulding, S., Edlund, M. (2008b) *Tetracyclus*. In Diatoms of North America. Retrieved March 25, 2020, from https://diatoms.org/genera/tetracyclus

1.7.72. Spaulding, S., Edlund, M. (2009) *Amphicampa*. In Diatoms of North America. Retrieved March 23, 2020, from https://diatoms.org/genera/amphicampa

1.7.73. Steinman, A.D. and Ladewski, T.B. (1987) Quantitative shape analysis of *Eunotia pectinalis* (Bacillariophyceae) and its application to season distribution patterns. *Phycologia* 26, 467-477.

1.7.74. Stidolph, S.R., Sterrenburg, F.A.S., Smith, K.E.L., and Kraberg, A. (2012) Stuart R. Stidolph Diatom Atlas: U.S. Geological Survey Open-File Report 2012–1163, available online at http://pubs.usgs.gov/of/2012/1163/.

1.7.75. Stoermer, E.F. and Ladewski, T.B. (1982) Quantitative analysis of shape variation in type and modern populations of *Gomphonema herculeana*. *Beih. Nova Hedwigia* 73, 347-386.

1.7.76. Stoermer, E.F., Qi Y.-Z., Ladewski, T.B. (1986) A quantitative investigation of shape variation in *Didymosphenia* (Lyngbye) M. Schmidt (Bacillariophyta). *Phycologia* 25, 494-502.

1.7.77. Tall, L., Cloutier, L., Cattaneo, A. (2006) Grazer-diatom size relationships in an epiphytic community. *Limnology Oceanography* 51, 1211-1216.

1.7.78. Tarasov, V.E. (2016) Some identities with generalized hypergeometric functions. *Appl. Math. Inf. Sci.* 10(5), 1729-1734.

1.7.79. Theriot, E. and Ladewski, T.B. (1986) Morphometric analysis of shape of specimens from the neotype of *Tabellaria flocculosa* (Bacillariophyceae). *American Journal of Botany* 73, 224-229.

1.7.80. Vincent, F. and Bowler, C. (2020) Diatoms are selective segregators in global ocean planktonic communities. *MSystems* 5:e00444-19. https://doi.org/10.1128/mSystems.00444-19.

1.7.81. Weinberger, H.F. (1965) *A First Course in Partial Differential Equations with Complex Variables and Transform Methods*. Dover Publications, Inc., New York, New York, USA.

1.7.82. Weisstein, E.W. (ed.) (2002) *Concise Encyclopedia of Mathematics, 2nd edition*. Chapman and Hall/CRC, New York, USA. https://doi.org/10.1201/9781420035223

1.7.83. Willmore, K.E. (2012) The body plan concept and its centrality in evo-devo. *Evo Edu Outreach* 5, 219-230.

1.7.84. Wishkerman, A. and Hamilton, P.B. (2018) Shape outline extraction software (DiaOutline) for elliptic Fourier analysis application in morphometric studies. *Appl. Plant Sci.* 6(12):e01204.

1.7.85. Witkowski, J., Sims, P.A., Williams, D.M. (2017) Typification of Eocene-Oligocene diatom taxa proposed by Grove & Sturt (1886-1887) from the Oamaru diatomite. *Diatom Research* 32(4), 363-408.

1.7.86. Wolfram Research, Inc. (2016) Mathematica, Version 11.0.0, Champaign, Illinois.

1.7.87. Yu, M., Ashworth, M.P., Hajrah, N.H., Khiyami, M.A., Sabir, M.J., Alhebshi, A.M., Al-Malki, A.L., Sabir, J.S.M., Theriot, E.C., Jansen, R.K. (2018) Evolution of the plastid genomes in diatoms. *Advances in Botanical Research* 85, 129-155.

2

Comparative Surface Analysis and Tracking Changes in Diatom Valve Face Morphology

Abstract

Digital images and 3D surface models are important in diatom morphological and morphometric analyses. While a digital image represents a real object analytically and algorithmically, a 3D surface model represents the same real object implicitly and geometrically. *Arachnoidiscus ehrenbergii* is used as an exemplar in valve surface morphology analysis. From a digital SEM, pixel values are converted geometrically to vector field values representing surface features. A parametric 3D surface model is devised, then converted to an image format where vector field values are extracted. Displacement and velocity vector field values in concert with the Jacobian (i.e., Jacobian matrix) are used in measuring image matching. The Jacobian determinant indicates how well surface features are matched. Adding an *Arachnoidiscus ornatus* SEM induced differentiating this taxon from *A. ehrenbergii*. The 3D surface model matched *A. ornatus* better than it matched *A. ehrenbergii*. Image matching of surface features may be instituted using many images, and the potential as an aid in taxonomic and morphologically-based phylogenetic studies is evident.

Keywords: Vector surface, Jacobian, image matching, crest line, extrema, displacement field, velocity field, visual servoing

2.1 Introduction

Surface morphology is a proxy for the phenotype [2.9.30.]. By studying organism surfaces, a better appreciation of the connection between morphology and function of biological structures may be obtained. Inferences about the evolution of such structures may lead to a better understanding of the way in which evolutionary processes occur [2.9.30.].

Sources of surface morphological data include 3D surface models and digital images. A 3D model reflects an implicit approximation of a real object and a geometric interpretation of morphology, while a digital image reflects an analytic approximation of a real object and an algorithmic interpretation of morphology. Both renditions of a real object may represent the same object, but it is not immediately evident how these two sets of data are related. Digital images provide immediate access to the 3D surface and may be mathematically represented via techniques resulting from computer vision and pattern recognition applications.

With a 3D model and digital image of the same real object, surface features may be extractable as a result of analyses of each set of data. Characteristics of the data may be changes in height in the z-direction of the surface with respect to the x, y-plane in 3D models, while changes in pixel or voxel intensity over the image surface occur in digital

Janice L. Pappas. *Mathematical Macroevolution in Diatom Research*, (39–80) © 2023 Scrivener Publishing LLC

images. The relation between changes in z-height and the surface geometry determined from intensity values determine the connection between a 3D model and digital image for the same real object.

Parameterized surfaces have been used to construct 3D models of organisms, including diatoms, and one extractable measurement of those 3D surfaces is the Jacobian (i.e., Jacobian matrix) [2.9.25.-2.9.29.]. Systems of parametric 3D equations are solved to devise 3D renditions of organisms, and first partial derivatives of the parameterized variables are used to represent 3D surface morphology with regard to changes in height via gradients [2.9.25.-2.9.29.]. Elements of the Jacobian consist of one of the measurements of ensemble surface features representing groups of points [2.9.29.]. These elements consist of directional changes on the surface where corresponding 3D points in one space are projected onto new corresponding 2D points in another space. Relation between one set of ensemble surface features as one Jacobian on one part of a surface and another group of ensemble surface features as another Jacobian on a different part of that same surface are measurable as the time derivatives of each Jacobian. The Jacobian is a linear map from one tangent space to another one.

Changes in intensity is inferentially recorded via pixel or voxel values from a digital image. On a digital image, the Jacobian may be used to measure the relation between motion of points at different locations on a surface (e.g., [2.9.23.]), or measurement of the displacements between points from a starting position on a surface to an ending position on another surface (e.g., [2.9.14.]). The image Jacobian is calculated using servoing (e.g., [2.9.6.]) techniques in robotics (e.g., [2.9.31.]) to determine the relational matching among the greater 3D space coordinates, camera motion spatially, and coordinates from one object to another one via pixel or voxel value changes in the resultant images (e.g., [2.9.12.]). That is, 2D points located on an image are moved from their original position to another position as a motion in 3D space so that 3D points in space are projected onto new 2D points on the image [2.9.17.]). The image Jacobian has been called a feature sensitivity matrix [2.9.42.]. Similar techniques are used in medical imaging in which successive magnetic resonance images need to be matched (e.g., [2.9.33., 2.9.40.]). Conceptually, the notion of motion occurring between observing one point on a surface to observing a comparable point on another surface may be used to represent the observed morphological change in comparable image surfaces from biological organisms. The motion between two images is infinitesimal as is the change with regard to the greater 3D space with rotational changes being negligible. Most importantly, because the matching of two images can be accomplished without reference to the greater 3D space, alignment of images is more easily accomplished.

2.1.1   Image Matching of Surface Features

Image matching methods may be applied to assessment of diatom digital images. Jacobian-based methods may be used to record the change in surface topography between two points on a given diatom valve surface or between points on two (or more) different diatom valve surfaces. Displacement is the change in shape of one object, such as a surface feature, with respect to another one during matching. Linear and angular velocities are changes in motion between the selected points on one object and another one. Displacement and velocity fields comprise the changes between images where the points with displacements are related to directions of motion as velocities. Columns of the Jacobian signify velocities [2.9.20.], and linear combinations of the columns indicate motion [2.9.36.] between image surface feature points.

The Jacobian is a map that is a transformation between different locations on surfaces representing local changes such as stretching, shearing or rotation (e.g., [2.9.18.]). The motion between two (or more) surfaces is relative to the locations chosen for analysis. Surface changes are recoverable not only on digital images, but also on comparable 3D surface models. Similarity between Jacobians measured on surfaces generated by different means is feasible via matrix operations (e.g., [2.9.10.]). For an image matrix, convolution produces a response curve for each row of the location of corners and edges, while summing each of the columns produces a response curve of the maximum offset of corners and edges. Edge orientation may be found by maximizing the aggregation of the maximum values of response curves. Three response curves are aggregated by summing along columns. Angles are found by aggregating along the diagonals (e.g., [2.9.10.]).

The location and position of a selected surface feature may be isolated as a window, where a window is defined as a polygonal shape so that matching is dependent on the same shape on two or more images rather than image resolution [2.9.7.]. Using a reference window, only the center of n-windows on n-surfaces is considered. Tracking surface changes from one image to another or tracking a series of images may occur by using a central or mean pixel value [2.9.12.]). Translated or rotated motion between two surfaces may occur from a location on one surface to a location on another surface [2.9.12.]. Surface markings definable as points, corners or edges are obtained via placing a window around the areas of interest on both surfaces, and the motion between the two surfaces records comparison of the areas' surface markings. The window is the neighborhood of the surface feature of interest and is typically a rectangle. The size of the rectangle dictates length of computational time and degree of complexity; the larger the rectangle, the more time expended for more computational complexity. New values are calculated from each rectangle representing the features in each window between images.

2.1.2   Image Matching: Diatoms

The organism of interest in this study is the diatom. These unicells have siliceous scaffolded structure with a multitude of geometric features that make them amenable for study as 3D surface models. Digital scanning electron (SEM) or light micrographs (LM) enable analyses of diatom surfaces with respect to pixel intensity with inferences about height of surface features via geometric transformation. Image surface features, $x_1, ..., x_n$, represent n-patterns, where each pattern represents the quantified surface of a given diatom valve face. Both 3D surface models and digital images enable measuring Jacobians to determine changes in surface morphology on a single cell or between cells.

On the surface of a diatom valve, there are different features that exist in a given proximity to one another. These morphological features may be distinct characters typically used in phylogenetic analysis, or groups of surface height changes as ensemble surface features that may be recognized [2.9.29.]. Proximity of morphological features is measurable as displacements on the surface. A given start of displacement to a given end of displacement is a function of changes in positions and orientations. Orientation may be preserved or reversed in which reversal means that the end position is considered first with the start position being considered last. The time derivatives of start positions and orientations and end positions and orientations may be calculable as displacements per unit time and velocities. These velocities are first partial derivatives as a function of parameterization of the 3D coordinates used to determine positions and orientations, and displacements and velocities are related to each other by the Jacobian.

Analytically, the Jacobian is a gateway to accessing the entire diatom valve surface so that morphology is quantifiable implicitly recovering information about changes in the surface without reference to particular morphological characteristics. Valve faces are compared via the changes throughout the surfaces within the valve margin. As a result, a group of taxa, even at the genus-level, may be quantified and compared for morphological variation with regard to classification correctness. The degree to which a classification scheme is correct is obtained via techniques available from optimization and regression theories.

Analyses of morphological characters and cellular structures of functional utility in a diatom cell enable study of evolutionary acquisitional morphology, including spines or other protrusions and jagged valve surfaces that may indicate predation resistance, pores and their shape and spacing that may be important in nutrient uptake, and valve thickness, pore distribution and smooth/flat areas of the valve surfaces as indicators of cell strength or stability. Jacobians and their related constituents obtained from 3D surfaces or digital images expands our ability to quantify and study morphological variation, and potentially used in applications to study speciation and morphological evolution in diatoms.

2.2   Purpose of this Study

Arachnoidiscus is used as an exemplar genus to illustrate the matching of a 3D surface model to SEMs using concepts from differential geometry, medical image processing, computer vision, and robotics techniques. Analyses are centered on Jacobian-based methods. Analyses will include how to define surface features and the fidelity of image matching. Specifically, SEM images of *A. ehrenbergii* and *A. ornatus* are used in the analyses. The techniques used in this study are applicable to taxonomic, developmental, phylogenetic, ecologic, and evolutionary studies, including speciation determinations.

2.3   Background on Image and Surface Geometry

2.3.1   The Geometry of the Digital Image and the Jacobian

Images consist of pixels if 2D and voxels if 3D, and the intensity of those pixels or voxels drive the capacity to match one image to another one. From pixels which are 2D planes and the exterior faces of voxels which are 3D cubes, values may be extracted and transformed to represent edges on the image surface that eventually may be summarized by the Jacobian. Whether using a digital image or 3D surface model, the Jacobian represents surface features.

Each feature may be defined as a crest surface [2.9.40.] whereby maximal points of curvature are the edges of interest of that feature. Surface edges of a given feature will have the highest intensities when compared to flatter parts of the surface so that these intensities represent the crest surfaces.

Crest surfaces at different intensities (Figure 2.1) are iso-surfaces with crest lines representing maximal crests [2.9.40., 2.9.43.]. The crest line corresponds to the extrema of two planes of principal curvatures, κ_1 and κ_2, (Figure 2.2): one aligned with the crest line and the other perpendicular to the crest line (e.g., [2.9.9., 2.9.15., 2.9.40., 2.9.45.]). The planes are crest surfaces which are also implicit surfaces [2.9.40.]. Principal curvatures and the original crest line define the shape of three implicit surfaces as extremal surfaces [2.9.38.] that make up the surface feature.

On the crest surface, perpendicular vectors in the plane and a normal vector to that plane of the crest line for a given point, P, form a Serret-Frenet trihedron (Figure 2.1) (e.g., [2.9.9., 2.9.15.]). The Serret-Frenet trihedron defines the principal directions and a unit normal at P on the surface, and curvature at P via κ_1 and κ_2 may be evaluated via the Gaussian and mean curvatures, $K = \kappa_1\kappa_2$ and $H = \dfrac{\kappa_1 + \kappa_2}{2}$, respectively [2.9.9.]. At a given P, κ_1 and κ_2 are extremal curvatures of all possible curvatures on the surface, unless P induces κ_1 and κ_2 to be equal, then the points are umbilics [2.9.9.].

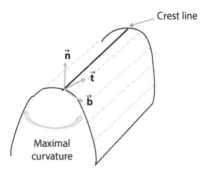

Figure 2.1 Crest line, Serret-Frenet trihedron (blue arrows) with $\vec{n}$ (unit normal), $\vec{b}$ (bi-normal) and $\vec{t}$ (tangent) as principal directions, and maximal curvature of an iso-surface as a surface feature.

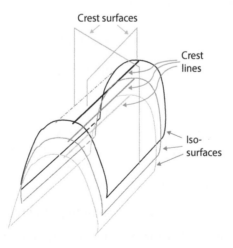

Figure 2.2 At a crest line, crest surfaces (planes of principal curvatures) intersect with iso-surfaces. The crest lines are extremal lines, and the iso-surfaces and crest surfaces are implicit surfaces as well as extremal surfaces.

From the crest line, maximal curvature of a surface feature is attained via principal curvatures and directions [2.9.40., 2.9.45.]. Derivatives define extremal points on crest lines, and the maximum is given as $e_{max} = \dfrac{\partial \kappa_{max}}{\partial r_{max}} < 0$, where r represents corresponding principal direction along corresponding curvatures [2.9.45.]. Crest lines may be convex or concave, so an extremal point may be defined as $e_{min} = \dfrac{\partial \kappa_{min}}{\partial r_{min}} > 0$ as well [2.9.45.].

At each point on the surface, there is a gradient, and the gradients of extremal points are associated to the maximal curvatures. A coordinate system is defined via points having coordinate pairs (u, v) that correspond to lines of curvature [2.9.46.]. From the Serret-Frenet trihedron, normal vectors along lines of curvature are associated to the gradients. The direction of the gradients is indicative of the directions of curvature [2.9.46.].

The first and second fundamental forms play a significant role in the connection between curvature, direction of curvature and gradients. The first fundamental form coefficients indicate tangent lines, while the second fundamental form coefficients indicate tangent planes via change in surface normals with respect to the tangents, inducing changes in gradients [2.9.9., 2.9.15.]. Gaussian and mean curvature may be expressed in terms of the first and second fundamental forms [2.9.9.]. From K, general shape of a surface is determined [2.9.15.], and from H, overall surface shape embedded in 3D space is determined [2.9.9.].

Consider the intersection of two surfaces (Figure 2.2). A unit normal vector, $\vec{n}$, at P may be expressed in a set of coordinates $\mathbf{r}(u, v)$ as $\vec{n}_{u,v}(P) = \dfrac{\mathbf{r}_u \wedge \mathbf{r}_v}{\|\mathbf{r}_u \wedge \mathbf{r}_v\|} = \dfrac{\mathbf{r}_u \times \mathbf{r}_v}{\|\mathbf{r}_u \times \mathbf{r}_v\|}$, where $\mathbf{r}$ represents tangent vectors in the u and v directions. The norm of the cross product is $(\mathbf{r}_u \cdot \mathbf{r}_u)(\mathbf{r}_v \cdot \mathbf{r}_v) - (\mathbf{r}_u \cdot \mathbf{r}_v)^2 = EG - F^2$, where E, F and G are coefficients of the first fundamental form, and $E = \mathbf{r}_u \cdot \mathbf{r}_u$, $F = \mathbf{r}_u \cdot \mathbf{r}_v$ and $G = \mathbf{r}_v \cdot \mathbf{r}_v$. For another set of coordinates $\vec{r}(u,v)$, $\overline{\mathbf{r}}_u \wedge \overline{\mathbf{r}}_v = (\mathbf{r}_u \wedge \mathbf{r}_v)\left(\dfrac{\partial(u,v)}{\partial(\overline{u},\overline{v})}\right)$ [2.9.9.]. The quotient of the first partial derivatives is the change from one set of coordinates to another one and is the Jacobian, J. The parameters u, v are used to express the connection between points on the surface via the inner product $J^T J = \begin{bmatrix} \mathbf{r}_u^T \mathbf{r}_u & \mathbf{r}_u^T \mathbf{r}_v \\ \mathbf{r}_u^T \mathbf{r}_v & \mathbf{r}_v^T \mathbf{r}_v \end{bmatrix}$. Because $\vec{n}_{u,v}$ is self-adjoint, a quadratic form is associated to the derivative of $\vec{n}_{u,v}$ via the second fundamental form [2.9.9.] with coefficients $L = \mathbf{r}_{uu} \cdot \mathbf{n}_{u,v}$, $M = \mathbf{r}_{uv} \cdot \mathbf{n}_{u,v}$ and $N = r_{vv} \cdot \mathbf{n}_{u,v}$.

For three implicit surfaces of an iso-surface of an image defined as $\Phi(\varphi_x, \varphi_y, \varphi_z)$, the parameterization of a curve in space is $\varphi_x = f(s)$, $\varphi_y = g(s)$ and $\varphi_z = h(s)$, where s is the arc length along a curve. Displacements from P along the crest line of the curve are $d\varphi_x = f'(s)ds$, $d\varphi_y = g'(s)ds$ and $d\varphi_z = h'(s)ds$ [2.9.13.]. The intersecting crest surfaces (Figure 2.2) are parametrically defined as $\dfrac{\partial \varphi_{x1}}{\partial x}dx + \dfrac{\partial \varphi_{y1}}{\partial y}dy + \dfrac{\partial \varphi_{z1}}{\partial z}dz = 0$ and $\dfrac{\partial \varphi_{x2}}{\partial x}dx + \dfrac{\partial \varphi_{y2}}{\partial y}dy + \dfrac{\partial \varphi_{z2}}{\partial z}dz = 0$ as the corresponding differentials. For a crest surface in vector form, the normal at P is the gradient vector $\nabla \Phi 1 = \dfrac{\partial \varphi_{x1}}{\partial x}\mathbf{i} + \dfrac{\partial \varphi_{y1}}{\partial y}\mathbf{j} + \dfrac{\partial \varphi_{z1}}{\partial z}\mathbf{k}$, and similarly, there is a gradient vector defined for $\Phi 2$. The crest line expressed as

$\dfrac{dx}{\dfrac{\partial(\varphi_{x1},\varphi_{x2})}{\partial(y,z)}} = \dfrac{dy}{\dfrac{\partial(\varphi_{y1},\varphi_{y2})}{\partial(z,x)}} = \dfrac{dz}{\dfrac{\partial(\varphi_{z1},\varphi_{z2})}{\partial(x,y)}}$ are Jacobians for x, y and z. The direction in which

the gradient is at a maximal magnitude is $|\nabla\Phi1| = \sqrt{\left(\dfrac{\partial\varphi_{x1}}{\partial x}\right)^2 + \left(\dfrac{\partial\varphi_{y1}}{\partial y}\right)^2 + \left(\dfrac{\partial\varphi_{z1}}{\partial z}\right)^2}$ and

similarly defined for the case of $\Phi2$. From this, the Jacobian is the compilation of the gradients of the surface.

2.3.2   The Geometry of the Diatom 3D Surface Model and the Jacobian

The valve face of a diatom is represented by the change in the outline of the face of a 3D cylinder (e.g., [2.9.28.-2.9.29.]). Parametric 3D surfaces with implicit functions are used to create 3D surface models that provide appropriate topography replicating a diatom valve surface (e.g., [2.9.25.-2.9.29.]). Pointwise changes in geometry on the diatom 3D surface model are changes in slopes on the surface that represent morphological features [2.9.28., 2.9.29.]. Any point on the surface has a slope, and changes from point to point are measurable as motion from one slope to another. Slopes are calculated as first partial derivatives from the equations used to create the parametric 3D surface with defined boundary conditions. The Jacobian consists of first partial derivatives of the surface and is numerically solved at maximum boundary conditions (e.g., [2.9.28.-2.9.30.]).

For the exemplar genus, *Arachnoidiscus*, the system of parametric 3D equations with u, $v = [0, 2\pi]$ is

$$x = 16 \cos u \cos v \, (1 + \sin u) \tag{2.1}$$

$$y = 16 \cos u \sin v \, (1 + \sin u) \tag{2.2}$$

$z = v \cos (0.5u^2) + \rho \cos (80u^2) \sin (1.9u)^2 + \Xi \sin (1 + u)^3 + \tau \cos (10 + 74u^2) \sin (0.5 + 0.8u)^3 \sin (2 + 2.9u)^3 + \vartheta \cos u \sin (0.37u)^3 \sin (22v)^3 \hfill (2.3)$

From this, the plotted 3D model is convertible to any image format.

From parametric 3D equations, first derivatives are expressed in terms of first partial derivatives with respect to arc length, s. The differential of s is $ds^2 = dx^2 + dy^2 + dz^2$ on the surface, and $dx = \dfrac{\partial x}{\partial u} du + \dfrac{\partial x}{\partial v} dv$, $dy = \dfrac{\partial y}{\partial u} du + \dfrac{\partial y}{\partial v} dv$, and $dz = \dfrac{\partial z}{\partial u} du + \dfrac{\partial z}{\partial v} dv$. From this, the Jacobian is

$$J = \begin{bmatrix} \dfrac{\partial x}{\partial u} & \dfrac{\partial x}{\partial v} \\[2mm] \dfrac{\partial y}{\partial u} & \dfrac{\partial y}{\partial v} \\[2mm] \dfrac{\partial z}{\partial u} & \dfrac{\partial z}{\partial v} \end{bmatrix}$$ with first partial derivatives of x, y, z with respect to u, v. Numerical solution

of the Jacobian characterizes all the tangent lines and planes on the 3D surface.

For image matching of surface features, the 3D surface model must be converted to an image format. While the Jacobian of the whole valve face characterizes the overall surface geometry, the Jacobian calculated for image matching is accomplished on a point-by-point basis with a digital image from an actual specimen. In this vein, characterization of the Jacobian with respect to digital images is necessary.

2.3.3   The Image Gradient and Jacobian

The connection measurement between surface features on valves occurs on the planar surface of pixels or one of the outside planar sides of voxels, or the centroid of either pixels or voxels. Image matching involves inclinations and declinations between points on respective images that are measurable as image gradients. Finite differences from a Taylor expansion gives the image gradient as a linear combination of neighborhood pixel values (e.g., [2.9.11.]).

An image gradient may be extracted based on edge detection via convolution using Canny, Prewitt or Sobel filters [2.9.21.]. Directional change and the magnitude of that change are encompassed in the image gradient as first partial derivatives and are calculable based on changes in pixel intensity with respect to edges in an image. In 2D, the image gradient is given as $\nabla f_{image} = \left(\dfrac{\partial f_{image}}{\partial x}, \dfrac{\partial f_{image}}{\partial y} \right) = \begin{pmatrix} \dfrac{\partial f_{image}}{\partial x} \\ \dfrac{\partial f_{image}}{\partial y} \end{pmatrix}$ (e.g., [2.9.21.]) and is indicative of the spatial change between two images [2.9.32.] as the change in pixel intensity [2.9.21.].

The image gradient of surface features is scalar valued. A convolution of the image gradient with a Gaussian function, $G(x, y)$, induces $\dfrac{\partial (f_{image} G)}{\partial x \partial y} = \left(f_{image} \dfrac{\partial G(x, y)}{\partial x}, f_{image} \dfrac{\partial G(x, y)}{\partial y} \right)$ [2.9.21.]. From the image gradient via filtering, horizontal (x) and vertical (y) changes across an image involve changes in displacement and direction of position. Each gradient at each point is a row vector of a Jacobian [2.9.13.]. The image gradients for surface features are related to the Gaussian derivative, and in terms of the Jacobian, a grayscale image is summarized as

$$J_{grayscale} = \left(\frac{\partial f_{grayscale}(x, y)}{\partial x \partial y} \right) = (f_{grayscale\,x}(x, y) \quad f_{grayscale\,y}(x, y)) \tag{2.4}$$

and a color image [2.9.21.] is

$$
J_{Color} = \begin{pmatrix} \dfrac{\partial f_{Color\,L}(x,y)}{\partial x \partial y} \\[2mm] \dfrac{\partial f_{Color\,a}(x,y)}{\partial x \partial y} \\[2mm] \dfrac{\partial f_{Color\,b}(x,y)}{\partial x \partial y} \\[2mm] \dfrac{\partial f_{Color\,Depth}(x,y)}{\partial x \partial y} \end{pmatrix} = \begin{pmatrix} f_{Color\,L\,x}(x,y) & f_{Color\,L\,y}(x,y) \\ f_{Color\,a\,x}(x,y) & f_{Color\,a\,y}(x,y) \\ f_{Color\,b\,x}(x,y) & f_{Color\,b\,y}(x,y) \\ f_{Color\,Depth\,x}(x,y) & f_{Color\,Depth\,y}(x,y) \end{pmatrix} \tag{2.5}
$$

where color is defined by the CIE(L*a*b) color space with three color channels, (L*a*b), and *Depth* connotes a fourth channel for depth image to ensure edge detection [2.9.21.]. The Jacobian enables a connection between small displacements among disparate spaces [2.9.14., 2.9.23.].

2.4   Image Matching Kinematics via the Jacobian

2.4.1   Position and Motion: The Kinematics of Image Matching

In order to match images via surface geometry, the conceptual framework must be established. Initially, a reference system is defined via a coordinate system in which to make geometric measurements. Position is the coordinates defining the point at which measurements are made from one reference frame to another one. The initial and final positions are defined as displacement, and the quantity that results is a displacement vector field. Displacement may occur linearly or at an angle where orientation is defined by a rotation matrix.

Local and global reference frames are needed. Reference frames are defined by the x, y, z-axes of Cartesian space. Characteristics of motion involve displacement, with the understanding that some amount of time has passed for motion to have occurred. Change over time in linear displacement is linear velocity, and change over time in angular displacement is angular velocity. For image matching, change in position of one image to the matching position in another image is defined to be linear.

Displacement occurs separately along the x, y and z axes. For example, during image matching, displacement may occur linearly in a constant direction along the x-axis with no rotation. That leaves displacement along the y and z-axes which occurs with respect to some change in orientation with respect to rotation. Displacement is a vector quantity with respect to change in position coordinates whether that change is linear or in an angular motion [2.9.8.]. Prismatic change is a linear sliding motion along one axis or in parallel when matching positions between two images. With the z-axis designated as the axis of motion, for the displacement vector having zero value elements for x and y, the only value of displacement would be z so that motion is limited to rectilinear translation [2.9.8.].

2.4.2   Displacement and Implicit Functions

Displacement rows and columns are represented parametrically as implicit functions $q = f(x, y)$ and $\omega = g(x, y)$, respectively. An arbitrary set of m equations in $m + n$ unknowns are used to represent pixel values with respect to the q rows of the matrix with ω columns. Arbitrary m equations in $m + n$ unknowns are defined as

$$F_1(q_1, \ldots, q_m, \omega_1, \ldots, \omega_n) = 0$$
$$\vdots$$
$$F_m(q_1, \ldots, q_m, \omega_1, \ldots, \omega_n) = 0$$

(2.6)

with

$$q_1 = f_1(\omega_1, \ldots, \omega_n)$$
$$\vdots$$
$$q_m = f_m(\omega_1, \ldots, \omega_n)$$

(2.7)

as implicit functions. From this, m differentiable functions are sought that would satisfy m equations [2.9.13.]. The m differentiable functions are summations of q and ω terms in m linear equations in m unknowns, thereby linearizing the relation among the pixel values. The m differentials define each row of a matrix as first partial derivatives in terms of q and ω.

At a given point with x, y coordinates on the image, a system of simultaneous linear equations is a linear approximation of an unknown implicit function as a total differential [2.9.13.]. The differentials of the functions are

$$F_{1y_1}\, dy_1 + \ldots + F_{1y_m}\, dy_m + F_{1x_1}\, dx_1 + \ldots + F_{1x_n}\, dx_n = 0$$
$$\vdots$$
$$F_{my_1}\, dy_1 + \ldots + F_{my_m}\, dy_m + F_{mx_1}\, dx_1 + \ldots + F_{mx_n}\, dx_n = 0$$

(2.8)

which can be written as $\mathbf{F_y}\, d\mathbf{y} + \mathbf{F_x}\, d\mathbf{x} = \mathbf{0}$ [2.9.13.]. $\mathbf{F_y}$ and $\mathbf{F_x}$ are the Jacobians

$$\mathbf{F_y} = \begin{bmatrix} F_{1y_1} & \cdots & F_{1y_n} \\ \vdots & & \vdots \\ F_{my_1} & \cdots & F_{my_m} \end{bmatrix} \text{ and } \mathbf{F_x} = \begin{bmatrix} F_{1x_1} & \cdots & F_{1x_n} \\ \vdots & & \vdots \\ F_{mx_1} & \cdots & F_{mx_m} \end{bmatrix}$$

(2.9)

with the assumption that their determinants are not equal to $\mathbf{0}$. The inverse of the Jacobians gives all the first partial derivatives. With respect to x_j

$$F_{1y_1} \frac{\partial y_1}{\partial x_j} + \ldots + F_{1y_m} \frac{\partial y_m}{\partial x_j} + F_{1x_j} = 0$$

$$\vdots \qquad\qquad .$$

$$F_{my_1} \frac{\partial y_1}{\partial x_j} + \ldots + F_{my_m} \frac{\partial y_m}{\partial x_j} + F_{mx_j} = 0$$

(2.10)

From Cramer's Rule more generally,

$$\frac{\partial y_i}{\partial x_j} = \frac{\dfrac{\partial(F_1,\ldots,F_m)}{\partial(y_1,\ldots,y_{i-1},x_j,y_{i+1},\ldots,y_m)}}{\dfrac{\partial(F_1,\ldots,F_m)}{\partial(y_1,\ldots,y_m)}}$$

(2.11)

where $i = 1, \ldots, m, j = 1, \ldots, n$ and the denominator is the Jacobian determinant [2.9.13.].
Implicit functions, ω in terms of x and y are

$$\omega_1 = g_1(q_1,\ldots,q_n)$$

$$\vdots$$

$$\omega_m = g_m(q_1,\ldots,q_n)$$

(2.12)

and a similar treatment for $G(x, y, q, \omega) = 0$ as it was used for $F(x, y, q, \omega)$ yields first partial derivatives in terms of G. As a result, $G_y\, dy + G_x\, dx = \mathbf{0}$, and G_y and G_x are Jacobians.

For implicit functions, q and ω, in terms of two equations with respect to x and y, F and G are given in four unknowns as $\dfrac{\partial q}{\partial x} = -\dfrac{\dfrac{\partial(F,G)}{\partial(x,\omega)}}{\dfrac{\partial(F,G)}{\partial(q,\omega)}}$, $\dfrac{\partial q}{\partial y} = -\dfrac{\dfrac{\partial(F,G)}{\partial(y,\omega)}}{\dfrac{\partial(F,G)}{\partial(q,\omega)}}$, $\dfrac{\partial\omega}{\partial x} = -\dfrac{\dfrac{\partial(F,G)}{\partial(q,x)}}{\dfrac{\partial(F,G)}{\partial(q,\omega)}}$, and

$\dfrac{\partial\omega}{\partial y} = -\dfrac{\dfrac{\partial(F,G)}{\partial(q,y)}}{\dfrac{\partial(F,G)}{\partial(q,\omega)}}$, where the denominator is the Jacobian. That is, the Jacobian for the pixel

values of a digital image is implicitly $\dfrac{\partial(F,G)}{\partial(q,\omega)}$.

2.4.3   Displacement and Motion: Position and Orientation

In a x, y plane, the relation between changing points between surfaces may be characterized via infinitesimal motion. Motion entails changes in position and orientation as displacement over time as velocity from one point to another between surfaces. Velocity means changes in both linear and angular directions with respect to position and orientation.

From a digital image, pixel coordinates from a x, y plane form a matrix of the position and orientation so that each pixel may become the endpoint of a matching with regard to another digital image. The Jacobian represents position and orientation from one image to another one. In 3D space, the correspondence between the displacement of positions and orientations over some small increment of time characterizes the Jacobian between two images.

Vector quantities are used to represent position and orientation where the displacements in position are $\mathbf{q}$ and in orientation, the displacements are $\boldsymbol{\omega}$. For displacement over time, $\mathbf{q} = \boldsymbol{\omega} + \mathbf{u}(\mathbf{x}, t)$ for a displacement vector $\mathbf{u}$ for some point x_k at time t. Motion between images is $\mathbf{q}$ as a function of $\boldsymbol{\omega}$ so that $\boldsymbol{\omega} = f(\mathbf{q})$. The displacement field is the change in 3D from one image to another one [2.9.2.]. A velocity field, $\omega(\mathbf{x}, t) = \dfrac{\partial \mathbf{u}(\mathbf{x}, t)}{\partial t}$ for some point at x_k at t, is the motion of each point on a planar surface where a displacement vector, $\mathbf{u}$, occurs at each point. Linear velocities are recorded in x, y, angular velocities are recorded by $\boldsymbol{\omega}$, and $\mathbf{q}$ records the joint velocities between positions and orientations on one image and another image.

The Jacobian relates displacements of position and orientation, and time derivatives of $\mathbf{q}$ and $\boldsymbol{\omega}$ are the relation between velocities $\dot{\mathbf{q}}$ and $\dot{\boldsymbol{\omega}}$ that comprise the Jacobian. Between groups of surface points on a diatom valve or between diatom valve surfaces

$$\dot{\omega}_{mx1} = J(q)_{mxn}\,\dot{\mathbf{q}}_{nx1} \tag{2.13}$$

where $\dot{\omega}$ is the time derivative vector of end displacements in terms of position and orientation of an end point, J is the Jacobian, and $\dot{\mathbf{q}}$ is the time derivative vector of start displacements in terms of angles at a start point so that $\dot{\omega}$ and $\dot{\mathbf{q}}$ become the linear and angular velocities, respectively. This equation is independent of the frame used for the variables [2.9.14., 2.9.23.], but a reference frame is necessary to obtain the original position vectors [2.9.8.]. For each different start and end displacement vectors, different Jacobians will result. That is, different start-end vectors will produce different Jacobians. By using Jacobians, the relation between small displacements between different surfaces is established and can be used in terms of rigid registration.

2.4.4   Surface Feature Matching via the Jacobian

The Jacobian with respect to surface features between two different images produces

$$\dot{\mathbf{f}}_{Between\ surfaces} = J_{Image_1} \cdot J_{Image_2} \cdot \dot{\mathbf{q}} = J_{Start\ position} \cdot J_{End\ position} \cdot \dot{\mathbf{q}} \tag{2.14}$$

where $\dot{f}$ is the vector of first partial derivatives that represent the joint velocity between surface features, $\dot{\mathbf{q}}$ (e.g., [2.9.20.]). As a result, $\dot{\mathbf{f}}$ represents the 2D gradient Jacobians via x, y coordinates for feature positions, $\mathbf{s}$, as

$$\delta\mathbf{s} = \begin{bmatrix} \delta x \\ \delta y \end{bmatrix} = \begin{bmatrix} \dfrac{\partial f_1}{\partial q_1} & \cdots & \dfrac{\partial f_1}{\partial q_n} \\ \vdots & & \vdots \\ \dfrac{\partial f_m}{\partial q_1} & \cdots & \dfrac{\partial f_m}{\partial q_n} \end{bmatrix} \delta\mathbf{q} \tag{2.16}$$

for each δs feature position and for each δq motion expressed as the matrix equation

$$\delta \mathbf{s} = J(\mathbf{q})\, \delta \mathbf{q} \qquad (2.16)$$

[2.9.14.] where $J(\mathbf{q}) \equiv \dfrac{\partial \mathbf{f}}{\partial \mathbf{q}}$, and the Jacobian via the time derivative equation is

$$\dot{\mathbf{s}} = J(\mathbf{q})\dot{\mathbf{q}} = J(\mathbf{q})\begin{pmatrix} J_v \\ J_\omega \end{pmatrix} \qquad (2.17)$$

where the feature position vector becomes the displacement velocity. The vector, $\begin{pmatrix} J_v \\ J_\omega \end{pmatrix} = \begin{pmatrix} \mathbf{v} \\ \omega \end{pmatrix}$, contains the image surface features at n-points on the surface calculated as the gradient magnitude, $\mathbf{v}$, and direction, ω, obtained from the Jacobian with respect to time [2.9.20.], J_v is the linear velocity with respect to gradient magnitude, and J_ω is the angular velocity with respect to gradient direction. Changes in gradient magnitude and direction over time are represented in the motion between surface features from one image to another one. If the gradient is measured at a static time, t, then t is eliminated and only the gradient magnitude and direction are used.

To determine the relation between feature points from one image to another, the context space in which the image matching of features is happening consists of a task space and a feature space (e.g., [2.9.8.]). Task space is the end position and orientation of the feature selected for matching on the image of the 3D surface model. The feature space is the place where joint displacements between both images with respect to the feature selected for matching occurs. The displacement between images is infinitesimal so that rotation change is negligible. Feature space is the local displacement between a selected feature found on both images with respect to position and orientation, i.e., $\mathbf{s}$ and $\mathbf{q}(s)$, while task space is the global space describing the setting in which local feature displacements occur, i.e., x, y, z in Cartesian coordinates in Euclidean space [2.9.41.].

Displacement over time produces velocity, and the velocities to consider are those at each juncture of a displacement and characterize the motion between features, between the spaces occupied by the images, and joint velocities connecting the features and spaces. The joint velocities, $\mathbf{q}$, are related to the spatial velocities, $\mathbf{r}$, via the Jacobian, and

$$\dot{\mathbf{r}} = J(\mathbf{r})_{spatial \cdot joint}\,\dot{\mathbf{q}} = \begin{bmatrix} \dfrac{\partial r_1}{\partial q_1} & \cdots & \dfrac{\partial r}{\partial q_n} \\ \vdots & & \vdots \\ \dfrac{\partial r_m}{\partial q_1} & \cdots & \dfrac{\partial r_m}{\partial q_n} \end{bmatrix}\dot{\mathbf{q}} \qquad (2.18)$$

with the relation between feature velocities, $\mathbf{s}$, and $\mathbf{r}$ given as

$$\dot{\mathbf{s}} = J(\mathbf{s})_{feature\cdot spatial}\dot{\mathbf{r}} = \begin{bmatrix} \dfrac{\partial s_1}{\partial r_1} & \cdots & \dfrac{\partial s}{\partial r_n} \\ \vdots & & \vdots \\ \dfrac{\partial s_m}{\partial r_1} & \cdots & \dfrac{\partial s_m}{\partial r_n} \end{bmatrix}\dot{\mathbf{r}} \qquad (2.19)$$

the total Jacobian between feature and joint velocities is obtained via $\mathbf{s}$ and $\mathbf{r}$ with respect to $\mathbf{q}$ through

$$\dot{\mathbf{s}} = J(\mathbf{s})_{feature\cdot joint}\dot{\mathbf{q}} = \begin{bmatrix} \dfrac{\partial s_1}{\partial q_1} & \cdots & \dfrac{\partial s}{\partial q_n} \\ \vdots & & \vdots \\ \dfrac{\partial s_m}{\partial q_1} & \cdots & \dfrac{\partial s_m}{\partial q_n} \end{bmatrix}\dot{\mathbf{q}} \qquad (2.20)$$

[2.9.32., 2.9.37.]. That is, $J_{feature\cdot joint} = J_{feature\cdot spatial}J_{spatial\cdot joint}$. For infinitesimal motion between one image and another one, displacements record this motion for constant time [2.9.42.].

2.4.5   The Jacobian of Whole Surface Matching

To obtain the composite Jacobian for n-configuration space and m-feature space, concatenated elements comprise matrices of sequential differences in each row in each matrix [2.9.37.] of joint and feature velocities (or displacements) so that

$$Q_{velocities} = \begin{bmatrix} q_1^{j-n+1} \cdots q_1^{j-1}\, q_1^{j} \\ q_2^{j-n+1} \cdots q_2^{j-1}\, q_2^{j} \\ \vdots \quad \ddots \quad \vdots \quad \vdots \\ q_n^{j-n+1} \cdots q_n^{j-1}\, q_n^{j} \end{bmatrix} \qquad (2.21)$$

and

$$S_{velocities} = \begin{bmatrix} s_1^{j-n+1} \cdots s_1^{j-1}\, s_1^{j} \\ s_2^{j-n+1} \cdots s_2^{j-1}\, s_2^{j} \\ \vdots \quad \ddots \quad \vdots \quad \vdots \\ s_m^{j-n+1} \cdots s_m^{j-1}\, s_m^{j} \end{bmatrix} \qquad (2.22)$$

where the transpose of each matrix produces n- and m-column vectors, resulting in Q as a n x n matrix and S as a m x n matrix, and the product matrix gives the Jacobian [2.9.37.] as

$$\mathbf{J_q} = \mathbf{S} \cdot \mathbf{Q}^{-1}. \tag{2.23}$$

2.5   Methods

2.5.1   Fiducial Outcomes of Image Matching of Surface Features

On a digital image of a SEM and 3D surface model converted to the same image format, a protocol is developed to enable image matching of surface features. The 3D surface model is geometrically devised, and the expectation is that matching of surface features in a point-wise fashion entails matching pixel intensities of a delineated surface feature of *Arachnoidiscus*. Data points are collected as pixel coordinates via standard image processing techniques. A window is selected that surrounds the surface feature of interest on the digital SEM. The window is a rectangular x, y plane representing a pixel and its neighborhood or the planar face of a voxel if a 3D image is used. For rigid registration, crest lines may be used to find extremal points in the rectangular x, y plane. Alternatively, the centroid of the window as a pixel or voxel planar face may be used to find non-extremal points for non-rigid registration.

Surface feature matching is determined by displacement as velocities with respect to the motion from one image to the next. From this, the Jacobian is calculated which records the change in displacements or velocities from one image to the next. Image matching utilized a prismatic set-up (Figure 2.3). Velocities were calculated via Eq. (2.21), and the mean joint velocity was calculated for the first and second images using Eq. (2.22). Using matrix algebra, the Jacobian was calculated according to Eq. (2.23).

The differences in image order when commencing to perform surface feature matching is determined by four set-ups: 1) *Arachnoidiscus ehrenbergii* SEM image is image 1, 3D surface model is image 2; 2) 3D surface model is image 1, SEM image is image 2; SEM image is analyzed alone; 3D surface model is analyzed alone. A second set of analyses are accomplished by adding *A. ornatus*. Image ordering involves SEM of *A. ehrenbergii* as image 1,

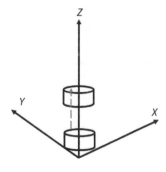

Figure 2.3 Prismatic set-up of image matching schema for *Arachnoidiscus* valves. Surface feature matching indicated by red dashed arrow.

SEM of *A. ornatus* as image 2, and 3D surface model as image 3. A permutation of the three images is used in analyses as is analyses of the individual images.

Columns of the Jacobian matrix are indicators of orientation preservation or reversal. If the column is negative, the orientation of space is reversed; if the column is positive, the orientation of space is preserved [2.9.36.]. Stability or instability may be assessed using the eigenvalues of the Jacobian. Critical points are indicated by the eigenvalues of the Jacobian, and a negative eigenvalue means stability, a positive eigenvalue means instability, and a mix of negative and positive eigenvalues may also indicate instability [2.9.35.] or be indeterminant.

The Jacobian determinant indicates how good the image matching has been. The determinant is calculated to represent the closeness of the image to the model with respect to the amount of expansion (positive Jacobian determinant) or contraction (negative Jacobian determinant) present as the amount of transformation necessary to match images. That is, if the Jacobian determinant is zero, then the image matching is perfect, and mismatching of images is determined as the degree that the value calculated deviates from zero. The smaller the difference, the more alike are the image and the model. The sign of the Jacobian determinant is an indicator of orientation preservation if positive or reversal if negative. The goal is to try to get Jacobian determinant to be as close as possible to zero. The Jacobian determinant at a zero value also signifies that matching between images was accomplished using critical points, including extrema.

2.6 Results

LMs (row a) and SEMs (row b) of *Arachnoidiscus ehrenbergii* interior valves are illustrated in Figure 2.4. The second image in row b is used in analyses. A 3D surface model (Figure 2.5, first image) was used for image matching. The digitized SEM was pre-treated to remove the background (Figure 2.6). The 3D model default did not have a background and was converted to the same image format as the SEMs.

Edge detection tests were conducted on the SEM and 3D surface model image windows using various filtering treatments (Figure 2.7), and as a result, a ridge filter (e.g., [2.9.19.]) was used from which to extract extremal points that constitute crest line data of pixel coordinates for SEM image (Figure 2.7e) and 3D surface model image (Figure 2.7f). In all analyses with SEMs and 3D surface model, ridge filtered images were used.

Morphological components were extracted from the whole valve images of the SEM (Figure 2.8a) and 3D surface model (Figure 2.8b). Examples of extracted ensemble surface features that could be used for image matching are illustrated in Figure 2.9. Each component could be used as an ensemble surface feature in image matching. To illustrate image matching, a double rib was chosen via a rectangular window encompassing the surface feature and neighborhood pixels from the SEM (Figure 2.10) for matching with the 3D surface model image.

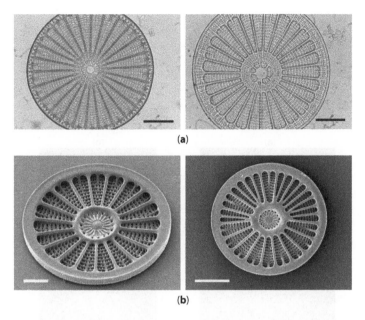

Figure 2.4 *Arachnoidiscus ehrenbergii*. a, Light micrographs. Scale bars = 50 μm. b, Scanning electron micrographs of internal valves. Left micrograph, scale bar = 20 μm; right micrograph, scale bar = 50 μm. All micrographs taken by Mary Ann Tiffany and used with permission.

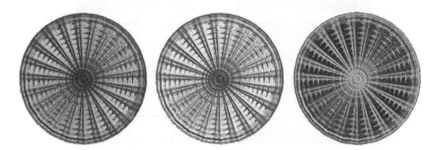

Figure 2.5 *Arachnoidiscus* 3D surface model. From left to right: the default model, illumination of the interior of the valve, and illumination of the central area.

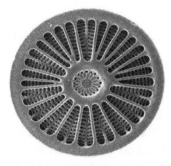

Figure 2.6 *Arachnoidiscus ehrenbergii* digitized SEM without background. Micrograph by Mary Ann Tiffany used with permission.

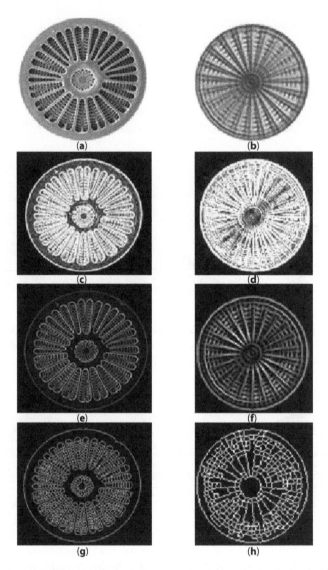

Figure 2.7 *Arachnoidiscus ehrenbergii* valve face surfaces with SEM images in the first column and 3D *Arachnoidiscus* model images in the second column: a, original SEM by Mary Ann Tiffany used with permission and b, 3D surface model; c, d, gradient Prewitt filter; e, f, crest lines; g, h, binarized.

2.6.1   Surface Feature Image Matching and the Jacobian

Surface feature selected for analysis was a double rib via the window (Figure 2.10) and matching points from the 3D surface model to the SEM of *A. ehrenbergii* (Figure 2.11). That double rib was isolated from each image (Figure 2.12), and matching was made from SEM to 3D surface model, then from 3D surface model to SEM. Calculation of displacements between images was depicted as optical flow diagrams where an almost mirror image was depicted between diagrams (Figure 2.13). Joint velocities were depicted as vector stream plots, where the comparison between SEM and 3D surface model showed an almost uniform velocity magnitude and twice broken velocity direction (Figure 2.14).

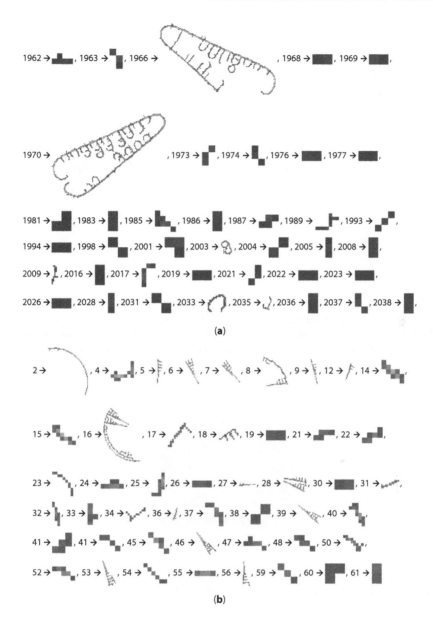

Figure 2.8 Samples of eccentricity morphological components extracted after treatment with a ridge filter to obtain crest line images: a, SEM; b, 3D surface model.

No vectors were plotted for matching from 3D surface model to SEM because the result was a scalar.

Jacobians for the double rib surface feature were calculated for each rendition of image matching. From *A. ehrenbergii* SEM to 3D surface model, the second and third columns were positive preserving orientation, but the first column of the Jacobian was negative, indicating orientation reversal (Table 2.1). From 3D surface model to SEM, the first Jacobian column was negative, while the second column was positive (Table 2.1). Magnitude of each Jacobian was similar. The Jacobian determinant was a very small positive number, meaning

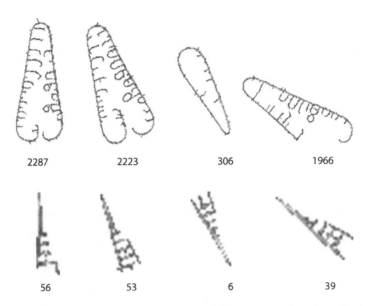

Figure 2.9 Examples of morphological components extracted for image matching from the *Arachnoidiscus ehrenbergii* SEM and 3D surface model after treatment with a ridge filter to obtain crest line images. Top row are components from the SEM, and bottom row are the components from the 3D surface model. Numbers refer to the component numbers extracted.

Figure 2.10 Rectangular window (yellow box) of selected rib pairs from the interior valve of *Arachnoidiscus ehrenbergii*. Micrograph by Mary Ann Tiffany used with permission.

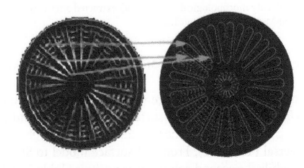

Figure 2.11 Motion from 3D surface model to SEM of *Arachnoidiscus ehrenbergii* showing three matching points between extremal points on crest lines.

(a)   **(b)**

Figure 2.12 Images of double rib surface feature: a, *Arachnoidiscus ehrenbergii* SEM window selection; b, 3D surface model window selection.

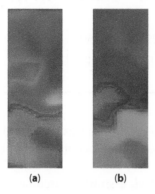

(a)   **(b)**

Figure 2.13 Optical flow of displacements from image matching of double rib: a, SEM to 3D surface model; b, 3D surface model to SEM.

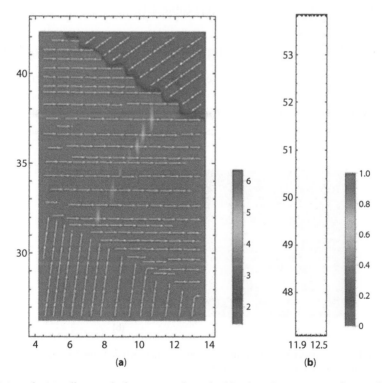

(a)   **(b)**

Figure 2.14 Joint velocities illustrated of image matching double rib: a, SEM to 3D surface model; b, 3D surface model to SEM. Highest velocities are red, and lowest velocities are blue.

Table 2.1 *Arachnoidiscus* image matching of double ribs. Jacobians, Jacobian determinants and non-zero Jacobian eigenvalues are reported.

Surface feature	Image matching	Jacobian			Jacobian determinant	Jacobian eigenvalues
Double ribs	*A. ehrenbergii* to 3D surface model	$\begin{bmatrix} -51.5889 & 83.4562 & 270.164 \\ -41.3075 & 58.3793 & 226.868 \\ -27.7191 & 27.0845 & 167.338 \end{bmatrix}$			9.18×10^{-12}	155.358, 18.7696
Double ribs	3D surface model to *A. ehrenbergii*	$\begin{bmatrix} -111.291 & 44.7502 \\ -103.18 & 42.2492 \end{bmatrix}$			-84.6393	-70.247, 1.20488

that the matching of extrema via crest lines was successful and orientation was preserved (Table 2.1). Non-zero eigenvalues of the Jacobian were positive, and therefore indicated instability in the matching.

From 3D surface model to SEM, a similar outcome for the Jacobian occurred (Table 2.1). The Jacobian determinant was a large negative number, indicating contraction and orientation reversal was necessary in order to image match extrema via crest lines. The first eigenvalue of the Jacobian was negative, while the second one was positive, signifying a mix of stability and instability in image matching (Table 2.1).

2.6.2   Whole Valve Images, Matching of Crest Lines and the Jacobian

Whole valves were compared for all surface features. Displacement pattern from *A. ehrenbergii* SEM to 3D surface model was approximately the opposite of that for matching the 3D surface model to the SEM (Figure 2.15). Joint velocities were somewhat similar in magnitude and somewhat generally mirror images regardless of the matching order (Figure 2.16).

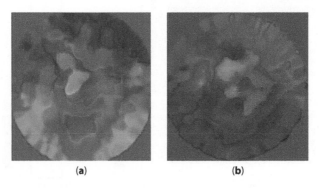

(a)   (b)

Figure 2.15 Optical flow of displacements from image matching of whole valves: a, *Arachnoidiscus ehrenbergii* SEM to 3D surface model; b, 3D surface model to SEM.

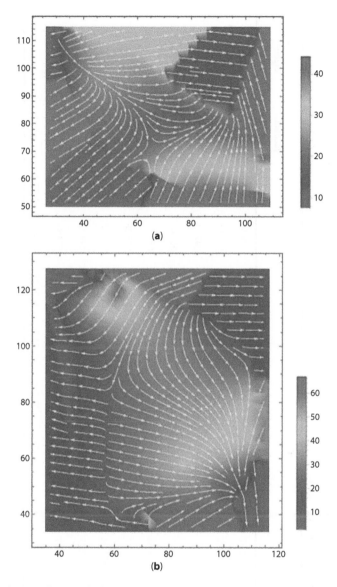

Figure 2.16 Joint velocities illustrated of image matching whole valves: a, *Arachnoidiscus ehrenbergii* SEM to 3D surface model; 3D surface model to SEM. Highest velocities are red, and lowest velocities are blue.

Image matching of whole valves sometimes produces Jacobians with a large number of elements. As a result, the Jacobian for image matching of *A. ehrenbergii* to the 3D surface model and vice versa is presented as an illustration. All other Jacobians will not be reported, but instead, the specifications of the Jacobian and number of negative- and/or positive-valued columns will be reported.

The Jacobians for image matching of whole valves were similar in magnitude (Table 2.2). More negative columns occurred in matching from 3D surface model to SEM, indicating

Table 2.2 *Arachnoidiscus* image matching of *A. ehrenbergii* SEM to 3D surface model and 3D surface model to SEM. Jacobians, Jacobian determinants and non-zero Jacobian eigenvalues are reported.

Surface feature	Image matching	Jacobian										Jacobian determinant	Jacobian eigenvalues
Whole valve	SEM to 3D surface model	2349.77	-471.037	1704.37	-2021.99	429.827	2099.48	-1989.64	3597.84	5169.34	5099.16	7.34×10^{-96}	13450.9, -2589.61
		1956.14	-369.908	1307.65	-1739.26	236.455	2114.69	-1765.66	3496.53	4822.85	4940.01		
		2701.89	-573.555	2119.57	-2244.56	668.628	1886.91	-2130.72	3416.59	5197.61	4864.62		
		1964.37	-406.804	1490	-1657.55	430.456	1540.11	-1599.24	2713.91	4017.07	3855.5		
		3082.16	-708.586	2689.66	-2423.68	1059.31	1255.89	-2163.47	2672.27	4659.79	3850.87		
		3780.18	-909.185	3499.58	-2871.51	1518.34	877.888	-2456.07	2372.26	4777.28	3468.14		
		2187.05	-491.351	1851.24	-1748.64	689.143	1080.19	-1591.48	2154.5	3574.13	3090.59		
		3494.79	-848.109	3273.23	-2635.67	1445.03	686.711	-2233.43	2022.49	4239.78	2969.71		
		2852.71	-687.457	2647.67	-2163.61	1153.14	640.347	-1846.87	1759.95	3573.81	2575.27		
		3292.77	-809.335	3135.32	-2457.49	1417.49	477.754	-2053.9	1674.28	3755.06	2477.41		

(Continued)

Table 2.2 *Arachnoidiscus image matching of A. ehrenbergii SEM to 3D surface model and 3D surface model to SEM. Jacobians, Jacobian determinants and non-zero Jacobian eigenvalues are reported. (Continued)*

Surface feature	Image matching	Jacobian										Jacobian determinant	Jacobian eigenvalues
Whole valve	3D surface model to SEM	-2642.98	-9432.78	-164.004	-315.773	-2235.56	866.288	-7923.91	1532.77	-537.223	2825.49	9.09×10^{-107}	-17614.7, 3773.63
		-2530.25	-9000.66	-157.655	-307.759	-2137.21	832.629	-7558.42	1482.09	-511.367	2702.17		
		-1911.96	-8109.79	-90.8031	6.92055	-1746.57	484.807	-6920.18	474.697	-515.639	2165.49		
		-2035.59	-7460.8	-122.076	-207.376	-1741.49	645.608	-6283.75	1083.99	-433.098	2194.66		
		-1949.91	-7960.79	-99.3165	-49.6752	-1750.07	528.631	-6771.2	636.979	-495.261	2179.19		
		-3490.21	-9343.92	-283.959	-986.636	-2639.12	1487.37	-7588.77	3558.92	-402.035	3437.16		
		-3409.81	-8025.6	-301.296	-1165.78	-2467.38	1574.8	-6395.01	4020.83	-283.836	3253.76		
		-2548.89	-6067.3	-223.751	-858.99	-1851.25	1169.69	-4843.22	2972.1	-218.89	2438.67		
		-2990.04	-6086.79	-284.787	-1196.27	-2068	1485.82	-4729.46	3994.68	-154.993	2763.37		
		-2716.69	-5601.53	-257.21	-1073.88	-1886.1	1342.13	-4362.85	3594.38	-147.855	2517.47		
		-1451.31	-3576.1	-124.772	-466.871	-1066.29	652.605	-2869.86	1632.39	-136.628	1400.02		

more orientation reversal when compared to matching from SEM to 3D surface model (Table 2.2). The Jacobian determinants were very small positive numbers, indicating successful image matching of crest lines and orientation preservation, while the non-zero eigenvalues were a mix of negative and positive values for a mix of stability and instability (Table 2.2).

Arachnoidiscus ornatus (Figure 2.17) was introduced into image matching analyses. Ridge filtering was used to obtain crest lines of extrema (Figure 2.18). Displacements as optical flow diagrams of image matching between *A. ornatus* SEM and 3D surface model and vice versa indicated similar geometric pattern of the ribs, yet differences in coloration patterns between the diagrams (Figure 2.19). Vector flows in joint velocity diagrams were only vaguely mirror images, with magnitudes of high velocities being perpendicular to one another with regard to image matching order (Figure 2.20).

Jacobians involving *A. ornatus* and 3D surface model matching had many more elements than matching involving *A. ehrenbergii* and the 3D surface model. Image matching from SEM to 3D surface model produced a 37 by 37 Jacobian with more negative or mixed negative-positive columns than positive ones. The reverse matching produced a 43 by 42 Jacobian with more negative or mixed columns than positive ones. Either matching meant more orientation reversals than not. Magnitude of the elements were similar for both renditions of image matching. Jacobian determinants were very small positive numbers from *A. ornatus* to 3D surface model and vice versa for successful image matching and orientation preservation (Table 2.3). Non-zero eigenvalues were negative for image matching from *A. ornatus* to 3D surface model as stability and positive for the reverse image matching for instability (Table 2.3).

Somewhat similar results occurred for image matching of SEMs of *A. ehrenbergii* and *A. ornatus*. Approximate similar geometric patterns of concentric circles and opposite color patterns were produced for displacement optical flow diagrams (Figure 2.21). Joint velocities were quasi-linear from *A. ehrenbergii* to *A. ornatus*, while joint velocities for *A. ornatus* to *A. ehrenbergii* were more convective (Figure 2.22). Velocity magnitudes were higher from *A. ehrenbergii* to *A. ornatus* than vice versa (Figure 2.22).

Jacobian for *A. ehrenbergii* to *A. ornatus* was a 9 by 9 matrix with 7 negative and 2 positive columns. For the reverse matching, a 33 by 31 Jacobian was produced with

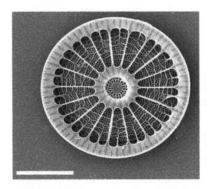

Figure 2.17 *Arachnoidiscus ornatus* SEM, scale bar = 50 μm. Micrograph by Mary Ann Tiffany used with permission.

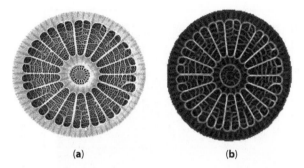

Figure 2.18 *Arachnoidiscus ornatus*: a, SEM without background; b, crest lines via ridge filtering. Micrograph by Mary Ann Tiffany used with permission.

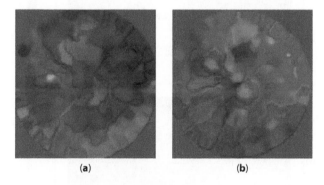

Figure 2.19 Optical flow of displacements from image matching of whole valves: a, *Arachnoidiscus ornatus* SEM to 3D surface model; b, 3D surface model to SEM.

17 negative and 14 positive columns. Again, more orientation reversals were found, and magnitudes were similar. Jacobian determinant was a very small positive number in successful image matching of crest lines for *A. ehrenbergii* to *A. ornatus*, but a very small negative number for the reverse successful matching with orientation reversal (Table 2.3). Jacobian eigenvalues for *A. ehrenbergii* to *A. ornatus* were negative for stability and the reverse matching produced complex-valued positive results for instability (Table 2.3).

2.6.3   Image Matching of more than Two Images

The next group of matchings involved ridge filtered SEM images of *A. ehrenbergii* and *A. ornatus* as well as the 3D surface model image (Figure 2.23). When mentioned, the base image is the first image to which other images are matched in a specified order. A permutation of the three images produced 6 displacement optical flow diagrams (Figure 2.24). Geometric and color patterns were somewhat similar for *A. ehrenbergii* as the base image (Figure 2.24 a, d), while this was not the case when the 3D surface model or *A. ornatus* was

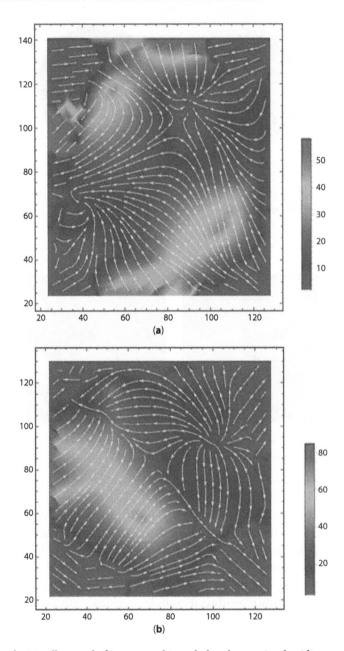

Figure 2.20 Joint velocities illustrated of image matching whole valves: a, *Arachnoidiscus ornatus* SEM to 3D surface model; 3D surface model to SEM. Highest velocities are red, and lowest velocities are blue.

the base image (Figure 2.24 b, c). Geometrically, displacement optical flow from 3D surface model to *A. ornatus* to *A. ehrenbergii* was similar to that from *A. ornatus* to 3D surface model to *A. ehrenbergii*, but the color pattern varied (Figure 2.24 e, f).

Almost a unidirectional, central flowline could be traced for *A. ehrenbergii* as the base image (Figure 2.25 a, d) or for the 3D surface model being the base image followed by

Table 2.3 *Arachnoidiscus* image matching of whole valves. Jacobians, Jacobian determinants and non-zero Jacobian eigenvalues are reported.

Image matching	Jacobian determinant	Jacobian eigenvalues
A. ornatus to 3D surface model	9.03×10^{-430}	$-16636.7 + 3600.16i$, $-16636.7 - 3600.16i$
3D surface model to *A. ornatus*	1.04×10^{-500}	19176.3, 8466.29
A. ehrenbergii to *A. ornatus*	1.07×10^{-84}	$-8408.86, -1281.89$
A. ornatus to *A. ehrenbergii*	-9.74×10^{-394}	$1096.79 + 302.165i$, $1096.79 - 302.165i$
A. ehrenbergii to 3D surface model to *A. ornatus*	2.28×10^{-98}	10882.3, -1516.62
3D surface model to *A. ehrenbergii* to *A. ornatus*	-3.11×10^{-109}	$-15310, 3580.16$
A. ornatus to *A. ehrenbergii* to 3D surface model	-2.20×10^{-395}	$-3353.07 + 1202.57i$, $-3353.07 - 1202.57i$
A. ehrenbergii to *A. ornatus* to 3D surface model	-4.10×10^{-86}	$-4970.74, -3686.04$
3D surface model to *A. ornatus* to *A. ehrenbergii*	4.75×10^{-504}	13503, 7781.66
A. ornatus to 3D surface model to *A. ehrenbergii*	2.15×10^{-433}	$-18232.3, -1412.26$

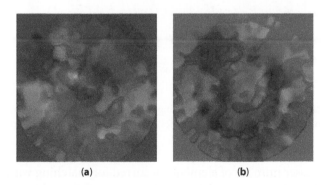

(a)   **(b)**

Figure 2.21 Optical flow of displacements from image matching of *Arachnoidiscus* whole valves: a, *A. ehrenbergii* to *A. ornatus*; b, *A. ornatus* to *A. ehrenbergii*.

A. ehrenbergii then *A. ornatus* (Figure 2.25 b) for joint velocities. Somewhat convective flows marked the joint velocities when *A. ornatus* was the base image (Figure 2.25 c, f) or when the 3D surface model was the base image followed by *A. ornatus* then *A. ehrenbergii* (Figure 2.25 e).

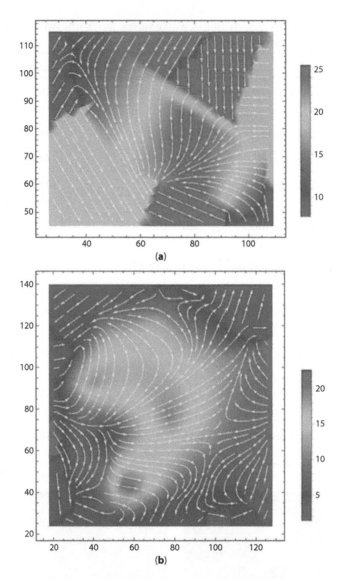

Figure 2.22 Joint velocities illustrated of image matching *Arachnoidiscus* whole valves: a, *A. ehrenbergii* to *A. ornatus*; *A.ornatus* to *A. ehrenbergii*. Highest velocities are red, and lowest velocities are blue.

Jacobians with a lesser number of elements occurred for matching with *A. ehrenbergii* as the base image or the 3D surface model as the base image followed by *A. ehrenbergii* as 10 by 10, 11 by 11 and 9 by 9 matrices. Large numbers of elements occurred for Jacobians when the base image was *A. ornatus* or the 3D model followed by *A. ornatus* with matrices of 33 by 33, 43 by 43 and 37 by 37. More positive than negative or mixed valued elements per column were found in Jacobians with image matching order *A. ehrenbergii* to 3D surface model to *A. ornatus*, 3D surface model to *A. ornatus* to *A. ehrenbergii*, and *A. ornatus*, to 3D

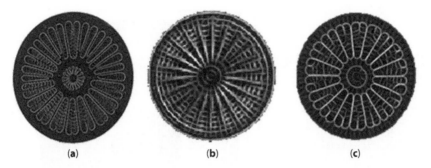

(a)   (b)   (c)

Figure 2.23 *Arachnoidiscus*: a, *A. ehrenbergii* image crest lines; b, 3D surface model image crest lines; c, *A. ornatus* image crest lines.

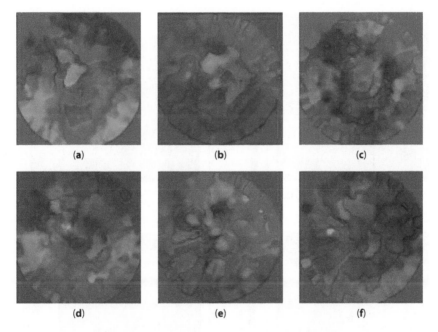

(a)   (b)   (c)

(d)   (e)   (f)

Figure 2.24 Optical flow of displacements from image matching of *Arachnoidiscus* whole valves: a, *A. ehrenbergii* to 3D surface model to *A. ornatus*; b, 3D surface model to *A. ehrenbergii* to *A. ornatus*; c, *A. ornatus* to *A. ehrenbergii* to 3D surface model; d, *A. ehrenbergii* to *A. ornatus* to 3D surface model; e, 3D surface model to *A. ornatus* to *A. ehrenbergii*; f, *A. ornatus* to 3D surface model to *A. ehrenbergii*.

surface model to *A. ehrenbergii*, assuring orientation preservation, while the opposite was found for other matching combinations of the images.

Jacobian determinants were positive in successful image matching of crest lines from *A. ehrenbergii* to 3D surface model to *A. ornatus*, from 3D surface model to *A. ornatus* to *A. ehrenbergii*, and from *A. ornatus* to 3D surface model to *A. ehrenbergii* for orientation preservation (Table 2.3). The remaining successful image matchings had negative Jacobian

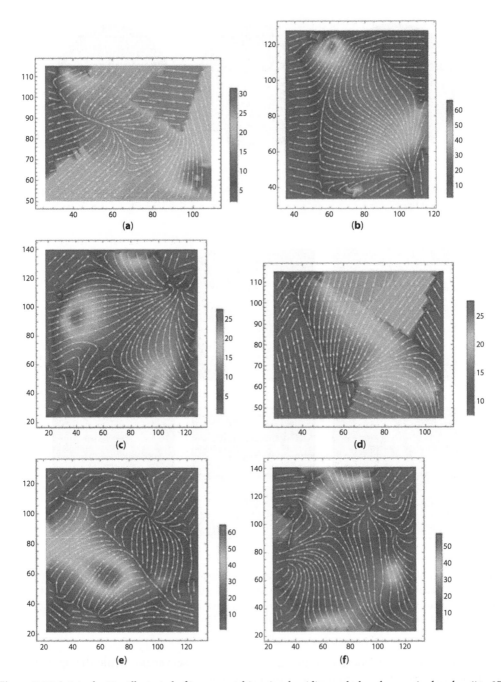

Figure 2.25 Joint velocities illustrated of image matching *Arachnoidiscus* whole valves: a, *A. ehrenbergii* to 3D surface model to *A. ornatus*; b, 3D surface model to *A. ehrenbergii* to *A. ornatus*; c, *A. ornatus* to *A. ehrenbergii* to 3D surface model; d, *A. ehrenbergii* to *A. ornatus* to 3D surface model; e, 3D surface model to *A. ornatus* to *A. ehrenbergii*; f, *A. ornatus* to 3D surface model to *A. ehrenbergii*.

determinants, indicating orientation reversal (Table 2.3). Non-zero eigenvalues were similar in magnitude, but differed by sign. Both values were negative for image matching of *A. ornatus* to *A. ehrenbergii* to 3D surface model (complex-valued), *A. ehrenbergii* to *A. ornatus* to 3D surface model, and *A. ornatus* to 3D surface model to *A. ehrenbergii*, indicating stability (Table 2.3). From 3D surface model to *A. ornatus* to *A. ehrenbergii*, Jacobian eigenvalues were positive, signifying instability (Table 2.3). Other image matching orders produced mixed valued Jacobian eigenvalues (Table 2.3).

2.7 Discussion

Measuring surface morphology in terms of the Jacobian and image matching is not dependent on specific morphological characters. To get the most out of using the methodology outlined in this study, some considerations are in order regarding the digital images used in analyses. Care must be exercised in the taking of digital images whether they are SEMs or LMs. It should be evident that any digital image regardless of the source could be used, as long as the resolution is impeccable. Any rendered diatom model may be used in matching to a digital image to facilitate experiments, such as matching surface changes in a 3D model to those seen in SEMs and LMs, or verifying that a model is sufficiently representative of the taxon being studied. As shown in this study, 3D surface models are suitable for comparative analysis with images. Data may be combined from each source, because models may be converted to the same image format, and geometry is used via implicit surfaces that are defined such that pixel intensity is converted to crest lines.

Jacobians and their determinants may be complex-valued [2.9.35.], but only the real part is needed in image matching and evaluation. To determine how good the image matching is between two or more images, the Jacobian determinant and Jacobian eigenvalues were used. Because the Jacobian is a quadratic, an error function based on the quadratic may be used to assess image matching as well (e.g., [2.9.20.]). Because of the wide-ranging applicability of the Jacobian, various assessments may be devised to evaluate the Jacobian in image matching.

At the outset, image matching using *Arachnoidiscus* relieved the necessity of matching disparate shaped forms as the outline of a circle was used. Surface features are boxed in by a window of the same size and shape on images. In this way, any group of taxa in a lineage could be analyzed using image Jacobian-based methods. Reference points would need to be determined in matching images with differently shaped valve outlines. However, because windows on the images are used and not valve shapes per se, determination of such reference points are reduced to potential geometric solutions.

From the results, the Jacobians with the largest number of elements means that more points were necessary to produce a successful matching, and that more points matched are an indication of closeness of surface features. This result is tempered by the number of columns indicating orientation reversals or a mix of orientation directions when larger Jacobians occur. Mostly, orientation preservation or reversal via the Jacobian was mixed.

Orientation preservation or reversal via the Jacobian determinant is a summary of the order of images in successful matching, and when comparing the matching of *Arachnoidiscus ehrenbergii* to the 3D surface model versus *A. ornatus* to the same model, orientation preservation held for *A. ornatus*. As a result, *A. ornatus* is a better match to the 3D surface model than *A. ehrenbergii*. This result is reinforced by the stability outcome of matching from *A. ornatus* to the 3D surface model (Table 2.3) and may be visually interpreted as such (Figure 2.23).

Image matching *Arachnoidiscus ehrenbergii* to *A. ornatus* produced orientation preservation, and the reverse matching produced orientation reversal (Table 2.3), indicating that *A. ehrenbergii* is a different taxon from *A. ornatus*. Stability in the eigenvalues of the first matching versus instability in the second matching reinforces this outcome.

The results so far have implications for matching outcomes of three images. Matching order where the base image was the 3D surface model followed by *Arachnoidiscus ornatus* or the base model was *A. ornatus* followed by the 3D surface model induced orientation preservation (Table 2.3). However, when the 3D surface model was the base image, instability resulted in contrast to *A. ornatus* being the base model. The only threesome that produced orientation preservation and stability was *A. ornatus* to 3D surface model to *A. ehrenbergii* (Table 2.3). *Arachnoidiscus ornatus* was a better match to the 3D surface model and different from *A. ehrenbergii* in outcomes of all permutations of matching three images.

2.7.1   Utility of Jacobian-Based Methods and Image Matching

Jacobian-based methods show that there is an additional utility to LMs and SEMs, even those that have been in storage, and perhaps, never used. LMs and SEMs that are printed and scanned may not have sufficient resolution to be useful, but that depends on the quality of the initial photographic results in these cases.

Potential applications for image matching include determinations of environmentally-influenced differences in taxa, indicator status ranges for a given taxon, morphological variation, or possibly, morphological constraints with respect to habitat changes. Longer term morphological changes may be determined within a genus or lineage with regard to evolutionary time or comparisons with fossil taxa and changes over geological time. Cell tracking (e.g., [2.9.5.]) using image matching and Jacobian-based methods enable the documenting of small shape and surface changes in a valve image sequence undergoing size diminution, or girdle development during vegetative reproduction. The ability to document changes in taxa may be applicable from the species to lineage level, and perhaps, considering body plan, the documentation of salient ensembles of surface features. By quantifying surface morphology using Jacobian-based methods and image matching, many avenues of research await.

2.7.2   The Image Jacobian and Rotation in A Reference Frame: Potential Application to Diatom Images

There may be times when rotation vectors between images are necessary in order to conduct 3D image matching. This situation might be applicable to taxa such as *Entomoneis*, *Cylindrotheca*, or colonies of *Asterionella* that are in a helix. Such 3D structures when treated as stereo pairs may induce the use of more involved visual servoing techniques, but such techniques could still be applied in image matching.

Translation and rotation may be considered with respect to linear and angular velocities in terms of frames. Displacements remain the same from one frame to the next in translation. Rotation involves one frame with a fixed vector and another one as a moving frame containing a rotating vector. A point on a rotating body is a linear velocity that is dependent on the rate of the rotation vector and the location of a point on a fixed vector. The angular velocity is a cross product that is the rate of the rotation vector.

At the end point, let the angle positions be $\mathbf{q} = (\theta_1, \theta_2) = [q_0, q_1]^T$ and orientation positions be $\mathbf{s} = [x, y, 0]^T$. Then, with respect to changes in positions and angles, the Jacobian is

$$J \equiv \begin{pmatrix} \dfrac{\partial x}{\partial \theta_1} & \dfrac{\partial x}{\partial \theta_2} \\[2mm] \dfrac{\partial y}{\partial \theta_1} & \dfrac{\partial y}{\partial \theta_2} \end{pmatrix} \tag{2.24}$$

On a 3D surface, define a point (x, y, z) and relate it to the angles made by displacement points in motion with joint velocities (q_0, q_1). We use a transformation matrix to relate the two frames of reference, 0 and 1, for (x, y, z) and (q_0, q_1). Transformation of position includes rotation and translation [2.9.14.]. For rotation,

$$_0^1\mathbf{R} = \begin{bmatrix} \cos(q_0) & -\sin(q_0) & 0 \\ \sin(q_0) & \cos(q_0) & 0 \\ 0 & 0 & 1 \end{bmatrix} \tag{2.25}$$

Length is $L_0 = \sqrt{[r\cos(q)]^2 + [r\sin(q)]^2 + 0}$, where the x and y positions cover the movement, while the z position is zero. From this, the translation is

$$_0^1\mathbf{D} = \begin{bmatrix} L_0 \cos(q_0) \\ L_0 \sin(q_0) \\ 0 \end{bmatrix} \tag{2.26}$$

so that the transformation matrix with rotation and translation [2.9.14.] from reference frame 0 to 1 is

$$
{}_{0}^{1}\mathbf{T} = \begin{bmatrix} \cos(q_0) & -\sin(q_0) & 0 & L_0\cos(q_0) \\ \sin(q_0) & \cos(q_0) & 0 & L_0\sin(q_0) \\ 0 & 0 & 1 & 0 \\ 0 & 0 & 0 & 1 \end{bmatrix}. \tag{2.27}
$$

For a new position in reference frame 1 with L_1,

$$
\mathbf{s} = \begin{bmatrix} L_1\cos(q_1) \\ L_1\sin(q_1) \\ 0 \\ 1 \end{bmatrix} \tag{2.28}
$$

and the new transformation matrix [2.9.14.] becomes

$$
{}_{0}^{1}\mathbf{T} = \begin{bmatrix} \cos(q_0) & -\sin(q_0) & 0 & L_0\cos(q_0) \\ \sin(q_0) & \cos(q_0) & 0 & L_0\sin(q_0) \\ 0 & 0 & 1 & 0 \\ 0 & 0 & 0 & 1 \end{bmatrix} \begin{bmatrix} L_1\cos(q_1) \\ L_1\sin(q_1) \\ 0 \\ 1 \end{bmatrix}
$$

$$
= \begin{bmatrix} L_1\cos(q_0)\cos(q_1) - L_1\sin(q_0)\sin(q_1) + L_0\cos(q_0) \\ L_1\sin(q_0)\cos(q_1) + L_1\cos(q_0)\sin(q_1) + L_0\sin(q_0) \\ 0 \\ 1 \end{bmatrix} \tag{2.29}
$$

with the position of **s** now

$$
\begin{bmatrix} L_0\cos(q_0) + L_1\cos(q_0 + q_1) \\ L_0\sin(q_0) + L_1\sin(q_0 + q_1) \\ 0 \end{bmatrix} \tag{2.30}
$$

Angular rotation around the z-axis is $\omega = \begin{bmatrix} \omega_x \\ \omega_y \\ \omega_z \end{bmatrix} = \begin{bmatrix} 0 \\ 0 \\ q_0 + q_1 \end{bmatrix}$. For rotation around the x- and

y- axes, angular rotation would be $\begin{bmatrix} q_0 \\ 0 \\ q_1 \end{bmatrix}$ and $\begin{bmatrix} 0 \\ q_0 \\ q_1 \end{bmatrix}$, respectively [2.9.14.].

2.7.3   Deformation and Registration of Image Surfaces: An Alternative Jacobian Calculation

On a 3D image, the relation among edges on the surface are displacements [2.9.33.], and those displacements in motion signify velocity. The motion between points on the surfaces via pixels are $P_i(t) = [x_i(t) \quad y_i(t)] \in \mathbb{R}^2$, where $x_i(t)$ and $y_i(t)$ are pixel coordinates with respect to t, time [2.9.23.]. Pixels are identified with respect to surface features where points are found on x, y planar surfaces, and pixel centroids are determined. The same technique may be used with voxels. That is, while 2D points are identifiable, 3D points using voxels may also be used [2.9.33.].

Using the 3D surface model and digital image in the same image format, points are extracted so that the shape of the extraction window is identical from one image to the other one. Surface feature points may be extremal points that are maximal. When images are superimposed on one another, maximal points of interest may be determined to define features of interest found by this approximate alignment of one image to the other one. The motion and displacement between images characterize the alignment procedure, and the fitting of the surface features of interest may involve a deformation procedure between images.

The Jacobian resulting from visual servoing techniques is produced as a result of rigid registration. For non-rigid registration, joint tracking between two surfaces may also be accomplished via the deformation gradient as it relates to the Jacobian [2.9.33.-2.9.34.]. The deformation gradient could be used for point matching in isolated clusters of pixel values to allow the usage of edges as relative extrema of curvature from crest lines.

A displacement gradient tensor is calculated as the change in displacement with respect to change in position between the first and second images. From this, a deformation gradient tensor is the identity tensor plus the displacement gradient tensor. More formally, the deforma-

tion gradient [2.9.2.] is expressed as $\left| J(\Phi) \right| = \det \nabla \Phi = \begin{vmatrix} \dfrac{\partial \varphi_1}{\partial x} & \dfrac{\partial \varphi_1}{\partial y} & \dfrac{\partial \varphi_1}{\partial z} \\[2mm] \dfrac{\partial \varphi_2}{\partial x} & \dfrac{\partial \varphi_2}{\partial y} & \dfrac{\partial \varphi_2}{\partial z} \\[2mm] \dfrac{\partial \varphi_3}{\partial x} & \dfrac{\partial \varphi_3}{\partial y} & \dfrac{\partial \varphi_3}{\partial z} \end{vmatrix}$ [2.9.33.], and

in terms of the displacement field, $J_{|\det \nabla \Phi|} = \begin{vmatrix} \dfrac{\partial u_1}{\partial x} + 1 & \dfrac{\partial u_1}{\partial y} & \dfrac{\partial u_1}{\partial z} \\[2mm] \dfrac{\partial u_2}{\partial x} & \dfrac{\partial u_2}{\partial y} + 1 & \dfrac{\partial u_2}{\partial z} \\[2mm] \dfrac{\partial u_3}{\partial x} & \dfrac{\partial u_3}{\partial y} & \dfrac{\partial u_3}{\partial z} + 1 \end{vmatrix} = \det(\mathbf{I} + \nabla u)$,

where ∇u is the displacement gradient between images, $u = (u_1, u_2, u_3)$ at P, and $J_{|\det \nabla \Phi|}$ is the Jacobian of the deformation gradient [2.9.2., 2.9.3., 2.9.33.]. From this, $\nabla \Phi^T \nabla \Phi$ is the metric tensor, and $|\nabla \Phi| > 0$ [2.9.33.]. The deformation determinant is $|J_\Phi| = |\nabla \Phi(\varphi_x, \varphi_y, \varphi_z)|$, where J is the Jacobian [2.9.33.]. At P, the deformation of the volume is related to the Jacobian as $J(\Phi) > 1$ for local expansion, $J(\Phi) < 1$ for local contraction, and $J(\Phi) = 1$ for no deformation [2.9.33.-2.9.34.].

From the initial position of the first image used, the position vector is defined as **x**, and the final position vector on the second image is **y**. A displacement vector, **u**, is calculated as the change in **x** with respect to a transformation, φ. A velocity vector is the change in initial position with respect to a transformation that results in displacement with respect to that transformation.

Changes from feature velocities on one image to feature velocities on another image will involve matching one image's window feature content to its counterpart on another image. The velocity gradient is the rate of deformation and is given as

$$\nabla \mathbf{q} = \begin{bmatrix} \dfrac{\partial q_x}{\partial x} & \dfrac{\partial q_x}{\partial y} & \dfrac{\partial q_x}{\partial z} \\[2mm] \dfrac{\partial q_y}{\partial x} & \dfrac{\partial q_y}{\partial y} & \dfrac{\partial q_y}{\partial z} \\[2mm] \dfrac{\partial q_z}{\partial x} & \dfrac{\partial q_z}{\partial y} & \dfrac{\partial q_z}{\partial z} \end{bmatrix} \tag{2.31}$$

so that the rate of change in deformation is a rate of change in the Jacobian [2.9.24.].

From deformation, the Jacobian characterizes the motion via velocity and displacement between images as the change in area or volume with respect to pixels or voxels, respectively, and is robust to small misalignments with respect to rigid registration of the images [2.9.7., 2.9.33.]. That is, image matching during non-rigid registration occurs with minimal misalignment with respect to the Jacobian (e.g., [2.9.33.]). For large misalignments or unusual deformations with image alignment, additional procedures invoking Jacobian-based methods or other optimization schema may be instituted to obtain better outcomes (e.g., [2.9.10., 2.9.47.]). Alignment between sufficiently similar images usually does not entail large angular changes or exclusively disparate surfaces in image matching calculations. Deformation also entails diffusion of one image blurred to match another image (e.g., [2.9.1.]) so that extremal points and curves may not necessarily be included (e.g., [2.9.46.]), in contrast to rigid registration where extremal points and curves as used in geometric computer vision techniques (e.g., [2.9.45.]) may be applied to image matching in visual servoing studies. Unlike visual servoing methods, deformation gradients need not be grounded in reference frame-based methods (e.g., [2.9.4., 2.9.14.]).

As an extension of Jacobian-based methods, second partial derivatives and the Hessian could be used so that point matching occurs with respect to a hypercube (e.g., [2.9.22.]). As another extension, the dot product of the norm and divergence of the displacement field in terms of deformation between images may be used so that deformation is related to the Jacobian [2.9.39.]. Deformation and visual servoing techniques both utilize the Jacobian as a measurable conduit to image matching.

Computationally, the Jacobian may be expensive and complicated to institute with large data sets, whether deformation or visual servoing techniques are used in image matching. There are approximation methods that may be instituted to estimate the Jacobian (e.g., [2.9.16., 2.9.37.]) that would facilitate obtaining useful results in these situations.

2.8 Summary and Future Research

With the plethora of diatom digital images as SEMs or LMs that reside in many research settings, a way to utilize this resource is in the offing. Jacobian-based methods and image matching, whether from models or micrographs, enables comparisons for purposes, ranging from taxonomic and classification treatments to other applied evolutionary studies. Morphological data from such images has been underutilized, and the Jacobian provides a computational means to productively exploit such data.

It may be possible to apply image surface analytical techniques as an aid in morphological character definition similarly to mathematical morphology techniques (e.g., [2.9.44.]). Combinations of surface features have as much morphological meaning as individual characters such as length, width or diameter. Diatom valve surface features are a rich resource of information, and using the Jacobian enables comparisons of diatom surfaces without regard to size or 2D shape of the valve face. The height of surface features or z-features is readily computed via the Jacobian and may be used to define ensemble surface features (e.g., [2.9.29.]), as the 3D-ness of protrusions, for example, on the valve surface are recoverable in size and shape. From a surface feature window, particular ensemble surface features may be obtained and recovered via the Jacobian [2.9.29.]. By using surface features as ensembles, any part of a surface may be used to match the same location in two or more images. In this way, new definitions of morphological characters may be made for studies in diatom morphological evolution, phylogenetic and speciation studies.

In diatom developmental studies, quantitative comparisons among stages in size diminution series from vegetative reproduction may be made using the Jacobian. The connection between successive stages and the change in surface features may be accounted for as size reduction occurs. Comparisons among species in a genus or characterization of lineages may inform evolutionary pathways of diatoms over short to longer time scales. The Jacobian in many forms has great utility in characterizing morphological surfaces and applications in morphological studies.

2.9 References

2.9.1.   Andresen, P.R. and Mielsen, M. (2001) Non-rigid registration by geometry-constrained diffusion. *Medical Image Analysis* 5, 81-88.

2.9.2.   Bower, A.F. (2008) *Applied Mechanics of Solids, Chapter 2: Governing Equations – 2.1 Deformation*, accessed on August 26, 2021, http://solidmechanics.org/text/Chapter2_1/Chapter2_1.htm

2.9.3.   Bro-Nielsen, M. (1997) Medical image registration and surgery simulation IMM-PHD-1996-25 thesis, Department of Mathematical Modelling, Technical University of Denmark.

2.9.4.   Cao, Y., Lu, K., Li, X., Zang, Y. (2011) Accurate numerical methods for computing 2D and 3D robot workspace. *International Journal of Advanced Robotic Systems* 8(6), *Special Issue Robotic Manipulators*, 1-13.

2.9.5.   Chen, M. (2021) Chapter 5: Cell tracking in time-lapse microscopy image sequences. In: *Computer Vision and Pattern Recognition - Computer Vision for Microscopy Image Analysis*, Chen, M. (ed.), Academic Press, London, United Kingdom: 101-129.

2.9.6.   Cowan, N.J. and Chang, D.E. (2005) Geometric visual servoing. *IDDD Transactions on Robotics* 21(6), 1128-1138.

2.9.7.   Davatzikos, C., Vaillant, M., Resnick, S., Prince, J.L., Letovsky, S., Bryan, R.N. (1996) Morphological analysis of brain structures using spatial normalization. In: *Visualization in Biomedical Computing, Lecture Notes in Computer Science, Vol. 1131*, Höhne, K.H., Kikinis, R. (eds.), Springer, Hamburg, Germany: 355-360.

2.9.8.   Denavit, J. and Hartenberg, R.S. (1964) *Kinematic Synthesis of Linkages*, McGraw-Hill Book Company, New York.

2.9.9.   do Carmo, M.P. (2016) *Differential Geometry of Curves and Surfaces, Revised & Updated Second Edition*, Dover Publications, Inc., Mineola, New York, USA.

2.9.10. Dulęba, I. and Opałka, M. (2013) A comparison of Jacobian-based methods of inverse kinematics for serial robot manipulators. *Int. J. Appl. Math. Comput. Sci.* 23(2), 373-382.

2.9.11. Hosseini, M.S. and Plataniotis, K.N. (2017) Finite differences in forward and inverse imaging problems: MaxPol design. *SIAM J. Imaging Sci.* 10(4), 1963-1996.

2.9.12. Hutchinson, S., Hager, G.D., Corke, P.I. (1996) A tutorial on visual servo control. *IEEE Transactions on Robotics and Automation* 12(5), 651-670.

2.9.13. Kaplan, W. (2003) *Advanced Calculus, 5th edition*. Addison-Wesley, Boston, Massachusetts, USA.

2.9.14. Khatib, O. (2020) *Chapter 4 - The Jacobian*, accessed on April 7, 2020, http://robotics. itee.uq.edu.au/~metr4202/2013/tpl/Chapter%204%20-%20Jacobain%20-%20from%20 Khatib%20-%20Introduction%20to%20Robotics.pdf

2.9.15. Koenderink, J.J. (1990) *Solid Shape*, MIT Press, Cambridge, Massachusetts USA.

2.9.16. Kosmopoulos, D.I. (2011) Robust Jacobian matrix estimation for image-based visual servoing. *Robotics and Computer-Integrated Manufacturing* 27, 82-87.

2.9.17. Kyung, M-H., Kim, M-S., Hong, S.J. (1995) Through-the-lens camera control with a simple Jacobian matrix. In *Proceedings of Graphics Interface '95*, Prusinkiewicz, P. (ed.), Canadian Human-Computer Communications Society, Toronto, Ontario, Canada: 171-178.

2.9.18. Leow, A.D., Yanovsky, I., Chiang, M.-C., Lee, A.D., Klunder, A.D., Lu, A., Becker, J.T., Davis, S.W., Toga, A.W., Thompson, P.M. (2007) Statistical properties of Jacobian maps and the realization of unbiased large-deformation nonlinear image registration. *IEEE Transactions on Medical Imaging* 26(6), 822-832.

2.9.19. Lindeberg, T. (2008) Scale-space. In: *Encyclopedia of Computer Science and Engineering, Volume IV*, Wah, B. (ed.), John Wiley and Sons, Hoboken, New Jersey: 2495-2504.

2.9.20. Mao, S., Huang, X., Wang, M. (2012) Image Jacobian matrix estimation based on online support vector regression. *International Journal of Advanced Robotic Systems* 9, 111:2012.

2.9.21. Mefteh, S., Kaâniche, M.-B., Ksantini, R., Bouhoula, A. (2019) A novel multispectral lab-depth based edge detector for color images with occluded objects. In: *Proceedings of the 14th International Joint Conference on Computer Vision, Imaging and Computer Graphics Theory and Applications (VISIGRAPP 2019*, SCITEPRESS – Science and Technology Publications, Lda., Prague, Czech Republic: 272-279.

2.9.22. Monga, O. and Benayoun, S. (1992) Using partial derivatives of 3D images to extract typical surface features. In: *Proceedings of the Third Annual Conference of AI, Simulation, and Planning in High Autonomy Systems 'Integrating Perception, Planning and Action', 1992, Perth, WA, Australia*, IEEE, 225-236.

2.9.23. Navarro-Alarcón, D. (2013) Model-free visually served deformation control of elastic objects by robot manipulators. *IEEE Transactions on Robotics* 29(6), 1457-1468.

2.9.24. Ohkitani, K. (2010) Dynamics of the Jacobian matrices arising in three-dimensional Euler equations: application of Riccati theory. *Proceedings of the Royal Society A: Mathematical, Physical and Engineering Sciences* 466, 1429-1439.

2.9.25. Pappas, J.L. (2005) Geometry and topology of diatom shape and surface morphogenesis for use in applications of nanotechnology. *Journal of Nanoscience and Nanotechnology* 5(1), 120-130.

2.9.26. Pappas, J.L. (2005) Theoretical morphospace and its relation to freshwater gomphonemoid-cymbelloid diatom (Bacillariophyta) lineages. *Journal of Biological Systems* 13(4), 385-398.

2.9.27. Pappas, J.L. (2008) More on theoretical morphospace and its relation to freshwater gomphonemoid-cymbelloid diatom (Bacillariophyta) lineages. *Journal of Biological Systems* 16(1), 119-137.

2.9.28. Pappas, J.L. (2016) Multivariate complexity analysis of 3D surface form and function of centric diatoms at the Eocene-Oligocene transition. *Marine Micropaleontology* 122, 67-86.

2.9.29. Pappas, J.L. (2021) Quantified ensemble 3D surface features modeled as a window on centric diatom valve morphogenesis. In: *Diatom Morphogenesis [DIMO, Volume in the series: Diatoms: Biology & Applications, series editors: Richard Gordon & Joseph Seckbach]*. V. Annenkov, J. Seckbach and R. Gordon, (eds.) Wiley-Scrivener, Beverly, MA, USA: 158-193.

2.9.30. Pappas, J.L. and Miller, D.J. (2013) A generalized approach to the modeling and analysis of 3D surface morphology in organisms. *Plos One* 8(10), #e77551.

2.9.31. Paquin, V., Maldague, X., Akhloufi, M. (2012) Avision-based method for improving absolute positioning accuracy of industrial robots. In: *Colloque Vision et Robotique Appliquées 2012, Proceedings of CVRA2012 (Applied Vision and Robotics Workshop 2012)*, Akhloufi, M (ed.), Montreal, Québec, Canada: 58-67.

2.9.32. Pari, L., Sebastián, J.M., Traslosheros, A., Angel, L. (2008) Image based visual servoing: estimated image Jacobian using fundamental matrix vs analytic Jacobian. In: *Image Analysis and Recognition, 5th International Conference, ICIAR 2008 Póvoa de Varzim, Portugal, June 2008 Proceedings*, Campilho, A., Kamel M. (eds.), Springer-Verlag, Berlin Heidelberg, Germany: 706-717.

2.9.33. Rey, D., Subsol, G., Delingette, H., Ayache, N. (2002) Automatic detection and segmentation of evolving processes in 3D medical images: application to multiple sclerosis. *Medical Image Analysis* 6, 163-179.

2.9.34. Riyahi, S., Choi, W., Liu, C-J., Zhong, H., Wu, A.J., Mechalakos, J.G., Lu, W. (2018) Quantifying local tumor morphological changes with Jacobian map for prediction of pathologic tumor response to chem-radiotherapy in locally advanced esophageal cancer. *Phys Med Biol* 63(14), 145020.

2.9.35. Roussel, M.R. (2019) Stability analysis for ODEs, Ch. 3. In: *Nonlinear Dynamics: A Hands-On Introductory Survey*, Morgan and Claypool Publishers, San Rafael, California: 3-12.

2.9.36. Selig, J.M. (1992) *Introductory Robotics*, Prentice Hall, New York.

2.9.37. Sutanto, H., Sharma, R., Varma, V. (1998) The role of exploratory movement in visual servoing without calibration. *Robotics and Autonomous Systems* 23, 153-169.

2.9.38. Thirion, J.-P. (1993) New feature points based on geometric invariants for 3D image registration. [Research Report] RR-1901, INRIA, inria-00077266.

2.9.39. Thirion, J.-P. and Calmon, G. (1999) Deformation analysis to detect and quantify active lesions in 3D medical image sequences. *IEEE Trans. Med. Imaging* 18(5), 429-441.

2.9.40. Thirion, J.-P. and Gourdon, A. (1992) The 3D marching lines algorithm and its application to crest lines extraction. [Research Report] RR-1672, INRIA, inria-00074885.

2.9.41. Watanabe, T. (2018) Background: dexterity in robotic manipulation by imitating human beings. In: *Human Inspired Dexterity in Robotic Manipulation*, Watanabe, T., Harada, K., Tada, M. (eds.), Academic Press, Elsevier, Cambridge, Massachusetts: 1-7.

2.9.42. Weiss, L.E., Sanderson, A.C., Neuman, C.P. (1987) Dynamic sensor-based control of robots with visual feedback. *IEEE Journal of Robotics and Automation* RA-3(5), 404-417.

2.9.43. Whitaker, R.T. (2004) 6-Isosurfaces and level-sets. In: *The Visualization Handbook*, Hansen, C.D. and Johnson, C.R. (eds.), Elsevier, Amsterdam, The Netherlands: 97-123.

2.9.44. Wilkinson, M., Urbach, E., Jalba, A., Roerdink, J. (2013) Diatom identification with mathematical morphology. In: *Mathematical Morphology: From Theory to Applications*, Najman, L. and Talbot, H. (eds.), Wiley-ISTE, London, UK: 357-366.

2.9.45. Yoshizawa, S., Belyaev, A., Yokota, H., Seidel, H.-P. (2008) Fast, robust, and faithful methods for detecting crest lines on meshes. *Computer Aided Geometric Design* 25, 545-560.

2.9.46. Yuille, A. and Leyton, M. (1990) 3D symmetry-curvature duality theorems. *Computer Vision, Graphics, and Image Processing* 52, 124-140.

2.9.47. Zhu, J. (2017) Image gradient-based joint direct visual odometry for stereo cameras. *Proceedings of the Twenty-Sixth International Joint Conference on Artificial Intelligence (IJCAI-17), Melbourne, Australia*, Sierra, C. (ed.), AAAI Press, Palo Alto, California, USA: 4558-4564.

3

Diatom Valve Morphology, Surface Gradients and Natural Classification

Abstract

Digital images play a key role in diatom morphology studies. From a digital image, pixel values are obtained via measurement of surface gradient magnitude and direction using the histogram of oriented gradients (HOG). Changes in these values among different specimen images reveal changes in edges on valve surfaces to enable a natural inter-taxon assessment of all valve morphological characters simultaneously. These changes record valve 3D surface and shape differences. Support Vector Regression (SVR) of HOG data is used to elucidate binning priorities via minimization of classification errors while maximizing the marginal classifier boundaries. The methodology provides a way to analyze light micrographs. The method can be used to analyze scanning electron micrographs as well. SVR using a kernel function provided morphological analysis of *Navicula* and other naviculoid taxa using HOG data. SVR is useful in predictive classification of diatom taxa. The protocol elucidated in this study enables the application of machine learning algorithms to multidimensional morphological data to better understand potential synapomorphic assessment and natural classification at the diatom generic level.

Keywords: *Navicula*, morphometrics, support vector regression, histogram of oriented gradients, pattern recognition, machine learning, synapomorphy, natural classification

3.1 Introduction

Diatoms morphology studies are often accomplished based on counting particular features or quantifying the outline of the organism. In either case, each measure is a portion of the total morphology of the whole diatom. To gain a more complete measure of morphology, surfaces contain more morphological information because the surface is a three-dimensional (3D) feature of the organism. Surface morphology is a proxy for the phenotype [3.8.76.], and measurement of the surface is necessary to acquire a more complete quantification of the morphology of any organism such as a diatom.

Sources of surface morphological data include digital images. Surface features may be extractable as a result of measuring changes in height in the z-direction of the surface with respect to the x, y-plane via changes in pixel intensity over the image surface. The relation between changes in z-height and pixel intensity are encapsulated in measuring surface edges in digital images. Measuring such changes from one digital image to the next is measuring 3D surfaces, and such measurements are useful in attaining measurement of total morphology of organisms including diatoms.

Janice L. Pappas. Mathematical Macroevolution in Diatom Research, (81–114) © 2023 Scrivener Publishing LLC

Diatoms have siliceous scaffolded structure with a multitude of geometric features that make them amenable for surface morphology analyses. Digital scanning electron micrographs (SEM) or light micrographs (LM) enable analyses of diatom surfaces with respect to pixel intensity differences of edges as surface features. As a result, a group of taxa, even at the genus-level, may be quantified and compared for surface morphological differences with regard to classification schema and correctness. The degree to which a classification scheme is correct is obtainable via techniques available from pattern recognition, optimization and statistical learning (e.g., [3.8.104.-3.8.106.]) theories.

Using digital images to measure 3D surface morphology expands our ability to quantify and study morphological variation within a genus, is potentially useful in applications regarding speciation, and more broadly, is applicable to morphological evolution studies of diatoms. In this regard, measurement of 3D surface morphology from digital images may be accomplished using techniques from image processing and computational computer vision and pattern recognition studies.

One well-established problem in diatom research is that some genera are bins with a mix of members. Unless a genus is succinctly defined in terms of morphological characters, cellular organelles or internal cellular characters, life cycle attributes, or genetic/molecular identifiers, species cannot be assuredly assigned to that genus. As a result, some diatom genera become bins in which to dump species that only qualify because they are bilaterally symmetric, lanceolate, round, or some other such broadly earmarked descriptor. Much work is needed to reassign misplaced species to the appropriate genus or a newly created genus.

3.2   Purposes of this Study

Using digital images, what kinds of analyses would aid in reallocating misassigned taxa to other or newly created diatom generic bins? There are limits to morphological variation with respect to defining species, but what are the limits for genera or any other classification level? Genetics plays a role in species designations, phyla are differentiated based on body plan, but on a macroevolutionary scale, what delineates genera? For guidance, the concept of monophyly is useful in determining which generic bin a species should be assigned to [3.8.59.]. Yet, being dependent on synapomorphies to define a given genus membership means being dependent on knowing the ancestor-descendent shared characters. Even when not knowing what those characters are, it is possible to discover such characters with a more wholistic approach.

Genera having polyphyletic origins, such as the slime mold *Physarum* [3.8.72.] or green alga *Dictosphaerium* [3.8.64.], indicate that achieving natural classification schema is an ongoing occupation for biologists. For diatoms, polyploidy is evident (e.g., [3.8.46., 3.8.57., 3.8.58.]), potentially indicating one aspect of the polyphyletic application of classification schema in attempts to understand diatom evolution. In ancient thalassiosiroid and some pennate lineages, allopolyploidy had occurred, making polyploidy a predominant feature of their genomic diversity [3.8.78.]. Diatom hybridization occurs as well, and *Pseudonitzschia pungens* x *cingulata* is one case where hybrid taxa are morphologically similar to *P. pungens* rather than having an intermediate morphology [3.8.41.]. Because of polyploidy and hybridization, it is evident that concept of monophyly must be applied with a carefulness and an exactness to achieve each level of a diatom natural classification system.

For monophyly as a delineator of diatom genera, applicable methods should encompass the utility of using such a descriptor, all the while knowing that confounding exceptions may exist or lack of complete information or data will occur. Methodologies with the concept of monophyly in mind should reflect the whole diatom whether using morphological or genomic data or a combination of both.

Often, diatom morphological studies are conducted using strewn mounts for light microscopy or stub-coated mounts for scanning electron microscopy or digital images from both kinds of mounts. With the availability of microphotographic images, how can we use this data effectively? To study morphological characters usually means counting identifiable structures. Quantifiable analyses proceed as if each character is discrete rather than a part of a continuous whole. To study whole diatom morphology, the surface contains all the morphological characters. In this case, the diatom valve face is most often used for identification purposes with its most identifiable attributes, and the whole diatom valve face is represented by its surface.

Digital images of a commonly found diatom genus, *Navicula* Bory de Saint-Vincent 1822 [3.8.98.], and other naviculoid taxa form the generic bins of study of how to approach the problem of genus assignment. From strewn slide or stub preparations, morphological data may be obtained, and digital images provide a record of this data of the taxa viewed. For naviculoid diatoms, digital images are used as the sources of surface information that will be used in analyses to determine the degree to which classification of these taxa belong to the genus *Navicula*. From analyses of taxa found to be non-inclusive, the results of this study will provide an initial impetus to re-evaluate such taxa concerning the genus *Navicula*.

3.2.1 The Genus *Navicula*

What is *Navicula*? According to Bory de Saint-Vincent [3.8.40.], a rough translation of the French describes *Navicula* as 'animalcules with a body that looked like a weaver's shuttle'. Although much work has been done on the genus *Navicula* since Bory de Saint-Vincent's time (e.g., [3.8.62.-3.8.63., 3.8.79., 3.8.93.), much more work is needed. Generally speaking, the genus *Navicula* has been used as a bin for bilaterally symmetric raphid diatom species that were not assigned to other pennate diatom genera. Many species remain in *Navicula sensu lato*, and there is a general consensus that there is a long way to go to achieve a genus *Navicula sensu stricto* [3.8.57., 3.8.61., 3.8.93.]. As Kociolek *et al.* [3.8.61.] lament, "*Navicula*…is largely still a repository for almost any symmetrical bi-raphid diatom, which may or may not share common descent."

A general characteristic of *Navicula* is that species are lanceolate in shape (e.g., [3.8.57., 3.8.61., 3.8.93.]). However, even this descriptor is inadequate as a starting point in determining what constitutes *Navicula* (e.g., [3.8.60., 3.8.108.]). Because *Navicula* comprises the largest botanical genus according to the International Code of Botanical Nomenclature, separating taxa appropriately to obtain a more natural classification is a monumental task. As an aid in this direction, digital images of naviculoid taxa possess a plethora of morphological data that may be gleaned in the quest for a more natural classification of *Navicula*. With digital imaging analysis, pattern recognition techniques, optimization theory, and statistical learning methodologies, morphological information from the digital images of diatom valve face surfaces may be used to start to disentangle the current members of the genus *Navicula*.

3.3   Methods

3.3.1   Naviculoid Diatom Surface Analysis

Digital grayscale light micrographs (LMs) of 67 naviculoid diatom valves were obtained from the Diatom of North America website ([3.8.99.]; Table 3.1). Taxa were chosen if they were listed as a *Navicula* species. Scrolling down the entire list of genera, non-*Navicula* genera were chosen if, upon cursory inspection, they were approximately lanceolate, both valves have raphes, were approximately apically symmetrical, or were named as "*Navicula*" at one time. Two of the taxa chosen were *Biremis* and *Brebissonia*, even though they are not quite within the loose boundary of what is being used as "naviculoid" for this study. No attempt was made to obtain specific taxa. The images were chosen because they exhibited no debris on the valve surface, and the backgrounds were fairly uniform in grayscale shade. If background markings were present, including white, black or other prominent markings such as micron bars, the background in that place on the LM was replaced with the same shade of grayscale nearby in the rest of the background. No attempt was made to change any feature, brightness or contrast of the diatom valve face image.

Of the 67 naviculoid taxa, 17 were from non-*Navicula* genera with the remaining 50 taxa were currently named *Navicula* species (Table 3.1). Surface analysis of the digital naviculoid LMs was accomplished using the image processing technique of measuring histograms of oriented gradients (HOG).

3.3.2   Gradients of Digital Image Surfaces

Any object possesses a surface. The texture of a surface determines its topography. If we take a photo of an object, the angle at which we take the picture enables us to determine the three-dimensionality of the object's surface. We can take a photo of an object face-forward or perpendicular to the object to see it from its side. For diatoms, the valve face is critical in our assessment of its identity, and the diatom side or girdle is additional, useful information of morphology. Photographing a diatom valve face, resulting in a LM or SEM, records the surface as well as shape.

Digital imaging of diatom valve faces provides access to information in a LM or SEM. Image processing techniques enable quantifying attributes of a digital image as a summary of the diatom valve surface and shape. Quantifying many diatom valve face images provides for numerical categorization and comparison for the sake of summarizing morphological differences at a species or genus level. Some diatom genera that contain a large number of species have historically been used as a dump for sketchily similar looking species. Sometimes subsequently, some of the species in a genus dump have been removed, only to find their placement in their own genus. Assignment in a given genus requires careful work by those with a meticulous ability to scrutinize each digital LM and/or SEM to make an expert determination.

To augment the expert work of a diatom researcher, digital LMs and SEMs may be read using the changes in grayscale or color as recordings of surface changes. Such surface changes are easily understood as changes in pixel intensity, recording an ascent or descent from pixel to pixel on the surface. That is, as pixel intensity increases in a grayscale image, the more whiteness is evident, signifying a peak; as pixel intensity decreases, the

Table 3.1 Sixty-seven naviculoid taxa used in HOG and SVR analyses. Authority information and images used are obtained from each authored page cited in [3.8.99.].

Taxon	Image	Citation
Aneumastus tusculus D.G. Mann and Stickle in Round *et al.* 1990		[3.8.14.]
Anomoeoneis sphaerophora Pfitzer 1871		[3.8.36.]
Biremis circumtexta (F. Meister *ex* Hust.) Lange-Bert. and Witkowski, Lange-Bert. and Metzeltin 2000		[3.8.85.]
Brebissonia lanceolata (Agardh) Mahoney and Reimer 1986		[3.8.38.]
Cavinula cocconeiformis (Gregory *ex* Grev.) D.G. Mann and Stickle in Round *et al.* 1990		[3.8.43.]
Cosmioneis citriformis A.R. Sherwood and R.L. Lowe in R.L. Lowe and A.R. Sherwood 2010		[3.8.67.]
Craticula buderi (Hust.) Lange-Bert. in Rumrich *et al.* 2000		[3.8.34.]
Decussata placenta (Ehrenb.) Lange-Bert. and Metzeltin in Lange-Bert. 2000		[3.8.52.]
Encyonopsis montana L.L. Bahls 2013		[3.8.28.]
Envekadea metzeltinii Sylvia S. Lee, Tobias, and Van de Vijer 2013		[3.8.66.]
Hippodonta capitata (Ehrenb.) Lange-Bert., Metzeltin and Witkowski 1996		[3.8.86.]
Kurtkrammeria subspicula Krammer 1997		[3.8.29.]
Lacustriella lacustris (W. Greg.) Lange-Bert. and Kulikovskiy 2012		[3.8.43.]
Navicula angusta Grunow 1862		[3.8.87.]
Navicula aurora Sovereign 1958		[3.8.2.]

(Continued)

Table 3.1 Sixty-seven naviculoid taxa used in HOG and SVR analyses. Authority information and images used are obtained from each authored page cited in [3.8.99.]. (*Continued*)

Taxon	Image	Citation
Navicula cari Ehrenb. 1863		[3.8.15.]
Navicula caroliniae L.L. Bahls 2012		[3.8.16.]
Navicula caterva M.H. Hohn and Hellerman 1963		[3.8.44.]
Navicula cryptocephaloides Hust. 1937		[3.8.3.]
Navicula cryptofallax Lange-Bert. and Kusber 2019		[3.8.4.]
Navicula duerrenbergiana Hust. In A. Schmidt *et al.* 1934		[3.8.17.]
Navicula eidrigiana Carter 1979		[3.8.31.]
Navicula elsoniana R.M. Patrick and Freese 1961		[3.8.48.]
Navicula erifuga Lange-Bert. in Krammer and Lange-Bert. 1985		[3.8.81.]
Navicula escambia (R.M. Patrick) Metzeltin and Lange-Bert. 2007		[3.8.69.]
Navicula flatheadensis L.L. Bahls 2011		[3.8.5.]
Navicula freesei R.M. Patrick and Freese 1961		[3.8.46.]
Navicula galloae L.L. Bahls 2011		[3.8.6.]
Navicula harmoniae L.L. Bahls 2014		[3.8.32.]
Navicula kotschyi Grunow 1860		[3.8.103.]
Navicula lanceolata (C. Agardh) Ehrenb. 1838		[3.8.82.]
Navicula leptostriata E.G. Jørg. 1948		[3.8.18.]
Navicula longicephala Hust. 1944		[3.8.88.]

Table 3.1 Sixty-seven naviculoid taxa used in HOG and SVR analyses. Authority information and images used are obtained from each authored page cited in [3.8.99.]. (*Continued*)

Taxon	Image	Citation
Navicula ludloviana A. Schmidt 1876		[3.8.35.]
Navicula lundii E. Reichardt 1985		[3.8.19.]
Navicula margalithii Lange-Bert. 1985		[3.8.33.]
Navicula metareinhardtiana Lange-Bert. and Kusber 2019		[3.8.89.]
Navicula oppugnata Hust. 1945		[3.8.20.]
Navicula peregrina (Ehrenb.) Kütz. 1844		[3.8.7.]
Navicula perotii R.M. Patrick and Freese 1961		[3.8.47.]
Navicula pseudolanceolata Lange-Bert. 1980		[3.8.8.]
Navicula radiosa Kütz. 1844		[3.8.90.]
Navicula recens Lange-Bert. in Krammer and Lange-Bert. 1985		[3.8.83.]
Navicula reinhardtii (Grunow) Grunow 1880		[3.8.9.]
Navicula rhynchocephala Ehrenb. 1844		[3.8.53.]
Navicula rhynchotella Lange-Bert. 1993		[3.8.10.]
Navicula salinarum Grunow 1880		[3.8.54.]
Navicula schweigeri L.L. Bahls 2012		[3.8.21.]
Navicula slesvicensis Grunow in Van Heurck 1880		[3.8.92.]
Navicula sovereignii L.L. Bahls 2011		[3.8.11.]
Navicula staffordiae L.L. Bahls 2012		[3.8.22.]

(*Continued*)

Table 3.1 Sixty-seven naviculoid taxa used in HOG and SVR analyses. Authority information and images used are obtained from each authored page cited in [3.8.99.]. (*Continued*)

Taxon	Image	Citation
Navicula subconcentrica Lange-Bert. 2001		[3.8.12.]
Navicula supleeorum L.L. Bahls 2013		[3.8.23.]
Navicula trilatera L.L. Bahls 2013		[3.8.24.]
Navicula trivialis Lange-Bert. 1980		[3.8.94.]
Navicula vaneei Lange-Bert. in Witkowski, Lange-Bert. and Stachura 1988		[3.8.13.]
Navicula venerablis M.H. Hohn and Hellerman 1963		[3.8.84.]
Navicula veneta Kütz. 1844		[3.8.91.]
Navicula vulpina Kütz. 1844		[3.8.55.]
Navicula weberi L.L. Bahls 2012		[3.8.25.]
Navicula whitefishensis L.L. Bahls 2012		[3.8.30.]
Navicula wildii Lange-Bert. 1993		[3.8.26.]
Navicula winona L.L. Bahls 2012		[3.8.27.]
Parlibellus delognei (Van Heurck) Cox 1988		[3.8.97.]
Placoneis gastrum (Ehrenb.) Mereschkowsky 1903		[3.8.109]
Prestauroneis integra (W. Sm.) Bruder 2008		[3.8.73.]
Pseudofallacia monoculata (Hust.) Liu, Kociolek and Wang 2012		[3.8.37.]

more blackness is evident, signifying a valley. Shades of gray tending toward one side of the scale or the other are transitional points. The direction of the changes is recorded as well, and the magnitude and direction or orientation of the changes are quantifiable as gradients. Changes in pixel intensity are changes in edges, and the image gradient is a measure of edge detection changes. These changes are derivatives of the horizontal and vertical directions at every point in an image with a continuous image function being defined over the whole surface. At a given point in a digital image, the gradient is a vector in the direction of greatest intensity increase, and the length of that vector is the rate of change in that direction. Measurement of gradient magnitude of intensity and the rate of change in that intensity is accomplished using HOG of digital images.

3.3.3   Histogram of Oriented Gradients and Surface Representation

Digital images whether in color or grayscale exhibit variations in pixel intensity. The image gradient defines the level of intensity with respect to edge detection in a digital image. The gradient is definable at each point in a digital image. Components of the gradient are derivatives in a vertical and horizontal direction from an image intensity function defined over the entire digital image. From this, the gradient as a vector quantity at each point measures the direction of greatest intensity as well as the rate of change in the greatest intensity direction (Table 3.2). The gradient magnitude is $\left\| \nabla f_{x,y} \right\| = \sqrt{\left(\dfrac{\partial f}{\partial x} \right)^2 + \left(\dfrac{\partial f}{\partial y} \right)^2}$, and the gradient direction or orientation is $\theta = \tan^{-1}\left(\dfrac{\partial f}{\partial y} \Big/ \dfrac{\partial f}{\partial x} \right)$, where f is an image intensity function with respect to x and y [3.8.102.].

Table 3.2 Image gradient components for each pixel intensity in a digital image.

Image gradient	Gradient direction	Gradient equation
	➡	$\nabla f_x = \left[\dfrac{\partial f}{\partial x}, 0 \right]$
	⬇	$\nabla f_y = \left[0, \dfrac{\partial f}{\partial y} \right]$
	↗	$\nabla f_{x,y} = \left[\dfrac{\partial f}{\partial x}, \dfrac{\partial f}{\partial y} \right]$

3.3.4   Application to Diatom Valve Face Digital Images

From computer vision and image processing techniques, HOG [3.8.42.] may be used to analyze and quantify diatom valve surfaces. As the name implies, gradient magnitude and direction are the measured quantities used in developing the content of the histogram representing the surface.

HOG is used to define feature characteristics of digital images. Pixel changes in grayscale or color are measured as gradient magnitude and direction changes in n-8x8 "cells" that are sliding windows covering a digital image. Feature extraction occurs over the grid of cells in an overlapping fashion where all positions and scales of the image are scanned.

To detect surface features, HOG is a technique that uses contrast normalization of overlapping cells to determine local contrasts within an image, where each cell may be a sliding window. Pixels in each window are analyzed for intensity of edges as gradient magnitudes while gradient directions are also obtained simultaneously. Each pixel is weighted by its gradient magnitude and is binned according to its corresponding local direction or orientation of gradients. The bars of the histogram are gradient direction or orientation bins for gradient magnitudes. HOG encapsulates a distribution of edge directions as intensity gradients. Because local contrast normalization is used, backgrounds are not influential in the resultant HOG for each diatom valve face analyzed.

HOG is a robust feature extraction method with respect to illumination variation, photometric transformation, and geometric transformation [3.8.101.]. Measurement of local edge features and contrast normalization provides invariance to geometric transformation [3.8.42.]. For many diatoms, image orientation is not relevant because valve features are symmetric. For diatoms that have directional or asymmetric features, orientation might matter, depending on the amount of translation or rotation relative to spatial and orientation sampling. If the sampling of gradients is coarse, and sampling of orientations is finely controlled, then transformation is not a necessary consideration. Illumination variation and photometric transformation are taken care of via contrast normalization so that preprocessing of images is unnecessary [3.8.42.].

Using HOG, surface feature detection is applicable to digital images of diatom valve faces, and the extracted valve surface features are used as the input data for morphological comparative analyses. Valve surface feature extraction is used with n-images across all the images. From n-images, n-HOG values represent n-feature vectors representing all the digital images. Any number of HOG bins may be calculated. Because local contrast normalization is used in collecting HOG data, digital image backgrounds do not influence the values obtained. HOG values for each taxon were assembled in a matrix for morphological analysis.

3.3.5   Support Vector Regression and Classification

Diatom morphological data is often non-linear and multidimensional, especially concerning morphometric measurements of the plethora of diatom shapes and morphological characters. Acquiring such data and transforming it to be a linear problem to solve makes diatom morphological and morphometric analyses tractable. Making morphological comparisons among diatoms valve surfaces may be accomplished using support vector machines (SVM) which are supervised learning techniques applied in classification and

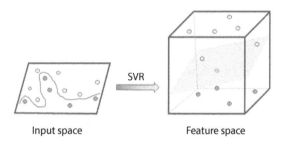

Input space   Feature space

Figure 3.1 Input data that are not separable with a straight line become separable with a hyperplane in n-dimensional space via a kernel function in SVR.

regression problems ([3.8.1., 3.8.50., 3.8.75., 3.8.95., 3.8.96., 3.8.104.-3.8.106.]). SVM is a convex optimization technique that uses kernel functions to analytically discriminate among inherently non-linear input data via mapping into a higher dimensional space where a best-fit hyperplane separator (Figure 3.1) is achieved [3.8.1., 3.8.50., 3.8.93.]. As a result, supervised learning occurs on a linear discriminant surface to produce the resultant classification space [3.8.1.].

To implement comparative surface morphology measurement among diatom valve digital images, HOG is used as input data in support vector regression (SVR) [3.8.50., 3.8.68., 3.8.70.]. As a regression technique, minimization of error is determined with an added constraint that takes the form of maximizing marginal separators [3.8.50., 3.8.68., 3.8.103-3.8.104]. Applied only to the training sets, the error function is an optimization criterion as a loss, cost or penalty function that is used to measure a minimization threshold [3.8.50., 3.8.68.].

SVR also induces a classification [3.8.70.], because a hyperplane is calculated over n-dimensions to produce feature separation in feature space with respect to support vectors. The support vectors provide the upper and lower bounds as the outside marginal feature classifiers. The loss, cost or penalty function serves to determine the width of the classification margins, so the support vectors contribute to the loss, cost or penalty induced [3.8.50.].

SVR and classification requires maximizing marginal separators set by the support vectors [3.8.50., 3.8.104.-3.8.105.] for the regression hyperplane as the best possible classification of data tested with respect to a training set. SVR is used for the prediction capability of a training set to correctly validate and classify test data. For diatom valve surfaces, training and validation test data may be used to produce a predictability assessment for new data introduced into the classification schema.

3.3.6   Using HOG as Combination Gradient Magnitude and Direction Input Data for SVR

Image surface features, $x_1, ..., x_n$, represent n-patterns, where each pattern represents the quantified surface of a given diatom valve face. Those patterns that are the most difficult to classify are the support vectors and occur farthest from the patterns that are the most similar, which are closer together, in feature space. The position of a given pattern is a prediction of where that pattern belongs relative to the closest support vectors. The support vectors influence the optimal resultant configuration of the feature space defined by the hyperplane separating the features (or patterns, equivalently). In some cases, a linear hyperplane suffices, and a generalized hyperplane is

$$\{\mathbf{x} \in S | \mathbf{w} \cdot \mathbf{x} = 0\}, \mathbf{w} \in S, b \in R \tag{3.1}$$

where x surface features occur in feature space, S, as a result of the dot product $\mathbf{w} \cdot \mathbf{x} = \sum_{i=1}^{n} w_i x_i$ with w weights, and b is the intercept [3.8.50.].

In other cases, a non-linear hyperplane must be used, and the resultant pattern is induced with respect to computation of a kernel function. The dot product is represented via linear SVR, while a kernel function represents the dot product with respect to non-linear SVR. The dot product of an unknown feature function that forms the training set is the basis of the kernel used. Concomitantly, SVR induces a classification of the prediction variables.

The feature function is one of a number of kernel functions that enable calculation of the dot product as

$$K(\mathbf{x}_i, \mathbf{x}_j) = \phi(\mathbf{x}_i) \cdot \phi(\mathbf{x}_j) = \sum_{i-1}^{\infty} \lambda_i \phi(\mathbf{x}_i) \phi(\mathbf{x}_j) \tag{3.2}$$

with weights, λ_i, and $K(\mathbf{x}_i, \mathbf{x}_j)$ is symmetric [3.8.50., 3.8.95.]. As a result, the kernel function replaces the dot product, and this is the hard margin version of SVR [3.8.50.].

The output vector comprises correlated features that "match" among surfaces. The regression analysis is a constrained quadratic optimization problem using

$$f(x_i) = y_i = w\phi(x_i) + b \tag{3.3}$$

where w is a weight vector, $\phi(\cdot)$ is a mapping as a non-linear feature function from input to an n-dimensional output feature space [3.8.50., 3.8.95.], and (w, b) are factors affecting the regression hyperplane separating the constituents of the higher dimensional feature space [3.8.50.]. The weight vector is expressible as a linear combination of training examples [3.8.39].

For dual input and regression

$$f(\mathbf{x}) = \sum_{i=1}^{n} \beta_i y_i \phi(\mathbf{x}_i) \phi(\mathbf{x}_j) + b = \sum_{i=1}^{n} \alpha_i K(\mathbf{x}_i, \mathbf{x}_j) + b \tag{3.4}$$

where non-linear Φ is mapped into feature space via α, which serves as a linear regressor variable, and the regression is a linear combination of $x_1, \ldots, x_n$ that serve as training points for prediction of newly added data via expansion of a kernel function, K [3.8.95.].

Kernel functions commonly used include the polynomial kernel [3.8.1., 3.8.50., 3.8.95.]. Higher order polynomials may be used cautiously because overfitting may occur. The higher the degree of the polynomial, the better chance of overfitting occurring. The general form of the polynomial kernel function is

$$K(\mathbf{x}_i, \mathbf{x}_j) = (1 + \mathbf{x}_i \cdot \mathbf{x}_j)^{poly} \tag{3.5}$$

where *poly* is the degree of the polynomial used [3.8.50.]. An alternative expression of the polynomial kernel function is

$$f(x) = \sum_{i=1}^{t} \theta_i \, K(x_i, x) \qquad (3.6)$$

where θ_i is a combination of dual input variables, and K is a polynomial kernel function [3.8.95.]. For degree = 1, the polynomial is a straight line.

In the event that certain conditions cannot be met, other variables are introduced to offset a penalty for errors in classification. This is the soft margin version of SVR [3.8.50.], and the penalty function to be minimized is given as

$$f(x)_{Penalty} = \frac{\|\mathbf{w}\|^2}{2} + C \sum_{i=1}^{m} \xi_i \qquad (3.7)$$

where C is a penalty capacity to classify imposed on $\|w\|$ with respect to slack variable, ξ_i, with constraints $y_i(\mathbf{w} \cdot \mathbf{x}_i + b) \geq +1 - \xi_i$, $\xi_i \geq 0$, and $i = 1, \ldots, m$, defining the hyperplanes with the penalty error, ϵ, taken into account via slack variable, ξ_i ($\mathbf{w}$, b) [3.8.50., 3.8.95., 3.8.105.]. That is, we want to minimize $\|w\|$, given constraints dictated by C and ϵ via ξ [3.8.50.].

The constrained quadratic minimization problem is expressed in terms of Lagrange multipliers (e.g., [3.8.50., 3.8.68.]) that represent the support vectors. The regression problem is converted into the Lagrangian function, L, as the dual optimization problem

$$L_{f(x)}(w, b, \Lambda, \mathbf{M}) = \frac{\|\mathbf{w}\|^2}{2} + C \sum_{i=1}^{m} \xi_i - \sum_{i=1}^{m} \lambda_i [y_i(\mathbf{w} \cdot \mathbf{x}_i + b) - 1 + \xi_i] - \sum_{i=1}^{m} \mu_i \xi_i \qquad (3.8)$$

where

$$\sum_{i=1}^{m} \gamma_i y_i - \epsilon \sum_{i=1}^{m} |\gamma_i| - \frac{1}{2} \sum_{i=1}^{m} \sum_{j=1}^{m} \gamma_i \gamma_j (\mathbf{x}_i \cdot \mathbf{x}_j) - \frac{1}{2C} \sum_{i=1}^{m} \gamma_i^2 \qquad (3.9)$$

is maximized, and

$$\sum_{i=1}^{m} \gamma_i = 0 \qquad (3.10)$$

where $i = 1, \ldots, m$ [3.8.50., 3.8.95.]. In the Lagrangian function, $\Lambda = (\lambda_1, \lambda_2, \ldots, \lambda_m)$ are the Lagrange multipliers for constraint $y_i(\mathbf{w} \cdot \mathbf{x}_i + b) \geq +1 - \xi_i$ and $\mathbf{M} = (\mu_1, \mu_2, \ldots, \mu_m)$ are the Lagrange multipliers for constraint $\xi_i \geq 0$ with $i = 1, \ldots, m$ [3.8.50., 3.8.95., 3.8.105.]. To achieve a better starting point for Lagrangian multipliers, rescaling may be applied with respect to training with different values of C [3.8.93.].

SVR is subject to Karush-Kuhn-Tucker (KKT) constraints [3.8.51., 3.8.65.] that gener-
alize the Lagrange multipliers for dual optimization, slack variables, and the gradient of
the Lagrangian function [3.8.1.]. SVR is used to solve non-linear as well as linear problems
[3.8.96.] via expression as a dual optimization problem [3.8.50.]. The dual problem is generally described as minimization of a cost while maximizing separation in feature space of
the features based on the dot product. That is, ϵ determines the maximal marginal separators, while C determines the slope of the fitted regression line. To achieve this outcome, the
primal and dual versions of the optimization problem achieve equivalency via utilization of
a Lagrangian function [3.8.50.]. When the product between dual variables is equal to zero,
or equivalently, the derivatives of the Lagrangian are equal to zero, the KKT conditions are
met [3.8.96.].

Using the Lagrangian, the dual optimization problem in terms of the kernel function is
given as

$$L_{dual}(w,b,\Lambda,\mathbf{M}) = -\frac{1}{2}\sum_{i=1}^{m}\sum_{j=1}^{m}(\lambda_i^- - \lambda_i^+)(\lambda_j^- - \lambda_j^+)K(\mathbf{x}_i,\mathbf{x}_j) \tag{3.11}$$

where

$$-\epsilon\sum_{i=1}^{m}(\lambda_i^- - \lambda_i^+) + \sum_{i=1}^{m}y_i(\lambda_i^- - \lambda_i^+) \tag{3.12}$$

is maximized with respect to constraint

$$\sum_{i=1}^{m}(\lambda_i^- - \lambda_i^+) = 0 \tag{3.13}$$

and $\lambda_i^-, \lambda_i^+ \in [0,C]$ [3.8.50.]. The non-linear SVR has

$$\mathbf{w} = \sum_{i=1}^{m}(\lambda_i^- - \lambda_i^+)\phi(\mathbf{x}_i) \tag{3.14}$$

with the ϵ-insensitive loss function being used [3.8.50., 3.8.106.], and the slack variable
constraint [3.8.50.] is

$$|\xi|_\epsilon = \begin{cases} 0 & \text{if } |\xi| \le \epsilon \\ |\xi| - \epsilon & \text{Otherwise.} \end{cases} \tag{3.15}$$

As computation of the kernel function is iterated, minimization of the error, ϵ, in the
updated output occurs to produce maximized marginal separators. The output is the update
of the institution of the kernel function with respect to change in surface morphology, and
therefore a measure of surface morphological differences for classification purposes.

The goal is to find a feature space that is the classification of n-diatom valve surfaces with optimal regression hyperplane separation with minimization of the loss function whereby the number of support vectors are sufficient to be used as a good prediction model for the testing of input diatom valve surface patterns. The classifier is

$$\text{class}(\mathbf{x}_k) = \text{sign}\left[\sum_{i-1}^{m} \lambda_i y_i \mathbf{x}_i \cdot \mathbf{x}_k + b\right] \tag{3.18}$$

where λ_i are Lagrange multipliers, and $i = 1, \ldots, m$ [3.8.50.]. For the maximum error boundary spread to produce the smoothest regression hyperplane separation, the best predictive model will result without overfitting, using the lowest order polynomial kernel function [3.8.50.].

3.3.7   Computational Efficiency and Cost

SVR as a dual optimization technique is a highly efficient and computationally cost-resistant method [3.8.1.]. However, the input data from HOG is potentially n-dimensional, and a cut-off point must be determined with regard to computational considerations using SVR. As input data increases, the kernel SVR matrix becomes so large to the point that matrix values must be calculated in real time or be retrieved from a pool of commonly determined values in a cache. The objective in SVR has a dual purpose so that a decision needs to be made as to whether the outcome will produce a true optimal solution or a generalization that has sufficiently low error, ϵ, relative to the value of the penalty classifier, C [3.8.50.]. Computational cost may be large at the outset when dealing with non-linear input data, but the outcome produces a lower computational cost as the n-dimensionality of the problem is reduced to a linear separator solution [3.8.39., 3.8.50.].

There is also consideration of the number of support vectors to be calculated in using kernel SVR, KKT conditions to be met, values of ϵ, and whether ξ is included in the calculations. Given such factors, computational time varies depending on whether training time or testing time is considered. Solution to training is a quadratic optimization, while testing for prediction purposes is linear. For linear SVR and a large training set, computational cost diminishes. For non-linear SVR, the kernel function matrix determines the computational cost. If loss function used is other than the ϵ-insensitive loss function, the quadratic problem may become a matrix inversion problem increasing computational cost. For asymptotic efficiency, the computational cost is $O(n^2)$ for small values of C and $O(n^3)$ for large values of C, signifying a lower and upper bound, respectively [3.8.39.] that can be used as general guidance for computational cost in using SVR.

3.4   Diatom Valve Surface Morphological Analysis

3.4.1   SVR Model Fit of Naviculoid Taxa

SVR was conducted using four HOG values of LMs per taxon to determine the goodness-of-model fit of a polynomial kernel function and reducing the initial n-dimensional problem to a linear feature space. Multivariate scatterplots of the four HOG values analysis were devised. Training and validation sets were obtained and used in model assessment. A plot of actual versus predicted values was devised as well as a plot of residuals versus predicted value. Root-mean-square-error (RMSE) plots were devised for C and ϵ as well.

The maximum number of HOG values used was four to qualify as n-dimensional data, yet minimize the cost of potentially extending computational time to complete each step in SVR analyses. Computational times for all runs were tabulated and reported.

3.4.2   Valve Surface Morphological Classification and Regression of Naviculoid Diatoms

A total of 67 naviculoid taxa digital LMs were used. HOG data of 50 *Navicula* species was obtained from the LMs for use in SVR analysis to determine the nature of the input data. SVR was conducted using two HOG values per taxon at the outset to classify *Navicula* species. Subsequently, SVR was conducted using the *Navicula* HOG data along with HOG data from 17 non-*Navicula* taxa to assess the classification efficacy of *Navicula* species used in analysis. Classification efficacy and efficiency were determined for the dataset of all naviculoid taxa.

For SVR analyses, nth-order polynomial kernels were used to find the optimal solution to the regression and classification problem using HOG data. A trade-off between maximal marginal error boundaries and the best-fit polynomial were obtained. The optimal outcome would be that all non-*Navicula* taxa were support vectors, and *Navicula* species were classified as non-support vectors. Those *Navicula* species that were correctly classified were eliminated from further analysis because those taxa have $\epsilon = 0$. Those *Navicula* species that were misclassified as support vectors were further analyzed with another SVR as either upper or lower boundary groups in combination with those non-*Navicula* taxa that occurred as support vectors in each group. From this, additional *Navicula* species that were correctly classified with $\epsilon = 0$ were eliminated from further SVR analysis.

The same procedure was used until no *Navicula* species were found to be support vectors, having only non-*Navicula* taxa as the support vectors, or, in lieu of this, when the outcome was at a standstill. The final product was evaluated by identifying those *Navicula* species that remain as misclassifications, still being support vectors. The percent of misclassified *Navicula* species is reported.

3.5   Results

3.5.1   HOG Data Analysis

HOG data may be visualized via inverse HOG functions (e.g., [3.8.107.]) as the differences in surface gradient direction (or orientation) weighted by gradient magnitude. Although some taxa look similar, inverse HOG rendering exhibits the differences among *Navicula* species with similar morphology. Some examples are given in Figure 3.2.

Gradient magnitudes and directions of valve surfaces differed among taxa. In Figure 3.3, examples are given of the gradient direction by arrows, and the gradient magnitude is recorded by the size of the arrow in each LM. Even subtle differences in valve surface morphology of *Navicula* species are recorded by gradient magnitude and direction.

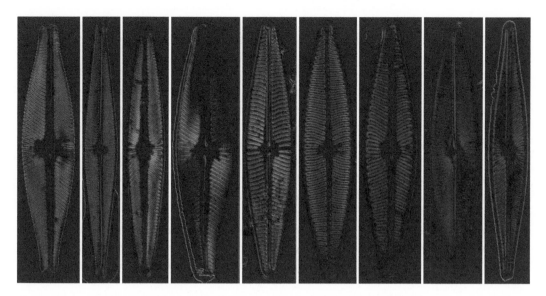

Figure 3.2 Visualization of surface morphology using inverse HOG functions of gradient direction weighted by gradient magnitude from LMs of *Navicula aurora*, *N. galloae*, *N. harmoniae*, *N. ludoviana*, *N. peregrina*, *N. schweigeri*, *N. supleeorum*, *N. vulpina*, and *N. wildii*.

Figure 3.3 Gradient direction as arrows and gradient magnitude as size of arrows obtained from LMs of *Navicula eidrigiana*, *N. freesia*, *N. harmoniae*, *N. peregrina*, *N. perotii*, *N. radiosa*, *N. schweigeri*, and *N. subconcentrica*.

From 67 naviculoid LMs, nine HOG bins were calculated per taxon. Any number of HOG bins could have been calculated for comparable results. From the nine bins per taxon the first four HOG bins were chosen as the *n*-dimensional data to be used in developing the SVR model. This represents the simplest multidimensional case to use in SVR with regard to computational time and efficiency. Each histogram is a probability density function

(PDF) plotted on the interval $\left[-\dfrac{\pi}{2}, \dfrac{\pi}{2} \right]$. By inspecting the histogram plots, similarities in distributions are evident among taxa.

In the central part of the histogram, some taxa exhibited a large portion of maximal probability, including *Aneumastus tusculus*, *Biremis circumtexta*, *Brebissonia lanceolata*, *Cavinula cocconeiformis*, *Craticula buderi*, *Hippodonta capitata*, *Parlibellus delognei*, *Prestauroneis integra*, *Pseudofallacia monoculata* (Figure 3.4), *Navicula angusta*, *N. cari*, *N. caroliniae*, *N. elsoniana*, *N. escambia*, *N. freesia*, *N. lanceolata*, *N. longicephala*, *N. margalithii*, *N. metareichardtiana*, *N. oppugnata*, *N. peregrina*, *N. radiosa*, *N. recens*, *N. reinhardtii*, *N. schweigeri*, *N. slesvicensis*, *N. supleeorum*, *N. trilateria*, *N. venerablis*, and *N. vulpina* (Figure 3.5).

Taxa with maximal probability largely on the edges of the distributions include *Anomoeoneis sphaerophora*, *Encyonopsis montana*, *Kurtkrammeria subspicula* (Figure 3.4), *Navicula caterva*, *N. cryptocephaloides*, *N. duerrenbergiana*, *N. erifuga*, *N. flatheadensis*, *N. galloae*, *N. kotschyi*, *N. leptostriata*, *N. perotii*, *N. rhynchocephala*, *N. rhynchotella*,

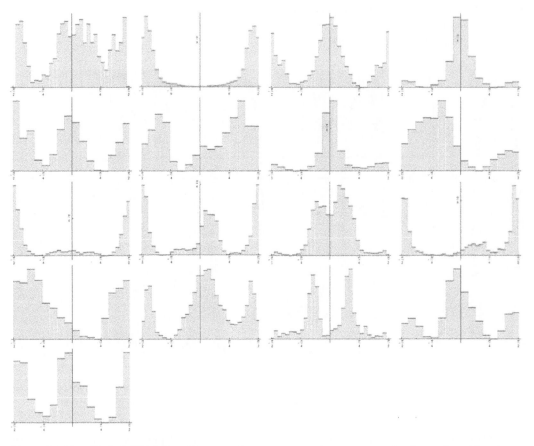

Figure 3.4 Non-*Navicula* HOG plots. First row: *Aneumastus tusculus*; *Anomoeoneis sphaerophora*; *Biremis circumtexta*; *Brebissonia lanceolata*. Second row: *Cavinula cocconeiformis*; *Cosmioneis citriformis*; *Craticula buderi*; *Decussata placenta*. Third row: *Encyonopsis montana*; *Envekadea metzeltinii*; *Hippodonta capitata*; *Kurtkrammeria subspicula*. Fourth row: *Lacustriella lacustris*; *Parlibellus delognei*; *Placoneis gastrum*; *Prestauroneis integra*. Fifth row: *Pseudofallacia monoculata*.

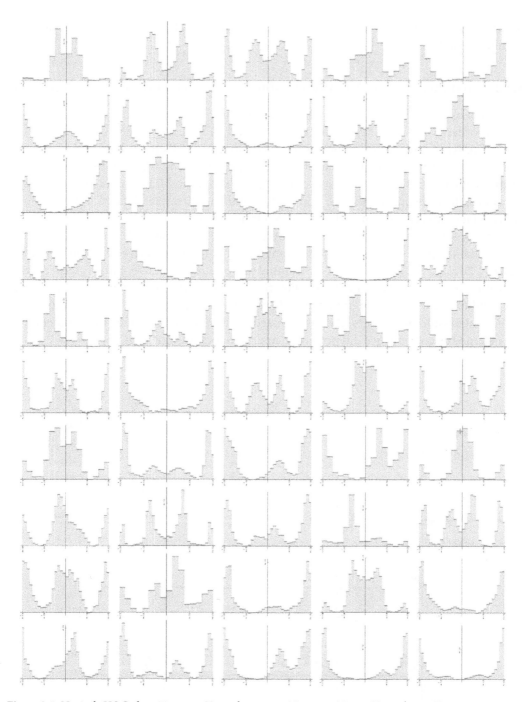

Figure 3.5 *Navicula* HOG plots. First row: *Navicula angusta*; *N. aurora*; *N. cari*; *N. caroliniae*; *N. caterva*. Second row: *N. cryptocephaloides*; *N. cryptofallax*; *N. duerrenbergiana*; *N. eidrigiana*; *N. elsoniana*. Third row: *N. erifuga*; *N. escambia*; *N. flatheadensis*; *N. freesia*; *N. galloae*. Fourth row: *N. harmoniae*; *N. kotschyi*; *N. lanceolata*; *N. leptostriata*; *N. longicephala*. Fifth row: *N. ludoviana*; *N. lundii*; *N. margalithii*; *N. metareichardtiana*; *N. oppugnata*. Sixth row: *N. peregrina*; *N. perotii*; *N. pseudolanceolata*; *N. radiosa*; *N. recens*. Seventh row: *N. reinhardtii*; *N. rhynchocephala*; *N. rhynchotella*; *N. salinarum*; *N. schweigeri*. Eighth row: *N. slesvicensis*; *N. sovereignii*; *N. staffordidae*; *N. subconcentrica*; *N. supleeorum*. Ninth row: *N. trilatera*; *N. trivialis*; *N. vaneei*; *N. venerablis*; *N. veneta*. Tenth row: *N. vulpina*; *N. weberi*; *N. whitefishensis*; *N. wildii*; *N. winona*.

N. staffordiae, N. vaneei, N. veneta, N. wildii, and *N. winona* (Figure 3.5). All other taxa fall somewhere in between.

3.5.2 Goodness-of-Fit SVR Model Using 4D HOG Data from Naviculoid LMs

Goodness-of-fit of SVR using 4D HOG data from all 67 naviculoid LMs was determined using n-order polynomial kernel functions. HOG data were pre-treated by using a factor of 10^4. Pre-treatment via rescaling was necessary to achieve distinguishable values greater than one, because grayscale LMs have very small pixel intensity values in the range of 10^{-6} to 10^{-2}. All HOG data were pre-treated identically so that no bias or undue influence was introduced to subsequent analyses. Scatterplots of the multivariate dataset of the first four HOG values for each naviculoid LM are given in Figure 3.6.

Training sets of three, four, five, and six were used in goodness-of-fit tests of C and ϵ for polynomial support vector regression. Computational time for each run using 4D HOG data was 20, 25, 31, and 43 minutes for training sets of three, four, five, and six, respectively, indicating computational cost of $O(n^2)$ (Figure 3.7). A randomization versus initial test run was conducted for one of the training sets for comparative purposes. At computation time of three hours and 30 minutes, the randomized solution to the optimization problem was not obtained as convergence did not occur, and therefore was not used.

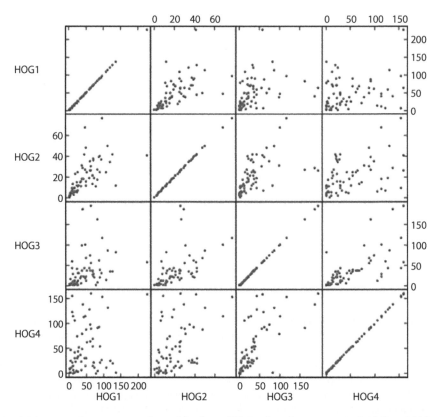

Figure 3.6 Multivariate dataset scatterplots of the first 4 HOG values from 67 naviculoid digital LMs.

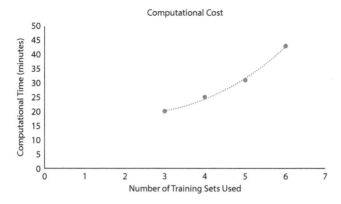

Figure 3.7 Computational time for different number of training sets used in SVR. Curve fit is a second-order polynomial or $O(n^2)$.

Actual (or Null) versus predicted model was plotted for all SVR test runs (Figure 3.8). The model best-fits most of the points to the regression line with some points distantly plotted being potentially not well-fitted. Points above the best-fit line are underestimates of the predicted value. The best-fit model residuals were plotted for all SVR test runs (Figure 3.9). RMSE for the plot was 30.5878, and the fitted curve is a polynomial.

Best-fit models for C and ϵ were plotted from RMSEs for each training set (Figure 3.10). RMSE for C was reduced from 81.9256 for three training sets to 34.9498 for four training sets to 33.2972 for five training sets to 30.7621 for six training sets. For ϵ, RMSE was

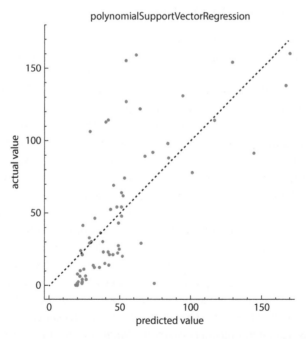

Figure 3.8 Comparison of the SVR model versus the Null model for all test runs. Points near the regression dashed line are closely fitted by the model. Points distant from the regression dashed line are not well fitted by the model and possible outliers. The distant points above the line are underestimating the predicted value.

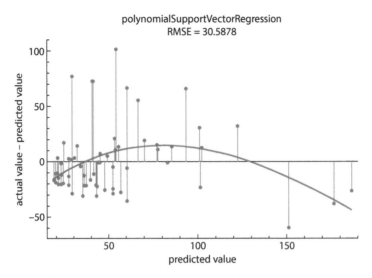

Figure 3.9 Plot of actual-predicted value of residuals versus predicted value of polynomial SVR for all test runs. The plot is a polynomial fit of the curve to the residuals.

43.9179, 31.533, 28.4445, and 29.2026 for training sets three, four, five, and six, respectively. To obtain the optimal values of **w** and b of the dual optimization solution, C with a value of 50 and ϵ with a value of zero from six training sets provided the overall best fit (Figure 3.11).

3.5.3   SVR of Naviculoid 2D HOG Data

Although original HOG data obtained do record differences in surface images, naviculoid digital LMs do not depict a wide range of grayscale tones, resulting in HOG values that are less than one. To augment the near-zero values recorded for grayscale, data were pretreated with a multiplied factor of ten to ensure better performance in SVR. The pretreated HOG data of HOG1 and HOG2 were input to SVR analysis.

To determine the relation among all naviculoid taxa, the initial test of the dual optimization problem was constrained by the following: The optimal maximum marginal separators would produce all non-*Navicula* taxa as support vectors, and all *Navicula* species would be non-support vectors. This outcome would occur if all *Navicula* were actual species of the genus. Initial values of C and ϵ from the goodness-of-fit tests were modified so that maximum marginal separators and slope of the hyperplane with the lowest possible order polynomial kernel resulted in all non-*Navicula* taxa as support vectors.

The SVR plot of the 67 naviculoid taxa is depicted in Figure 3.12. A fourth-order polynomial kernel function was used with $C = 100$ and $\epsilon = 0.006$. The distance from each support vector to the maximum marginal separator is the value of the slack variables, ξ_i with respect to the support vectors. The 15 upper support vectors are *Aneumastus tusculus*, *Anomoeoneis sphaerophora*, *Cosmioneis citriformis*, *Decussata placenta*, *Envekadea metzeltinii*, *Lacustriella lacustris*, *Parlibellus delognei*, *Placoneis gastrum*, *Navicula aurora*, *N. cari*, *N. flatheadensis*, *N. freesia*, *N. harmoniae*, *N. peregrina*, and *N. rhychotella* (Figure 3.12).

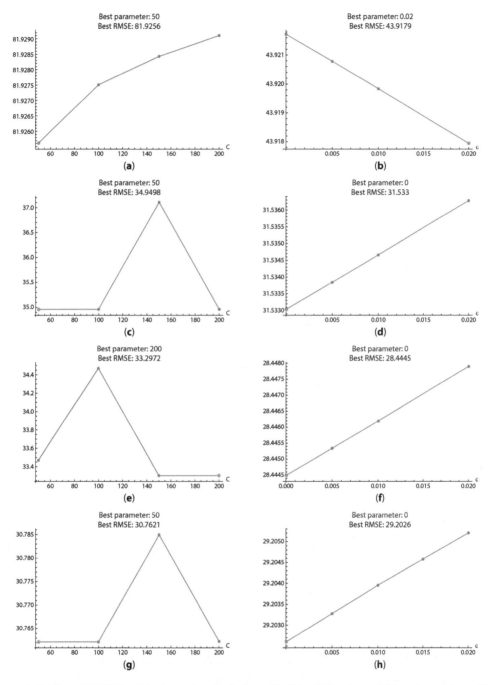

Figure 3.10 Plots of RMSE for C (column 1) and ϵ (column 2) of the SVR polynomial kernel model. a and b, Three training sets; c and d, Four training sets; e and f, Five training sets; g and h, Six training sets.

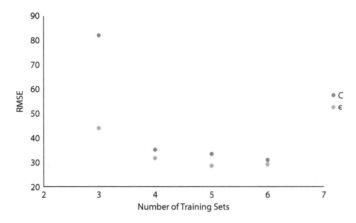

Figure 3.11 Best-fit values for C and ϵ. Six training sets provided the overall best fit to the dual optimization solution for SVR.

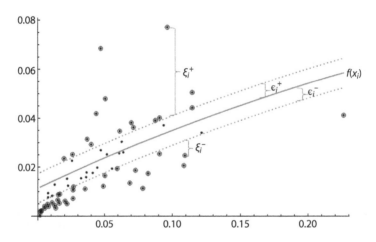

Figure 3.12 Plot of 67 naviculoid taxa HOG1 vs. HOG2 data. Error variable, ϵ, and slack variable, ξ, for regression curve $f(x)$ for i-taxa are labeled. All non-*Navicula* taxa are SVs as well as some *Navicula* species. There are 20 *Navicula* within the maximal marginal boundaries (green dashed lines). Fitted curve is a fourth-order polynomial with $\epsilon = 0.006$ and $C = 100$.

The 32 lower support vectors are *Biremis circumtexta, Brebissonia lanceolata, Cavinula cocconeiformis, Craticula buderi, Encyonopsis montana, Hippodonta capitata, Kurtkrammeria subspicula, Prestauroneis integra, Pseudofallacia monoculata, Navicula caroliniae, N. caterva, N. cryptocephaloides, N. duerrenbergiana, N. erifuga, N. galloae, N. kotschyi, N. lanceolata, N. leptostriata, N. longicephala, N. lundii, N. margalithii, N. metareichardtiana, N. oppugnata, N. recens, N. rhynchocephala, N. salinarum, N. staffordiae, N. supleeorum, N. trivialis, N. veneta, N. whitefishensis,* and *N. wildii* (Figure 3.12).

Twenty *Navicula* species with $\epsilon = 0$ and are not support vectors are *N. angusta, N.cryptofallax, N. eidrigiana,, N. elsoniana, N. escambia, N. ludloviana, N. perotii, N. pseudolanceolata, N. radiosa, N. reinhardtii, N. schweigeri, N. slesvicensis, N. sovereignii, N. subconcentrica, N. trilatera, N. vaneei, N. venerablis, N. vulpina, N. weberi,* and *N. winona* (Figure 3.12).

3.5.4   Second Round of SVR Analysis of Remaining Naviculoid 2D HOG Data

From the results of SVR for 67 naviculoids, the upper and lower support vectors were analyzed separately in SVR. With *Navicula* species and non-*Navicula* taxa as support vectors in each analysis, the resultant *Navicula* species that were not support vectors in this second SVR analysis were tabulated and removed from further analysis.

Using 15 taxa that were upper support vectors from the original SVR, the second SVR analysis with $C = 100$ and $\epsilon = 0.002$ with a second-order polynomial kernel function has upper support vectors *Aneumastus tusculus, Decussata placenta, Lacustriella lacustris,* and *Placoneis gastrum* (Figure 3.13). *Anomoeoneis sphaerophora, Cosmioneis citriformis, Envekadea metzeltinii, Parlibellus delognei, Navicula aurora, N. cari, N. flatheadensis, N. freesia, N. harmoniae,* and *N. rhychotella* comprise the new lower support vectors (Figure 3.13). Only *N. peregrina* is not a support vector (Figure 3.13).

Using 32 taxa from the original SVR lower support vectors, the second SVR analysis with $C = 0.28$ and $\epsilon = 0.0069$ and a seventh-order polynomial kernel function produced *Encyonopsis montana, Kurtkrammeria subspicula, Navicula galloae,* and *N. wildii* as the new upper support vectors (Figure 3.14). The new lower support vectors are *Biremis circumtexta, Brebissonia lanceolata, Cavinula cocconeiformis, Craticula buderi, Hippodonta capitata, Prestauroneis integra, Pseudofallacia monoculata, Navicula caroliniae, N. caterva, N. longicephala, N. metareichardtiana, N. oppugnata, N. recens, N. salinarum,* and *N. veneta* (Figure 3.14). *Navicula* that are not support vectors from the second SVR analysis are *N. cryptocephaloides, N. duerrenbergiana, N. erifuga, N. kotschyi, N. lanceolata, N. leptostriata, N, lundii, N. margalithii, N. rhynchocephala, N. staffordiae, N. supleeorum, N. trivialis, N. whitefishensis* (Figure 3.14).

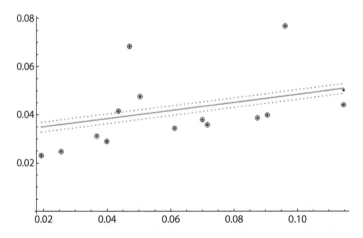

Figure 3.13 Plot of 15 naviculoid taxa HOG1 vs. HOG2 data of the original SVR upper support vectors. Fitted curve is a second-order polynomial kernel function, $C = 100$ and $\epsilon = 0.002$.

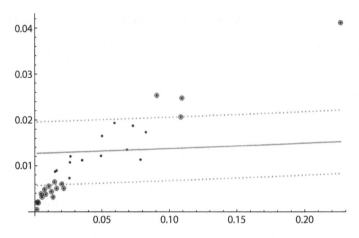

Figure 3.14 Plot of 32 naviculoid taxa HOG1 vs. HOG2 data of the original SVR lower support vectors. Fitted curve is a seventh-order polynomial kernel function, $C = 0.28$ and $\epsilon = 0.0069$.

3.5.5   Last Round of SVR Analysis of Remaining Naviculoid 2D HOG Data

For the last round, *Navicula* species with $\epsilon = 0$ from the second SVRs were eliminated, and separate SVR analyses for the remaining upper and lower support vectors was accomplished once more. Last SVR for the remaining upper and lower support vectors had no *Navicula* species with $\epsilon = 0$; therefore, solution to the problem is not attainable. The remaining 19 lower support vectors were analyzed by the last SVR. A solution to the problem is not attainable here as well. Maximum marginal separators for all combinations of C, ϵ and nth-order polynomial do not produce all non-*Navicula* as support vectors.

3.5.6   Classification Results from SVR Analysis of Naviculoid Taxa

From results of SVR analyses, given the non-*Navicula* taxa as support vectors, the *Navicula* species that remained as support vector are *Navicula aurora, N. cari, N. caroliniae, N. caterva, N. flatheadensis, N. freesia, N. galloae, N. harmoniae, N. longicephala, N. metareichardtiana, N. oppugnata, N. recens, N. rhychotella N. salinarum, N. veneta,* and *N. wildii.* Of all *Navicula* species, this is a 32% rate of misclassification with a 68% correct classification. Of all naviculoid taxa, the *Navicula* species that were support vectors comprise approximately 24%. Because misclassified *Navicula* species occurred with non-*Navicula* taxa, they may require re-examination with respect to their assignment to the genus *Navicula.*

3.6   Discussion

Usually, deformation models are used to obtain variation in the appearance of a given morphological entity. However, deformable models may not convey the subtle variation among a collection of images from which to recover relevant detailed information for a given analysis. To fully obtain the rich surface and pattern information available in such a collective, HOG data fits this bill. Using HOG, important features of each digital image were extracted to recover useful surface information for comparative analysis. For this study, a small number of naviculoid diatom images and their HOG data were used to elucidate and illustrate the methodology.

Extraction of HOG data from digital LMs of naviculoid diatoms provided a way to measure whole valve face surface morphology that is size and transformation, translation, and rotationally invariant. This is true provided that all digital images were oriented in the same way; that is, either vertically or horizontally. For this study, all LMs were oriented vertically. Contrast normalization of digital LMs ensured that backgrounds were not influential on the outcome of SVR analyses. No matter how similar taxa looked to the viewing eye, HOG data for each taxon were unique as confirmed by HOG plots. Measurement in this way is not dependent on the taxa chosen or the sample size used in the subsequent SVR analysis. Any number of taxa could be used and analyzed for classification tests and regression predictive value. By measuring diatom valve surfaces using HOG data, all morphological characters were included from each taxon's LM. If a diatom researcher wanted to be even more careful in using HOG data, triplicate acquisition of HOG data from the same species could be obtained, averaged and used in SVR. The general point of using HOG data in SVR is to show that such measurement and analyses may be useful in discerning synapomorphic characteristics of diatom morphological data such as those obtained for naviculoid taxa.

LMs of naviculoid taxa were chosen because grayscale images have subtle differences in shades of gray that make it difficult to distinguish taxa on first view, especially for those without a trained eye, having impeccable observational skills and a knowledge base on diatom identification. Because HOG data recover surface morphology in all the changes in pixel intensity of a digital LM, and this data from edge detection includes specific, individual morphological attributes such as shape, striae, spines or central area, distinguishing naviculoid taxa was accomplished via comparison of whole valve surface features. HOG data were subject to contrast normalization so that slight changes in tilting of specimens or variable background shading does not affect the accuracy of data acquired per digital LM, enabling comparison among all taxa. That is, acquisition of HOG data from digital LMs was sensitive enough to enable distinguishing taxa using grayscale LMs, providing unique HOG data per taxon.

From the results of this study, some of the species currently identified as *Navicula* may need to be reassessed. If the goal of delineating a diatom genus is based on synapomorphy, then SVR results may be used to augment re-examination of currently assigned species to *Navicula*. Given the vague criteria that qualifies assignment to *Navicula*, HOG data analyzed using SVR may be a way to advance research toward a more natural classification.

Using SVR with a polynomial kernel function enabled application of regression to a classification problem. Although 3 training and validation sets were used with the input HOG data, 4, 5 or 6 training sets could have been used as well. The number of training sets to use is a matter of some trial and error to achieve the best possible outcome. Regardless, as more input data are used, more training and validation sets may be used.

3.6.1   Characteristics of SVM and SVR

SVR is an extension of SVM and has advantages in its utility in classification analysis. SVM is a sparse technique, a kernel technique and maximum margin separator. SVM is robust, has good generalization ability, unique global optimum solutions, a type of supervised learning, maps non-linear data into high dimensional space via kernel functions with a linear separator to discriminate classes, has optima that are found analytically rather than heuristically, and has training done on a sample of the data set while prediction is performed on the yet-to-be-seen data via convex optimization and structural risk minimization [3.8.1.].

SVM compared favorably with 16 classifiers by Meyer, Leisch and Hornik [3.8.71.] and was found to be one of the best classifiers. Other kernel machines include kernel principal component analysis, kernel Gram-Schmidt analysis, kernel Fisher discriminant analysis, Gaussian process analysis, and Bayes point machines. Other alternative methods include partial least-squares and artificial neural networks. All of these methods did not perform as well as SVM and SVR methods [3.8.1., 3.8.50.].

SVR may be accomplished using different penalty functions which may improve results such as the Huber insensitive function [3.8.1., 3.8.49.]. However, computational cost may increase making the utility of such loss functions less attractive. Generally, SVR is a cost-efficient dual optimization problem solver because multidimensional problems are reduced to tractable linear or low order polynomial solutions [3.8.50., 3.8.95.].

3.6.2   Advantages in Using HOG Data and SVR

By using HOG data in measuring 3D surface morphology, there is no dependence on specific morphological characters. The entire diatom valve face contains morphological information and by using HOG, comprehensive morphological quantification is not limited to pre-defined or devised characters.

Another advantage to using HOG is that any LM is the basis for data acquisition. It should be evident that any digital image, such as SEM, could be used, provided that either all LMs or SEMs or whatever the format used are analyzed with regard to the same scale. Quality of LMs and SEMs is key, and because of contrast normalization, LMs and SEMs with as little background "noise" as possible as well as using replicate digital images and data averaging enables and promotes confidence in results from SVR analysis.

3.6.3   Potential Utility of HOG Data and SVR in Diatom Research

Kociolek and Williams [3.8.59.] stated that diatom classification levels be based on the concept of monophyly. So, a genus must contain ancestor and all descendants. As a goal of natural classification, this is the desired result. The utility of purely morphological studies toward achievement of this goal is of the utmost importance concerning diatom natural classification. Because of the sheer numbers of diatom species, fossilized or embedded as strewn or stub-coated mounts, morphology will continue to be a rich source of necessary data in the quest for a diatom natural classification.

Using SVR on whole valve surface morphology may be a way to augment and further the goal of a natural classification. This study on naviculoid taxa has been a useful exercise in alerting a diatom researcher about whether currently named taxa should be tested further on their generic membership. Because digital images were used, the methodology is applicable to any group of diatom taxa not just genera. Groupings based on major morphological characteristics such as araphids, monoraphids, eucentrics, or eccentrics [3.8.77.] as well as major clades based on molecular data such as melosiroids, coscinodiscoids, biddulphioids, rhizosolenioids, fragilarioids, eunotioids, bacillarioids, neidioids, cymbelloids, stauroneoids, amphoroids, rhopalodioids, or surirelloids [3.8.56., 3.8.100.] could be subject to analysis using HOG data in SVR. Initial divisions of such groups may prove useful in developing a more natural classification, given the pain-staking, time-consuming nature of diatom taxonomic analysis and the sheer number of diatom taxa to study.

HOG data may be combined with size data as a constraint and analyzed with SVR. Other constraints could be introduced in combination with HOG such as environmental data, habitat preference data, indicator status data, geographic or location data in terms of depth, substrate preference data, colonial or single cell status, or vegetative reproductive stage. Such tests could be conducted not only on diatom groups of convenience, but also on size reduction series. Developmental trajectories may be analyzed to determine how size reduction affects valve surface morphology. Because SVR is a multivariate technique, constrained SVR could be useful in a variety of diatom studies.

More broadly, SVR and other SVM methods may be applicable in studies of diatom morphological evolution. With the ability to obtain predictive results, SVR used in conjunction with measurement of 3D surface morphology of diatom valve faces, applications of changes in morphology over time are in the offing via implementation in computationally sophisticated programs using this machine learning technique.

3.7   Summary and Future Research

This study of naviculoids used a small number of taxa to illustrate the methodology in using HOG and SVR. Because SVR is a machine learning technique, the current study could be expanded by using thousands of LMs of naviculoid taxa. Using many more images of naviculoid taxa, the percentage of predicted correct classification should increase. Such an application is plausible algorithmically in light of similar applications in, e.g., facial recognition (e.g., [3.8.80.]). Because of the ubiquitous nature of machine learning methods and availability of diatom LMs and SEMs, diatom morphological analyses could be aided considerably.

Studying *Navicula* and other naviculoid taxa using digital images and computationally sophisticated methods enables understanding what a given genus definition of diatoms means as well as providing the impetus to delineate and verify the assignment of taxa to that genus. Using whole valve surface morphology is not viewed as a substitute, but rather a way to augment traditional careful microscopic studies. Being able to analyze digital LMs and SEMs provides access to more information about morphology in the face of incomplete or unavailable information. HOG data and SVR may be useful in devising more natural diatom classifications.

Using HOG and SVR, studies in morphological evolution, including speciation and morphological character definition may be in the offing. The application of such methods may provide a way to answer the following questions: What is a genus? What is a level of classification? What is a natural classification?

This study illustrated that morphology is not just shape, external individual characters, or any other partial rendition of a diatom cell. Morphology is all of these descriptors, and using whole valve diatom surfaces as the basis of external morphological assessment is one way to combine all possible descriptors. Using HOG data in SVR provides a useful way to quantify and assess diatom morphological groups in the quest to develop a more natural classification.

3.8   References

3.8.1.   Awad, M. and Khanna, R. (2015) *Efficient Learning Machines – Theories, Concepts, and Applications for Engineers and System Designers*. Apress Media, LLC, Springer, New York.

3.8.2.   Bahls, L. (2011) *Navicula aurora*. In Diatoms of North America. Retrieved June 30, 2020, from https://diatoms.org/species/navicula_aurora

3.8.3.   Bahls, L. (2011) *Navicula cryptocephaloides*. In Diatoms of North America. Retrieved June 30, 2020, from https://diatoms.org/species/navicula_cryptocephaloides

3.8.4.   Bahls, L. (2011) *Navicula cryptofallax*. In Diatoms of North America. Retrieved June 30, 2020, from https://diatoms.org/species/navicula_cryptofallax

3.8.5.   Bahls, L. (2011) *Navicula flatheadensis*. In Diatoms of North America. Retrieved June 30, 2020, from https://diatoms.org/species/navicula_flatheadensis

3.8.6.   Bahls, L. (2011) *Navicula galloae*. In Diatoms of North America. Retrieved June 30, 2020, from https://diatoms.org/species/navicula_galloae

3.8.7.   Bahls, L. (2011) *Navicula peregrina*. In Diatoms of North America. Retrieved June 30, 2020, from https://diatoms.org/species/navicula_peregrina

3.8.8.   Bahls, L. (2011) *Navicula pseudolanceolata*. In Diatoms of North America. Retrieved June 30, 2020, from https://diatoms.org/species/navicula_pseudolanceolata

3.8.9.   Bahls, L. (2011) *Navicula reinhardtii*. In Diatoms of North America. Retrieved June 30, 2020, from https://diatoms.org/species/navicula_reinhardtii

3.8.10.   Bahls, L. (2011) *Navicula rhynchotella*. In Diatoms of North America. Retrieved June 30, 2020, from https://diatoms.org/species/navicula_rhynchotella

3.8.11.   Bahls, L. (2011) *Navicula sovereignii*. In Diatoms of North America. Retrieved June 30, 2020, from https://diatoms.org/species/navicula_sovereignii

3.8.12.   Bahls, L. (2011) *Navicula subconcentrica*. In Diatoms of North America. Retrieved June 30, 2020, from https://diatoms.org/species/navicula_subconcentrica

3.8.13.   Bahls, L. (2011) *Navicula vaneei*. In Diatoms of North America. Retrieved June 30, 2020, from https://diatoms.org/species/navicula_vaneei

3.8.14.   Bahls, L. (2012) *Aneumastus tusculus*. In Diatoms of North America. Retrieved June 30, 2020, from https://diatoms.org/species/aneumastus_tusculus

3.8.15.   Bahls, L. (2012) *Navicula cari*. In Diatoms of North America. Retrieved June 30, 2020, from https://diatoms.org/species/navicula_cari

3.8.16.   Bahls, L. (2012) *Navicula caroliniae*. In Diatoms of North America. Retrieved June 30, 2020, from https://diatoms.org/species/navicula_caroliniae

3.8.17.   Bahls, L. (2012) *Navicula duerrenbergiana*. In Diatoms of North America. Retrieved June 30, 2020, from https://diatoms.org/species/navicula_duerrenbergiana

3.8.18.   Bahls, L. (2012) *Navicula leptostriata*. In Diatoms of North America. Retrieved June 30, 2020, from https://diatoms.org/species/navicula_leptostriata

3.8.19.   Bahls, L. (2012) *Navicula lundii*. In Diatoms of North America. Retrieved June 30, 2020, from https://diatoms.org/species/navicula_lundii

3.8.20.   Bahls, L. (2012) *Navicula oppugnata*. In Diatoms of North America. Retrieved June 30, 2020, from https://diatoms.org/species/navicula_oppugnata

3.8.21.   Bahls, L. (2012) *Navicula schweigeri*. In Diatoms of North America. Retrieved June 30, 2020, from https://diatoms.org/species/navicula_schweigeri

3.8.22.   Bahls, L. (2012) *Navicula staffordiae*. In Diatoms of North America. Retrieved June 30, 2020, from https://diatoms.org/species/navicula_staffordiae

3.8.23.   Bahls, L. (2012) *Navicula supleeorum*. In Diatoms of North America. Retrieved June 30, 2020, from https://diatoms.org/species/navicula_supleeorum

3.8.24.   Bahls, L. (2012) *Navicula trilatera*. In Diatoms of North America. Retrieved June 30, 2020, from https://diatoms.org/species/navicula_trilatera

3.8.25.   Bahls, L. (2012) *Navicula weberi*. In Diatoms of North America. Retrieved June 30, 2020, from https://diatoms.org/species/navicula_weberi

3.8.26.   Bahls, L. (2012) *Navicula wildii*. In Diatoms of North America. Retrieved June 30, 2020, from https://diatoms.org/species/navicula_wildii

3.8.27.   Bahls, L. (2012) *Navicula winona*. In Diatoms of North America. Retrieved June 30, 2020, from https://diatoms.org/species/navicula_winona

3.8.28.   Bahls, L. (2013) *Encyonopsis montana*. In Diatoms of North America. Retrieved June 30, 2020, from https://diatoms.org/species/encyonopsis_montana

3.8.29.   Bahls, L. (2013) *Kurtkrammeria subspicula*. In Diatoms of North America. Retrieved June 30, 2020, from https://diatoms.org/species/kurtkrammeria_subspicula

3.8.30.   Bahls, L. (2013) *Navicula whitefishensis*. In Diatoms of North America. Retrieved June 30, 2020, from https://diatoms.org/species/navicula_whitefishensis

3.8.31.   Bahls, L. (2016) *Navicula eidrigiana*. In Diatoms of North America. Retrieved June 30, 2020, from https://diatoms.org/species/navicula_eidrigiana

3.8.32.   Bahls, L. (2016) *Navicula harmoniae*. In Diatoms of North America. Retrieved June 30, 2020, from https://diatoms.org/species/navicula_harmoniae

3.8.33.   Bahls, L. (2016) *Navicula margalithii*. In Diatoms of North America. Retrieved June 30, 2020, from https://diatoms.org/species/navicula_margalithii

3.8.34.   Bahls, L. and Kociolek, P. (2012) *Craticula buderi*. In Diatoms of North America. Retrieved June 30, 2020, from https://diatoms.org/species/craticula_buderi

3.8.35.   Bahls, L. and Potapova, M (2015) *Navicula ludloviana*. In Diatoms of North America. Retrieved June 30, 2020, from https://diatoms.org/species/navicula_ludloviana

3.8.36.   Bahls, L., Morgan, L., Bishop, I , Edlund, M. (2018) *Anomoeoneis sphaerophora*. In Diatoms of North America. Retrieved June 30, 2020, from https://diatoms.org/species/anomoeoneis-sphaerophora

3.8.37.   Bishop, I. (2016) *Pseudofallacia monoculata*. In Diatoms of North America. Retrieved June 30, 2020, from https://diatoms.org/species/fallacia_monoculata

3.8.38.   Bogan, D. (2013) *Brebissonia lanceolata*. In Diatoms of North America. Retrieved June 30, 2020, from https://diatoms.org/species/brebissonia_lanceolata1

3.8.39.   Bordes, A., Ertekin, S., Weston, J., Botou, L. (2005) Fast kernel classifiers with online and active learning. *Journal of Machine Learning* 6, 1579-1619.

3.8.40.   Bory de Saint-Vincent, J.B.M. & coll. (1822-1831) Dictionnaire Classique d'Histoire Naturelle. Paris. Rey & Gravier, libraires-éditeurs; Baudouin Frères, libraires-éditeurs., vol. 1 to 17.

3.8.41.   Casteleyn, G., Adams, N.G., Vanormelingen, P., Debeer, A.-E., Sabbe, K., Vyverman, W. (2009) Natural hybrids in the marine diatom *Pseudo-nitzschia pungens* (Bacillariophyceae): genetic and morphological evidence. *Protist* 160(2), 343-354.

3.8.42.   Dalal, N. and Triggs, b. (2005) Histograms of oriented gradients for human detection. International Conference on Computer Vision & Pattern Recognition (CVPR '05), Jun 2005, San Diego, Califormia, U.S.A.: 886-893.

3.8.43.   Edlund, M. and Burge, D. (2017) *Lacustriella lacustris*. In Diatoms of North America. Retrieved June 30, 2020, from https://diatoms.org/species/cavinula_lacustris

3.8.44.   Edlund, M. and Potapova, M., Spaulding, S. (2010) *Navicula caterva*. In Diatoms of North America. Retrieved June 30, 2020, from https://diatoms.org/species/navicula_caterva

3.8.45.   Giri, B.S. (1992) Nuclear cytology of naviculoid diatoms. *Cytologia* 57, 173-179.

3.8.46.   Hamilton, P. (2011) *Navicula freesei*. In Diatoms of North America. Retrieved June 30, 2020, from https://diatoms.org/species/navicula_freesei

3.8.47.   Hamilton, P. (2011) *Navicula perotii*. In Diatoms of North America. Retrieved June 30, 2020, from https://diatoms.org/species/navicula_perotii

3.8.48.   Hamilton, P. (2012) *Navicula elsoniana*. In Diatoms of North America. Retrieved June 30, 2020, from https://diatoms.org/species/navicula_elsoniana

3.8.49.   Huber, P.J. (1964) Robust estimation of a location parameter. *The Annals of Mathematical Statistics* 35(1), 73-101.

3.8.50.   Ivanciuc, O. (2007) Applications of support vector machines in chemistry. In: *Reviews in Computational Chemistry, Volume 23*, Lipkowitz, K.B. and Cundari, T.R. (eds.), Wiley-VCH, Weinheim: 291-400.

3.8.51.   Karush, W. (1939) Minima of functions of several variables with inequalities as side constraints. unpublished Master's thesis, Department of Mathematics, University of Chicago.

3.8.52.   Kociolek, P. (2011) *Decussata placenta*. In Diatoms of North America. Retrieved June 30, 2020, from https://diatoms.org/species/decussata_placenta

3.8.53.   Kociolek, P. (2011) *Navicula rhynchocephala*. In Diatoms of North America. Retrieved June 30, 2020, from https://diatoms.org/species/navicula_rhynchocephala

3.8.54.   Kociolek, P. (2011) *Navicula salinarum*. In Diatoms of North America. Retrieved June 30, 2020, from https://diatoms.org/species/navicula_salinarum

3.8.55.   Kociolek, P. (2011) *Navicula vulpina*. In Diatoms of North America. Retrieved June 30, 2020, from https://diatoms.org/species/navicula_vulpina

3.8.56.   Kociolek, J.P. (2018) A worldwide listing and biogeography of freshwater diatom genera: a phylogenetic perspective. *Diatom Research* 33(4), 509-534.

3.8.57.   Kociolek, J.P. and Spaulding, S.A. (2003) Symmetrical naviculoid diatoms, Chapter 17. In: *Freshwater Algae of North America: Ecology and Classification*, J.D., Sheath, R.G. (eds.), Elsevier Science, San Diego, California: 637-653.

3.8.58.   Kociolek, J.P. and Stoermer, E.F. (1989) Chromosome numbers in diatoms: a review. *Diatom Research* 4(1, 47-54.

3.8.59.   Kociolek, J.P. and Williams, D. (2015) How to define a diatom genus? Notes on the creation and recognition of taxa, and a call for revionary studies of diatoms. *Acta Botanica Croatica* 74(2), 195-210.

3.8.60.   Kociolek, J.P., Spaulding, S.A., Kingston, J.C. (1998) Valve morphology and systematic position of *Navicula walkeri* (Bacillariophyceae), a diatom endemic to Oregon and California. *Nova Hedwigia* 67, 235-245.

3.8.61.   Kociolek, J.P., Spaulding, S.A., Lowe, R.L. (2015) Bacillariophyceae: the raphid diatoms, Chapter 16. In *Freshwater Algae of North America: Ecology and Classification 2nd edition*, J.D., Sheath, R.G., Kociolek, J.P. (eds.), Academic Press, San Diego, California: 709-772.

3.8.62.   Krammer, K., Lange-Bertalot, H. (1985) Naviculaceae. Neue und wenig Taxa, neue Kombinationen und Synonyme sowie Bemerkungen zu einigen Gattung. *Bibl. Diatomol.* 9, 1–230.

3.8.63.   Krammer, K., Lange-Bertalot, H. (1986) Naviculaceae. In *Bacillariophyceae, Part 1. Süsswasserflora von Mitteleuropa, Vol. 2.*, Ettl, H., Gerloff, J., Heynig, H., Mollenhauer, D. (eds.), Gustav Fischer, Stuttgart, Germany.

3.8.64.   Krienitz, L., Bock, C., Luo, W., Pröschold, T. (2010) Polyphyletic origin of the *Dictyosphaerium* morphotype within Chlorellaceae (Trebouxiophyceae). *Journal of Phycology* 46, 559-563.

3.8.65.   Kuhn, H.W. and Tucker, A.W. (1951) Nonlinear programming. In *Proceedings 2nd Berkeley Symposium on Mathematical Statistics and Probabilistics*, Berkeley, California, University of California Press: 481-492.

3.8.66.   Lee, S. and Van de Vijver, B. (2013) *Envekadea metzeltinii*. In Diatoms of North America. Retrieved June 30, 2020, from https://diatoms.org/species/envekadea_metzeltinii

3.8.67.   Lowe, R. (2011) *Cosmioneis citriformis*. In Diatoms of North America. Retrieved June 30, 2020, from https://diatoms.org/species/cosmioneis_citriformis

3.8.68.   Ma, J., Theiler, J. Perkins, S. (2003) Accurate online support vector regression. *Neural Computation* 15(11), 2683-2703.

3.8.69.   Manoylov, K. and Hamilton, P. (2010) *Navicula escambia*. In Diatoms of North America. Retrieved June 30, 2020, from https://diatoms.org/species/navicula_escambia

3.8.70.   Martin, M. (2002) On-line support vector machine regression. In *Machine Learning: ECML 2002. ECML 2002. European Conference on Machine Learning, Lecture Notes in Computer Science, vol. 2430*, Elomaa, T., Mannila, H., Toivonen, H. (eds.), Springer, Berlin, Heidelberg: 282-294.

3.8.71.   Meyer, D., Leisch, F., Hornik, K. (2003) The support vector machine under test. *Neurocomputing* 55, 169-186.

3.8.72.   Nandipati, S.C.R., Haugli, K., Coucheron, D.H., Haskins, E.F., Johansen, S.D. (2012) Polyphyletic origin of the genus *Physarum* (Physarales, Myxomycetes) revealed by nuclear rDNA mini-chromosome analysis and group I intron synapomorphy. *BMC Evolutionary Biology* 12, 166.

3.8.73.   Neil, K. (2014) *Prestauroneis integra*. In Diatoms of North America. Retrieved June 30, 2020, from https://diatoms.org/species/prestauroneis_integra

3.8.74.   Otu, M. and Spaulding, S. (2011) *Cavinula cocconeiformis*. In Diatoms of North America. Retrieved June 30, 2020, from https://diatoms.org/species/cavinula_cocconeiformis

3.8.75.   Paláncz, B., Völgyesi, L., Popper, Gy. (2005) Support vector regression via Mathematica. *Periodica Polytechnica Civ. Eng.* 49(1), 59-84.

3.8.76.   Pappas, J.L. and Miller, D.J. (2013) A generalized approach to the modeling and analysis of 3D surface morphology in organisms. *Plos One* 8(10), #e77551.

3.8.77.   Pappas, J.L. and Stoermer, E.F. (1995) Great Lakes Diatoms, http://umich.edu/~phyto-lab/GreatLakesDiatomHomePage/groups/majorgroups.html, accessed on 6 July 2020.

3.8.78.   Parks, M.B., Nakov, T., Ruck, E.C., Wickett, N.J., Alverson, A.J. (2018) Phylogenomics reveals an extensive history of genome duplication in diatoms (Bacillariophyta). *American Journal of Botany* 105(3), 330-347.

3.8.79.   Patrick, R.M., Reimer, C.W. (1966) *The Diatoms of the United States. Monograph 13*, Academy of Natural Sciences of Philadelphia.

3.8.80.   Patwary, M.J.A., Parvin, S., Akter, S. (2015) Significan HOG-Histogram of oriented gradient feature selection for human detection. *International Journal of Computer Applications (0975-8887)* 132(17), 20-24.

3.8.81.   Potapova, M. (2009) *Navicula erifuga*. In Diatoms of North America. Retrieved June 30, 2020, from https://diatoms.org/species/navicula_erifuga

3.8.82.   Potapova, M. (2009) Navicula *lanceolata*. In Diatoms of North America. Retrieved June 30, 2020, from https://diatoms.org/species/navicula_lanceolata

3.8.83.   Potapova, M. (2009) *Navicula recens*. In Diatoms of North America. Retrieved June 30, 2020, from https://diatoms.org/species/navicula_recens

3.8.84.   Potapova, M. (2009) *Navicula venerablis*. In Diatoms of North America. Retrieved June 30, 2020, from https://diatoms.org/species/navicula_venerablis

3.8.85.   Potapova, M. (2011) *Biremis circumtexta* In Diatoms of North America. Retrieved June 30, 2020, from https://diatoms.org/species/biremis_circumtexta

3.8.86.   Potapova, M. (2011) *Hippodonta capitata*. In Diatoms of North America. Retrieved June 30, 2020, from https://diatoms.org/species/hippodonta_capitata

3.8.87.   Potapova, M. (2011) *Navicula angusta*. In Diatoms of North America. Retrieved June 30, 2020, from https://diatoms.org/species/navicula_angusta

3.8.88.   Potapova, M. (2011) *Navicula longicephala*. In Diatoms of North America. Retrieved June 30, 2020, from https://diatoms.org/species/navicula_longicephala

3.8.89.   Potapova, M. (2011) *Navicula metareichardtiana*. In Diatoms of North America. Retrieved June 30, 2020, from https://diatoms.org/species/navicula_metareichardtiana

3.8.90.   Potapova, M. (2011) *Navicula radiosa*. In Diatoms of North America. Retrieved June 30, 2020, from https://diatoms.org/species/navicula_radiosa

3.8.91.   Potapova, M. (2011) *Navicula veneta*. In Diatoms of North America. Retrieved June 30, 2020, from https://diatoms.org/species/navicula_veneta

3.8.92.   Potapova, M. and Bahls, L. (2011) *Navicula slesvicensis*. In Diatoms of North America. Retrieved June 30, 2020, from https://diatoms.org/species/navicula_slesvicensis

3.8.93.   Round, F.E., Crawford, R.M. and Mann, D.G. (1990) *The Diatoms, Biology & Morphology of the Genera*. Cambridge University Press, Cambridge, U.K.

3.8.94.   Rushforth, S. and Spaulding, S. (2010) *Navicula trivialis*. In Diatoms of North America. Retrieved June 30, 2020, from https://diatoms.org/species/navicular_trivialis

3.8.95.   Ruskeepää, H. (2017) Support vector regression and other prediction methods: a competition with Mathematica, https://library.wolfram.com/infocenter/MathSource/9548/.

3.8.96.   Smola, A.J. and Schölkopf, B. (2004) A tutorial on support vector regression. *Statistics and Computing* 14, 199-222.

3.8.97.   Spaulding, S. (2012) *Parlibellus*. In Diatoms of North America. Retrieved June 30, 2020, from https://diatoms.org/genera/parlibellus

3.8.98.   Spaulding, S. and Edlund, M. (2008) *Navicula*. In Diatoms of North America. Retrieved June 30, 2020, from https://diatoms.org/genera/navicula

3.8.99.   Spaulding, S.A., Bishop, I.W., Edlund, M.B., Lee, S., Furey, P., Jovanovska, E. and Potapova, M. (2019) Diatoms of North America. https://diatoms.org/

3.8.100.   Theriot, E.C., Ashworth, M.P., Nadov, T., Ruck, E., Jansen, R.K. (2015) Dissecting signal and noise in diatom chloroplast protein encoding genes with phylogenetic information profiling. *Molecular Phylogenetics and Evolution* 89,28-36.

3.8.101.   Tian, S., Lu, S., Su, B., Tan, C.L. (2013) Scene text recognition using co-occurrence of histogram of oriented gradients. *12th International Conference on Document Analysis and Recognition, 2013*, 912-916, doi: 10.1109/ICDAR.2013.186.

3.8.102.   Tomasi, C. (2020) Histograms of oriented gradients. *Computer Vision Sampler*, 1-6.

3.8.103.   Tyree, M. and Bishop, I. (2015) *Navicula kotschyi*. In Diatoms of North America. Retrieved June 30, 2020, from https://diatoms.org/species/navicula_kotschyi

3.8.104.   Vapnik, V. (1995) *The Nature of Statistical Learning Theory*, 2nd edition, Springer Science+Business Media, New York.

3.8.105.   Vapnik, V. (1999) An overview of statistical learning theory. *IEEE Transactions on Neural Networks* 10(5), 988-999.

3.8.106.   Vapnik, V. and Lerner, A. (1963) Pattern recognition using generalized portrait method. *Automation and Remote Control* 24, 774-780.

3.8.107.   Vondrick, C., Khosla, a., Malisiewicz, T., Torralba, A. (2013) HOGgles: visualizing object detection features. *IEEE International Conference on Computer Vision* 1-9.

3.8.108.   Witkowski, A., Lange-Bertalot, H., Stachura, K., (1998) New and confused species in the genus *Navicula* (Bacillariophyceae) and the consequences of restrictive generic circumscription. *Cryptogam. Algol.* 19, 83-108.

3.8.109.   Woodell, J. (2015) *Placoneis gastrum*. In Diatoms of North America. Retrieved June 30, 2020, from https://diatoms.org/species/placoneis_gastrum

Part II
MACROEVOLUTIONARY SYSTEMS
ANALYSIS OF DIATOMS

4

Probabilistic Diatom Adaptive Radiation in the Southern Ocean

Abstract

Diatoms are found worldwide as a result of high diversification. They are major influential drivers of the carbon and silica cycles and are ecologically tolerant of a wide range of environmental conditions, including nutrient availability. In the Southern Ocean (SO), two diatom lineages account for a high degree of diversity, namely, Chaetocerotales with *Chaetoceros*, and Bacillariales with *Fragilariopsis* and *Pseudo-nitzschia*. Adaptive radiation of genera represented by phenotypic traits is constrained by iron, nitrate, and silica availability, among other ecological factors with respect to niche filling. Adaptive radiation trajectories for each genus and lineage were modeled using Ornstein-Uhlenbeck (OU), Lyapunov modified OU and Hamilton-Jacobi (HJ) processes. Slice sampling was used to verify OU-based outcomes. Exit probabilities and time described the relation between volatility and outcomes. Niche filling with regard to iron, nitrate and silica availability in preferred SO habitats of sea ice, meltwaters and pelagic waters indicated that the adaptive advantage is sometimes with the Bacillariales despite the earlier evolution of the Chaetocerotales. The differences between the Bacillariales and Chaetocerotales are enough to indicate different trajectories for the adaptive radiation successes of these lineages in the Southern Ocean.

Keywords: Adaptive radiation, Antarctic, Hamilton-Jacobi equation, niche filling, Ornstein-Uhlenbeck process, Lyapunov function, nutrient availability, short term evolution

4.1 Introduction

Diatoms have been in existence since at least the Jurassic [4.8.19., 4.8.50., 4.8.76., 4.8.134.], and according to molecular clock estimates and biogeochemical assessments, diatoms may have originated just after the end Permian extinction event [4.8.9., 4.8.20., 4.8.134.]. With centric diatoms already in existence by the Jurassic and araphid diatoms commencing in the Lower Cretaceous, raphid diatoms were present in the Upper Cretaceous, according to molecular clock estimates [4.8.134.].

Diatom diversification occurred rapidly from the Late Cretaceous [4.8.50.] concomitantly with changes in ecological conditions (e.g., [4.8.9., 4.8.64.) throughout the Cenozoic, although there were fluctuations in diatom diversification [4.8.119.] with events such as the Paleocene-Eocene Thermal Maximum (PETM) [4.8.76., 4.8.97.], Eocene-Oligocene transition (EOT) [4.8.9., 4.8.70., 4.8.100., 4.8.119., 4.8.137.], and the Middle Miocene Climatic Optimum (MMCO) [4.8.9., 4.8.24., 4.8.76., 4.8.80.]. Major diversification of raphid diatoms occurred as early as the Paleocene ([4.8.134.]) with speciation proliferating since the Eocene (e.g., [4.8.19., 4.8.64., 4.8.94.]).

Janice L. Pappas. *Mathematical Macroevolution in Diatom Research*, (117–158) © 2023 Scrivener Publishing LLC

Diatoms account for around 45% of primary production and control the marine silica cycle as well as holding a 20% influence over the carbon cycle (e.g., [4.8.9.]). With high productivity, diatoms are highly adapted to various habitats and environments worldwide (e.g., [4.8.120.]). Despite physiologically-challenging conditions, diatoms have evolved strategies to enable adaptation to limiting nutrients such as iron (Fe) by forming resting stages (e.g., [4.8.64.]) or by having an ornithine-urea cycle [4.8.4.] that enables rapid nitrogen uptake during upwelling events for metabolism during nitrogen deficient conditions [4.8.64.]. Adaptive strategies enabling diatom sinking or suspension include cytoplasmic density regulation (e.g., [4.8.146.]) or chitin threads (e.g., [4.8.120.]) to control position with respect to light, nutrients and predation avoidance strategies (e.g., [4.8.47.]). At the MacDonald-Pfitzer limit of size reduction during the diatom life cycle, sexual reproduction occurs to induce genetic diversity, and is therefore an adaptive strategy for which timing of such an event occurs during optimal ecological conditions (e.g., [4.8.34.]). Diatoms exhibit high diversity and are the result of many lines of adaptive strategies enabling high adaptive radiation [4.8.64.].

Adaptation is a result of the influence of natural selection (e.g., [4.8.129.]) and may be described with respect to different aspects of selection. Evolution may proceed as directional change, random walks or stasis (e.g., [4.8.54.]). Adaptive change is associated with directional selection as a rare occurrence, and potentially, punctuated equilibrium (e.g., [4.8.54.]). Genetic drift is associated with randomness without boundary exemplifying neutral evolution (e.g., [4.8.40.]), and net to almost zero phenotypic change is associated with stasis (e.g., [4.8.35., 4.8.54.]) as evidence of stabilizing selection (e.g., [4.8.15., 4.8.32., 4.8.48.]).

Directional change is less frequently seen in calcareous planktonic versus benthic microfossils, while size is less frequently indicative of stasis in such planktonic versus benthic microfossils [4.8.54.]. Shape is less frequently indicative of stasis in calcareous benthic rather than planktonic microfossils [4.8.54.]. Evolution in siliceous microfossil size and shape may resemble an inverse trend with regard to diatoms, because diatoms contribute to the biological carbon pump in contrast to the alkalinity pump to which calcareous microfossils contribute [4.8.119.].

Geography may contribute to stasis [4.8.35.]. Continental masses or ocean currents at different scales may contribute to geographic barriers to reproduction (e.g., [4.8.36.]), flows (e.g., [4.8.30.]), and may contribute to stasis as well (e.g., [4.8.24.]). The Southern Ocean (SO) is bounded by the Polar Front [4.8.24.] (Figure 4.1) where meridional transport defines temperature, salinity and nutrient gradients (e.g., [4.8.116.]) and is generally the largest area of high nutrient and low chlorophyll content (e.g., [4.8.51.]). For diatoms in the SO, there have been five instances of extinctions and originations occurring since the Middle Miocene [4.8.24.], but despite such fluctuations, marine planktonic diatoms have continued to diversify, evidenced by large opal deposits in the SO [4.8.119.].

Various definitions of adaptive radiation have been used, including speciation via divergent selection, speciation via genetic drift, speciation via polyploidy, and ecological speciation [4.8.126., 4.8.127.]. Adaptation may occur with respect to the whole organism or to particular characters [4.8.40.]. The phenotypic diversification of lineage members increasing their occupation of available ecological niches over evolutionary time [4.8.113., 4.8.128.] is a form of ecological speciation. Filled niche spaces induces branching of a given lineage (e.g., [4.8.42., 4.8.113.]), and adaptive radiation may result from ecological

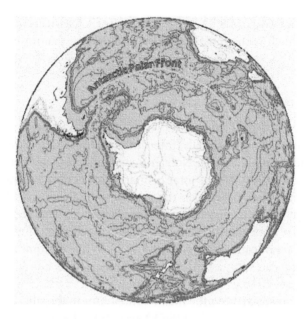

Figure 4.1 Map of the Southern Ocean.

speciation. Characteristics that may change throughout a lineage include body size and shape (e.g., [4.8.113.]), and tolerances to ecological conditions may span wide gradients (e.g., [4.8.55.]). For diatoms, ecological speciation (e.g., [4.8.64.]) and polyploidy [4.8.107.] may induce adaptive radiation, and their multitude of sizes and shapes are representative of lineage speciation (e.g., [4.8.93.]).

Of the general patterns of dynamical adaptive radiation resulting in speciation, oceans may be viewed spatially as two-dimensional geographic areas [4.8.42.]. These dimensions involve the horizontal or open ocean and vertical or coastal areas. Surface or deep ocean currents along with coastal or open ocean areas at multiple spatial scales and may induce constraints on organismal movements concerning biological reproductive activity, influencing speciation [4.8.42.].

For marine diatoms, the Antarctic represents a large biogeographic barrier [4.8.24.]. Evidence of diatom sexual reproduction occurring despite long geographical distances (e.g., [4.8.59.]) indicates that gene flow and speciation as a result of adaptive radiation occurs for diatoms where selective pressures and ecological conditions may dictate restrictions (e.g., [4.8.22.]). Despite barriers, diatoms have taken advantage of ecological opportunities (e.g., [4.8.136.]), resulting in colonization (e.g., [4.8.57.]), both temporally on geologic time scales and spatially worldwide [4.8.119.]. Diatom speciation has occurred at the clade level [4.8.76.], especially concerning rapid speciation of raphid diatom lineages [4.8.19.].

Adaptive radiation of pennate diatoms throughout the Cenozoic coincides with large ecological changes with greenhouse conditions existing prior to the Eocene and ice house conditions occurring from the Middle Eocene to the present (e.g., [4.8.91., 4.8.150.]). Adaptive radiation is analyzable at higher classification levels, including genera in lineages [4.8.135.]. For diatoms, the lineage of Bacillariales is important as a representative of the origination of the raphe as a novel phenotypic characteristic used in motility and sediment burial (e.g., [4.8.19., 4.8.62.]) and as representational of adaptative radiation with respect to ecological conditions (e.g., [4.8.22.]). The lineage of Chaetocerotales [4.8.19.] has the genus, *Chaetoceros*, which is the most abundant and diverse diatom in the oceans worldwide [4.8.83.].

4.1.1   Diatoms in the SO

Diatoms are a dominant feature in the SO (e.g., [4.8.21., 4.8.124.]). They are physically restricted by sea ice occurring seasonally, intermittently and as a permanent feature along continental Antarctica (e.g., [4.8.24.]). Nutrient concentrations are controlled by diatoms in the SO as they account for 40% of total primary production and carbon cycling (e.g., [4.8.124.]) and the interrelation between the carbon and silica cycles [4.8.133., 4.8.141.]. Fe concentrations may act as an accelerant to diatom growth when temperatures increase [4.8.11.]. However, Fe concentrations may influence reduced silicification of diatom frustules as a result of acidification of the SO [4.8.112.]. While diatoms remove carbon from the atmosphere via sequestration as oozes in the SO, Fe acts as a link between the carbon and silica cycles via diatoms (e.g., [4.8.124.]).

Since the EOT, diatoms have been the major contributor to the silica cycle [4.8.119.]. The interaction of Fe, nitrate (NO_3) and silica (SiO_2) availability has influenced diatom physiology and growth [4.8.123., 4.8.149.] in the SO [4.8.124.]. Biofacies from the Holocene indicate that seasonal changes in light and amount of sea ice constrain which diatom species are present [4.8.13.], and the availability of such nutrients as Fe, NO_3 and SiO_2 have an impact as well. As one example, highest diatom biomass occurs in the Weddell Sea as a result of stabile conditions from sea ice melting, with NO_3 and SiO_2 concentrations being proportionately low (e.g., [4.8.58.]).

Of the SO phytoplankton, diatoms are one of the most highly abundant and diverse groups (e.g., [4.8.13.]). *Chaetoceros* ([4.8.122.]) from the Chaetocerotales, along with Bacillariales lineage members, *Fragilariopsis* [4.8.123.] and *Pseudo-nitzschia* [4.8.2., 4.8.6.] are some of the dominant diatoms (e.g., [4.8.21.]) (Figure 4.2), as they constitute a major portion of algal communities in coastal high latitude regions [4.8.83.]. All three genera occur in the waters around the Drake Passage and in the Weddell Sea (e.g., [4.8.58.]) as well as being circumpolar Antarctic in distribution (4.8.131.). These genera represent distinct phenotypes that have adapted to varying sea ice conditions in the SO (e.g., [4.8.88.]).

4.1.2   Chaetocerotales and Bacillariales Speciation Rates

The diatom lineages of Chaetocerotales [4.8.19.] and Bacillariales (e.g., [4.8.19., 4.8.62., 4.8.82.]) have been established as monophyletic groups. Representative genera from these lineages are *Chaetoceros*, *Fragilariopsis* and *Pseudo-nitzschia* [4.8.19.]. According to the phylogeny established by Castro-Bugallo *et al.* [4.8.19.], *Fragilariopsis* has the highest speciation rate with first appearance estimated to occur in the Pliocene at around 5 Ma, *Pseudo-nitzschia* has the second highest speciation rate with first appearance at approximately 28 Ma during the Oligocene, and the slowest speciation rate belongs to *Chaetoceros* with first appearance at around 70 Ma in the Late Cretaceous.

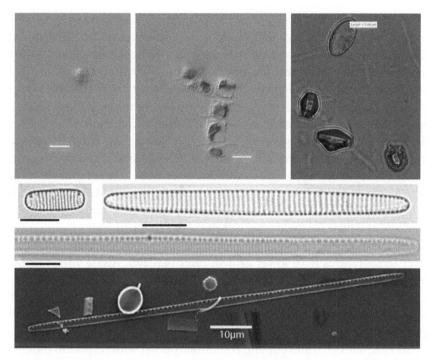

Figure 4.2 Examples of Southern Ocean diatoms: a, *Chaetoceros socialis* (valve view, differential interference contrast (DIC) LM by Isabel Dove, University of Rhode Island); b, *Chaetoceros socialis* (girdle view, DIC LM by Isabel Dove, University of Rhode Island); c, *Chaetoceros* sp. resting spores (LM by Isabel Dove, University of Rhode Island); d, *Fragilariopsis curta* (LM by Amy Leventer, Colgate University); e, *Fragilariopsis sublinearis* (LM by Amy Leventer, Colgate University); f, *Pseudo-nitzschia turgiduloides* (LM by Amy Leventer, Colgate University); g, *Pseudo-nitzschia turgiduloides* (SEM by Gastón Almandoz, Universidad Nacional de La Plata). Scale bar = 10 μ. All micrographs used with permission.

4.1.3   Chaetocerotales and Bacillariales: Fe, NO_3 and SiO_2 Availability in the SO

Primary production may be determined by Fe availability with regard to silica utilization by diatoms which affects growth [4.8.124.]. Fe in the SO is important in carbon fixation, nitrogen fixation and NO_3 assimilation [4.8.28.]. In the SO, high silica concentrations with low chlorophyll conditions exist in which Fe limitation is present to influence diatom growth [4.8.149.]. NO_3 is influenced by Fe as well as SiO_2 where the Si (silicon) to N (nitrogen) ratio under Fe limiting conditions may induce an increase in surface to volume ratio of the diatom cell ([4.8.84.]). NO_3 and SiO_2 uptake are affected by Fe concentrations (e.g., [4.8.90.]) along vertical and horizontal gradients [4.8.68.].

Fe fertilization in the waters of the Kerguelen Plateau region induce smaller, lightly silicified cells of *Chaetoceros* to proliferate as blooms, in contrast to Fe limited areas which have larger, more heavily silicified cells such as *Fragilariopsis* [4.8.69.]. Smaller heavily silicified cells of *Fragilariopsis* may occur under Fe limitation as well, resulting in lesser availability for grazers [4.8.60., 4.8.61.]. *Chaetoceros* response to low Fe concentrations is growth inhibition [4.8.108.], while *Pseudo-nitzschia* acclimates to Fe limited conditions via increasing their accumulation of Si on their valve surfaces [4.8.84.].

High NO_3 concentrations along with Fe fertilization fuel increasing phytoplankton growth in the SO [4.8.85.]. *Pseudo-nitzschia* growth proliferates under this nutrient regime, and the degree to which *Pseudo-nitzschia* can tolerate Fe depletion and still thrive in the open ocean as well as in coastal regions may be the result of the ability to store Fe [4.8.85.]. Along with the wide-ranging adaptive ability of *Pseudo-nitzschia* [4.8.84.], *Fragilariopsis* exhibits phenotypic plasticity with respect to Fe enrichment in the SO [4.8.123.].

Chaetoceros may occupy many niches whereby adaptive radiation of other species may be curtailed (e.g., [4.8.137.]). In general, *Chaetoceros*, along with *Fragilariopsis*, have high abundances in the SO, with *Chaetoceros* occurring in both coastal and open ocean settings [4.8.21., 4.8.83.], but preferring pelagic conditions in terms of nutrient availability [4.8.111.]. By contrast, *Pseudo-nitzschia* is highly abundant across all oceans [4.8.6., 4.8.83.]. High surface water productivity for *Chaetoceros* and lower productivity for *Fragilariopsis* may be indicators of low and high seasonal ice cover in the SO [4.8.13.].

Chaetoceros has undergone cryptic speciation in which phenotypic forms occurring at lower temperatures have lower growth rates [4.8.29.]. Along with vegetative cells, *Chaetoceros* forms resting spores in the SO [4.8.65., 4.8.75.], contributing to carbon export [4.8.118.]. *Fragilariopsis* is found in the cold waters of the SO and is important with regard to the silica cycle when Fe is limited [4.8.21., 4.8.27.]. During the Pliocene, *Fragilariopsis* had undergone adaptive radiation with respect to increasing sea ice and cooling conditions in the SO [4.8.111., 4.8.131.], and cooling conditions persist with respect to modern species [4.8.21.]. The resultant speciation produced new forms that are, by inference from the stratigraphy, adapted to different environmental regimes [4.8.131.].

Pseudo-nitzschia is a prolific marine diatom that occurs worldwide (e.g., [4.8.143.]) and has gained notoriety as a vector for domoic acid contamination of marine organisms (e.g., [4.8.6., 4.8.72., 4.8.143.]). Upwelling events contribute to *Pseudo-nitzschia* species diversity in coastal regions, especially with respect to those taxa that produce toxic blooms [4.8.81.]. Directional selection may play a role in adaptive radiation by *Pseudo-nitzschia*, and the consequent speciation may be influenced by changes in specific nutrients in the SO. *Pseudo-nitzschia* is found in Fe and NO_3 rich as well as SiO_2 limited coastlines [4.8.84.]. Although preference is for meltwaters of coastal regions, *Pseudo-nitzschia* have been documented to be present in the open oceans where limiting nutrients is the dominating scenario (e.g., [4.8.6.]).

The underlying mechanisms of adaptive radiation are expressible as mathematical statements that encompass empirical and theoretical considerations [4.8.42.]. Sympatric speciation is assumed so that factors controlling reproductive isolation are not included for the sake of simplicity. Regardless of homoplasy (e.g., [4.8.62.]), adaptive radiation and the consequent of speciation are about processes and the whole organism phenotype as well as particular phenotypic characteristics.

4.1.4   Modeling Diatom Adaptive Radiation

For diatom lineages, adaptive radiation may be modeled and analyzed via nutrient availability constraints. Diatom growth may induce adaptive radiation with regard to nutrients in a given geographic oceanic setting. The SO is one of the most productive concerning diatoms, and the diatom lineages of the Chaetocerotales and Bacillariales have members that are ubiquitous in this part of the world. *Chaetoceros*, *Fragilariopsis* and *Pseudo-nitzschia* exhibit high productivity and biomass in the SO and form an important component of the ecology

[4.8.122.]. All three genera exhibit phenotypic plasticity [4.8.122.], producing phenotypes adapted to various ecological conditions as well as selection for particular phenotypes (e.g., [4.8.39.]) that have diversified in the SO (e.g., [4.8.45.]).

Adaptive radiation via niche filling is a result of ecological opportunity (e.g., [4.8.130., 4.8.136., 4.8.148.]). Niche filling may result because of extinction, dominant clade prevention, or other promoting or inhibitory factors in the environment, and such a scenario may be used to model diatom adaptive radiation in the SO. Table 4.1 is a compilation of ecological opportunities as niche preferences, photosynthesis efficiency, and nutrient availability for SO Chaetocerotales and Bacillariales.

4.2   Purposes of this Study

Modeling diatom adaptive radiation is accomplished via stochastic differential (SDEs) and stochastic partial differential (SPDEs) equations in the form of the Ornstein-Uhlenbeck (OU) process, Lyapunov modified OU process, and the Hamilton-Jacobi (HJ) process. Outcomes from these equations provide general trends for adaptive radiation. Short time scales are used to model and simulate the earliest time in the adaptive radiation process. The intent is to model the initial process of diatom adaptive radiation relative to stochastic, deterministic and deterministic chaos properties of two lineages: Bacillariales, represented by *Fragilariopsis* and *Pseudo-nitzschia*; Chaetocerotales, represented by *Chaetoceros*, Adaptive radiation completion is analyzed as the relation between short term evolution and exit times and probabilities.

Adaptive radiation of SO Bacillariales and Chaetocerotales is assessed using reliability analyses and survival functions and probabilities. Diatom niche filling resulting from ecological opportunity is used to characterize SO adaptive radiation. That is, diatom adaptive radiation with respect to an optimal phenotypic trait is modeled in terms of SO ecological conditions. For *Chaetoceros*, *Fragilariopsis* and *Pseudo-nitzschia*, phenotypic traits are determined via digital images and quantified for use in adaptive radiation models. SO ecological opportunities are represented by niche preference, photosynthetic efficiency, Fe, NO_3 and SiO_2 enrichment or limitation and used as parameters affecting modeling outcomes.

4.3   Mathematical Modeling of Adaptive Radiation

4.3.1   Quantitative Phenotypic Trait Measurement and Adaptive Radiation

Changes in diatom size during the Cenozoic is associated with changes in diversity, with increasing size culminating in the Eocene, and decreasing size commencing in the Miocene [4.8.38.]. Since the Neogene, diatom size has remained the same [4.8.38.], indicating the potential for stasis in diatoms, regardless of fluctuations present in speciation in the SO since the Middle Miocene [4.8.24.].

For adaptive radiation analysis, body size and shape have been used as measurable quantities of phenotype [4.8.49.]. For diatom adaptive radiation, measurement of body size or shape may be obtained from multivariate analysis. Typically, principal components analysis (PCA) is used to obtain a size variation gradient on PC1 with a shape variation gradient

Table 4.1 Southern Ocean diatom lineages represented by *Chaetoceros*, *Fragilariopsis* and *Pseudo-nitzschia* with relative, general ecological opportunity conditions for maximal (3) to minimal (1) adaptive radiation.

	Ecological niche preference	Photosynthesis efficiency: light response and oxygen production[b] – Ranked from maximum (1) to minimum (3)			Sea ice[c,g]: ↑Nutrient enriched ↓Nutrient limited			Meltwaters[c,d,f]: ↑Nutrient enriched ↓Nutrient limited			Pelagic[e,f]: ↑Nutrient enriched ↓Nutrient limited		
	Temperature (°C)[b]	Sea ice	Meltwaters	Pelagic	Fe	NO$_3$	SiO$_2$	Fe	NO$_3$	SiO$_2$	Fe	NO$_3$	SiO$_2$
Chaetocerotales[a]													
Chaetoceros[e,f,m] (*socialis*)	5.0 (pelagic)	2	2	3	→	↑	→	↑	→	→	→	↑	↑
Bacillariales[a]													
Fragilariopsis[f,h,k] (*curta*)	-1.5 (ice)	1	3	1	→	↑	→	↑	→	→	→	↑	↑
Pseudo-nitzschia[f,j] (*turgiduloides*)	2.0 (meltwaters)	3	1	2	→	↑	→	↑	→	→	→	↑	↑

From: [a][4.8.19.]; [b][4.8.111.]; [c][4.8.33.]; [d][4.8.67.]; [e][4.8.5.]; [f][4.8.77.]; [g][4.8.51.]; [h][4.8.28.]; [j][4.8.84.]; [k][4.8.21.]; [m][4.8.37.].

on PC2 (e.g., [4.8.49., 4.8.105.]). Other measurements of size include cell volume [4.8.53.], surface to volume ratio (e.g., [4.8.38.]), or apical length. Regardless of the size measurement used, shape is related to size as a multivariate quantity (e.g., [4.8.105.]), and simplified geometric models that may be useful for cell volume calculations (e.g., [4.8.53.]) do not adequately address the morphogenetic aspects of diatom valve formation (e.g., [4.8.104.]) or distribution of valve shapes [4.8.102.] with respect to ecological niches and adaptive radiation as an evolutionary process.

Size reduction during diatom vegetative reproduction may complicate using either size or shape measures as discrete phenotypic indicators of adaptation, especially across a broad spectrum of lineages and/or ecological conditions in contrast to comparisons of marine versus freshwater taxa in a single lineage [4.8.93.]. One alternative is to use phenotypic surface traits that may be obtained as groupings of multiple features via image processing techniques. Digitized diatom images from light or scanning electron microscopy may be used to extract phenotypic features as groups of pixels in areas of the valve surface. Such groupings may include edge and corner detection of features such as striae, punctae, valve outline, or any other feature of the valve surface. Such measures include aspects of surface features as well as shape and size [4.8.103.] but are not exclusively bound by a particular feature. Ensemble surface features [4.8.101.] may be indicative of important continuous phenotypic traits that may have significance phylogenetically or ecologically. Feature extraction of the whole valve surface may be used as a composite phenotypic trait measurable via image analytical techniques as well.

Feature detection of phenotypic surface traits is measurable in terms of edges and corners which are comprised of groups of pixels at varying values. Different thresholds may be set and used across a number of digital images to ensure unbiased measuring. Grayscale images may be used, and the amount of lighting, shadowing, brightness, and contrast may be normalized among the images used in analysis. Background pixels may be removed using various image processing techniques. Length of the maximum threshold of extracted features is the sum of the surface traits for each image that is a measure of the extracted edges and corners of phenotypic characteristics of each valve surface.

4.3.2 Adaptive Radiation: Implicit Stochastic Models

Recognition of adaptive radiations has been thought to proceed according to changes in the accumulation of lineages (e.g., [4.8.44., 4.8.49.]). However, a plethora of criteria, suppositions, or other specifications are not comprehensive, generally applied, or in agreement about what entails adaptive radiation and recognition of it with respect to a given lineage [4.8.49.]. Alternatively, adaptive radiation as it reflects Simpson's [4.8.129.] original model is diversification of a given lineage that occurs according to the filling of available niches as adaptive zones [4.8.49.]. Distribution of species in a given group and their relations in a phylogenetic tree have provided a basis for measurement of adaptive radiation (e.g., [4.8.49., [4.8.79.]).

Measurement of adaptive radiation may follow the models of Brownian motion (B), the early burst (EB) [4.8.49.], or the OU process [4.8.7., 4.8.15., 4.8.48.]. While the B model does not result in a pattern, the EB and OU models are more complex, each producing noticeable patterns.

For the B model, $f(x)_B = \sum_{k=1}^{n} x_k$, where x is a trait series that is summed from k to n-species as the net distance walked for each x at shared path lengths [4.8.49.].

No constraints are specified. The expected value is zero, and the covariance matrix is defined as $CV_{B_{ij}} = \sigma^2 s_{ij}$, where σ^2 is the net divergence rate (e.g., [4.8.1.]), and s_{ij} is the ith, jth shared path length of taxa in the lineage covariance matrix, CV (after [4.8.49.]).

For the EB model, an exponential rise is followed by an asymptotic finish whereby many niches are initially available, with adaptive radiation tailing off over time as niches become filled and stasis ensues. The EB model is given generally as $f(x)_{EB} = 1 - e^{-kx}$, where x is the variable used to measure adaptive radiation of n-species, and k is the rate of decay of that radiation over time. The expected value is Gaussian, and the covariance matrix is

$$CV_{EB_{ij}} = \sigma_0^2 \left(\frac{e^{r\, s_{ij}} - 1}{r} \right) \tag{4.1}$$

where σ_0^2 is the initial value for the net divergence rate, s_{ij} is defined as above, and r is a pattern rate change (after [4.8.49.]). EB with respect to body size and shape are considered to be rare events in adaptive evolution [4.8.49.].

The OU model is given as the SDE

$$dx_k(t) = \alpha[\beta_k(t) - x_k(t)]\, dt + \sigma dB_k(t) \tag{4.2}$$

where x is the kth quantitative phenotypic trait, t is time with $t \geq 0$, α is the strength of selection ([4.8.32., 4.8.130.]), β is an optimum trait in a lineage occurring via selection (e.g., [4.8.7., 4.8.32., 4.8.48.]) and is the long-term mean, σ is a measure of volatility (as the amplitude of randomness) (e.g., [4.8.144.]), and B is Brownian motion [4.8.7.] or Gaussian white noise [4.8.26.]. The bracketed quantity in the first term of the right-hand side of the equation is also the drift term. If considering stabilizing selection, ω^2, $\alpha = \omega^2 / 2N_{effective}$, where $N_{effective}$ is the effective size of the population [4.8.66.]. The net divergence rate is also the rate of stochastic evolution [4.8.121.], and is σ^2 for the short term and α for long term evolution [4.8.48.]. The simplest form of σ^2 in terms of a probability density function, p, is

$$\frac{\partial p}{\partial t} = \frac{f_{max}(1 - f_{max})}{2N} \frac{\partial^2 p}{\partial f^2} + s\, f_{max}(1 - f_{max}) \frac{\partial p}{\partial f} + \mu(2f_{max} - 1) \frac{\partial p}{\partial f} \tag{4.3}$$

where the first term on the right-hand side of the equation is random drift, the second term is selection, and the third term is mutation for allele frequency f, and population size N [4.8.121.]. As α approaches zero, the OU process becomes the B model (e.g., [4.8.32.]), but over long-term time scales, where presumably, the rate of mutation becomes known, α dictates the course of evolution as deterministic selection [4.8.121.].

For the OU model, the expected value is Gaussian, and the covariance matrix is given as

$$CV_{OU_{ij}} = \frac{\sigma^2}{2\alpha}(e^{-2\alpha\,(t - s_{ij})})(1 - e^{-2\alpha\,(s_{ij})}) \tag{4.4}$$

where σ^2, s_{ij} and α are defined as above, and t is the time at the deepest divergence in a phylogenetic tree (after [4.8.49.]).

The variance of an OU process with respect to time is an exponential decrease [4.8.32.]. The exponential OU process is

$$dx_k(t) = \alpha[\beta_k(t) - ln(x_k(t))] \, x_k(t) \, dt + \sigma \, x_k(t) \, dB_k(t) \tag{4.5}$$

where the optimum trait is equilibrated between it and the natural logarithm of the quantitative trait with respect to the strength of selection. Analytically, applying Itô's lemma (as a geometric Brownian process, e.g., [4.8.140.]),

$$d\left[ln(x_k(t))e^{\theta t} \right] = \left(\theta\beta_k(t) - \frac{\sigma^2 \, e^{\theta t}}{2} \right) dt + \sigma e^{\theta t} dB_k(t) \tag{4.6}$$

where θ is the rate of mean reversion (e.g., [4.8.144.]).

The OU model is stationary and normally distributed (e.g., [4.8.10.]) and is sensitive to initial conditions because of the Brownian motion term [4.8.23.]. Bias toward accepting a fitted OU model may occur when using likelihood ratio tests [4.8.23.]. Evaluation and modification of the OU model may include statistical testing or rescaling using total branch length (e.g., [4.8.26.]) as a pretreatment concerning initial conditions. However, rescaling that is dependent on total branch length may be biased because of dependence on the particular phylogeny used.

To offset the potential shortfalls of the OU model and the possibility of autocorrelation in sampling during the implementation of the OU process, the slice distribution as a Markov chain Monte Carlo (MCMC) method is used (e.g., [4.8.95.]). The OU process is Markovian [4.8.10.], and because the slice distribution has the Markov property, over the long run, it will converge to the stationary probability density distribution uniformly [4.8.95.]. Slice sampling involves obtaining coordinate values from beneath a distribution curve in which MCMC induces convergence of the perpendiculars as the density of the distribution. A horizontal slice is obtained through the distribution at the perpendicular position of a given coordinate value, indicating the OU process at a given time. The slice distribution represents the stationary probability density distribution of the steps of the OU process and ensures fidelity of the resultant OU fitted model [4.8.3.].

4.3.3   Adaptive Radiation Models: Time Evolution of a Stochastic System

Consider the OU process to exhibit positive recurrence with respect to adaptive radiation, and additionally, determine stability in terms of the OU model. With respect to stochastic stability inducing positive recurrence (e.g., [4.8.96.]), a Lyapunov function, $V_t(x_k(t))$, is inserted in the OU model as time evolution of the quantitative phenotypic trait so that the Lyapunov modified OU model is given as

$$dV_t(x_k(t)) = \left(-2\mu[V_t(x_k(t))] + \frac{\sigma^2}{2} \right) dt + \sigma\sqrt{2V_t(x_k(t))}dB_k^H(t) \tag{4.7}$$

where $1/2 < H < 1$, and at equilibrium, $V_t(x_0(t)) \equiv 0$ when positive definite, or V is negative definite if $-V$ is positive definite [4.8.151.]. We want a positive definite Lyapunov function $V_t(x_k(t)) \geq \mu(|x|), \forall(x,t) \in \mathbb{S}_h = \{x \in \mathbb{R}^m_+ : |x| < h\} \times [t_0, \infty)$ for the mth exponent [4.8.151.].

For $V_t(x_k(t)) = \dfrac{[x_k(t)]^2}{2}$, $\mu \equiv \alpha$, and $V_t(x_k(t)) = x_k(t)$. By using a Lyapunov function, the deterministic term becomes a deterministic chaos term, and a mixed term results from a combination of fractional Brownian motion and deterministic chaos. The Lyapunov modified OU process is a stochastic Lyapunov function of pseudo-random (self-similar) and random terms, where the product is a random term. The overall process is deterministic if $\sigma \to 0$.

The Lyapunov modified OU model expresses adaptive radiation in terms of stochastic stability despite deterministic chaos where the goal is convergence to a stationary probability distribution [4.8.96.] of phylogenetic patterns, and this stability is understood to reflect resultant niche filling, and ultimately, speciation. Phenotypic divergence may range from a linear to exponential rate (e.g., [4.8.32.]). Expressing the Lyapunov modified OU model in terms of exponential functions,

$$d[lnV_t(x_k(t))e^{\lambda t}] = \left(\lambda \beta_k(t) - \frac{\sigma^2}{2} e^{\lambda t} \right) dt + \sigma e^{\lambda t} dB_k(t) \qquad (4.8)$$

where the Lyapunov exponent, λ, indicates exponential stability, depending on the sign of the exponent (e.g., [4.8.96., 4.8.151.]). The maximum Lyapunov exponent is $\lambda_{max} = \lim\limits_{t \to \infty} \left[\max\limits_k \sup \frac{1}{t} \ln V_k(x_k(t)) \right]$ [4.8.138.]. If λ_{max} is negative, then asymptotic Lyapunov stability exists, and if λ_{max} is positive, then chaos or instability is present (e.g., [4.8.96.]). Lyapunov exponents are indicators of divergence if positive or convergence if negative (e.g., [4.8.99.]). The rate of mean reversion may be highly volatile given by a positive Lyapunov exponent or highly stable, given by a negative Lyapunov exponent (e.g., [4.8.63.]).

Initial conditions under the Lyapunov OU model may reflect deterministic chaotic inputs rather than likelihood inferences with respect to Brownian motion of the adaptive radiation process. Deterministically, because tracing a single adaptive radiation within a lineage may be theoretically recoverable as a pattern, but not necessarily predictable, deterministic chaos becomes part of the adaptive radiation process as a possible outcome. That is, variability in adaptive radiations and ecological conditions may produce patterns or gradients that may be discernable but not necessarily predictable (e.g., [4.8.117.]). Small changes in initial conditions may have a large effect on the outcome of adaptive radiations or ecological conditions, affecting speciation as the result of stochastic, deterministic or deterministic chaotic processes.

4.3.4   Adaptive Radiation as an Optimal Control Problem

Time evolution of the infinitesimal generator, $\mathcal{L}$, [4.8.139.] of a Lyapunov modified OU process is given as

$$\mathcal{L}V_t(x) = \frac{\partial V_t(x)}{\partial t} + \nabla_x V_t(x) + \frac{1}{2} \langle \sigma, \nabla^2_x V_t(x) \sigma \rangle \qquad (4.9)$$

where $\langle \cdot, \cdot \rangle$ is the Frobenius inner product [4.8.96.]. The Lyapunov modified OU model formulated in terms of optimal control theory involves the ith volatility, σ_i, that is optimally controlled via an input value on the interval $[-\sigma, \sigma]$ [4.8.96.]. Maximizing the probability, $\mathbb{P}$, that the quantitative phenotypic trait, x, becomes optimal and is definable as the stochastic reachability problem is

$$\sup_{\sigma_i} \mathbb{P}_x \left[\sup_{0 \le t \le T} x_t \ge \tau \right] \tag{4.10}$$

where sup is the supremum, x_t is the quantitative phenotypic trait that follows Eq. (4.7), first exit time, τ, is the level reached by x in $V_t(x_k(t))$, and $|\sigma_i| \le \sigma$, where σ is the maximum value [4.8.96.].

Because mean reversion is assumed to be part of the adaptive radiation process, time reversal may be used as a way to recognize potential change in the opposite direction as convergence in the mean occurs. The result is to obtain an optimum despite changes in the phenotypic trait with non-directional time. For a stochastically controlled adaptive radiation process with time reversal, $|\sigma_i| = \sigma$, and applying Itô's lemma, the optimal value function that is continuously differentiable is the Hamilton-Jacobi (HJ) partial differential equation (PDE)

$$\frac{\partial \Phi^\sigma}{\partial t} = \frac{\sigma_t^2}{2} \frac{\partial^2 \Phi^\sigma}{\partial x^2} - \theta x \frac{\partial \Phi^\sigma}{\partial x} \tag{4.11}$$

with $(t, x) \in [0,T] \times [-\infty,\tau]$, which becomes the optimal value function, $\Phi^*(t, x)$, when $\sigma^* \in \arg \sup_{\sigma} \dfrac{\sigma^2}{2} \dfrac{\partial^2 \Phi^*}{\partial x^2}$ for the optimal input, $\sigma^*(t, x)$, and initial and boundary conditions are $\Phi(0, x) = 0$ for $x \in [-\infty,\tau]$ and $\Phi(t, \tau) = 1$ for $t \in [0,T]$, respectively [4.8.96.]. Φ^* is stochastic optimal control via a Lyapunov function, $V_t(x_k(t)) = \dfrac{[x_k(t)]^2}{2}$, in which the maximum value is based on the parameters at a given time-state for quantitative phenotypic trait, x. In Eq. (4.11), the first term on the right serves as the characterization of the optimal phenotypic trait, and the second term on the right serves as the characterization of the optimum converging in the mean. Eq. (4.11) serves as the solution to reaching adaptive radiation at $\tau \le T$. That is, adaptive radiation is achieved as the quantitative phenotypic trait approaches the level reached as defined in Eq. (4.10).

Eq. (4.11) may be reformulated with respect to the OU process via a stochastic control system given as $dX_t = \beta(t, X_t, \theta_t)\, dt + \sigma(t, X_t, \theta_t)\, dB_t$, with $X_0 = x$ and $T \in (0,\infty)$ [4.8.115.] so that the deterministic HJ equation [4.8.109.]

$$-\frac{\partial \Phi(x,T)}{\partial t} = -\frac{\partial \Phi}{\partial t} = \sup_{v \in \Phi} \left\{ \frac{1}{2} \langle D^2 \Phi \sigma(x,a), \sigma(x,a) \rangle + \langle D\Phi, b(x,a,t) \rangle + f(x,a,t) \right\} \tag{4.12}$$

becomes the backward SPDE

$$-d\Phi = \sup_{v \in \Phi} \left\{ \frac{1}{2} \langle D^2\Phi\sigma(x,a,t), \sigma(x,a,t) \rangle + \langle D\Phi, b(x,a,t) \rangle \right.$$

$$\left. + \langle Dv, \sigma(x,a) \rangle + f(x,a,t)dt - v(x,t)dB \right\} \tag{4.13}$$

where D is the partial derivative, f is a deterministic function, and the other parameters are defined as above. As with Φ, v is a value function with $v(x,t) = \sigma^T(x,\Phi,v)\nabla\Phi$ and $(x,t) \in \mathbb{R}^n \times [0,T]$ and is necessary to satisfy the backward SPDE [4.8.109.].

4.3.5   Exit Probabilities as Boundaries for Completion of Adaptive Radiation

Computation of boundaries for the OU process is a formidable problem, given the multitude of possible values for the n-variables involved. Using a Lyapunov function with the Markov property in mind, a probability space based on the OU process, T, and τ is devised.

For two quantitative phenotypes either within a lineage or between lineages, the exit probabilities of the adaptive radiation process are

$$\mathbb{P}_x \left[\sup_{0 \le t \le T} x_t^1 \ge \tau \right] \le \mathbb{P}_x \left[\sup_{0 \le t \le T} x_t^2 \ge \tau \right] \tag{4.14}$$

where x_t^1 and x_t^2 are from two different adaptive radiations with the superscripts describing quantitative traits 1 and 2, and

$$\mathbb{P}_x \left[\sup_{0 \le t \le T} x_t^1 \ge \tau \right] \le \mathbb{P}_{x\frac{\sqrt{\theta}}{\sigma}} \left[\sup_{0 \le \theta t \le \theta T} x_t^2 \frac{\sqrt{\theta}}{\sigma} \ge \tau \frac{\sqrt{\theta}}{\sigma} \right] \tag{4.15}$$

are exit probabilities for an OU process with $OU_t = x_t^2 \dfrac{\sqrt{\theta}}{\sigma}$ with $\theta = \sigma = 1$ [4.8.52., 4.8.96.]. Lyapunov stability via the exit probability from the OU process represents an upper bound on V, and

$$\mathcal{L}V_t(x) \le -\alpha V_t(x) + c_t \tag{4.16}$$

where $c_t = 0$ with $\|\nabla V(x)\, e(x)\|_2 \le \sigma$ to ensure stochastic stability. Then, the exit probabilities are

$$\mathbb{P}_x \left[\sup_{0 \le t \le T} V_t(x_k(t)) \ge \frac{c_t}{\alpha}\tau \right] \le \mathbb{P}_{OU} \left[\sup_{0 \le t \le T} OU_t \ge \tau \right] \tag{4.17}$$

[4.8.96.].

For the HJ equation expressed via the infinitesimal operator (modified after [4.8.12., 4.8.96.]) with Dirichlet boundary $\partial\Omega = 0$, and $u(x) = \Phi^\sigma(x)$,

$$\inf_{v \in V}\{\mathcal{L}V_t(x(t))u(x) + f(x, u(x), \nabla u(x)\sigma V_t(x(t))), v\} = 0, x \in D_{\bar{\Omega}}, \qquad (4.18)$$

where $D_{\bar{\Omega}}$ is the time-dependent domain of the open bounded subset Ω and closed bounded subset $\bar{\Omega}$, inf is the infimum and is upper semicontinuous [4.8.25.], and first exit time is $\tau := \inf\{t \geq 0 : x_t^0 \notin \bar{D}_{\bar{\Omega}}\}$. For a local maximum x of $(u(x) - \varphi)$, there is a viscosity subsolution of HJ ≥ 0, and for a local minimum x of $(u(x) - \varphi)$, there is a viscosity supersolution of HJ ≤ 0 [4.8.12.]. Exit probabilities (modified after [4.8.52., 4.8.96.]) are expressed as

$$\mathbb{P}\left[\inf_{[0,\tau]} x_t^1 \leq 0\right] \leq \mathbb{P}\left[\sup_{[0,\tau]} V_t(x_1(t)) \geq 0\right] \qquad (4.19)$$

for x and $V(x)$, and sup is lower semicontinuous [4.8.25.]. Exit probabilities concerning optimality related to HJ for the OU process are

$$\mathbb{P}_x\left[\sup_{0 \leq t \leq T} x_t^1 \geq \beta + \tau\right] \leq \mathbb{P}_x\left[\sup_{0 \leq t \leq T} x_t^2 \geq \beta + \tau\right] \qquad (4.20)$$

where $\tau > 0$ on the upper bound with initial values of x at $t_0 = 0$ [4.8.96.].

From all of the probability inequalities presented with respect to the OU, Lyapunov modified OU and HJ processes, exit probabilities were calculated for Lyapunov modified OU simulations in view of $\sigma^* \in \arg\sup_\sigma \dfrac{\sigma^2}{2}\dfrac{\partial^2 \Phi^*}{\partial x^2}$, and that from Eq. (4.7), $2V_t(x_k(t)) = [x_k(t)]^2$ when $\alpha = \theta = \sigma = 1$.

4.3.6   Exit Times for the Adaptive Radiation Process

The probability that one genus (or lineage) completes adaptive radiation (e.g., just prior to speciation) before another is countable as an exit time. Exit probabilities are in concert with exit times as time evolution occurs during adaptive radiation. If $\sigma = 0$, an exit time of $\log 2 + \epsilon$ occurs, resulting in $\mathbb{P}_\tau = 1$; if $|\sigma_i| \leq \sigma$ for all $t \in [0, T]$, then all $\tau > 0$, and $\mathbb{P}_\tau < 1$ [4.8.96.]. The state of a given x occurs at an instant of time, τ, and because the adaptive radiation process occurs via the OU process containing Brownian motion and having the Markov property, the average exit time is $\sqrt{\log(1+\tau)}$ [4.8.46.]. From Eq. (4.17), the Lyapunov function has exponential stability, inducing an exit time of the form $\tau_{exit\ time} = \dfrac{\log(\tau + 1)}{2\theta}$, which satisfies $\tau = \{\inf t > 0 : B_t \notin [tboundary_-, tboundary_+]\}$ [4.8.52.] for the completion of adaptive radiation via an OU process. From Eq. (4.15) and $OU_t = x_t^2 \dfrac{\sqrt{\theta}}{\sigma}$ with $\theta = \sigma = 1$, the modified threshold for an OU process is $\tau \dfrac{\sqrt{\theta}}{\sigma}$, and the modified exit time equation is

$$\tau_{OU\ exit\ time} = \frac{\log\left(t\dfrac{\sqrt{\theta}}{\sigma}+1\right)}{2\theta} \tag{4.21}$$

where log is base 10 for $\tau_k \leq T$ time steps.

4.3.7   Adaptive Radiation: A Study of Southern Ocean Diatoms and Niche Filling

Until now, constraints to adaptive radiation have been implicit. For SO diatoms, habitat and nutrient constraints may be incorporated into the adaptive radiation model. An exponential distribution is used in reliability analysis to model the interaction of habitat and nutrients Fe, NO_3 and SiO_2 and their effect on the probability of success in adaptive radiation for each genus and each lineage. An exponential distribution has the Markovian property [4.8.71.] and is suitable for relatively short time span or rapid events such as adaptive radiations (e.g., [4.8.136.]), and optimal environmental influences are testable via Gaussian demographics in which the exponential distribution describes organismal responses (e.g., [4.8.98.]).

An exponential distribution models a constant event (or risk) rate with respect to a survival function that serves as the means to calculate survival probability (e.g., [4.8.43.]) and probability of adaptive radiation success. The probability that a lineage will achieve successful adaptive radiation is dependent on environmental stress levels and rescue mutations [4.8.8.]. Environmental stress levels explicitly and rescue mutations implicitly serve as a basis since such mutations are linked to adaptive ability with respect to severity of environmental stresses [4.8.8.].

We do not know the exact trajectories over time for each genus with regard to adaptive radiation. However, we do know the end result, that each genus used in analysis is a dominant form in the SO, and therefore censoring of the data is unnecessary [4.8.74.]. From Table 4.1, rankings for genera from the categories of sea ice, meltwaters and pelagic are used to calculate a probability matrix, and nutrient enriched or limited conditions are scored as well. Photosynthesis efficiency as a habitat condition is emphasized in terms of each nutrient for each genus and is weighted so that assignment of a high value indicates high environmental adaptiveness. A much lower score indicates deemphasizing the particular habitat condition, and for a given genus, environmental stress is present, indicating lack of adaptiveness.

For each genus, niche filling scenarios are devised. From one category to the next, a disjunction of niche filling probabilities is defined as

$$nf_{max} = \max_i\{nf_1, nf_2, \ldots, nf_n\} = nf_1 \vee nf_2 \vee \ldots \vee nf_n = 1 - \prod_{i=1}^{n}(1 - nf_i) \tag{4.22}$$

for parallel niche filling by genera, or a conjunction as

$$nf_{min} = \min_i\{nf_1, nf_2, \ldots, nf_n\} = nf_1 \wedge nf_2 \wedge \ldots \wedge nf_n = \prod_{i=1}^{n} nf_i \qquad (4.23)$$

for sequential or series [4.8.73.] niche filling by genera. For incomplete niche filling, a warm standby system is used [4.8.73.]. Mixed systems were used to define each genus, depending on the rank scores for ecological parameters given in Table 4.1.

For each genus, 1000 permutations of the scores via an exponential distribution is used. Survival functions for the niche filling configurations are calculated, and scenarios are devised and plotted to illustrate the adaptive radiation successfulness of the genera. Probabilities of successful adaptive radiation are calculated as survival probability products with respect to different times as indicated via Eqs. (4.22) and (4.23).

4.4 Methods

The phylogeny in Castro-Bugallo *et al.* [4.8.19.] is used as a basis for study, with lineages Chaetocerotales represented by *Chaetoceros* and Bacillariales represented by *Fragilariopsis* and *Pseudo-nitzschia* (Figure 4.3). Petrou and Ralph's [4.8.111.] assessment of *Chaetoceros*, *Fragilariopsis* and *Pseudo-nitzschia* in terms of Antarctic ecological conditions affecting primary production is used as the structure for constraints on diatom adaptive radiation in the SO (Table 4.1). While all three genera have cosmopolitan members, and there are differences between east and west Antarctic oceanic environments, including influences from the Antarctic Circumpolar Current affecting sea ice formation and upwelling, some generalizations were made about each genus with regard to the assignment of ecological conditions for utility in modeling adaptive radiation.

Light micrographs (LMs) as grayscale digital images for each genus were used to obtain valve surface coordinate data via Laplacian of Gaussian filtering (LoG) [4.8.14.]. As a convolution, Gaussian smoothing removes noise first, then the Laplacian is used as an edge

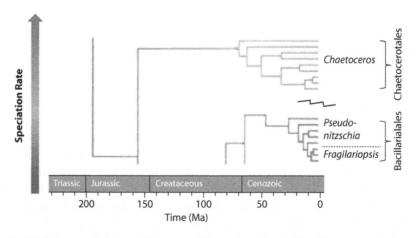

Figure 4.3 Partial phylogeny showing sublineages of Chaetocerotales and Bacillariales, speciation rate and time line adapted from [4.8.19.].

detector (e.g., [4.8.106.]). Pixel intensity values are treated via second partial derivatives using convolution methods to obtain remaining coordinates of non-smooth valve surface features. The surface feature values are summed to produce a quantity representing the topography of the valve surface. LoG is used on each digital image to obtain a series of image surface values between the minimum (approximately black) and maximum (approximately white). Each summed value represents a quantitative phenotypic surface trait, $x_1, ..., x_n$, and these values are averaged per genus, resulting in the quantity, β. Parameter values σ, θ, α, and t are kept constant throughout simulations.

Simulations using OU, Lyapunov modified OU and HJ models are used according to Eqs. (4.2), (4.7) and (4.11). The OU model is used for the classical analysis of adaptive radiation. The Lyapunov modified OU model is used to include stability in adaptive radiation. The optimal value function for adaptive radiation is calculated according to the HJ model.

Plots were constructed of simulation runs for each model to determine general trajectories of adaptive radiation for each genus. Means were calculated for OU and Lyapunov modified OU simulations, verified with slice sampling, and plotted as slice distributions. Exit times and probabilities were plotted as distribution σ-curves with $\sigma_i = \sigma_{0.1}, \sigma_{0.2}, ..., \sigma_{0.5}$. Exit probabilities were plotted versus exit times as distribution σ-curves with $\sigma_i = \sigma_{0.1}, \sigma_{0.2}, ..., \sigma_{0.5}$.

4.4.1   Niche Filling and Adaptive Radiation

For each genus, ecological niche preference, photosynthesis efficiency, Fe, NO_3 and SiO_2 were used to define ecological opportunities for successful adaptive radiation as niche filling. From Table 4.1, ecological niche preferences were identified and photosynthesis efficiencies and nutrient status were ranked based on the scientific literature. Probability analyses were used to characterize niche filling with respect to time via reliability analysis, survival functions and survival probabilities. An exponential distribution was used as a constant rate to indicate waiting time between events. Survival probabilities were calculated as well as the probability of successful adaptive radiation at a given time. Characterization was given at the generic and lineage levels specifically for photosynthesis efficiency as well as for all environmental parameters given in Table 4.1.

4.5   Results

For each genus, a series of LoG minima to maxima images were obtained on a normalized scale, and the sum of the image surface values as x_k at each LoG step were plotted for *Chaetoceros*, *Fragilariopsis* and *Pseudo-nitzschia* (Figure 4.4). Original LMs were converted to grayscale, if needed, and image samples were selected at the same coordinates and same size, making the input data size and shape invariant. Each LoG image sample is a "fingerprint" of each genus and served as input data in adaptive radiation simulations.

Parameters for OU and Lyapunov modified OU simulations were $\sigma = 1$, $\theta = 1$, $t = 4$ with 0.01 stepwise increments, $x = 1$, and β was the mean LoG value for each genus. Simulation conditions were 1000 iterations for each run, and 200 iterations for each mean calculation.

From OU simulation, non-discernable trends were evident for each genus (Figure 4.5), and with slice sampling, mean values ensured valid results (Table 4.2). By contrast,

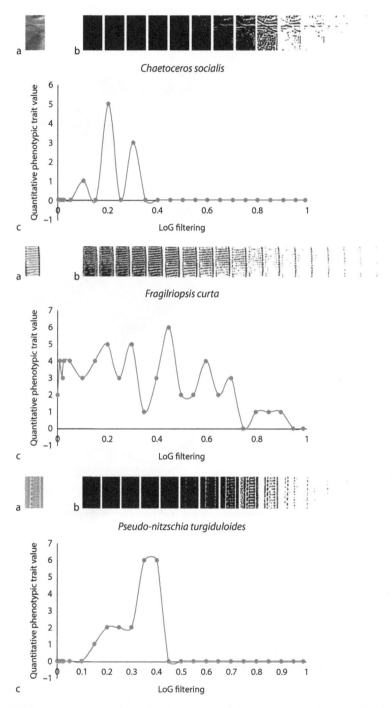

Figure 4.4 *Chaetoceros socialis* (top), *Fragilariopsis curta* (middle) and *Pseudo-nitzschia turgiduloides* (bottom). a, original LM; b, series of LoG images from minimum to maximum; c, plots of quantitative phenotypic trait values versus LoG filtering.

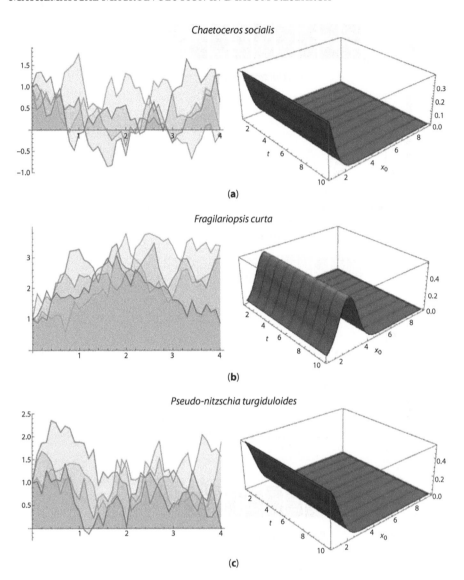

Figure 4.5 OU simulations (left) and slice sampling distributions (right): a, *Chaetoceros socialis*; b, *Fragilariopsis curta*; c, *Pseudo-nitzschia turgiduloides*. Parameters: $\theta = \sigma = 1$; $x = 1$; $t = 4$.

Lyapunov modified OU simulations produced overall negative exponential trends for each genus (Figure 4.6), and slice sampling indicated valid results as well (Table 4.2).

HJ modeling via initial values $\sigma = 1$, $\theta = 1$, $t = 4$ with 0.01 stepwise increments, $x = 1$, and β was the mean LoG value for each genus. Genera in the Bacillariales had similar distributions to each other in contrast to Chaetocerotales (Figure 4.7). Plots for each genus are the classic HJ inverted "horseshoe" in 3D with the two branches corresponding to the minimum and maximum, and the flat portion of the surface corresponding to the viscosity solution of HJ, bounded by the two branches (e.g., [4.8.16.]). The two branches are optimality ridges on the right and left sides of each plot, indicating discontinuity for the Chaetocerotales (Figure 4.7).

Table 4.2 Mean values for x and V, OU and Lyapunov modified OU, and slice samples.

Genus	Mean$_x$	Mean$_{OU}$	Mean$_{Slice\ OU}$
Chaetoceros	0.375	0.386447	0.368132
Fragilariopsis	2.625	2.59524	2.57692
Pseudo-nitzschia	0.791667	0.795483	0.777167
Genus	**Mean$_V$**	**Mean$_{Lyapunov\ OU}$**	**Mean$_{Slice\ Lyapunov\ OU}$**
Chaetoceros	0.72917	0.724973	0.715815
Fragilariopsis	4.8125	4.73351	4.72436
Pseudo-nitzschia	1.770833	1.74756	1.7384

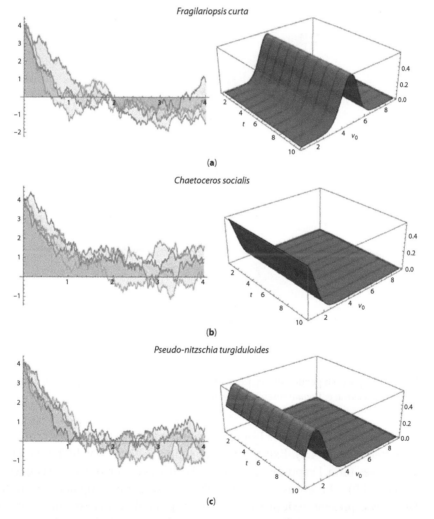

Figure 4.6 Lyapunov modified OU simulations (left) and slice sampling distributions (right): a, *Chaetoceros socialis*; b, *Fragilariopsis curta*; c, *Pseudo-nitzschia turgiduloides*. Parameters: $\theta = \sigma = 1$; $x = 1$; $t = 4$.

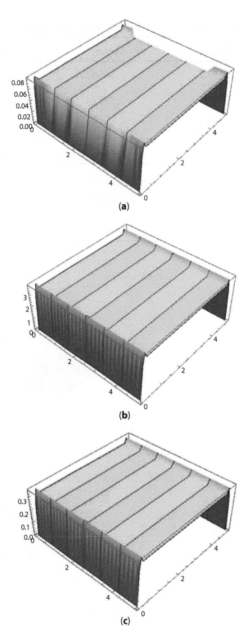

Figure 4.7 HJ modeling of each genus with regard to optimum quantitative phenotypic trait. a, *Chaetoceros socialis*; b, *Fragilariopsis curta*; c, *Pseudo-nitzschia turgiduloides*.

For a range of σ-curves for OU or Lyapunov modified OU processes, exit time, τ, versus time, t was plotted (Figure 4.8). Exit times were sooner if there was higher volatility. Exit probability σ-curves for a Lyapunov modified OU process over time, t, was devised with $\mathbb{P}(V_t \geq 2)$ (Figure 4.9). There was a higher probability of completing adaptive radiation if higher volatility was present. Exit probabilities were plotted versus exit times for σ-curves. (Figure 4.10). A higher probability to completing adaptive radiation occurs sooner with higher volatility.

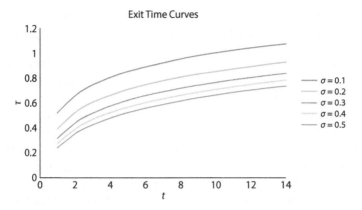

Figure 4.8 Exit time σ-curves for OU process over time, t. Higher volatility implies lower exit time to completion of adaptive radiation.

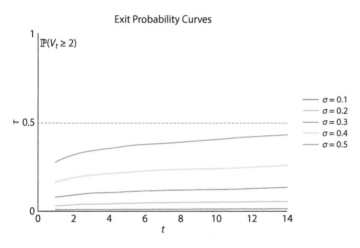

Figure 4.9 Exit probability σ-curves for Lyapunov modified OU process over time, t. All curves approach the 0.5 probability threshold for $V_t \geq 2$ which implies a 50% completion rate for adaptive radiation. Higher volatility implies a higher probability to completion of adaptive radiation.

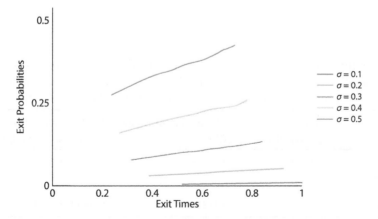

Figure 4.10 Exit probabilities versus exit times, τ. Higher volatility indicates less time and a higher probability to completion of adaptive radiation.

4.5.1   Ecological Niche Preference, Photosynthesis Efficiency, Nutrient Enrichment or Limitation, and Adaptive Radiation

From Table 4.1, the influence of ecological niche preference, photosynthesis efficiency, Fe, NO_3 and SiO_2 in the SO for diatom adaptive radiation was calculated as probabilities using survival functions. All probabilities were calculated to be at least 0.99 to differentiate times to successful niche filling. Weighting all environmental conditions in favor of *Chaetoceros* induced the probability of its successful adaptive radiation as niche filling at a time much sooner than that for Bacillariales genera (Table 4.3). When Bacillariales genera were favorably weighted in terms of all environmental conditions, time to successful niche filling was much sooner than it was for Chaetocerotales (Table 4.4). Between Bacillariales genera, the probability of successful niche filling was a much longer time with weighting for Chatocerotales (Table 4.3), with a shorter time with weighting favoring both Bacillariales genera (Table 4.4).

Table 4.3 Adaptive radiation as niche filling in terms of ecological niche preference, photosynthesis efficiency, Fe, NO_3, and SiO_2 enrichment/limitation in the Southern Ocean for *Chaetoceros*, *Fragilariopsis* and *Pseudo-nitzschia*, weighted toward Chaetocerotales.

Genus	Survival probability	Mean time to niche filling	Probability at time of successful niche filling
Pseudo-nitzschia	0.9958	11035.2	0.9998 at $t \geq 1000$
Chaetoceros		107.213	0.9958 at $t \geq 20$
Pseudo-nitzschia	0.9932	11035.2	0.9998 at $t \geq 1000$
Fragilariopsis			
Chaetoceros	0.9958	107.213	0.9958 at $t \geq 20$
Fragilariopsis		11035.2	0.9998 at $t \geq 1000$

Table 4.4 Adaptive radiation as niche filling in terms of ecological niche preference, photosynthesis efficiency, Fe, NO_3, and SiO_2 enrichment/limitation in the Southern Ocean for *Chaetoceros*, *Fragilariopsis* and *Pseudo-nitzschia*, weighted toward the Bacillariales.

Genus	Survival probability	Mean time to niche filling	Probability at time of successful niche filling
Pseudo-nitzschia *Fragilariopsis*	0.9966	110.352	0.9966 at $t \geq 20$
Chaetoceros		10721.3	0.9999 at $t \geq 1000$

4.5.2   Ecological Niche Preference and Photosynthesis Efficiency Rankings Representing Niche Filling as Adaptive Radiation

From Table 4.1, probabilities were calculated as a niche filling matrix (Table 4.5). Photosynthesis efficiency niche filling among genera was illustrated via plots of probability versus time (Figure 4.11). Comparisons within a lineage and between lineages were made. Incomplete niche filling by all genera were ordered with the highest probability for *Fragilariopsis*, followed by *Pseudo-nitzschia*, with *Chaetoceros* as the least probable (Figure 4.11a). Bacillariales incomplete niche filling is less probable than parallel niche filling for Chatocerotales (Figure 4.11b). Sequential niche filling by Chatocerotales is less probable than incomplete niche filling by Bacillariales (Figure 4.11c).

Probability versus time for photosynthesis efficiency niche filling between lineages indicated that parallel niche filling by Bacillariales had a higher probability than incomplete niche filling by Chatocerotales (Figure 4.12a), while the opposite was true when considering Bacillariales sequential niche to incomplete Chaetocerotales niche filling (Figure 4.12b). Parallel niche filling by one lineage had a higher probability than sequential niche filling by the other lineage (Figure 4.12c,d).

4.5.3   Niche Filling and the Lyapunov modified OU Adaptive Radiation Model

Niche filling using the Lyapunov modified OU model resulted in distinct outcomes for photosynthesis efficiencies (Figure 4.13). Parameter values for all simulations were $\sigma = \theta = 1$, $t = 4$. *Chaetoceros* indicated the most stability (Figure 4.13, top row) followed by *Pseudo-nitzschia* (Figure 4.13, bottom row) then *Fragilariopsis* (Figure 4.13, middle row). A combination plot of the reverse diagonal maximum ranks illustrated that each genus had a different signature with regard to photosynthesis efficiency (Figure 4.14).

Combined photosynthesis efficiencies for each genus indicated that *Chaetoceros* niche filling was consistent until a later time when niche filling in pelagic waters fell off (Figure 4.15a). For *Fragilariopsis*, combined niche filling had a negative exponential trend throughout (Figure 4.15b). *Pseudo-nitzschia* combined niche filling was mostly consistent from the beginning (Figure 4.15c). Overall, part of the Bacillariales matched the Chaetocerotales in completing adaptive radiation.

Photosynthesis efficiency category comparisons among the genera indicated that niche filling occurred initially by *Chaetoceros* and *Fragilariopsis*, but fell off as time progressed for

Table 4.5 Niche filling probability matrix of ecological niche preference and photosynthesis efficiency in the Southern Ocean for *Chaetoceros*, *Fragilariopsis* and *Pseudo-nitzschia*.

	Sea ice	Meltwaters	Pelagic
Chaetoceros	0.29	0.29	0.43
Fragilariopsis	0.20	0.60	0.20
Pseudo-nitzschia	0.50	0.17	0.33

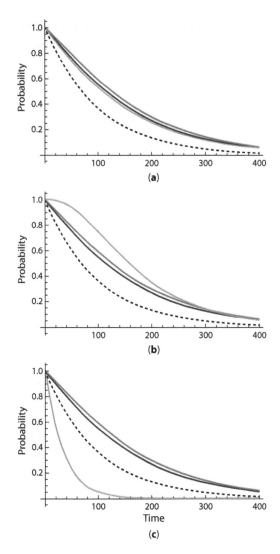

Figure 4.11 Niche filling plots of probability versus time for *Chaetoceros* (orange curve), *Fragilariopsis* (green curve) and *Pseudo-nitzschia* (purple curve). The dashed blue line is the demarcation between lineage niche filling with higher (above) and lower probabilities (below). a, Incomplete niche filling; b, Bacillariales with incomplete and Chaetocerotales with parallel niche filling; c, Bacillariales with incomplete and Chaetocerotales with sequential niche filling.

sea ice, while *Pseudo-nitzschia* maintained niche filling over time (Figure 4.16a). *Chaetoceros* and *Pseudo-nitzschia* had a higher meltwaters niche filling capacity at first, and subsequently, *Chaetoceros* increased in capacity, *Pseudo-nitzschia* decreased in capacity, while *Fragilariopsis* maintained capacity, albeit apparently lower than *Chaetoceros* and *Pseudo-nitzschia* (Figure 4.16b). *Fragilariopsis* started with the highest niche filling, but decreased while *Chaetoceros* and *Pseudo-nitzschia* maintained niche filling capacity for pelagic waters (Figure 4.16c). Adaptive radiation occurred for Bacillariales and Chaetocerotales regardless of the type of habitat in the SO.

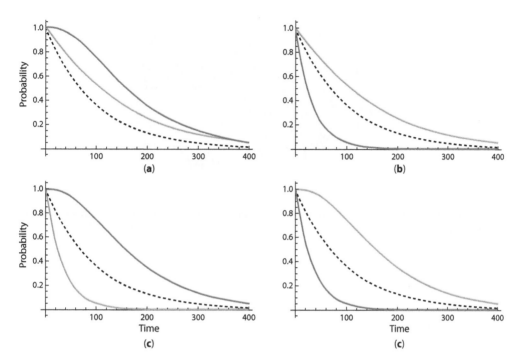

Figure 4.12 Niche filling plots of probability versus time for Chaetocerotales (orange curve) and Bacillariales (green curve). The dashed blue line is the demarcation between lineage niche filling with higher (above) and lower probabilities (below). a, Bacillariales parallel and Chaetocertales incomplete niche filling; b, Bacillariales sequential and Chaetocerotales incomplete niche filling; c, Bacillariales parallel and Chaetocerotales sequential niche filling; d, Bacillariales sequential and Chaetocerotales parallel niche filling.

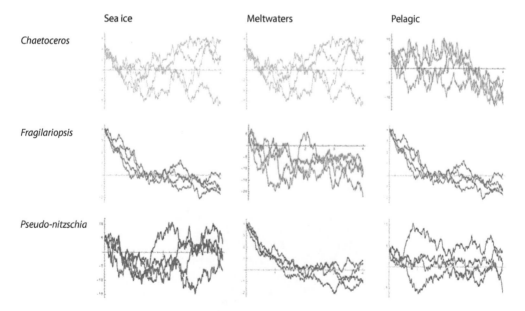

Figure 4.13 Lyapunov modified OU simulation of rank-ordered habitat preferences from Table 4.1 as lineage niche filling for Chaetocerotales (*Chaetoceros*, orange curves) and Bacillariales (*Fragilariopsis*, turquoise curves and *Pseudo-nitzschia*, blue curves). The brighter colored curves on the reverse diagonal are the maximum ranked niche filling outcomes.

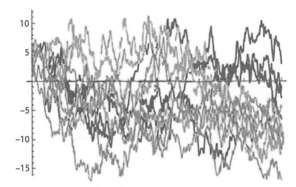

Figure 4.14 Lyapunov modified OU simulations of maximum ranks for Chatocerotales (*Chaetoceros*, orange curve) and Bacillariales (*Fragilariopsis*, turquoise curve and *Pseudo-nitzschia*, blue curve) niche filling.

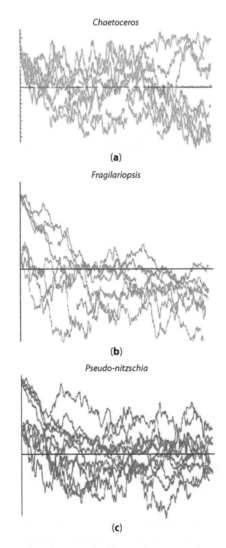

Figure 4.15 From Figure 4.13, combined row niche filling of sea ice, meltwaters and pelagic photosynthetic efficiencies for *Chaetoceros* (a), *Fragilariopsis* (b) and *Pseudo-nitzschia* (c).

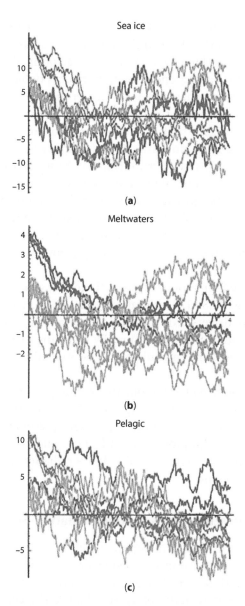

Figure 4.16 From Figure 4.13, combined column niche filling for *Chaetoceros* (orange curve), *Fragilariopsis* (turquoise curve) and *Pseudo-nitzschia* (blue curve) for photosynthesis efficiencies of sea ice (a), meltwaters (b) and pelagic (c).

4.6 Discussion

The intent of this study was to introduce modeling of adaptive radiation by looking at properties of this evolutionary process. Probabilistic outcomes at very short time intervals for diatom adaptive radiation in the SO was used for illustration. Low versus high adaptive probability of success was determined to occur because of many ecological factors. Adaptive radiation as a compilation of phenotypic evolutionary processes involved simulation of the random, deterministic and deterministically chaotic properties via the OU process as a basis for bounded time trajectories. The Lyapunov modified OU model induced the establishment of upper bounds on the OU adaptive radiation process as the solution to an optimal control problem via HJ and injected stability as a factor in adaptive radiation.

Interjection of stability via a Lyapunov function induces tighter bounds on the adaptive radiation process; i.e., when stochasticity is low, determinism is high, but that determinism may be tempered by chaos, and the Lyapunov function covers this case. Rather than depending solely on the stochastic part of adaptive radiation, other segments of this process must be taken into account. If the stochastic part of adaptive radiation is downplayed, then assessment of the deterministic and deterministically chaotic factors may be made.

Because of the emphasis on probabilistic outcomes, exit times and probabilities were useful characterizations of the short-term adaptive radiation process. By varying σ, short-term evolution was dictated by volatility, resulting in adaptive radiation. For longer term evolution, α, the deterministic aspects of adaptive radiation may be found with respect to generation time (e.g., [4.8.7.]) or perhaps even geologic time scales, regardless of the behavior of σ.

Adaptive dynamics have been modeled using OU and HJ where mutation (e.g., [4.8.87.]), competitive exclusion (e.g., [4.8.31.]), migration (e.g., [4.8.86.]), or dispersal and mutation (e.g., [4.8.110.]) were the drivers. For diatoms, dispersal rate, migration or local extinctions may play a key role in adaptive radiation. By matching niche filling with geographic location and speciation, resultant endemism may be studied as probabilistic outcomes of adaptive radiation events.

Adaptive radiation is modeled here implicitly as a jump or delayed differential process. As is the case in adaptive radiation modeling generally, each modeling schema is sensitive to initial input values. Despite this, the modeling schema presented indicated the flexibility of such an approach. While OU, Lyapunov modified OU and HJ simulations provided stochasticity, stability and extrema conditions of adaptive radiation, respectively, the interrelation of such models was evident.

4.6.1 More on Specifications for Adaptive Radiation Modeling

Usually, size is represented as a morphological evolutionary measure. Because of the vegetative reproductive process of diatoms involving size diminution, the approach here was taken to use measurable surface morphology as a "fingerprint" for a given genus, and at the level of a lineage, may be representative of a given "bauplan" specific to that lineage.

This study was about adaptive radiation as a short-term process, perhaps over ecological time. Diatoms have a high dispersal rate (e.g., [4.8.132.]), and because of their status as environmental indicators of specific conditions as well as their relative constancy in morphological forms present since the Neogene (e.g., [4.8.38.]), analyzing adaptive radiation as a short-term process induced the utility of using niche filling as one of the potential mechanisms. As a result, the focus was on the short-term evolution parameter, σ, and the probabilistic aspects of successful adaptive radiation at a given time (Figures 4.8-4.10) while still using the OU-based approaches to modeling adaptive radiation.

Given the divergence times for Bacillariales and Chaetocerotales (Figure 4.3), does Chatocerotales adaptively radiate before Bacillariales? From the OU and Lyapunov modified OU models, Chaetocerotales may adaptively radiate at a slightly earlier time

(Figures 4.5-4.6). However, according to the HJ model, Chaetocerotales may indicate growth fragmentation with the apparent discontinuous optimality ridge (Figure 4.7).

In this study, stabilizing selection, α, represented by a phenotypic trait, x, converging in the mean with respect to an optimal trait, β, was unchanged throughout simulations and analyses. The intent was to study short-term evolution and the effect that initial stochasticity had on an OU process. Introducing a Lyapunov modification while equating V with x provided the inclusion of implied stability. Potentially, negative, disruptive or other forms of selection may be included in adaptive radiation models. By looking at short-term evolution in this study, initial trends in adaptive radiation may be studied.

4.6.2   Diatom Adaptive Radiation Short-Term Trends as a Result of Niche Filling in the SO

This study was an illustration of including evolutionary ecological considerations such as niche filling in adaptive radiation models. Combinations of photosynthesis efficiencies and nutrients that are important in the SO were used as to infer general trends that would induce adaptive radiation. While the volatility of the trends may be a confounding factor, the short-term trends displayed would still be present at lower volatilities. From the simulations, trends may be determined despite the degree of volatility present. The intent is to show that multiple influences can be used to infer adaptive radiation short-term outcomes.

Fe dictated the SO nutrient regime, which in turn, is a proxy for temperature via the amount of sea ice present. Fe enrichment from islands in the SO enables diatom growth via utilization of available NO_3 and SiO_2 [4.8.92.]. In the SO, sea ice and meltwaters provide for a more constant or slightly positive trend compared to pelagic waters for the Chaetocerotales and Bacillariales (Figures 4.13, 4.15 and 4.16). Fe is associated with sea ice (e.g., [4.8.67.]), initially favoring *Fragilariopsis* over *Chaetoceros*, with the inference that Fe fertilization via ice melt is a contributing factor in contrast to Fe retention in or adsorption to sea ice [4.8.67.]. However, *Pseudo-nitzschia* ultimately maintains an almost constant trend concerning sea ice as well as in pelagic waters over *Chaetoceros* (Figures 4.13, 4.15 and 4.16). Fe depletion is tempered by the ability to store Fe in *Pseudo-nitzschia* [4.8.85.]. Overall, photosynthesis efficiency and Fe may favor Bacillariales adaptive radiation in sea ice and meltwater over Chaetocerotales.

Fragilariopsis in meltwaters had the lowest declining trend (Figure 4.13) which may be inferred as NO_3 limiting conditions (Table 4.1). At the edge of sea ice, nitrogen fixation readily occurs [4.8.128.]. NO_3 availability may be impacted by Fe as well, and may induce a more positive trend for *Chaetoceros* in shallow waters and a more positive trend for *Pseudo-nitzschia* in pelagic waters (Figure 4.13). Generally, photosynthesis efficiency and NO_3 may favor adaptive radiation in both lineages.

Although SiO_2 limiting conditions at depth may exist, *Chaetoceros* may produce an abundance of heavily silicified resting spores that end up in pelagic waters as possible "seeding" events for vegetative reproduction [4.8.37.]. In sea ice and meltwaters, the positive trends for *Chaetoceros* (Figure 4.13) may imply SiO_2 enrichment prior to a bloom event, yet enables vegetative reproduction at depth via resting spores when Fe conditions are right and $Si:NO_3$ is high [4.8.90.]. With *Pseudo-nitzschia* at a somewhat mixed trend with *Fragilariopsis* in terms of sea ice, meltwaters and pelagic waters and higher than *Chaetoceros* in pelagic waters, SiO_2 coupled with Fe availability may indicate adaptive radiation favorability for the Bacillariales over the Chateocerotales.

4.6.3 Other Potential Mathematical Modeling Regimes of Adaptive Radiation

Although the B model has martingale properties, the OU process does not [4.8.10.]. The Lyapunov function is rooted in supermartingale and ergodic theories of Markov processes (e.g., [4.8.96., 4.8.140.]). Using a Lyapunov approach is not amenable to generalization [4.8.140.], but it does enable the inclusion of potentially deterministic chaotic aspects of the adaptive radiation process into modeling. Because detailed, explicit information on phenotypic trait evolution is usually unavailable, parameters for the OU and Lyapunov modified OU models may be difficult to determine, along with determining whether the initial condition of stationarity or non-stationarity is present, or whether the Markov or martingale properties are appropriate. Adaptive radiation modeling depends not only on the properties of the equations used, but also on circumscribed, reasoned selection of parameter values as well as selection of a particular phylogeny that are the specific focus and objective of a given study.

Other modifications to the traditional OU model include other SDEs such as stochastic functional differential equations, pseudo-SDEs, or Riccati SDEs (e.g., [4.8.96.]). Other PDEs, including other stochastic PDEs (SPDEs) or pseudo-SPDEs, may be suitable as well. Various properties of these equations must be considered for use in adaptive radiation modeling. Different spatial and time scales may be incorporated as well, and other stochastic jump processes (e.g., [4.8.41.]) may be useful tools in adaptive radiation modeling (e.g., [4.8.10.]).

Blomberg *et al.* [4.8.10.] used the Fokker-Planck equation via a Kolmogorov forward equation to model evolution of traits that do not follow a Gaussian distribution such as sex ratios or life spans. However, such traits are not specifically descriptive of phenotypes with respect to adaptive radiation. The Fokker-Planck equation and/or Kolmogorov forward equation model the flux in probability and have the Markov property [4.8.10.]. Noteworthy is that a Fokker-Planck equation is a one-dimensional OU process, where the connection is found via Bernoulli-Laplace Urn models [4.8.56.]. Evolutionary jump models based on a Lévy process [4.8.41., 4.8.147.], or models that incorporate varying speeds at which evolutionary changes occur (e.g., [4.8.10.]) may be useful in simulating adaptive radiation over evolutionary time. Forward and backward problems associated with the OU process may be calculated according to the probability density or cumulative probability exit time distribution as well as the time scale considered (e.g., [4.8.78.]).

Calculating exit probabilities and times is a somewhat heuristic process. Determination of a barrier function or certificate (e.g., [4.8.114.]) or choice of boundary conditions is dependent on initial conditions, settings for parameters, and changes in settings throughout running simulations. Without more generalized bounds for adaptive radiation functions, and in spite of phylogenetic results, there are a multitude of ways to determine boundaries for adaptive radiation schema.

Using reliability analysis including survival functions and probabilities could be expanded to model more complex scenarios of adaptive radiation. Although niche filling was used in this study, other characteristics or features of adaptive radiation could be incorporated into reliability analysis to obtain probabilistic outcomes when comparing lineages.

4.7   Summary and Future Directions

Adaptive radiation was mathematically modeled based on two SDEs—OU and a Lyapunov modified OU process—as well as a PDE in the form of the HJ equation. In any case, an additive function that covers the deterministic, deterministic chaotic and random aspects of the adaptive radiation process is useful in modeling potential adaptive radiation trajectories. The limiting factors in modeling include the extent of a data set, robustness of the phylogeny considered, choice of input values, computational time, and convergence to solutions. There are also short versus longer time scales and spatial or geographic constraints that may be important factors, depending on how a given adaptive radiation study is defined. Other potential considerations include hybridization (e.g., [4.8.17., 4.8.18.]) or polyploidy versus mutation, the failure to adaptively radiate despite optimal conditions and ecological opportunity, multiple key innovations leading to adaptive radiation, invasives adaptively radiating, and changing optimality states for a given lineage.

Additional members of the Chaetocerotales and Bacillariales could be included in adaptive radiation analysis for the SO. Other lineages may be included or analyzed separately. Other oceanic, freshwater or continental environments could be analyzed. World-wide diatom adaptive radiation over geologic time could be studied. In terms of ecological parameters, nutrients in addition to or instead of Fe, NO_3 and SiO_2 and other parameters could be used to gauge ecological opportunity for adaptive radiation. Diatom adaptive radiation treated as an optimal control problem opens up the opportunity to analyze adaptive radiation as an energy transfer/flow process.

Until actual trajectories for adaptive radiation are demonstrably and historically traceable for diatom lineages with constancy, this evolutionary process will remain subject to multiple modeling regimes and input values. For adaptive radiation, the interaction of lineages may be considered as ecological trait changes at shorter time scales or at geologic time scales. Ecological opportunity expressed as niche filling was one way to study short term diatom adaptive radiation.

Aspects of diatom evolution other than the ecological may be considered in more complex schema for studying adaptive radiation, including reproduction and size reduction or phenotypic traits that may exhibit stabilizing selection (e.g., keel in some pennate diatoms), but may be suboptimal when considering common traits of the diatom bauplan (e.g., raphe in pennate diatoms). The relation between geographic distributions and niche filling dynamics (e.g., [4.8.98.]) as a result of ecological opportunity at different time and spatial scales (e.g., [4.8.148.]) may be interesting avenues of study with regard to diatom adaptive radiations. In light of this, the possibilities for amendments are in the offing as we learn more about the adaptive radiation process in diatom lineages.

4.7.1   Caveats in Adaptive Radiation Studies to be Considered

As longer time scales are considered, the role of the strength of selection in ecological adaptive radiation may be studied with respect to divergence times (e.g., [4.8.142.]). Older lineages, including non-interacting lineages, may induce convergence in phenotypic traits, and the role of selection here may be studied with respect to adaptive radiation. The relation between divergence times and adaptive radiation at multiple evolutionary time scales may provide insight into trait differences.

However, the relation between phylogenetic trees and adaptive radiation requires careful considerations. Because phylogenies are hypotheses in themselves, relying on them for the sole picture of evolution may be inadequate (e.g., [4.8.125.]). Techniques such as Bayesian analysis of macroevolutionary mixtures (BAMM) to determine diversification rates across branches of a phylogeny may be fraught with problems including the inability to estimate accurate priors and other parameters which in turn leads to incorrect likelihood calculations [4.8.89.]. Great care must be taken to implement Bayesian approaches (e.g., [4.8.145.]) in adaptive radiation studies.

Until such time that adaptive radiation trajectories are established, a variety of methods may be used to study different facets of this evolutionary process. Trends and probabilistic outcomes may be useful in furthering our understanding of the interrelation among adaptive radiation parameters, temporal and spatial scales, and the role that ecological and other evolutionary processes have in adaptive radiation.

4.8   References

4.8.1. Ackerly, D. (2009) Conservatism and diversification of plant functional traits: evolutionary rates versus phylogenetic signal. *Proc. Natl. Acad. Sci.* 106, 19699-19706.

4.8.2. Almandoz, G., Ferreyra, G.A., Schloss, I.R., Dogliotti, A.I., Rupolo, V., Paparazzo, F.E., Esteves, J.L., Ferrario, M.E. (2008) Distribution and ecology of *Pseudo-nitzschia* species (Bacillariophyceae) in surface waters of the Weddell Sea (Antarctica) *Polar Biology* 31, 429-442.

4.8.3. Andrieu, C., Doucet, A., Holenstein, R. (2010) Particle Markov chain Monte Carlo methods. *J.R. Statist. Soc. B* 72(Part3), 269-342.

4.8.4. Armbrust, E., Berges, J.A., Bowler, C., Green, B.R., Martinez, D., Putnam, N.H., *et al.* (2004) The genome of the diatom *Thalassiosira pseudonana*: ecology, evolution, and metabolism. *Science* 306, 79-86.

4.8.5. Assmy, P., Smetacek, V., Montresor, M., Klaas, C., Henjes, J., Strass, V.H., Arrieta, J.M., Bathmann, U., Berg, G.M., Breitbarth, E., Cisewski, B., *et al.* (2013) Thick-shelled, grazer-protected diatoms decouple ocean carbon and silicon cycles in the iron-limited Antarctic Circumpolar Current *PNAS* 110(51):20633-8.

4.8.6. Bates, S.S., Hubbard, K.A., Lundholm, N., Montresor, M., Leaw, C.P. (2019) *Pseudo-nitzschi*a, *Nitzschia*, and domoic acid: new research since 2011. *Harmful Algae* 79, 3-43.

4.8.7. Beaulieu, J.M., Jhwueng, D.-C., Boettiger, C., O'Meara, B.C. (2012) Modeling stabilizing selection: expanding the Ornstein-Uhlenbeck model of adaptive evolution. *Evolution* 66(8), 2369-2383.

4.8.8. Bell, G. and Collins, S. (2008) Adaptation, extinction and global change. *Evolutionary Applications* 1(2008), 3-16.

4.8.9. Benoiston, A.-S., Ibarbalz, F.M., Bittner, L., Guidi, L., Jahn, O., Dutkiewicz, S., Bowler, C. (2017) The evolution of diatoms and their biogeochemical functions. *Phil. Trans. R. Soc. B* 372: 20160397.

4.8.10. Blomberg, S.P., Rathnayake, S.I., Moreau, C.M. (2020) Beyond Brownian motion and the Ornstein-Uhlenbeck process: stochastic diffusion models for evolution of quantitative characters. *The American Naturalist* 195(2), 145-165.

4.8.11. Boyd, P.W. (2019) Physiology and iron modulate diverse responses of diatoms to a warming Southern Ocean. *Nature Climate Change* 9, 148-152.

4.8.12.   Buckdahn, R. and Nie, T. (2018) Generalized Hamilton-Jacobi-Bellman equations with Dirichlet boundary and stochastic exit time optimal control problem. https://arxiv.org/pdf/1412.0730.pdf.

4.8.13.   Buffen, A., Leventer, A., Rubin, A., Hutchins, T. (2007) Diatom assemblages in surface sediments of the northwestern Weddell Sea, Antarctic Peninsula. *Marine Micropaleontology* 62, 7-30.

4.8.14.   Burt, P.J. and Adelson, E.H. (1983) The Laplacian pyramid as a compact image code. *IEEE Transactions on Communications* COM-31(4), 532-540.

4.8.15.   Butler, M.A. and King, A.A. (2004) Phylogenetic comparative analysis: a modeling approach for adaptive evolution. *Am. Nat.* 164, 683-695.

4.8.16.   Calvez, V., Gabriel, P., Gaubert, S. (2014) Non-linear eigenvalue probalems arising from growth maximization of positive linear dynamical systems. *53rd IEEE Conference on Decision and Control, Los Angeles, California 2014*, 1600-1607.

4.8.17.   Casteleyn, G., Adams, N.G., Vanormelingen, P., Debeer, A.-E., Sabbe, K., Vyverman, W. (2009) Natural hybrids in the marine diatom *Pseudo-nitzschia pungens* (Bacillariophyceae): genetic and morphological evidence. *Protist* 160, 343-354.

4.8.18.   Casteleyn, G., Leliaert, F., Backelijau, T., Debeer, A.-E., Kotaki, Y., Rhodes, L., Lundholm, N., Sabbe, K., Vyverman, W. (2010) Limits to gene flow in a cosmopolitan marine planktonic diatom. *PNAS* 107(29), 12952-12957.

4.8.19.   Castro-Bugallo, A., Rojas, D., Rocha, S., Cermeño, P. (2017) Phylogenetic analyses reveal an increase in the speciation rate of raphid pennate diatoms in the Cretaceous. bioRxiv, https://doi.org/10.1101/104612.

4.8.20.   Cavalier-Smith, T. (2009) Megaphylogeny, cell body plans, adaptive zones: causes and timing of eukaryote basal radiations. *J. Eukaryot. Microbiol.* 56(1), 26-33.

4.8.21.   Cefarelli, A.O., Ferrario, M.E., Almandoz, G.O., Atencio, A.G., Akselman, R., Vernet, M. (2010) Diversity of the diatom genus *Fragilariopsis* in the Argentine Sea and Antarctic waters: morphology: distribution and abundance. *Polar Biology* 33, 1463-1484.

4.8.22.   Cermeño, P. and Falkowski, P.G. (2009) Controls on diatom biogeography in the ocean. *Science* 325, 1539-1541.

4.8.23.   Cooper, N., Thomas, G.H., Venditti, C., Meade, A., Freckleton, R.P. (2016) A cautionary note on the use of Ornstein Uhlenbeck models in macroevolutionary studies. *Biological Journal of the Linnean Society* 118, 64-77.

4.8.24.   Crampton, J.S., Cody, R.D., Levy, R., Harwood, D., McKay, R., Naish, T.R. (2016) Southern Ocean phytoplankton turnover in response to stepwise Antarctic cooling over the past 15 million years. *PNAS* 113(25), 6868-6873.

4.8.25.   Crandall, M.G., Ishii, H., Lions, P.-L. (1992) User's guide to viscosity solutions or second order partial differential equations. *AMS* 27(1), 1-67.

4.8.26.   Cressler, D.E., Butler, M.A., King, A.A. (2015) Detecting adaptive evolution in phylogenetic comparative analysis using the Ornstein-Uhlenbeck model. *Syst. Biol.* 64(6), 953-968.

4.8.27.   Cortese, G. and Gersonde, R. (2007) Morphometric variability in the diatom *Fragilariopsis kerguelensis*: implications for Southern Ocean paleoceanography. *Earth and Planet. Sci. Lett.* 257, 526-544.

4.8.28.   De Baar, H.J.W., Van Leeuwe, M.A., Scharek, R., Goeyens, L., Bakker, M.J., Fritsche, P. (1997) Nutrient anomalies in *Fragilariopsis kerguelensis* blooms, iron deficiency and the nitrate/phosphate ratio (A.C. Redfield) of the Antarctic Ocean. *Deep-Sea Research II* 44(1-2), 229-260.

4.8.29.   Degerlund, M., Huseby, S., Zincone, A., Sarno, D., Landfald, B. (2012) Functional diversity in cryptic species of *Chaetoceros socialis* Lauder (Bacillariophyceae). *Journal of Plankton Research* 34(5), 416-431.

4.8.30.   Deppleler, S.L. and Davidson, A.T. (2017) Southern Ocean phytoplankton in a changing climate. *Frontiers in Marine Science* 4:40.

4.8.31.   Diekmann, O., Jabin, Pl-E., Mischler, S., Perthame, B. (2005) The dynamics of adaptation: an illuminating example and a Hamilton-Jacobi approach. *Theoretical Population Biology* 67(4), 257-271.

4.8.32.   Diniz-Filho, J.A.F. (2001) Phylogenetic autocorrelation under distinct evolutionary processes. *Evolution* 556, 1104-1109.

4.8.33.   DuPrat, L., Corkill, M., Genovese, C., Townsend, A.T., Moreau, S., Meiners, K.M., Lannuzel, D. (2020) Nutrient distribution in east Antarctic summer sea ice: a potential iron contribution from glacial basal melt. *Journal of Geophysical Research: Oceans* 125, e2020JC016130.

4.8.34.   Edlund, M.B. and Stoermer, E.F. (1997) Ecological, evolutionary, and systematic significance of diatom life histories. *Journal of Phycology* 33, 897-918.

4.8.35.   Eldredge, N., Thompson, J.N., Brakefield, P.M., Gavrilets, S., Jablonski, D., Jackson, J.B.C., Lenski, R.E., Lieberman, B.S., McPeek, M.A., Miller III, W. (2005) The dynamics of evolutionary stasis. *Paleobiology* 31(2), 133-145.

4.8.36.   Faria, R., Johannesson, K., Stankowski, S. (2021) Speciation in marine environments: diving under the surface. *J. Evol. Biol.* 34, 4-15.

4.8.37.   Ferrario, M.E., Sar, E.A., Vernet, M. (1998) *Chaetoceros* resting spores in the Gerlache Strait, Antarctic Peninsula. *Polar Biology* 19, 286-288.

4.8.38.   Finkel, Z.V., Katz, M.E., Wright, J.D., Schofield, O.M.E., Falkowski, P.G. (2005) Climatically driven macroevolutionary patterns in the size of marine diatoms over the Cenozoic. *PNAS* 102(25), 8927-8932.

4.8.39.   Fox, R.J., Donelson, J.M., Schunter, C., Ravasi, T., Gaitán-Espitia, J.D. (2019) Beyond buying time: the role of plasticity in phenotypic adaptation to rapid environmental change. *Phil. Trans. R. Soc. B* 374, 20180174.

4.8.40.   Futuyma, D.J. (2010) Evolutionary constraint and ecological consequences. *Evolution* 64(7), 1865-1884.

4.8.41.   Gao, T., Duan, J., Li, X., Song, R. (2014) Mean exit time and escape probability for dynamical systems driven by Lévy noises. *SIAM J. Sci. Comput.* 36(3), A887-A906.

4.8.42.   Gavrilets, S. and Losos, J.B. (2009) Adaptive radiation: contrasting theory with data. *Science* 323, 732-737.

4.8.43.   Giorgini, L.T., Moon, W., Wettlaufer, J.S. (2020) Analytical survival analysis of the Ornstein-Uhlenbeck process. *Journal of Statistical Physics* 181, 2404-2414.

4.8.44.   Givnish, T.J. and Systma, K.J. (eds.) (1997) *Molecular Evolution and Adaptive Radiation*, Cambridge University Press, Cambridge, United Kingdom.

4.8.45.   Glemser, B., Kloster, M., Esper, O., Eggers, S.L., Kauer, G., Beszteri, B. (2019) Biogeographic differentiation between two morphotypes of the Southern Ocean diatom *Fragilariopsis kerguelensis*. *Polar Biology* 42, 1369-1376.

4.8.46.   Graversen, S.E. and Peskir, G. (2000) Maximal inequalities for the Ornstein-Uhlenbeck process. *Proceedings of the American Mathematical Society* 128(10), 3035-3041.

4.8.47.   Hamm, C.E., Merkel, R., Springer, O., Jurkojc, P., Maier, C., Prechtel, K., Smetacek, V. (2003) Architecture and material properties of diatom shells provide effective mechanical protection. *Nature* 421(6925), 841-843.

4.8.48.   Hansen, T.F. (1997) Stabilizing selection and the comparative analysis of adaptation. *Evolution* 51, 1341-1351.

4.8.49.   Harmon, L.J., Losos, J.B., Davies, T., Gillespie, R.G, Gittleman, J.L., Bryan Jennings, W., *et al.* (2010) Early burst of body size and shape evolution are rare in comparative data. *Evolution* 64, 2385-2396.

4.8.50. Harwood, D.M. and Nikolaev, V.A. (1995) Cretaceous diatoms; morphology, taxonomy, biostratigraphy. In: *Siliceous microfossils*. Paleontological Society Short Courses in Paleontology, Vol 8, Blome, D. (ed). Paleontology Society, University of Tennessee, Knoxville: 81–106.

4.8.51. Henley, S.F., Cavan, E.L., Fawcett, S.E., Kerr, R., Monteiro, T., Sherrell, R.M., Bowie, A.R., Boyd, P.W., Barnes, D.K.A., Schloss, I.R., Marshall, T., Flynn, R., Smith, S. (2020) Changing biogeochemistry of the Southern Ocean and its ecosystem implications. *Frontiers in Marine Science* 7:581.

4.8.52. Herrmann, S. and Massin, N. (2019) Exit problem for Ornstein-Uhlenbeck processes: a random walk approach. https://hal.archives-ouvertes.fr/hal-02143409v2

4.8.53. Hillebrand, H., Dürselen, C.-D., Kirschtel, D., Pollingher, U., Zohary, T. (1999) Biovolume calculation for pelagic and benthic microalgae. *J. Phycol.* 35, 403-424.

4.8.54. Hunt, G. (2007) The relative importance of directional change, random walks, and stasis in the evolution of fossil lineages. *PNAS* 104(47), 18404-18408.

4.8.55. Ingram, T. (2011) Speciation along a depth gradient in a marine adaptive radiation. *Proc. R. Soc. B* 278, 613-618.

4.8.56. Jacobsen, M. (1996) Laplace and the origin of the Ornstein-Uhlenbeck process. *Bernoulli* 2(3), 271-286.

4.8.57. Kaczmarska, I. and Ehrman, J.M. (2015) High colonization and propagule pressure by ship ballast as a vector for the diatom genus *Pseudo-nitzschia. Management of Biological Invasions* 6(1), 31-43.

4.8.58. Kang, S.-H., Kang, J.-S., Lee, S., Chung, K.H., Kim, D., Park, M.G. (2001) Antarctic phytoplankton assemblages in the marginal ice zone of the northwestern Weddell Sea. *Journal of Plankton Research* 23(4), 333-352.

4.8.59. Kim, J.H., Ajani, P., Murray, S.A., Kim, J.-H., Lim, H.C., Teng, S.T., Lim, P.T., Han, M.-S., Park, B.S. (2020) Sexual reproduction and genetic polymorphism within the cosmopolitan marine diatom *Pseudonitzschia pungens. Scientific Reports* 10: 10653.

4.8.60. Kloster, M., Kauer, G., Esper, O., Fuchs, N., Beszteri, B. (2018) Morphometry of the diatom *Fragilariopsis kerguelensis* from Southern Ocean sediment: high-throughput measurements show second morphotype occurring during glacials. *Marine Micropaleontology* 143, 70-79.

4.8.61. Kloster, M., Rigual-Hernández, A.S., Armand, L.K., Kauer, G., Trull, T.W., Beszteri, B. (2019) Temporal changes in size distributions of the Southern Ocean diatom *Fragilariopsis kerguelensis* through high-throughput microscopy of sediment trap samples. *Diatom Research* 34(3), 133-147.

4.8.62. Kociolek, J.P., Williams, D.M., Stepanek, J., Liu, Q., Liu, Y., You, Q., Karthick, B., Kulikovskiy, M. (2019) Rampant homoplasy and adaptive radiation in pennate diatoms. *Plant Ecology and Evolution* 152(2), 131-141.

4.8.63. Kříž, R. and Kratochvil, Š. (2014) Analyses of the chaotic behavior of the electricity price series. In: *ISCS 2013: Interdisciplinary Symposium on Complex Systems, Emergence, Complexity and Computation 8*, A. Sanayei *et al.* (eds.), Springer-Verlag, Berlin Heidelberg: 215-226.

4.8.64. Kuwata, A. and Jewson, D.H. (2015) Ecology and evolution of marine diatoms and Parmales. In: *Marine Protists*, Ohtsuka, S., *et al.* (eds.), Springer, Japan: 251-275.

4.8.65. Lafond, A., Leblanc, K., Legras, J., Cornet, V., Quéguiner, B. (2020) The structure of diatom communities constrains biogeochemical properties in surface waters of the Southern Ocean (Kerguelen Plateau). *Journal of Marine Systems* 212, 103458.

4.8.66. Lande, R. (1976) Natural selection and random genetic drift in phenotypic evolution. *Evolution* 30, 314-334.

4.8.67.　Lannuzel, D., Vancoppenolle, M., van der Merwe, P., de Jong, J., Meiners, K.M., Grotti, M., Nishioka, J., Schoemann, V. (2016) Iron in sea ice: review and new insights. *Elementa: Science of the Anthropocene* 4:000130.

4.8.68.　Lannuzel, D., Bowie, A.R., Remenyi, T., Lam, P., Townsend, A., Ibisanmi, E., Butler, E., Wagener, T., Schoemann, V. (2011) Distributions of dissolved and particulate iron in the sub-Antarctic and Polar Frontal Southern Ocean (Australian sector). *Deep-Sea Research II* 58, 2094-2112.

4.8.69.　Lasbleiz, M., Leblanc, K., Armand, L.K., Christaki, U., Georges, C., Obernosterer, I., Quéguiner, B. (2016) Composition of diatom communities and their contribution to plankton biomass in the naturally iron-fertilized region of Kerguelen in the Southern Ocean. *FEMS Microbiology Ecology* 92(11), fiw171.

4.8.70.　Lazarus, D., Barron, J., Renaudie, J., Diver, P., Türke, A. (2014) Cenozoic planktonic marine diatom diversity and correlation to climate change. *PLoS ONE* 9(1), e84857.

4.8.71.　Lehoczky, J.P. (2001) Distributions, statistical: special and continuous. In *International Encyclopedia of the Social & Behavioral Sciences, Reference Work*, Smelser, N.J. and Baltes, P.B. (eds.), Pergamon, Elsevier, Oxford, UK: 3787-3793.

4.8.72.　Lelong, A., Hégaret, H., Soudant, P., Bates, S.S. (2012) *Pseudo-nitzschia* (Bacillariophyceae) species, domoic acid and amnesic shellfish poisoning: revisiting previous paradigms. *Phycologia* 51(2), 168-216.

4.8.73.　Lenz, M. and Rhodin, M. (2011) Reliability calculations for complex systems. Masters Thesis, Linköping University, LiTH-ISY-EX-11(4441-SE), Sweden.

4.8.74.　Leung, K.-M., Elashoff, R.M., Afifi, A.A. (1997) Censoring issues in survival analysis. *Annu. Rev. Public Health* 18, 83-104.

4.8.75.　Leventer, A. (1991) Sediment trap diatom assemblages from the northern Antarctic Peninsula region. *Deep Sea Research* 38(8/9), 1127-1143.

4.8.76.　Lewitus, E., Bittner, L., Malviya, S., Bowler, C., Morlon, H. (2018) Clade-specific diversification dynamics of marine diatoms since the Jurassic. *Nature Ecology & Evolution* 2, 1715-1723.

4.8.77.　Lim, S.M., Moreau, Vancoppenolle, M., Deman, F., Roukaerts, A., Meiners, K.M., Janssens, J., Lannuzel, D. (2019) Field observations and physical-biogeochemical modeling suggest low silicon affinity for Antarctic fast ice diatoms. *Journal of Geophysical Research: Oceans* 124, 7837-7853.

4.8.78.　Lipton, A. and Kaushansky, V. (2018) On the first hitting time density of an Ornstein-Uhlenbeck process. arxiv.org:1810.02390v2 [q-fin.CP]

4.8.79.　Losos, J.B. and Miles, D.B. (2002) Testing the hypothesis that a clade has adaptively radiated: iguanid lizard clades as a case study. *The American Naturalist* 160(2), 147-157.

4.8.80.　Loughney, K.M., Hren, M.T., Smith, S.Y., Pappas, J.L. (2019) Vegetation and habitat change in southern California through the Middle Miocene Climatic Optimum: Paleoenvironmental records from the Barstow Formation, Mojave Desert, USA. *GSA Bulletin* 132(1-2), 113-129.

4.8.81.　Louw, D.C., Doucette, G.J., Lundholm, N. (2018) Morphology and toxicity of *Pseudo-nitzschia* species in the northern Benguela upwelling system. *Harmful Algae* 75, 118-128.

4.8.82.　Lundholm, N., Daugbjerg, N., Moestrup, Ø. (2002) Phylogeny of the Bacillariaceae with emphasis on the genus *Pseudo-nitzschia* (Bacillariophyceae) based on partial LSU rDNA. *Eur. J. Phycol.* 37, 115-134.

4.8.83.　Malviya, S., Scalco, E., Audic, S., Vincent, F., Veluchamy, A., Poulain, J., Wincker, P., Iudicone, D., deVargas, C., Bittner, L., Zingone, A., Bowler, C. (2016) Insights into global diatom distribution and diversity in the world's ocean. *PNAS* 113(11), E1516-E1525.

4.8.84. Marchetti, A., and Harrison, P.J. (2007) Coupled changes in the cell morphology and the elemental (C, N, and Si) composition of the pennate diatom *Pseudo-nitzschia* due to iron deficiency. *Limnol. Oceanogr.* 52(5), 2270-2284.

4.8.85. Marchetti, A., Maldonado, M.T., Lane, E.S., Harrison, P.J. (2006) Iron requirements of the pennate diatom *Pseudo-nitzschia*: comparison of oceanic (high-nitrate, low-chlorophyll waters) and coastal species. *Limnol. Oceangr.* 51(5), 2092-2101.

4.8.86. Mirrahimi, S. (2013) Migration and adaptation of a population between patches. *Discrete and Continuous Dynamical Systems-Series B (DCDS_B)* 18(3), 753-768.

4.8.87. Mirrahimi, S. (2017) A Hamilton-Jacobi approach to characterize the evolutionary equilibria in heterogeneous environments. *Mathematical Models and Methods in Applied Sciences* 27(13), 2425-2460.

4.8.88. Mock, T., Otillar, R.P., Strauss, J., McMullan, M., Paajanen, P., Schmutz, J., Salamov, A., et al. (2017) Evolutionary genomics of the cold-adapted diatom *Fragilariopsis cylindrus*. *Nature* 541, 20803.

4.8.89. Moore, B.R., Höhna, S., May, M.R., Rannala, B., Huelsenbeck, J.P. (2015) Critically evaluating the theory and performance of Bayesian analysis of macroevolutionary mixtures. *PNAS* 113(34), 9569-9574.

4.8.90. Mosseri, J., Quéguiner, B., Armand, L., Cornet-Barthaux, V. (2008) Impact of iron on silicon utilization by diatoms in the Southern Ocean: a case study of Si/N cycle decoupling in a naturally iron-enriched area. *Deep-Sea Research II* 55, 801-819.

4.8.91. Mudelsee, M., Bickert, T., Lear, C.H., Lohmann, G. (2014) Cenozoic climate changes: a review based on time series analysis of marine benthic $\delta18O$ records. *Rev. Geophys.* 52, 333-374.

4.8.92. Nair, A., Mohan, R., Manoj, M.C., Thamban, M. (2015) Glacial-interglacial variability in diatom abundance and valve size: implications for Southern Ocean paleoceanography. *Paleoceanography* 30, 1245-1260.

4.8.93. Nakov, T., Theriot, E.C., Alverson, A.J. (2014) Using phylogeny to model cell size evolution in marine and freshwater diatoms. *Limnol. Oceangr.* 59(1), 79-86.

4.8.94. Nakov, T., Beaulieu, J.M., Alverson, A.J. (2018). Accelerated diversification is related to life history and locomotion in a hyperdiverse lineage of microbial eukaryotes (Diatoms, Bacillariophyta) *New Phytologist* 219, 462-473.

4.8.95. Neal, R.M. (2003) Slice sampling. *The Annals of Statistics* 31(3), 705-767.

4.8.96. Nilsson, P. and Ames, A.D. (2020) Lyapunov-like conditions for tight exit probability bounds through comparison theorems for SDEs. *2020 American Control Conference (ACC), Denver, CO, USA, 2020*: 5175-5181.

4.8.97. Oreshkina, T. and Radionova, E.P. (2014) Diatom record of the Paleocene-Eocene Thermal Maximum in marine paleobasins of central Russia, Transuralia and adjacent regions. *Nova Hedwigia, Beiheft* 143, 307-336.

4.8.98. Pagel, J., Thrurnicht, M., Bond, W.J., Kraaij, T., Nottebrock, H., Schutte-Vlok, A., Tonnabel, J., Esler, K.J., Schurr, F.M. (2020) Mismatches between demographic niches and geographic distributions are strongest in poorly dispersed and highly persistent plant species. *PNAS* 117(7), 3663-3669.

4.8.99. Papaioannou, G.P., Dikaiakos, C., Dramountanis, A., Georgiadis, D.S., Papaioannou, P.G. (2017) Using nonlinear stochastic and deterministic (chaotic tools) to test the EMH of two electricity markets; the case of Italy and Greece. arXiv:1711.10552 [q-fin.ST]

4.8.100. Pappas, J.L. (2016) Multivariate complexity analysis of 3D surface form and function of centric diatoms at the Eocene-Oligocene transition. *Marine Micropaleontology* 122, 67-86.

4.8.101. Pappas, J.L. (2021) Quantified ensemble 3D surface features modeled as a window on centric diatom valve morphogenesis. In: *Diatom Morphogenesis [DIMO, Volume in the series: Diatoms: Biology & Applications, series editors: Richard Gordon & Joseph Seckbach].* V. Annenkov, J. Seckbach and R. Gordon, (eds.) Wiley-Scrivener, Beverly, MA, USA: 158-193.

4.8.102. Pappas, J.L. (2023) Diatom bauplan as modified 2D valve face shapes of a 3D capped cylinder and valve shape distribution. In: *Mathematical Macroevolution in Diatom Research [MMDR, Volume in the series: Diatoms: Biology & Applications, series editors: Richard Gordon & Joseph Seckbach].* Wiley-Scrivener, Beverly, MA, USA: this volume.

4.8.103. Pappas, J.L. (2023) Comparative surface analysis, the image Jacobian and tracking changes in diatom valve face morphology. In: *Mathematical Macroevolution in Diatom Research [MMDR, Volume in the series: Diatoms: Biology & Applications, series editors: Richard Gordon & Joseph Seckbach].* Wiley-Scrivener, Beverly, MA, USA: this volume.

4.8.104. Pappas, J.L. and Gordon, R. (2021) Buckling: a geometric and biophysical multiscale feature of centric diatom valve morphogenesis. In: *Diatom Morphogenesis [DIMO, Volume in the series: Diatoms: Biology & Applications, series editors: Richard Gordon & Joseph Seckbach].* V. Annenkov, J. Seckbach and R. Gordon, (eds.) Wiley-Scrivener, Beverly, MA, USA: 195-230.

4.8.105. Pappas, J.L., Kociolek, J.P., Stoermer, E.F. (2014) Quantitative morphometric methods in diatom research. In: Nina Strelnikova Festschrift. J.P. Kociolek, M. Kulivoskiy, J. Witkowski, and D.M. Harwood, D.M. (eds.), *Nova Hedwigia, Beihefte* 143, 281-306.

4.8.106. Paris, S., Hasinoff, S.W., Kautz, J. (2011) Local Laplacian filters: edge-aware image processing with a Laplacian pyramid. *ACM Trans. Graph. (Proc. SIGGRAPH)* 30(4), https://doi.org/10.1145/2010324.1964963.

4.8.107. Parks, M., Nakov, T., Ruck, E., Wickett, N.J., Alverson, A.J. (2018) Phylogenomics reveals an extensive history of genome duplication in diatoms (Bacillariophyta). *American Journal of Botany* 105(3), 330-347.

4.8.108. Pausch, F., Bischof, K., Trimborn, S. (2019) Iron and manganese co-limit growth of the Southern Ocean diatom *Chaetoceros debilis. PLoS ONE* 14(9), e0221959.

4.8.109. Peng, S. (1999) Open problems on backward stochastic differential equations. In: *Control of Distributed Parameter and Stochastic Systems – Proceedings of the IFIP WG 7.2 International Conference, June 19-22, 1998, Hangzhou, China,* Chen, S., Li, X., Yong, J., Zhou, X.Y. (eds.), Springer Science + Business Media, LLC, Berlin, Germany: 265-273.

4.8.110. Perthame, B. and Souganidis, P.E. (2016) Rare mutations limit of a steady state dispersal evolution model. *Mathematical Modelling of Natural Phenomena* 11(4), 154-166.

4.8.111. Petrou, K. and Ralph, P.J. (2011) Photosynthesis and net primary productivity in three Antarctic diatoms: possible significance for their distribution in the Antarctic marine ecosystem. *Marine Ecology Progress Series* 437, 27-40.

4.8.112. Petrou, K., Baker, K.G., Nielsen, D.A., Hancock, A.M., Schulz, K.G., Davidson, A.T. (2019) Acidification diminishes diatom silica production in the Southern Ocean. *Nature Climate Change* 9, 781-786.

4.8.113. Pincheira-Donoso, D., Harvey, L.P., Ruta, M. (2015) What defines an adaptive radiation? Macroevolutionary diversification dynamics of an exceptionally species-rich continental lizard radiation. *BMC Evolutionary Biology* 15:153.

4.8.114. Prajna, S., Jadbabaie, A., Pappas, G.J. (2007) A framework for worst-case and stochastic safety verification using barrier certificates. *IEEE Transactions on Automatic Control* 52(8), 1415-1428.

4.8.115. Qiu, J. (2018) Viscosity solutions of stochastic Hamilton-Jacobi-Bellman equations. *SIAM Journal on Control and Optimization* 56(5), 3708-3730.

4.8.116. Ramaiah, N., Jain, A., Meena, R.M., Naik, R.K., Verma, R., Bhat, M., Mesquita, A., Nadkarni, A., D'Souza, S.E., Ahmed, T., Bandekar, M., Gomes, J. (2015) Response of bacteria and phytoplankton from a subtropical front location Southern Ocean to micro-nutrient amendments *ex-situ*. *Deep Sea Research II* 118, 209-220.

4.8.117. Rego-Costa, A., Débarre, F., Chevin, L.-M. (2018) Chaos and the (un)predictability of evolution in a changing environment. *Evolution* 72(2), 375-385.

4.8.118. Rembauville, M., Manno, C., Tarling, G.A., Blain, S., Salter, I. (2016) Strong contribution of diatom resting spores to deep-sea carbon transfer in naturally iron-fertilized waters downstream of South Georgia. *Deep-Sea Research I* 115, 22-35.

4.8.119. Renaudie, J. (2016) Quantifying the Cenozoic marine diatom deposition history: links to the C and Si cycles. Biogeosciences 13, 6003-6014.

4.8.120. Round, F.E., Crawford, R.M., Mann, D.G. (1990) *The Diatoms – Biology and Morphology of the Genera*, Cambridge University Press, Cambridge, United Kingdom.

4.8.121. Rouzine, I.M., Rodrigo, A., Coffin, J.M. (2001) Transition between stochastic evolution and deterministic evolution in the presence of selection: general theory and application to virology. *Microbiology and Molecular Biology Reviews* 65(1), 151-185.

4.8.122. Sackett, O., Petrou, K., Reedy, B., De Grazia, A., Hill, R., Doblin, M., Beardall, J., Ralph, P., Heraud, P. (2013) Phenotypic plasticity of Southern Ocean diatoms: key to success in the sea ice habitat? *PLoS ONE* 8(11), e81185.

4.8.123. Sackett, O., Armand, L., Beardall, J., Hill, R., Doblin, M., Conelly, C., Howes, J., Stuart, B., Ralph, P., Heraud, P. (2014) Taxon-specific responses of Southern Ocean diatoms to Fe enrichment revealed by synchrotron radiation FTIR microspectroscopy. *Biogeosciences* 11, 5795-5808.

4.8.124. Sarthou, G., Timmermans, K.R., Blain, S., Tréguer, P. (2005) Growth physiology and fate of diatoms in the ocean: a review. *Journal of Sea Research* 53, 25-42.

4.8.125. Schenk, J.J. (2021) The next generation of adaptive radiation studies in plants. *Int. J. Plant Sci.* 182(4), 245-262.

4.8.126. Schluter, D. (2000) *The Ecology of Adaptive Radiation*, Oxford University Press, Oxford, United Kingdom.

4.8.127. Schluter, D. (2001) Ecology and the origin of species. *TRENDS in Ecology & Evolution* 16(7), 372-380.

4.8.128. Shiozaki, T., Fujiwara, A., Inomura, K., Hirose, Y., Hashihama, F., Harada, N. (2020) Biological nitrogen fixation detected under Antarctic sea ice. *Nature Geoscience* 13, 729-732.

4.8.129. Simpson, G.G. (1944) *Tempo and Mode in Evolution*, Columbia University Press, New York.

4.8.130. Simpson, G.G. (1953) *The Major Features of Evolution*, Columbia University Press, New York.

4.8.131. Sjunneskog, C., Riesselman, C.R., Winter, D., Scherer, R. (2012) *Fragilariopsis* diatom evolution in Pliocene and Pleistocene Antarctic shelf sediments. *Micropaleontology* 58(3), 273-289.

4.8.132. Soininen, J. and Teittinen, A. (2019) Fifteen important questions in the spatial ecology of diatoms. *Freshwater Biology* 64, 2071-2083.

4.8.133. Soppa, M.A., Völker, C., Bracher, A. (2016) Diatom phenology in the Southern Ocean: mean patterns, tremds and the role of climate oscillations. *Remote Sensing* 8, 420.

4.8.134. Sorhannus, U. (2007) A nuclear-encoded small-subunit ribosomal RNA timescale for diatom evolution. *Marine Micropaleontology* 65, 1-12.

4.8.135. Stadler, T. and Bokma, F. (2013) Estimating speciation and extinction rates for phylogenies of higher taxa. *Syst. Biol.* 62(2), 220-230.

4.8.136. Stroud, J.T. and Losos, J.B. (2016) Ecological opportunity and adaptive radiation. Ann. *Rev. Ecol. Evol. Syst.* 47, 507-532.

4.8.137. Suto, I. (2006) The explosive diversification of the diatom genus *Chaetoceros* across the Eocene/Oligocene and Oligocene/Miocene boundaries in the Norwegian Sea. *Marine Micropaleontology* 58, 259-269.

4.8.138. Temam, R. (1988) *Infinite Dimensional Dynamical Systems in Mechanics and Physics*, Springer-Verlag, Cambridge, United Kingdom.

4.8.139. Thiffeault, J.-L. (2018) Minicourse: Exit time problems for swimming microorganisms. Accessed on 2 June 2018. https://people.math.wisc.edu/~jeanluc/lecturing/exit-minicourse/exit-minicourse.pdf

4.8.140. Thygesen, U.H. (1997) A survey of Lyapunov techniques for stochastic differential equations. IMM Technical Report/Dept. Math. Modeling, Tech. Uni. Denmark 18-1997, 1-26.

4.8.141. Timmermans, K.R., Gerringa, L.J.A., debar, J.J.W., van der Wagt, B., Veldhuis, M.J.W., de Jong, J.T.M., *et al.* (2001) Growth rates of large and small Southern Ocean diatoms in relation to availability of iron in natural seawater. *Limnol. Oceanogr.* 46, 260-266.

4.8.142. Tobias, J.A., Cornwallis, C.K., Derryberry, E.P., Claramunt, S., Brumfield, R.T., Seddon, N. (2013) Species coexistence and the dynamics of phenotypic evolution in adaptive radiation. *Nature* 506, 359-363.

4.8.143. Trainer, V.L., Bates, S.S., Lundholm, N., Thessen, A.E., Cochlan, W.P., Adams, N.G., Trick, C.G. (2012) *Pseudo-nitzschia* physiological ecology, phylogeny, toxicity, monitoring and impacts on ecosystem health. *Harmful Algae* 14, 271-300.

4.8.144. Vega, C.A.M. (2018) Calibration of the exponential Ornstein-Uhlenbeck process when spot prices are visible through the maximum log-likelihood method. Example with gold prices. *Advances in Difference Equations* 2018:269.

4.8.145. van de Schoot, R., Depaoli, S., King, R., Kramer, B., Märtens, K., Tadesse, M.G., Vannucci, M., Gelman, A., Veen D., Willemsen, J., Yau, C. (2021) Bayesian statistics and modelling. *Nature Reviews Methods Primers* 1, 1 (2021).

4.8.146. Waite, A., Fisher, A., Thompson, P.A. (1997) Sinking rate vs volume relationships illuminate sinking control mechanisms in marine diatoms. *Mar. Ecol. Progr. Ser.* 157, 97-108.

4.8.147. Wei, P. and Duan, J. (2021) Lyapunov exponents for Hamiltonian systems under small Lévy perturbations. arXiv:2011.10735v2 [math.DS]

4.8.148. Wellborn, G.A. and Langerhans, R.B. (2015) Ecological opportunity and the adaptive diversification of lineages. *Ecology and Evolution* 5(1), 176-195.

4.8.149. Yool, A. and Tyrrell, T. (2003) Role of diatoms in regulating the ocean's silicon cycle. *Global Biogeochemical Cycles* 17(4), 1103-1123.

4.8.150. Zachos, J.C., Dickens, G.R., Zeebe, R.E. (2008) An early Cenozoic perspective on greenhouse warming and carbon-cycle dynamics. *Nature* 451, 279-283.

4.8.151. Zeng, C., Yang, Q., Chen, Y.Q. (2014) Lyapunov techniques for stochastic differential equations driven by fractional Brownian motion. *Abstr. Appl. Anal.* S135, 1-9, doi: 10.1155/2014/292653.

Cenozoic Diatom Origination and Extinction and Influences on Diversity

Abstract

Diatom origination, extinction and diversity have occurred throughout the Cenozoic. Using relative maxima ages from origination and extinction rate percent peak heights, cumulative frequency curves were established and used to determine the relative influence of origination and extinction on diatom diversity. Cumulative frequency curves were piecewise continuous, and the least-squares fitted curves as *n*th-order polynomials were converted to piecewise functions via Heaviside functions. Using this result and Laplace transforms of Heaviside functions, switching between diatom origination and extinction during the Cenozoic was analyzed. Switching order was used to determine influence on diatom diversity as well. The Poisson and Lyapunov Poisson processes were used to analyze probabilistic origination and extinction in terms of stability in the switching of these states throughout the Cenozoic. Combined origination-extinction exhibited instability after the MMCO and at the PETM, while diversity instability was exhibited at the EOT. Overall, origination was probabilistically more influential on diversity than was extinction for diatoms during the Cenozoic.

Keywords: Diversity, origination, extinction, hazard function, survival function, Laplace transform, Heaviside function, stability

5.1 Introduction

Diversity in biological organisms has been said to be a function of origination and extinction [5.7.2., 5.7.5.]. Diversity is not equal to origination, and loss of diversity is not equal to extinction. Origination is the turnover rate with respect to first appearances or per capita instantaneous rates [5.7.3.], while extinction is the turnover rate with respect to last appearances or per capita instantaneous rates [5.7.3.]. Diversity takes into account species richness and evenness (or relative abundance) and may be represented by various indexes, with or without constraints [5.7.40.]. Origination and extinction rates are exponential [5.7.2., 5.7.3.], while diversity rates may be characterized by power law or exponential functions [5.7.2.]. Other distributions may be used to characterize origination, extinction or diversity over time.

Phanerozoic diversity fluctuates with regard to origination and extinction [5.7.64.]. Diversity loss is reflected in extinction rates when such rates are cumulative [5.7.2.]. Origination rates are out of step with diversity gains unless diversity prior to extinction is taken into account [5.7.2.]. Diversity loss that exceeds extinction may be greater than origination, and diversity gain may exceed extinction when origination does so as well [5.7.5.].

Diversity loss as an event is important and best understood in the context of geologic time [5.7.5.] as is the case with origination and extinction.

Proportional influence of origination over extinction on diversity loss may have occurred at particular times of diversity depletion [5.7.5.]. Proportionally, the magnitude of genus extinction may correspond to that for species [5.7.3., 5.7.22.], and genus origination and extinction may be used to understand diversity [5.7.3., 5.7.5.]. Changes in lineage origination, extinction and diversification events are analyzable at the genus as well as species level (e.g., [5.7.5., 5.7.10.]).

Origination and extinction rates are variable but may correlate with each other [5.7.2.]. High origination rates occur after large extinction events, and the quickest recovery occurs after the largest extinction events [5.7.2.]. Continuous origination and extinction time rates may be expressed as separate processes that may or may not overlap at given age intervals from at least the Cambrian to the Neogene to the present [5.7.2.]. Over the Cenozoic, origination and extinction rates have generally declined [5.7.2., 5.7.75.].

Diversification rates have declined as well during the Cenozoic (e.g., [5.7.2.]), although for diatoms, diversity has increased (e.g., [5.7.37.]). Of origination or extinction, one or the other may have a more pronounced influence on diversity at a given time, and this influence may be tempered by environmental conditions. For example, sea level and temperature are major definers of time interval events throughout the Cenozoic, and are indicators of cumulative effects on the biota, especially concerning diatoms and their origination, extinction and diversity.

5.1.1   Cenozoic Diatoms and Environmental Conditions

The Cretaceous-Paleogene mass extinction event (K-Pg) took its toll on diatoms as species numbers plummeted [5.7.69.]. During the Cenozoic, different environmental regimes such as the Paleocene-Eocene Thermal Maximum (PETM), Eocene-Oligocene Transition (EOT), Mid-Miocene Climatic Optimum (MMCO) along with the Mid-Miocene Climatic Transition (MMCT), and Pliocene-Pleistocene boundary extinction (Pli-Ple) are notable for their effects on diatom origination and extinction. Cenozoic diatoms experienced distinct changes in diversity with respect to the PETM [5.7.51.], EOT (e.g., [5.7.6., 5.7.52.]), MMCO and MMCT (e.g., [5.7.27., 5.7.56.]), and the Pli-Ple boundary extinction [5.7.37.].

Sea level and temperature have been influential in diatom diversity and may affect the timing of origination and extinction events (e.g., [5.7.56.]). For example, increases in temperature that occurred over a short time frame characterize the PETM [5.7.25., 5.7.70.], and diatoms experienced a decline in diversity from mid to high latitudes and high turnover rates in diversity at the regional level [5.7.51.]. At the PETM, sea level increased at the Antarctic margins potentially as a result of melting ice, and this may be the case globally with eustatic rise [5.7.71.].

Diatom assemblages changed as a result of the EOT as diversity increased during cooling conditions of the early Oligocene via proliferation in the high latitudes [5.7.35., 5.7.37., 5.7.55., 5.7.61.]. While some genera and species became extinct (e.g., [5.7.52.]), others proliferated, especially in the Southern Ocean by the late Oligocene [5.7.61.]. This proliferation occurred specifically in terms of relative abundances [5.7.61.], as the EOT saw one of the largest increases in diatom abundances during the Cenozoic [5.7.35., 5.7.61.]. Sea level fell at the EOT [5.7.25., 5.7.44., 5.7.45., 5.7.57.] as a result of the formation of Antarctic ice [5.7.28., 5.7.57.].

Modern diatom assemblages commenced in the early Miocene with marine taxa (e.g., [5.7.27., 5.7.66.], and by the mid to late Miocene, freshwater diatoms were evident [5.7.27.]. This change is reflected in additional silica availability from volcanism and seasonal environmental influences via monsoons, and because eustasy, tectonic uplift and mountain formation that characterize the mid-Miocene were particularly influential factors [5.7.27.]. Meridional temperature gradients induced diatom diversity changes during the MMCO as warmer conditions gave way to the MMCT [5.7.25., 5.7.37.]. Sea level increased at the MMCO when Antarctic ice receded, and this is in contrast to the MMCT when sea level decreased [5.7.23., 5.7.42.]. However, overall, fluctuations in sea level occurred throughout the Miocene [5.7.23., 5.7.25.].

During the Pli-Ple, diatom diversity increased as temperatures decreased [5.7.30., 5.7.31., 5.7.35., 5.7.37.]. However, the level of diversification was less than what it was during the Miocene, especially in regard to the MMCT [5.7.56.]. Diatom diversification varied with respect to endemism in that species richness declined in Antarctica with global cooling [5.7.56.]. Sea level changed dramatically at the end Pliocene with alternating increased glaciation and eustatic events (e.g., [5.7.54.]), with overall increasing sea level with respect to Antarctica [5.7.77.]. For marine and freshwater diatoms across lineages, speciation was greater than extinction, and net turnover was greater than net diversification [5.7.48.].

5.1.2   Diatom Diversity during the Cenozoic

Unlike vertebrates, for regional floras and faunas as well as some insect groups, the relation between diversity and extinction is not well known, and this is the case with unicells, including algae such as diatoms [5.7.68.]. Plankton suffered diversity loss during the Cenozoic as evidenced at events such as the EOT, and such losses account for a greater diversity loss than that for marine macroinvertebrates and continental shelf dwellers [5.7.37.]. Up until recently, the marine realm dictated the extent of extinction and its effect on diversity (e.g., [5.7.5.]). However, diversity loss of plankton in contrast to the benthos has been more pronounced with high turnover rates, reflecting changing environmental and ecological conditions [5.7.37.].

Currently, siliceous plankton, including diatoms, have the greatest diversity at high latitudes around the Arctic and Antarctic poles, while calcareous plankton diversity dominates the tropical and subtropical marine environments [5.7.37.]. Because diatoms are able to form cysts, this may have induced their capacity to withstand instability in the water column and short periods of global darkness. Icehouse conditions that prevail today may be a sustaining factor in diatom increasing abundance at high latitudes [5.7.29. 5.7.37.]. Diatoms are thought to have evolved during the Mesozoic, specifically during the Jurassic [5.7.37.]. Diatoms are most notably known from the mid-Cretaceous to the present as the dominant silica-secreting microorganism (e.g., [5.7.58.]) and are the most important primary producers of marine food webs (e.g., [5.7.26., 5.7.37.]), contributors to the carbon and silica cycles (e.g., [5.7.7.]), and are carbon sequesterers (e.g., [5.7.67.]).

The Cenozoic diatom record is replete, even though microfossil recovery indicates less than well-preserved specimens (e.g., [5.7.21.]). This record enables constructing origination, extinction and diversity continuous time rates (e.g., [5.7.37., 5.7.62.]. Having continuous time rates enables conducting analyses to understand the relation among origination, extinction and diversity for Cenozoic diatoms.

Diversity is often quantified with regard to species richness and evenness or other quantification methods that enable a semblance of representing the result of originations and extinctions over time (e.g., [5.7.2.]). Environmental conditions may be influential in diatom diversity. For example, diatom lineage diversification has been documented with respect to environmental events during the late Eocene [5.7.36.], and expansion of diatom diversity, especially to the Southern Ocean and the tropics is evident today [5.7.36., 5.7.39.]. Diatom origination and extinction influence diversity that is tempered by environmental conditions present during the Cenozoic.

5.1.3   Diversity as a Result of the Frequency of Origination and Extinction Events

As a given group of organisms diversify, their lineage changes shape over time (e.g., [5.7.49., 5.7.59., 5.7.60.]), and that shape change may be characterized as an accumulation of diversified organisms over time. Diversity may be represented as a cumulative function just as with origination and extinction. However, one caveat to consider is that there is not necessarily an equivalence between origination and extinction taxa within the same bounded time intervals in the Cenozoic. Overlapping time bins are one problem to consider when assessing diversity in terms of origination and extinction (e.g., [5.7.2., 5.7.10., 5.7.37.]). The difference between origination and extinction events influences diversity assessment at any given age. Changes in diversity with lineage sizes increasing or decreasing over time induce lineage shifts over time (e.g., [5.7.47.]). As a result, diversity may be obtained as a net value, and one way to describe this value is net success.

Diversification rate is usually calculated with regard to speciation and extinction (e.g., [5.7.18.]). Within a lineage, speciation, as the number of new species per time interval, may be deemed to be a time segment of origination compiled as genera (e.g., [5.7.18.]) or species (e.g., [5.7.75.]). That is, speciation may be represented by origination per time interval. New species origination rate is proportional to the average speciation rate [5.7.75.], and proportional originations for a given time interval may be the result of allopatric speciation [5.7.9.]. By characterizing diversification rate in terms of extinction and origination rather than speciation, origination may be used per time interval bin. As a result, evolutionary rates may be obtained as the net difference between origination and extinction, representing net diversification rate (e.g., [5.7.2., 5.7.10.]).

Different time bins characterizing net diversification rates may indicate different speciation and origination rates which may induce different diversity outcomes over time. New diatom genera have been introduced more readily in recent times, and this induces reconfiguring phylogenetic relations among lineages [5.7.32.] which may induce differing outcomes concerning diversification rates. Net diversification rates have been documented for diatom lineages across the Cenozoic, with increases across the PETM and EOT (e.g., [5.7.10., 5.7.11.]). In the late Eocene, some diatom lineage diversification rates increased as a result of sea level changes and silica availability while others did not [5.7.36.]. Origination rates decreased during the Oligocene while extinction remained constant, but diatom diversity decreased during this time (e.g., [5.7.10.]). With origination potentially occurring rapidly and extinction occurring in a more constant fashion [5.7.10.], origination may be an influential facet of diversity (e.g., [5.7.41.]).

5.2   Purposes of this Study

The relation among diatom origination, extinction and diversity during the Cenozoic may be ascertained by using published data in analyses. Origination and extinction rate percents are used to analyze their accumulated affect in a stepwise manner as probabilistic. Piecewise continuous functions are established for origination and extinction, and switching between origination and extinction is analyzed for Cenozoic diatoms. Diatom origination and extinction with regard to events K-Pg, PETM, EOT, MMCO, and Pli-Ple are assessed in this context. The cumulative effect of origination and extinction on diatom diversity may be characterizable as deterministic, stochastic or chaotic, which is testable. Accumulated diatom origination and extinction may be indicative of the influence these processes have had on Cenozoic diatom diversity.

5.3   Methods and Background

5.3.1   Reconstructed Diatom Origination, Extinction and Diversity during the Cenozoic

Cenozoic diatom origination and extinction rate percent curves were reconstructed by extracting arbitrary coordinate data from Figure 2 in Lowery *et al.* [5.7.37.] and plotted by matching each peak value with the age at which it occurred (Figure 5.1, top and middle curves). The age at which peak values occurred are relative maxima and used in further data analysis. To compare the relative maxima ages by inspection for origination and extinction rate percents during the Cenozoic, a mirror image of the extinction curve is superimposed on the origination curve and presented (Figure 5.1, bottom).

For the Cenozoic, diatom diversity has been compiled per Spencer-Cervato [5.7.72.], Lazarus *et al.* [5.7.35.] and updated by Renaudie *et al.* [5.7.62.]. A Cenozoic diatom diversity curve based on species richness was reconstructed by extracting arbitrary coordinate data from Figure 4 in Renaudie *et al.* [5.7.62.] (Figure 5.2). There is a section of the curve that is filled in with a dashed line in Renaudie *et al.'s* [5.7.62.] figure, but for the purposes of this study, a solid line replaces the dashed line. The time line used is from approximately 65 Ma to the present, covering the Cenozoic. The age at which each peak diversity value is identified is used in further data analysis. The K-Pg as well as the PETM, EOT, MMCO, and Pli-Ple are specifically of interest in assessing diatom origination, extinction and diversity during the Cenozoic.

Reconstructed origination and extinction curves for Cenozoic diatoms provide the comparative data for assessing the contribution of origination and extinction to diversity. Origination is an indication of initial successful survival, while extinction is the failure to survive. Diversity is the enduring indication of successful survival. Using survival analysis, an origination rate percent curve is amenable to being converted into a survival function, while the extinction rate percent curve may be converted into a hazard function. To

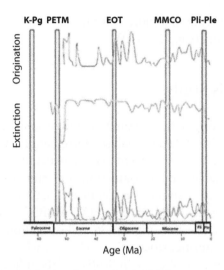

Figure 5.1 Reconstructed origination (top) and extinction (middle) rate curves for Cenozoic diatoms by extracting arbitrary coordinates from curves presented in [5.7.37.], Figure 2. The bottom superimposed curves are the reconstructed origination curve and a mirror-image of the extinction curve to the Age axis, showing the position of relative maxima (i.e., when peak values of origination and extinctions coincide, or not).

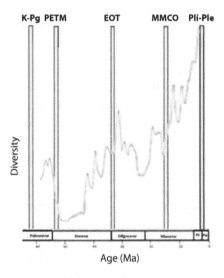

Figure 5.2 Reconstructed diversity curve for Cenozoic diatoms by extracting arbitrary coordinates from curve presented in [5.7.62.], Figure 4.

accomplish this, origination and extinction peak values were identified with respect to age (Ma), and the age at which these relative maxima occurred are used as input data to create survival and hazard functions. Diatom diversity with respect to origination and extinction as a time-dependent process may be compared with origination and extinction via survivorship curves (e.g., [5.7.10.]) or as survival distributions indicating the probability of origination, extinction or diversity throughout the Cenozoic.

5.3.2 Cumulative Functions and the Frequency of Cenozoic Origination and Extinction

To analyze the occurrence of originations and extinctions during the Cenozoic, probability density functions (PDFs) and cumulative density functions (CDFs) lie at the heart of the frequency of occurrence of relative maxima ages. To define these functions,

$$
\left.
\begin{aligned}
&\text{PDF}: f_{PDF}(t)\\
&\text{CDF}: F_{CDF}(t) = Prob(T \le t)
\end{aligned}
\right\}, F(0) = Prob(T = 0)
\tag{5.1}
$$

where t is time, and T, a continuous random variable, is the time of a given event [5.7.50.].

Origination as a survival function means that from origination as first appearance to some specified time t later, an organism's survival is ongoing [5.7.15.], and so origination like survival is representative of success. By surviving at time t, an organism's extended origination is captured by a survival function. An origination as a survival function is $S(t) = 1 - F_{CDF}(t)$ for $t > 0$ with

$$
F_{CDF}(t) = 1 - \exp\left[-\int_0^t h(t)dt\right] = 1 - \exp\left[\int_0^t \frac{d \ln S(t)}{dt}\right] = 1 - S(t)
\tag{5.2}
$$

where $h(t) = \dfrac{f_{PDF}(t)}{1 - F_{CDF}(t)} = \dfrac{f_{PDF}(t)}{S(t)}$ is a hazard function [5.7.15., 5.7.50.].

The cumulative survival function records the product of the probability of initial survival and the probability of surviving until a specified later time [5.7.73.]. For n-diatoms, origination times are initial survival times, and the accumulation of origination events during the Cenozoic is the cumulative origination function, $Surv$, of relative maxima ages. Using the Kaplan-Meier method (e.g., [5.7.63.]), the cumulative origination function is given as

$$
Surv(t_i) = S(t_{i-1})\left(1 - \frac{Orig_i}{Orig_{Total}}\right)
\tag{5.3}
$$

for $t > 0$ [5.7.15.] where the ith origination occurrence (i.e., a relative maximum age) is $Orig_i$ of the total originations (i.e., sum of the relative maxima ages), $Orig_{Total}$, from t_{i-1}, ..., t. The cumulative survival function is a PDF [5.7.63.] and is a step function [5.7.15.] so that the probability changes only at the time at which each origination occurs.

For the accumulation of diatom extinction events during the Cenozoic as relative maxima ages, the continuous cumulative hazard function is given as

$$
Haz(t) = \int_0^t h(t)dt = \int_0^t \frac{f_{PDF}(t)}{S(t)} dt = -\ln S(t)
\tag{5.4}
$$

where $S(t) = e^{-h(t)} = e^{Haz(t)}$ [5.7.50.] and $f_{PDF}(t) = h(t)\, e^{Haz(t)}$. Note that the hazard function is not a PDF, but it is conditioned on an actual event occurring and is an instantaneous failure rate to survive past a specified time t [5.7.15., 5.7.50.]. The discrete form of the hazard function induces a conditional probability. With regard to the survival function,

$$S_j(t) = (1-h_1(t))(1-h_2(t))\cdots(1-h_{j-1}(t)) = \prod_{1 \le j \le t}(1-h_j(t)) \tag{5.5}$$

[5.7.33., 5.7.65.], each jth failure to survive, $h_j, \ldots, h_{j-1}$, is a probabilistic stepwise event. The discrete hazard function is a bounded function so that

$$S_j(t) = \begin{cases} \displaystyle\prod_{t=0}^{j-1}(1-h(t)) & j \ge 1 \\ 1 & j = 0 \end{cases} \tag{5.6}$$

and the probability mass function (PMF) is

$$f_{PMF}(t) = h_j(t)\prod_{t=0}^{j-1}(1-h(t)) \tag{5.7}$$

(e.g., [5.7.46.]). For extinction, each jth event occurs as a probability independent of previous extinction events. From this,

$$h_j(t) = \ln(1 - h(t)) \tag{5.8}$$

and

$$h(t) = 1 - e^{-h_j(t)}. \tag{5.9}$$

Then, the cumulative extinction function becomes the discrete equation

$$Haz_j(t) = \sum_{j \le t}\ln(1-h_j(t)) \tag{5.10}$$

[5.7.33.] for the jth diatom extinction event recorded from relative maxima ages during the Cenozoic.

Cumulative origination and cumulative extinction functions as specified need to be altered with respect to the Cenozoic timeline used. The oldest age is at the start, while the youngest age is at the end, which is numerically contrary to the usual depiction of time in survival analysis. The adjustment is a reversal of time for the independent variable, and the relative maxima ages are recorded according to this order from the K-Pg to the present. Both the cumulative origination function and cumulative extinction function are used with

$t = t_{Reverse}$. However, in some analyses that follow, it suffices to leave the timeline from newest to oldest age.

5.3.3   Origination and Extinction: Heaviside Functions and Switching

Cumulative origination and extinction equations are indicative of frequencies of events. Origination and extinction events are indicated at particular ages throughout the Cenozoic timeline. These events are definable by piecewise intervals that form a continuous Cenozoic timeline. A piecewise continuous function has discontinuities at definable points, inducing a step function (e.g., [5.7.76.]).

For a given function, $f(t)$, the stepwise origination functions are

$$f(t) = \begin{cases} f_1(t) & 0 \leq t < c_1 \\ f_2(t) & c_1 \leq t < c_2 \\ f_2(t) & c_2 \leq t < c_3 \\ \vdots & \vdots \\ f_{n-1}(t) & c_{n-2} \leq t < c_{n-1} \\ f_n(t) & t \geq c_{n-1} \end{cases} \tag{5.11}$$

where c is a particular Cenozoic event, and a Heaviside function, H, is used to devise a schema so that

$$H(t)f(t) = \begin{cases} 0 & t < 0 \\ f(t) & t > 0 \end{cases} \tag{5.12}$$

meaning that $f(t)$ is switched on when $t > 0$ and switched off when $t < 0$ [5.7.43.]. To match relative maxima ages to Cenozoic events of the PETM, EOT, MMCO, and Pli-Ple, a shifted Heaviside function is used that is given as

$$1 - H(t - c) = \begin{cases} 1 - 0 = 1 & t < c \\ 1 - 1 = 0 & t \geq c \end{cases} \tag{5.13}$$

so that 1 signifies when the switch is off, and 0 signifies when the switch is on [5.7.17.]. In this way, 0 means that the relative maximum for a given age for either origination or extinction matches the age of the particular Cenozoic event.

To calculate f for the $f_1(t)$, ..., $f_n(t)$ origination terms with regard to the shifted Heaviside functions, the generalized equation for f is

$$f = H_0(f_1) + H_{c_1}(f_2 - f_1) + H_{c_2}(f_3 - f_2) + \cdots + H_{c_{n-1}}(f_n - f_{n-1}). \tag{5.14}$$

For extinction, a function $g(t)$ is similarly calculated. The number of Heaviside functions is equivalent to the number of switches, and switches as either on or off indicate whether origination or extinction relative maxima ages are more closely associated to an age of a particular Cenozoic event.

5.3.4   Origination and Extinction as a Sequence of Steps and Accumulated Switches during the Cenozoic

The total number and order of diatom origination and extinction relative maxima ages are identifiable and countable at the ages at which origination or extinction peak values occur in sequence from the K-Pg to the present during the Cenozoic. Comparison of relative maxima ages to the PETM, EOT, MMCO, and Pli-Ple events at their ages may be accomplished via characterizing changes from origination to extinction as changes in a sequence of steps during the Cenozoic. Changes in the steps as switching from origination to extinction may be analyzed with respect to diatom diversity.

Accumulating the switching intervals [5.7.78.] induces the overall timing relation between origination and extinction during the Cenozoic. Via the Heaviside functions, the origination or extinction switch count may be obtained at each Cenozoic event. For each event, $W_{Count} = W_{Switch} (1 - k)^n$ [5.7.78.], where W_{Count} is the total number of switches, W_{Switch} is a given switch expressed as the $f_1 (t), \cdots, f_5 (t)$ for origination or $g_1 (t), \cdots, g_5 (t)$ for extinction which are sums of Heaviside functions, k is the event step, and n is the time of switching for each event. Taylor expansion of the term $(1 - k)^n$ yields an approximation of $(1 - k)^n = 1 - nk$ [5.7.78.] so that

$$W_{Count} = W_{Switch} (1 - nk) \tag{5.15}$$

and this count is based on Heaviside functions. For $k = 1$ as the step, $W_{Count} = W_{Switch} (1 - n)$.

5.3.5   Piecewise Continuous Switching via the Laplace Transform of the Heaviside Functions

The Laplace transform of the shifted Heaviside functions enables recovering the sequence of switches as piecewise continuous (e.g., [5.7.76.]) because either origination or extinction at a given age is considered to be a discrete response during the Cenozoic. The Laplace transform, $\mathcal{L}$, of origination, $f(t)$ is

$$\mathcal{L}[f] \equiv \int_0^\infty e^{-st} f(t)dt \tag{5.16}$$

where $f(t) = 0$ for $t < 0$ for Laplace exponent, s, [5.7.74.] which is a complex-valued frequency parameter that is independent of t, and the Laplace transform for $f(t) = 1$ is $\frac{1}{s}$ [5.7.1., 5.7.76.].

The Laplace transform of the shifted Heaviside functions for origination is given in the form of

$$\mathcal{L}\{H_c(t)f(t-c)\} = \int_0^\infty e^{-st} f(t-c)dt = e^{-cs}F(s) \tag{5.17}$$

with $H_c(t) = \begin{cases} 0 & t < c \\ 1 & t \geq c \end{cases}$, and $F(s) = \mathcal{L}\{f(t)\}$ [5.7.17.]. The inverse Laplace transform is

$$\mathcal{L}^{-1}\{e^{-cs}F(s)\} = H_c(t)f(t-c) \tag{5.18}$$

where $f(t) = \mathcal{L}^{-1}\{F(s)\}$ with the switch being $H_c(t)f(t-c)$, and the Laplace transform as $F(s)$ [5.7.17.]. That is, for the shifted Heaviside functions, the Laplace transform is

$$\mathcal{L}[H] \equiv \int_0^\infty e^{-st} H(t)\, dt = \frac{e^{-cs}}{s} \tag{5.19}$$

[5.7.1., 5.7.76.]. Substituting $g(t)$ as the extinction function, the Laplace transform and the inverse Laplace transform are determined in a similar fashion. Because the Heaviside function is a step function, switching between origination and extinction at the PETM, EOT, MMCO, and Pli-Ple Cenozoic events is recovered as piecewise continuous via the Laplace transform.

5.3.6 Overlapping of Origination and Extinction: A Convolution Product

A convolution product indicates the overlap of origination and extinction. Define a switch from origination to extinction during the Cenozoic. For origination Heaviside functions, $f(t)$, and extinction Heaviside functions, $g(t)$, the convolution product [5.7.76.] is

$$f * g(t) = \begin{cases} \int_0^t f(t_q)g(t-t_q)dt_q & \text{when } t \geq 0 \\ 0 & \text{when } t \leq 0 \end{cases} \tag{5.20}$$

so that the Laplace transform is

$$\mathcal{L}[f * g] = \mathcal{L}[f]\,\mathcal{L}[g] \tag{5.21}$$

and

$$\mathcal{L}\left[\int_0^t f(t_q)dt_q\right] = \mathcal{L}[f * H] = \frac{1}{s}\mathcal{L}[f] \tag{5.22}$$

or

$$g * f(t) = \begin{cases} \int_0^t g(t_q)f(t-t_q)dt_q & \text{when } t \geq 0 \\ 0 & \text{when } t \leq 0 \end{cases} \tag{5.23}$$

and the Laplace transform is

$$\mathcal{L}\left[\int_0^t g(t_q)dt_q\right] = \mathcal{L}[g * H] = \frac{1}{s}\mathcal{L}[g] \tag{5.24}$$

The convolution product is the sum of switches during the Cenozoic for diatom origination and extinction.

5.3.7   Non-Overlapping Origination and Extinction: A Poisson Process

The exponential distribution figures prominently in origination and extinction cumulative frequency functions and is indicative of time between continuously occurring successive events as evidenced via solutions to the Laplace transform. The Laplace transform induces a probability measure via

$$\mathcal{L}\{f\}(s) = \mathbb{E}[e^{-sT}] = \left(\frac{b}{b+t}\right)^a \tag{5.25}$$

where $\mathbb{E}$ is the expected value for T, a random time variable that has a PDF, f, and for an exponential distribution, $a = 1$, and b is a scaling parameter [5.7.34.]. The PDF is obtained via the CDF, F_T [5.7.1.] which is given as

$$F_T(t) = \mathcal{L}^{-1}\left\{\frac{1}{s}[e^{-sT}]\right\}(t) = \mathcal{L}^{-1}\left\{\frac{1}{s}\mathcal{L}\{f\}(s)\right\}(t) = \int_{-\infty}^t f_T(s)ds. \tag{5.26}$$

The Laplace transform characteristic of a Poisson process provides probabilistic distribution of origination and extinction with respect to events K-Pg, PETM, EOT, MMCO, and Pli-Ple.

The number of relative maxima ages are discrete values occurring within non-overlapping bounded time intervals of the K-Pg and PETM, the PETM and EOT, the EOT and MMCO, and the MMCO and Pli-Ple. The PMF for real-valued numbers, $\mathbb{R}_T$, is defined as

$$f_{PMF}(t_T) = \begin{cases} \dfrac{e^{-\lambda}\lambda^{t_T}}{t_T!} & \text{for } t_T \in \mathbb{R}_T \\ 0 & \text{otherwise} \end{cases} \tag{5.27}$$

where parameter λ is the expected value of time t_T as a rate [5.7.50.]. The discrete values are characterizable by a Poisson process [5.7.16.] so that with respect to origination and extinction,

$$Poisson\{t_{Origination} > t_{Extinction}\} = e^{-\lambda t_T} \tag{5.28}$$

where λ is the expected value of time $t_T \geq 0$ as a rate, $t_{Origination}$ are the origination time intervals with respect to events K-Pg, PETM, EOT, MMCO, and Pli-Ple defined as 65 to 51 Ma, 51 to 34 Ma, 34 to 15 Ma, 15 to 3 Ma, and 3 to 0 Ma. For $t_{Extinction}$, this is the number of times in which extinctions occur within the origination time intervals. Origination is treated as interarrival time intervals, and extinction is treated as arrival times, according to a Poisson process.

5.3.8 Test of Switch Reversibility, Cenozoic Events and a Lyapunov Function

Switches from origination to extinction as piecewise values are probabilistic in a Poisson process context. Deterministic or stochastic characteristics of switches enable analysis of the reversibility of those switches. If the sequence of two switches in a row were either both on or off for origination or extinction, then either origination or extinction was considered to be non-reversible, and therefore deterministic. If the switches were singular, then either origination or extinction prevailed, but the switch was considered to be reversible and therefore, stochastic.

A Lyapunov function may be used to test for chaos. This function may take the form of a summation involving a logarithmic or quadratic function (e.g., [5.7.24.]). Lyapunov functions are indicative of the stability properties of a given process (e.g., [5.7.13.]), and as a result, may be used as indicators of chaos. Using a Lyapunov function, a test is devised for deterministic non-reversal switches from origination to extinction. That is, non-reversal deterministic switches are testable in terms of chaos.

Transformation of cumulative frequency functions to Poisson processes enables application of a Lyapunov function (e.g., [5.7.4.]) to characterize stability of origination and extinction during the Cenozoic. From Eq (5.28), λ becomes a Lyapunov function, $V(t)$, defined as

$$V(t) = \sum_i t_i (\ln t_i - \ln c_i - 1) + c_i \tag{5.29}$$

where c_i is stability with respect to a Lyapunov function [5.7.4.]. In this case, events K-Pg, PETM, EOT, MMCO, and Pli-Ple are c_i as stable events which are 65, 55, 34, 15, and 2.5 Ma, respectively. The new Poisson equation becomes

$$Poisson_V \{t_{V_{Origination}} > t_{V_{Extinction}}\} = e^{-Vt_V} \tag{5.30}$$

While Eq. (5.28) may cover stationarity with respect to stochasticity, Eq. (5.30) would cover the deterministic chaos of switches for origination and extinction in that the Lyapunov function is an indicator of stability.

For a Lyapunov function, interarrival and arrival times are identified as origination and extinction, respectively. Cenozoic intervals 65 to 55 Ma, 55 to 34 Ma, 34 to 15 Ma, 15 to 2.5 Ma, and 2.5 to 0 Ma are used with reference to the K-Pg, PETM, EOT, MMCO, and Pli-Ple. Non-reversal deterministic sequences are found for origination and extinction. The analysis is set up such that the deterministic component is being defined as interarrival time intervals in which origination switches, and the stochastic component is extinction as arrival times at the Cenozoic events. This set up ensures scaling of stationarity [5.7.4.]. With $V > 0$, interarrival times have the memoryless property as non-overlapping, independent variables, and like arrival times, interarrival times have the Markov property [5.7.38.].

The ages for non-reversal are recorded and compared to the lowest boundary value of each Cenozoic interval in order to be binned. Lyapunov rate values are calculated for each non-reversal according to Eq. (5.30), and the sum of the Lyapunov rate values for each Cenozoic interval are determined. From this, the probability that a deterministic non-reversal is chaotic is determined.

To merge the Poisson processes with respect to origination and extinction, the stationary distribution of the product function [5.7.4.] is

$$\pi(t) = \prod_{i=1}^{n} \frac{c_i^{t_i}}{t_i!} e^{-c_i} \tag{5.31}$$

for the state space of a stochastic process, $\mathbb{Z}$, for $t \in \mathbb{Z}_{\geq 0}^n$ [5.7.4.]. For a Lyapunov Poisson process, the stationary distribution [5.7.4.] becomes

$$\pi^V(t) = \prod_{i=1}^{n} \frac{(Vc_i)^{t_i}}{t_i!} e^{-Vc_i} \tag{5.32}$$

so that the Lyapunov rate function deterministically is related to the stochastic function. Inserting a Lyapunov function induces a state space of a chaotic process, and from this, a probabilistic interpretation of a Lyapunov function is used to determine chaos for deterministic origination and extinction.

5.3.9   Origination and Extinction: Relation to Diversity

Data for Cenozoic diatom diversity was extracted from Figure 5.2 as peak values for relative maxima and valley values for relative minima. Origination and extinction data were rate percents, and comparisons to each other could be made as a result of transforming the piecewise data to cumulative frequency. Likewise, diversity relative maxima and minima were used to obtain a cumulative frequency function, and this function forms the basis for comparison with a combination function of origination and extinction cumulative frequency.

Comparisons to be made were between combination of origination and extinction versus diversity with regard to piecewise continuity. Laplace transform of Heaviside functions was used as was Poisson process analysis and test of switchability to compare diversity to a combination of origination and extinction. During switching analysis, for two origination or extinction relative maxima ages in a row, diversity may be compared at the age interval covered. If diversity increases because of origination (two in a row) or extinction (two in a row), no reversal is present for origination or extinction, and either of these processes may be related to an increase in diversity. For a decrease in diversity, the same non-reversal may be present for originations or extinctions. No reversal may be construed as being in a deterministic state. If there is a switch from one to the other, i.e., a reversal, then either origination or extinction may coincide with but may not necessarily have anything to do with a diversity increase or decrease. Reversal could mean that a random state exists. If a deterministic state is present, deterministic chaos may be present.

5.4 Results

5.4.1 Cumulative Frequency of Cenozoic Diatom Origination and Extinction Events

Survival and hazard functions were used to depict cumulative frequencies for diatom origination and extinction during the Cenozoic. For origination relative maxima ages only, the survival function is a continuous function but may be depicted stepwise over the Cenozoic (Figure 5.3a). The survival function as piecewise continuous induces piecewise values and conditions (as intervals) (Table 5.1) for each step in the probability plot (Figure 5.3a). For extinction relative maxima ages only, the cumulative hazard function is continuous, but may be depicted stepwise over the Cenozoic (Figure 5.3b). The cumulative hazard function as piecewise continuous induces piecewise values and conditions (Table 5.2) for each step in the plot (Figure 5.3b). While the origination plot depicts probability, the extinction plot does not (Figure 5.3). Note that both plots give age from the present to the K-Pg.

All relative maxima ages for Cenozoic diatom origination and extinction induce continuous functions. Cumulative origination and cumulative extinction were obtained as least-squares fitted n^{th}-order polynomials (Table 5.3) and plotted (Figure 5.4). The cumulative origination curve is increasing from the K-Pg to the present (Figure 5.4 a) so that the higher the frequency of origination, the higher the probability of origination (Figure 5.4a). The cumulative extinction curve is decreasing from the K-Pg to the present (Figure 5.4b) so that the higher the frequency of extinction, the lower the probability of extinction (Figure 5.4b). The probability of origination exceeds the probability of extinction throughout the Cenozoic (Figure 5.4).

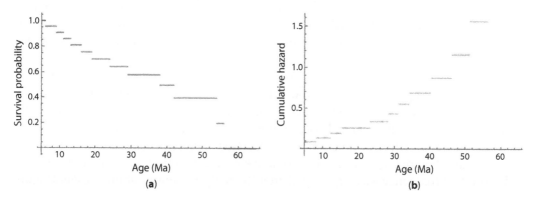

Figure 5.3 a, Origination relative maxima only plotted as a piecewise survival function. b, extinction relative maxima only plotted as a piecewise cumulative hazard function. See Tables 5.1 and 5.2 for piecewise values and conditions.

Table 5.1 Piecewise values and conditions obtained from expansion of the survival function calculated with Cenozoic diatom origination relative maxima.

Piecewise values	Piecewise conditions
0	$t = 56$
0.198212	$54 \leq t < 56$
0.396424	$42 \leq t < 54$
0.495529	$38 \leq t < 42$
0.578118	$29 \leq t < 38$
0.642353	$24 \leq t < 29$
0.700749	$19 \leq t < 24$
0.754652	$16 \leq t < 19$
0.808556	$13 \leq t < 16$
0.859091	$11 \leq t < 13$
0.906818	$9 \leq t < 11$
0.954545	$6 \leq t < 9$
1.	$t < 6$
0.	---

Identifier bars for the PETM, EOT, MMCO, and Pli-Ple events of the Cenozoic were superimposed on a plot of cumulative origination and cumulative extinction (Figure 5.5). From the plots, cumulative origination has a higher probability of being influential at the EOT, MMCO and Pli-Ple (Figure 5.5). Cumulative extinction has a higher probability of being influential at the PETM (Figure 5.5). Cumulative frequencies of origination probabilistically increased, while cumulative frequencies of extinction probabilistically declined (Figure 5.5). The probability of cumulative origination frequency equaled cumulative extinction frequency where the curves cross at 40 Ma (Figure 5.5).

Using a survival function, a combination of origination and extinction relative maxima ages produced piecewise values and conditions for the entire Cenozoic (Table 5.4). The probabilities were plotted versus age for the Cenozoic (Figure 5.6), and a piecewise continuous plot was depicted of a least-squares curve fitting of a nth-order polynomial for the combined cumulative origination and extinction frequency for the entire Cenozoic (Table 5.3 and Figure 5.7). Using the cumulative hazard function, combined origination and extinction frequency was highest just after the K-Pg and lowest at the present (Figure 5.8).

Table 5.2 Piecewise values and conditions obtained from expansion of the cumulative hazard function calculated with Cenozoic diatom extinction relative maxima.

Piecewise values	Piecewise conditions
0.210213	$51 \leq t < 56$
0.315319	$46 \leq t < 51$
0.420426	$40 \leq t < 46$
0.504511	$34 \leq t < 40$
0.576584	$31 \leq t < 34$
0.648657	$28 \leq t < 31$
0.713523	$23 \leq t < 28$
0.772983	$15 \leq t < 23$
0.824515	$12 \leq t < 15$
0.873016	$8 \leq t < 12$
0.916667	$5 \leq t < 8$
0.958333	$3 \leq t < 5$
1.	$t < 3$
Indeterminate	---

Table 5.3 Cumulative functions as least-squares fitted n^{th}-order polynomials for origination and extinction of Cenozoic diatoms.

Diatom evolution process	Least-squares fitted n^{th}-order polynomial	Coefficient of determination
Origination	$f(t)_{Origination} = -4 \times 10^{-6}t^3 + 0.0005t^2 - 0.0353t + 1.0699$	$R^2 = 0.9975$
Extinction	$g(t)_{Extinction} = 3 \times 10^{-9}t^6 - 5 \times 10^{-7}t^5 + 3 \times 10^{-5}t^4$ $- 0.0008t^3 + 0.0108t^2 + 0.0644t + 0.1709$	$R^2 = 0.9921$
Origination and Extinction	$fg(t)_{Origination\ and\ Extinction} = -0.0002t^2 + 0.002t + 0.09799$	$R^2 = 0.9972$
Diversity	$diversity(t) = -9 \times 10^{-10}t^6 + 1 \times 10^{-7}t^5 + 9 \times 10^{-6}t^4$ $+ 0.0002t^3 - 0.003t^2 + 0.0055t + 0.9874$	$R^2 = 0.9952$

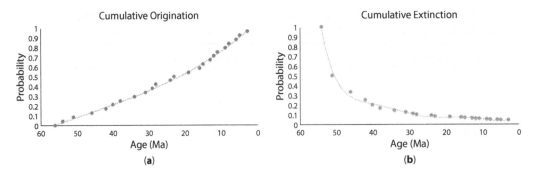

Figure 5.4 Polynomial least-squares fitted curves for: a, cumulative diatom origination (blue); b, cumulative diatom extinction (orange) for the Cenozoic. See Table 5.3 for least-squares equations and coefficients of determination.

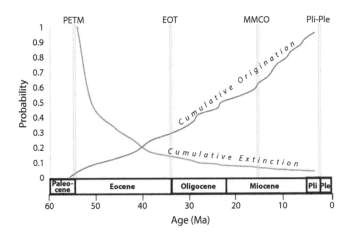

Figure 5.5 Polynomial least-squares fitted curves for cumulative diatom origination (blue) and cumulative diatom extinction (orange) for the Cenozoic. Identifier bars for the PETM, EOT, MMCO, and Pli-Ple are superimposed on the plots. See Table 5.3 for least-squares equations and coefficients of determination.

5.4.2 Switching from Diatom Origination to Extinction over the Cenozoic

During the Cenozoic, diatom origination and extinction may be equal to or switch times when relative maxima ages occur. For Cenozoic events, switching occurrences are analyzable via Heaviside functions. The Cenozoic ages for the events of the PETM, EOT, MMCO, and Pli-Ple are equal to $c = c_5, c_4, c_3, c_2, c_1$ for 55, 34, 15, 2.5 with units Ma. K-Pg and the present are boundaries 65 Ma and 0 Ma, respectively. For $t > 55$, c_5 covers the K-Pg, and $t < c_1$ covers the present. Step functions as shifted Heaviside functions are $H_{2.5}(t)$, $H_{15}(t)$, $H_{34}(t)$, and $H_{55}(t)$, and the switches are defined as $H(t) = \begin{cases} 1 - 0 & when\ t < c \\ 1 - 1 & when\ t \geq c \end{cases}$. That is, the switch is on at 0 until the value of c is reached, then the switch goes off at 1.

For origination, $t = 51, 51, 34, 15,$ and 3 for the ages at which relative maxima occur nearest to the c-values. Except for the youngest aged event, all t-values did not exceed the c-values. The time intervals were ordered from the present to the K-Pg with respect to

Table 5.4 Piecewise values and conditions obtained from expansion of the survival function calculated with Cenozoic diatom origination and extinction relative maxima.

Piecewise values	Piecewise conditions
0.067183	$55 \leq t < 56$
0.134367	$53 \leq t < 55$
0.179156	$52 \leq t < 53$
0.223945	$50 \leq t < 52$
0.261269	$49 \leq t < 50$
0.298593	$48 \leq t < 49$
0.335917	$47 \leq t < 48$
0.373241	$45 \leq t < 47$
0.407172	$44 \leq t < 45$
0.441103	$43 \leq t < 44$
0.475034	$41 \leq t < 43$
0.506703	$39 \leq t < 41$
0.536509	$37 \leq t < 39$
0.564747	$36 \leq t < 37$
0.592984	$35 \leq t < 36$
0.621221	$33 \leq t < 35$
0.648231	$32 \leq t < 33$
0.675241	$30 \leq t < 32$
0.701211	$27 \leq t < 30$
0.725391	$26 \leq t < 27$
0.749571	$25 \leq t < 26$
0.77375	$22 \leq t < 25$
0.796508	$21 \leq t < 22$
0.819265	$20 \leq t < 21$
0.842022	$18 \leq t < 20$
0.864181	$17 \leq t < 18$
0.886339	$14 \leq t < 17$

(Continued)

Table 5.4 Piecewise values and conditions obtained from expansion of the survival function calculated with Cenozoic diatom origination and extinction relative maxima. (*Continued*)

Piecewise values	Piecewise conditions
0.907443	$10 \leq t < 14$
0.92717	$7 \leq t < 10$
0.946092	$4 \leq t < 7$
0.964286	$2 \leq t < 4$
0.982143	$1 \leq t < 2$
1.	$t < 1$
Indeterminate	---

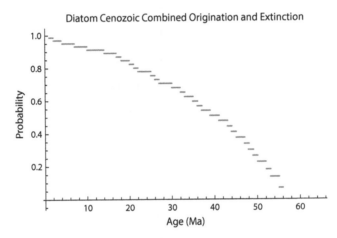

Figure 5.6 Survival function analysis of combined origination and extinction for Cenozoic diatoms with the age scale from the present to the K-Pg.

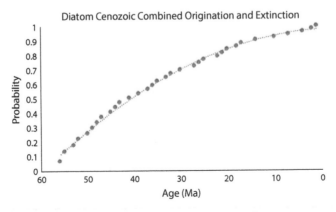

Figure 5.7 Piecewise continuous least-squares fitted polynomial of cumulative origination and extinction for Cenozoic diatoms. The age scale is from the K-Pg to the present. See Table 5.3 for least-squares equations and coefficients of determination.

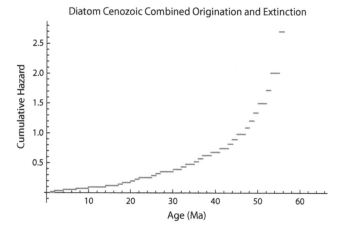

Figure 5.8 Cumulative hazard function analysis of combined origination and extinction for Cenozoic diatoms with the age scale from the present to the K-Pg.

Cenozoic origination events and represent the shift from each Cenozoic event to the next and are given as follows: the Pli-Ple to the present; the Pli-Ple to MMCO; the MMCO to EOT; the EOT to PETM; the last shift is from the Cretaceous to the K-Pg. For origination only, $f(t)$ is given as

$$f(t) = \begin{cases} 0 & t < 2.5 \\ 0 & 2.5 \leq t < 15 \\ 0 & 15 \leq t < 34 \\ 0 & 34 \leq t < 55 \\ -4 & t \geq 55 \end{cases} \tag{5.33}$$

where the first column behind the curly bracket is $f_1(t), f_2(t), f_3(t), f_4(t)$, and $f_5(t)$, and the second column from top to bottom represents the time event shifts as intervals. The values for $f_1(t)$ through $f_5(t)$ were calculated piecewise according to Eq. (5.14), and the origination function in terms of Heaviside functions is

$$f(t) = 0 + 0H_{2.5}(t) + 0H_{15}(t) + 0H_{34}(t) - 4H_{55}(t) = -4 \tag{5.34}$$

The on switches were with respect to $f_1(t), f_2(t), f_3(t)$, and $f_4(t)$, while there was an off switch for $f_5(t)$.

For extinction events, $t = 56, 54, 29, 13$, and 6 are the ages at which relative maxima occurred nearest to the c-values. Except for the youngest aged event, all t-values did not exceed the c-values for $g_2(t)$ through $g_4(t)$. The t-value for $g_5(t)$ did exceed the c-value to obtain the correct evaluation. The time intervals were ordered from the present to the K-Pg with respect to Cenozoic extinction events only so that $g(t)$ is given as

$$g(t) = \begin{cases} 0 & t < 2.5 \\ -2 & 2.5 \leq t < 15 \\ -7 & 15 \leq t < 34 \\ -10 & 34 \leq t < 55 \\ -10 & t \geq 55 \end{cases} \qquad (5.35)$$

where the first column behind the curly bracket is $g_1(t), g_2(t), g_3(t), g_4(t)$, and $g_5(t)$, and the second column from top to bottom represents the time event intervals. The values for $g_1(t)$ through $g_5(t)$ were calculated piecewise according to Eq. (5.14), and the extinction function in terms of Heaviside functions is

$$g(t) = 0 - 2H_{2.5}(t) - 7H_{15}(t) - 10H_{34}(t) - 10H_{55}(t) = -29 \qquad (5.36)$$

An on switch occurred with respect to $g_1(t)$ and $g_5(t)$, while the off switches were $g_2(t)$, $g_3(t)$ and $g_4(t)$. Total sums as solutions to $f(t)$ and $g(t)$ for all Heaviside functions are -4 and -29, respectively.

From Heaviside functions for origination and extinction, switching during the Cenozoic is graphically depicted as a probability plot (Figure 5.9). Origination occupies the highest probability from 14 to approximately 54 Ma, while extinction occupies the lowest probability from 0 to 14 Ma and approximately 54 to 65 Ma (Figure 5.9).

Heaviside functions are calculable for a combination of origination and extinction during the Cenozoic. The combined switching is the difference between origination and extinction of Cenozoic diatoms expressed as Heaviside functions, and the switches are given at ages

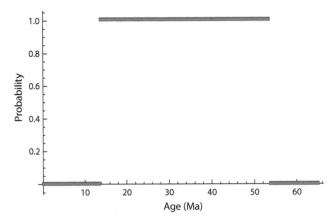

Figure 5.9 Heaviside function switches from origination to extinction. Highest probability is origination, and lowest probability is extinction.

indicated by subscripts of the Heaviside functions. The combined switching from origination to extinction, $fg(t)$, is given via the Heaviside functions as

$$fg_{Combined}(t) = 0H_0(t) + 1H_4(t) + 2H_7(t) + 3H_{10}(t) + 1H_{14}(t) + 2H_{17}(t) + 1H_{20}(t)$$
$$+ 2H_{25}(t) + 3H_{30}(t) + 4H_{32}(t) + 5H_{35}(t) + 4H_{39}(t) + 5H_{41}(t)$$
$$+ 4H_{43}(t) + 5H_{47}(t) + 6H_{52}(t) + 6H_{55}(t) + 1H_{56}(t) \tag{5.37}$$

for all ages during the Cenozoic. Time interval values as piecewise conditions are obtained from expansion of the survival function (Table 5.4).

Results from the combined equation may be applied to the particular Cenozoic events of the PETM, EOT, MMCO, and Pli-Ple. The time intervals are ordered from the present to the K-Pg, and the combined events induce $fg_{Combined}(t_{Events})$ to be

$$fg_{Combined}(t_{Events}) = \begin{cases} 0 & t_{Events} < 2.5 \\ 0 & 2.5 \le t_{Events} < 15 \\ -3 & 15 \le t_{Events} < 34 \\ -8 & 34 \le t_{Events} < 55 \\ -56 & t_{Events} \ge 55 \end{cases} \tag{5.38}$$

with $fg_{Combined}(t_{Events})$ given in terms of Heaviside functions as

$$fg_{Combined}(t_{Events}) = 0 - 0H_{2.5}(t_{Events}) - 3H_{15}(t_{Events}) - 8H_{34}(t_{Events}) - 56H_{55}(t_{Events}) = -67. \tag{5.39}$$

On switches occurred at $fg_{Combined}(t_{Events})$ 1 and 2, while off switches occurred at $fg_{Combined}(t_{Events})$ 3, 4, and 5. Total sum of Heaviside functions was -67.

5.4.3 Origination and Extinction Sequential Steps and Accumulated Switches during the Cenozoic

Relative maxima ages used in analyses were based on peak values at ages during the Cenozoic (Figures 5.1 and 5.2 and Table 5.5) and depicted in a diagram from the K-Pg to the present to show the peak value ordering (Figure 5.10). For origination, successive relative maxima occur at 51 and 46 Ma, 34 and 31 Ma, and 5 and 3 Ma (Figure 5.10 and Table 5.5). For extinction, successive relative maxima occur at 56 and 54 Ma, 19 and 16 Ma, and 11 and 9 Ma (Figure 5.10 and Table 5.5). These pairs of successive peak values represent piecewise continuity for both origination and extinction.

At the PETM, EOT, MMCO, and Pli-Ple, origination and extinction may be defined by position with regard to switching. Using Eq. (5.15) with values based on Heaviside functions, accumulated switch counts were calculated. For $k = 0$, accumulated switch counts were just the functions resulting in 0, 0, 0, 0, and -4 for origination and 0, -2, -7, -10, and

Table 5.5 Relative maxima ages for Cenozoic diatom origination and extinction rates from reconstructed plots based on [5.7.37.] Figure 2 and Cenozoic diatom diversity relative maxima and relative minima ages from reconstructed plot based on [5.7.62.], Figure 4.

Origination rate relative maxima age (Ma)	Extinction rate relative maxima age (Ma)	Diversity relative maxima age (Ma)	Diversity relative minima age (Ma)
3	6	3	0
5	9	7	2
8	11	9	8
12	13	17	10
15	16	20	16
23	19	24	18
28	24	26	23
31	29	27	25
34	38	33	29
40	42	37	31
46	54	41	35
51	56	42	39
		45	43
		55	46
		58	53
			59
			61

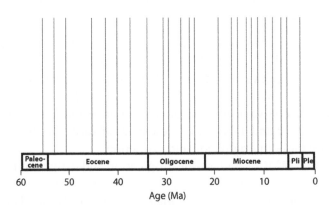

Figure 5.10 Diagram of the ages at which Cenozoic diatom origination (blue perpendiculars) and extinction (orange perpendiculars) relative maxima occur.

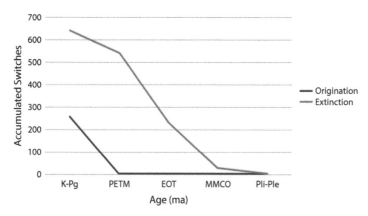

Figure 5.11 Accumulated switches for origination (blue curve) and extinction (orange curve) for Cenozoic diatoms.

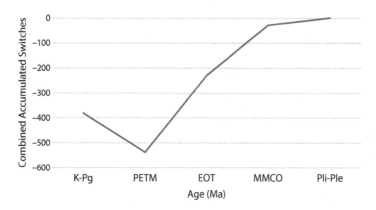

Figure 5.12 Combined accumulated switches of origination and extinction for Cenozoic diatoms.

-10 for extinction from Pli-Ple to K-Pg. For $k = 1$, accumulated switch counts were 0, 0, 0, 0, and 256 for origination and 0, 28, 231, 540, and 640 for extinction from Pli-Ple to K-Pg (Figure 5.11). Combined accumulated switches for origination and extinction were weighted toward extinction because zero was used as the on switch in Heaviside function calculations (Figure 5.12).

5.4.4 Overlapping via a Convolution Product of Origination and Extinction

For the cumulative frequency origination equation, the Laplace transform of the Heaviside functions was $\dfrac{1-e^{-53.4971s}}{s}$, while the counterpart for extinction was $\dfrac{e^{-13.8009s}}{s}$. From these solutions, the convolution product was calculated as coefficients and applied to events Pli-Ple, MMCO, EOT, and PETM with respect to origination and extinction switches, resulting in plots of convolution curves (Figure 5.13). Origination was more similar to the Cenozoic events than was extinction (Figure 5.13). Origination and extinction overlapped at approximately 4, 11, 18, 30, and 52 Ma (Figure 5.13).

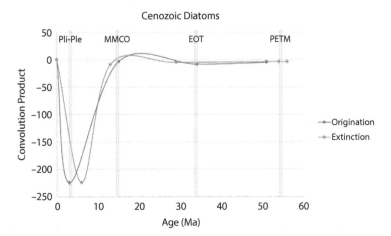

Figure 5.13 Convolution product plot for Cenozoic events Pli-Ple, MMCO, EOT, and PETM, origination switches (blue) curve, and extinction switches (orange) curve.

5.4.5 Origination and Extinction as Poisson Processes

Origination as the interarrival intervals and extinction as arrival times were obtained from Figure 5.10 and used in Poisson process analysis. The input data was defined according to Table 5.6. The arrival probabilities were calculated as 0, 0.224, 0.195, 0.271, and 0.271 for interarrival intervals 0 to 3 Ma, 3 to 15 Ma, 15 to 34 Ma, 34 to 51 Ma, and 51 to 65 Ma. That is, the probability of extinction was from 0.2 to 0.3 for Cenozoic diatoms at the Pli-Ple, MMCO, EOT, PETM, and K-Pg. The probability product was 0.0032.

5.4.6 Test of Origination and Extinction Switches: Stochastic or Deterministic Chaos?

A test of switches is invoked to determine whether the sequence of switches was stochastic or deterministically chaotic with regard to comparisons with Cenozoic events K-Pg, PETM, EOT, MMCO, and Pli-Ple. Initially, successive relative maxima ages may indicate deterministic conditions. However, the characteristic of successive on or off switches at those relative maxima ages as deterministically chaotic was tested via a Lyapunov function.

Table 5.6 Interarrival (origination) and arrival (extinction) values for Poisson process analysis.

Interarrival origination interval (Ma)	Arrival number of extinctions	Total time span of interarrival interval (Ma)	Rate value (λ)
0 to 3	0	3	0.00
3 to 15	3	12	0.25
15 to 34	4	17	0.24
34 to 51	2	17	0.12
51 to 65	2	14	0.14

Table 5.7 Non-reversal ages for origination and extinction and the Lyapunov rate values for Poisson process analysis.

Extinction (E) or Origination (O)	Cenozoic interval (Ma)	Non-reversal interval (Ma)	t_i (Ma)	c_i (Ma)	Lyapunov rate value (V)	Sum of the Lyapunov rates
---	0 to 2.5	0 to 3	0	0	0.000	0.000
O	2.5 to 15	3 to 5	3	2.5	0.018	
O	2.5 to 15	3 to 5	5	2.5	0.422	
E	2.5 to 15	9 to 11	9	2.5	1.533	
E	2.5 to 15	9 to 11	11	2.5	2.071	4.045
E	15 to 34	16 to 19	16	15	0.114	
E	15 to 34	16 to 19	19	15	0.460	
O	15 to 34	31 to 34	31	15	1.767	
O	15 to 34	31 to 34	34	34	0.000	2.340
O	34 to 55	46 to 51	46	34	0.855	
O	34 to 55	46 to 51	51	34	1.189	
E	34 to 55	54 (to 56)	54	34	1.383	3.427
E	55 to 65	56	56	55	0.055	0.055

From Eq. (5.32), origination on switches occurred for all successive relative maxima ages, and are therefore, deterministic. From Eq. (5.34), extinction off switches via $g_2(t)$ and $g_3(t)$ occurred for successive relative maxima 19 and 16 Ma as well as 11 and 9 Ma, and are therefore, deterministic. An on and off switch occurred for successive relative maxima 56 and 54 Ma, respectively, resulting in stochasticity.

For deterministic chaos analysis, a Lyapunov function was inserted into the Poisson process according to Eq. (5.30) and calculated for non-reversals (Table 5.7). The resultant probabilities were 0.0, 0.00001, 0.0038, 0.0004, and 0.052 for the presence of chaos for intervals 0 to 2.5 Ma, 2.5 to 15 Ma, 15 to 34 Ma, 34 to 55 Ma, and 55 to 65 Ma. The product probability was $8.51 \times 10^{-13} \approx 0$.

5.4.7 Diversity and Its Relation to Origination and Extinction for Cenozoic Diatoms

A diagram of Cenozoic diatom diversity was devised of peak value ages as relative maxima and relative minima ages as valleys (Figure 5.14). Using this diagram as the basis of diversity analyses, comparisons were made between diversity and combined origination and extinction results.

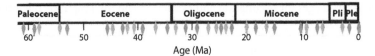

Figure 5.14 Diagram of the ages as which Cenozoic diatom diversity occurs as relative maxima (green diamonds) and relative minima (pink diamonds).

Values for diversity relative maxima and minima (Table 5.5) were used to construct a cumulative frequency curve using a survival function (Table 5.8). The result is depicted in a plot with the combination cumulative frequency function of origination and extinction (Figure 5.15). Comparison between combination origination and extinction and diversity was depicted as piecewise continuous least-squares fitted curves as well (Figure 5.16).

Table 5.8 Piecewise values and conditions obtained from expansion of the survival function calculated with Cenozoic diatom diversity relative maxima and minima.

Piecewise values	Piecewise conditions
0.	$t = 61$
0.154234	$59 \leq t < 61$
0.308467	$53 \leq t < 59$
0.385584	$46 \leq t < 53$
0.462701	$43 \leq t < 46$
0.528801	$39 \leq t < 43$
0.581682	$35 \leq t < 39$
0.630155	$31 \leq t < 35$
0.675166	$29 \leq t < 31$
0.720177	$25 \leq t < 29$
0.760187	$23 \leq t < 25$
0.798196	$18 \leq t < 23$
0.834478	$16 \leq t < 18$
0.869248	$10 \leq t < 16$
0.904018	$8 \leq t < 10$
0.9375	$2 \leq t < 8$
0.96875	$0 \leq t < 2$
1.	$t < 0$
0.	---

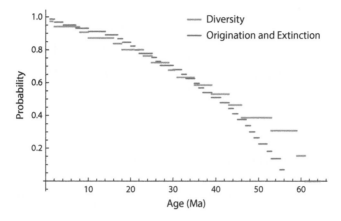

Figure 5.15 Survival function analysis of combined origination and extinction (blue curve) and diversity (green curve) for Cenozoic diatoms with the age scale from the present to the K-Pg.

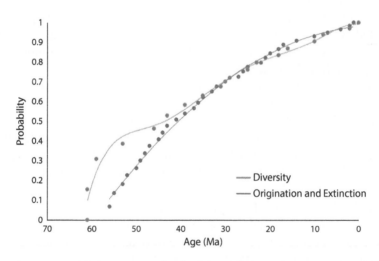

Figure 5.16 Piecewise continuous least-squares fitted polynomial of cumulative origination and extinction (blue curve) and diversity (green curve) for Cenozoic diatoms. See Table 5.3 for least-squares equations and coefficients of determination.

For diversity, the Heaviside functions are

$$
\begin{aligned}
Diversity(t) = {}& 0H_0(t) + 2H_2(t) + 8H_8(t) + 9H_{10}(t) + 3H_{16}(t) + 4H_{18}(t) + 2H_{23}(t) \\
& + 1H_{25}(t) + 2H_{29}(t) + 0H_{31}(t) - 2H_{35}(t) - 4H_{39}(t) - 5H_{43}(t) - 7H_{46}(t) \\
& - 14H_{53}(t) - 15H_{59}(t) - 17H_{61}(t)
\end{aligned}
$$

$$(5.40)$$

The Heaviside functions go from positive to negative at age 35 Ma. For Cenozoic events of the PETM, EOT, MMCO, and Pli-Ple, the time intervals are ordered from the present to the K-Pg, and diversity as $diversity(t_{Events})$ is

$$Diversity(t_{Events}) = \begin{cases} 0 & t_{Events} < 2.5 \\ 2 & 2.5 \le t_{Events} < 15 \\ 12 & 15 \le t_{Events} < 34 \\ -32 & 34 \le t_{Events} < 55 \\ -32 & t_{Events} \ge 55 \end{cases} \qquad (5.41)$$

and diversity in terms of the Heaviside functions is given as

$$Diversity(t_{Events}) = 0 + 2H_{2.5}(t_{Events}) + 12H_{15}(t_{Events}) - 32H_{34}(t_{Events}) - 32H_{55}(t_{Events}) = -50.$$
$$(5.42)$$

On switches occurred at $diversity(t_{Events})$ 1, while off switches occurred at 2, 3, 4, and 5. Total sum of Heaviside functions was -50.

Combined origination and extinction accumulated switches were compared with diversity accumulated switches between relative maxima and minima (Figure 5.17). The early Cenozoic saw diversity favored until the EOT with origination being more associated to diversity during this time than extinction (Figure 5.17).

For Poisson process analysis of Cenozoic diatom diversity, interarrival intervals were chosen from diversity relative maxima and minima ages that were the closest value to the Pli-Ple, MMCO, EOT, PETM, and K-Pg. Those values were 0, 2, 16, 35, and 53 Ma which were all relative minima ages. The input data was defined according to Table 5.9. The arrival probabilities were calculated as 0.0, 0.224, 0.161, 0.195, and 0.271 for interarrival intervals 0 to 2 Ma, 2 to 16 Ma, 16 to 35 Ma, 35 to 53 Ma, and 53 to 65 Ma. That is, the probability of diversity decreasing ranged from 0.2 to 0.3 for Cenozoic diatoms at the Pli-Ple, MMCO, EOT, PETM, and K-Pg. The probability product was 0.0019.

A Lyapunov Poisson process analysis was conducted to test for non-reversals concerning diversity increases or decreases and the presence of deterministic chaos according to Eq. (5.30). The probabilities were 0.0, 0.0019, 4.348×10^{-9}, 0.0002, and 0.2382 for the presence of chaos for intervals 0 to 2.5 Ma, 2.5 to 15 Ma, 15 to 34 Ma, 34 to 55 Ma, and 55 to 65 Ma. The product probability was 3.63×10^{-16}.

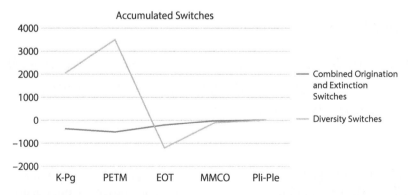

Figure 5.17 Accumulated switches for combined origination and extinction (blue curve) and diversity (green curve) for Cenozoic diatoms.

Table 5.9 Diversity interarrival (relative minima ages) and arrival (relative maxima ages) values for Poisson process analysis.

Interarrival origination interval (Ma)	Arrival number = relative maxima	Total time span of interarrival interval (Ma)	Rate value (λ)
0 to 2	0	2	0.00
2 to 16	3	14	0.21
16 to 35	6	19	0.32
35 to 53	4	18	0.22
53 to 65	2	12	0.17

A plot of the Poisson (Figure 5.19) and Lyapunov functions (Figure 5.20) of the rate values for the combination of origination and extinction (Tables 5.6 and 5.7) versus diversity (Tables 5.9 and 5.10) revealed the fluctuations associated with Cenozoic ages as maximum values just after the MMCO and around the EOT and PETM. For comparative purposes, the calculated results of the Poisson process (Eq. 5.28) and Lyapunov Poisson process (Eq. 5.30) for a combination of origination and extinction as well as diversity was compiled in Table 5.11. The Poisson process for combination origination and extinction was similar to that for diversity, while the Lyapunov Poisson process produced differing results for combination origination and extinction in contrast to diversity from the PETM to K-Pg (Table 5.11 and Figure 5.18).

5.5   Discussion

5.5.1   Diversity and the Effects from Origination and Extinction of Cenozoic Diatoms

Overall, diatom diversity as species richness has increased during the Cenozoic (e.g., [5.7.35., 5.7.62.], while originations and extinctions have decreased (e.g., [5.7.37.]). Increases in origination rates do not necessarily follow extinctions, and origination rates slowly declining do not necessarily lead to declining diversity at the clade level [5.7.75.]. Trying to find the interrelation among origination, extinction and diversity may not necessarily be straightforward or readily at hand without additional analyses.

A different perspective or approach was taken here in which the cumulative frequency of diatom origination and extinction during the Cenozoic was analyzed using methods involving piecewise continuous functions with respect to probability, and these results were related to diversity. With established data, diatom origination and extinction rate percents were converted into probability data, with the view that the peak value ages during the Cenozoic were of interest rather than the entire pattern of the rate percents. Cumulative frequencies were used as a conduit to determining probabilistic representations of diatom origination and extinction.

Table 5.10 Non-reversal ages for diversity relative maxima or relative minima and the Lyapunov rate values for Poisson process analysis.

Relative maxima (max) or Relative minima (min)	Cenozoic interval (Ma)	Non-reversal interval (Ma)	t_i (Ma)	c_i (Ma)	Lyapunov rate value (V)	Sum of the Lyapunov rates
---	0 to 2.5	0 to 3	0	0	0.000	0.000
max	2.5 to 15	3 to 7	3	2.5	0.018	
max	2.5 to 15	3 to 7	7	2.5	0.974	
min	2.5 to 15+	10 to 16	10	2.5	1.805	2.798
min	15 to 34	10 to 16	16	15	0.114	
min	15 to 34	23 to 25	23	15	0.913	
min	15 to 34	23 to 25	25	15	1.133	
max	15 to 34	27 to 33	27	15	1.349	
max	15 to 34	27 to 33	33	15	1.968	5.479
max	34 to 55	41 to 42	41	34	0.508	
max	34 to 55	41 to 42	42	34	0.578	
min	34 to 55	46 to 53	46	34	0.855	
min	34 to 55	46 to 53	53	34	1.318	3.260
max	55 to 65	55 to 58	55	55	0.000	
max	55 to 65	55 to 58	58	55	0.163	
max	55 to 65	59 to 61	59	55	0.216	
max	55 to 65	59 to 61	61	55	0.322	0.701

Table 5.11 Poisson process and Lyapunov process results for combination origination and extinction and diversity for Cenozoic diatoms.

Age	Poisson process – origination and extinction	Lyapunov Poisson process – origination and extinction	Poisson process – diversity	Lyapunov Poisson process – diversity
Pli-Ple	0.	0.	0.	0
MMCO	2.3364	0.0525	1.6142	0.1219
EOT	11.8551	1.4449	11.6674	0.0668
PETM	30.2263	1.1045	28.0258	1.3433
K-Pg	44.2108	48.2707	44.8643	26.3006

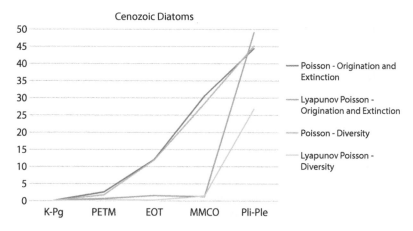

Figure 5.18 Combination origination and extinction Poisson and Lyapunov Poisson process (blue curves) compared to diversity (green curves) for Cenozoic diatoms.

For Cenozoic diatoms, origination cumulative frequency was a larger influence than extinction cumulative frequency over the Cenozoic (Figure 5.5) as was origination switches via Heaviside functions (Figure 5.9). Switching favored origination, with on values for the Cenozoic events of the PETM, EOT and MMCO, and for extinction, switches were on for the Pli-Ple and K-Pg. Extinction influenced diversity via the K-Pg and at the Pli-Ple, while origination ruled the PETM, EOT and MMCO for Cenozoic diatoms.

Accumulated switches between origination and extinction throughout the Cenozoic favored extinction (Figure 5.11). The combination of origination and extinction reduced the influence of extinction with the exception of the K-Pg (Figure 5.12). A convolution product of origination and extinction switches produced more similarity between origination and events K-Pg, PETM, EOT, MMCO, and Pli-Ple than did extinction (Figure 5.13). Accumulated switches were analyzed with respect to origination and extinction intervals, while convolution product switches were analyzed with respect to specified events, accounting for the differences in outcomes.

From Poisson analyses, the combination of origination and extinction compared to diversity were approximately the same (Figure 5.18), although the rates differed for all events except at the Pli-Ple (Figure 5.19). From Lyapunov Poisson analyses, diversity diverged from the combination of origination and extinction at the PETM, being much higher in value (Figure 5.18).

A test of switches was used to analyze whether chaos was present via the characteristic of non-reversals with respect to diatom origination and extinction, and then with diversity. Lyapunov function values were larger for the combination of origination and extinction around the MMCO and PETM, while the values were larger at the EOT for diversity (Figure 5.20). That is, deterministic chaos may be indicative of the MMCO and PETM regarding origination and extinction, while the EOT with respect to diversity may be a result of deterministic chaos.

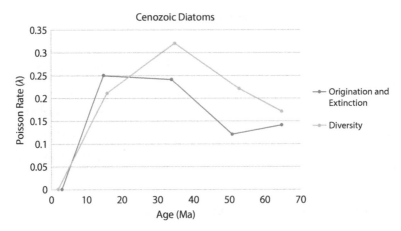

Figure 5.19 Poisson rates for combination of origination and extinction (blue curve) in contrast to that for diversity (green curve).

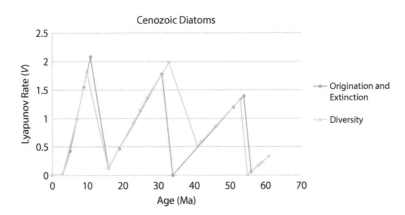

Figure 5.20 Lyapunov functions for combination of origination and extinction (blue curve) in contrast to that for diversity (green curve).

5.5.2 Cenozoic Events and Diatom Diversity, Origination and Extinction

Using cumulative frequency functions, Heaviside functions, Poisson and Lyapunov Poisson analyses, the relation between Cenozoic events of the K-Pg, PETM, EOT, MMCO, and Pli-Ple and diatom diversity, origination and extinction may be discerned. The combination of origination and extinction were most influenced by the PETM onward (Figures 5.11 and 5.12). Except for a slight dip at the EOT, diversity followed the same path (Figures 5.17 and 5.18). Overall, the results signify a decrease in diversity with respect to the PETM.

At the EOT, there is a divergence between the combination of origination and extinction and diversity (Figures 5.15 and 5.16), where the increase in diatom diversity just after the EOT coincides with this result. Subsequently, a decrease in diversity is accompanied by an increase in diversity to the MMCO as was the case for a combination of origination and extinction (Figures 5.12 and 5.17). From the MMCO to Pli-Ple, a slight decrease occurred for both a combination of origination and extinction as well as diversity (Figure 5.18), but the rate of change was greater for the combination of origination and extinction (Table 5.6 and Figure 5.19) in contrast to diversity (Table 5.9 and Figure 5.20).

Deterministic chaos has been found for a combination of origination and extinction just after the MMCO and at the PETM with maximum peak values (Figure 5.20). For diversity, deterministic chaos has been determined at the EOT with a maximum peak value (Figure 5.20). From the plot of the Lyapunov rates, there is a sawtooth pattern, meaning instability is present and is an indicator of deterministic chaos.

5.5.3   Origination and Extinction Related to Diversity: Markov Chain, Martingale, Ergodic Processes, and Lyapunov Functions

Although the analyses herein were based on piecewise continuous functions, continuous functions may be used, provided that the appropriate conditions are instituted. In that vein, the probability of the frequency of origination and extinction with respect to diversity during the Cenozoic as Poisson processes may be characterized via a Markov chain (stochastic, ordered, recursive with no dependence on history) as indicated throughout via the memoryless property. Other properties that may be present include the martingale (stochastic with non-stationarity, i.e., finishing at the same place as the start, but conditioned on history and ending up at the same place in the present), or in the longer term, ergodicity (dynamic or stochastic with stationarity, i.e., producing the same outcome over the long term).

The probabilistic representation of origination, extinction and diversity hinges on the cumulative frequency of each process over time. The nature of the probability may be characterized as each of the relative maxima ages occurring independently from the previously occurring one, or each of the relative maxima ages essentially being the same at any given time, regardless of the amount of time gone by. That is, whether Markovian in the first instance, or a martingale in the second instance, probability of origination, extinction or diversity may be assessed via these properties.

For example, a Markov chain may be representative of the step-by-step process with respect to origination, extinction or diversity. Alternatively, at each step, martingale conditions may be present (e.g., [5.7.53.]). The total cumulative result over a long period of time such as the Cenozoic may entail recurrence, where the time average equals the expected value, thereby exhibiting ergodicity (e.g., [5.7.19.]). Such functions may characterize change from one process to another at any given time, affecting origination, extinction or diversity outcomes so that Markov chain, martingale or ergodic properties may be evident throughout the Cenozoic.

To test for the Markov property, martingale property or ergodicity, cumulative functions could be converted to time series, e.g., using a seasonal autoregressive integrated moving average (SARIMA) model. Testing could be done with respect to Cenozoic events PETM, EOT, MMCO, and Pli-Ple, among others, using the seasonal aspect of SARIMA to define event time intervals. Other time series models could be implemented. The Markov property is testable, e.g., using a conditional characteristic function (e.g., [5.7.12.]). Testing for martingale properties could be accomplished via regression and goodness-of-fit tests such as the Cramér-von Mises or Kolmogorov-Smirnov tests with conditional expected sequential values taken into consideration (e.g., [5.7.53.]). Ergodicity is testable, e.g., using Monte Carlo methods (e.g., [5.7.19.]).

All possible tests concerning Markovian, martingale or ergodic properties may be subject to equilibrium or stability conditions. Testing for such conditions using Lyapunov functions may result in evidence of ergodicity while preserving the Markov property (e.g., [5.7.14.]). Quadratic Lyapunov functions may be used with regard to martingales (e.g., [5.7.8.]). A test for stochasticity versus deterministic chaos may be devised using stochastic Lyapunov functions (e.g., [5.7.79.]). No matter which properties are individually or simultaneously governing origination, extinction and diversity over geologic time, there interrelatedness with regard to stochasticity, ergodicity or chaos is testable.

5.6   Summary and Future Research

Studying Cenozoic diatom origination, extinction and diversity in the context of cumulative frequency as piecewise continuous induces other possibilities to expand the results of this study. Late Oligocene and Miocene-Pliocene boundaries are other Cenozoic events that could be included. The opening of the Drake Passage, closure of the Panama Seaway, or volcanism events may form the basis of study in the frequency of origination and extinction events and their relation to diversity regarding diatoms.

Using Heaviside functions, initial value problems (IVPs) (e.g., [5.7.20.]) may be set up and applied to origination, extinction and diversity functions. That is, combining derivatives of cumulative origination and extinction functions with shifted Heaviside functions define the IVP. The derivatives of the cumulative functions are explicitly made piecewise continuous by inserting the shifted Heaviside functions. However, if Lyapunov functions are eventually included, sensitivity to initial conditions may be a consideration.

Other methods regarding starting and even ending positions may be identified in accordance with geologic time benchmarks as boundary conditions concerning origination or extinction. Solutions to these problems may be informative with regard to deterministic studies. Queueing theory (e.g., [5.7.8.]) is another avenue worth applying to origination, extinction and diversity processes in view of the application of the Poisson process in this study.

The relation among origination, extinction and diversity requires further exploration. Diatom diversity which is calculable by many avenues and the measurement of origination and extinction need to be better integrated so that the interrelation of these concepts is better understood.

5.7   References

5.7.1.   Abramowitz, M. and Stegun, I.A. (eds.) (1972) *Handbook of Mathematical Functions with Formulas, Graphs, and Mathematical Tables*, 9th edition, Dover Publications, Inc., New York, USA.

5.7.2.   Alroy, J. (2008) Dynamics of origination and extinction in the marine fossil record. *PNAS* 105 (suppl. 1), 11536-11542.

5.7.3.   Alroy, J. (2014) Accurate and precise estimates of origination and extinction rates. *Paleobiology* 40(3), 374-397.

5.7.4. Anderson, D.F. Craciun, G., Gopalkrighnan, M., Wiuf, C. (2015) Lyapunov functions, stationary distributions, and non-equilibrium potential for reaction networks. *Bull. Math Biol.* 77(9), 1744-1767.

5.7.5. Bambach, R.K., Knoll, A.H., Wang, S.C. (2004) Origination, extinction, and mass depletions of marine diversity. *Paleobiology* 30(4), 522-542.

5.7.6. Behrenfeld, M.J., Halsey, K.H., Boss, E., Karp-Boss, L., Milligan, A.J., Peers, G. (2021) Thought on the evolution and ecological niche of diatoms. *Ecological Monographs* 91(3), e01457.

5.7.7. Benoiston, A.-S., Ibarbalz, F.M., Bittner, L., Guidi, L., Jahn, O., Dutkiewicz, S., Bowler, C. (2017) The evolution of diatoms and their biogeochemical functions. *Phil. Trans. R. Soc. B* 372: 20160397.

5.7.8. Brémaud, P. (1999) *Markov Chains: Gibbs Fields, Monte Carlo Simulation, and Queues*, Springer Science+Business Media, New York.

5.7.9. Castiglione, S., Mondanaro, A., Melchionna, M., Serio, C., Di Febbraro, M., Carotenuto, F., Raia, P. (2017) Diversification rates and the evolution of species range size frequency distribution. *Frontiers in Ecology and Evolution* 5, 147. doi: 10.3389/fevo.2017.00147

5.7.10. Cermeño, P. (2011) Marine planktonic microbes survived climatic instabilities in the past. *Proc. R. Soc. B* doi:10.1098/rspb.2011.1151

5.7.11. Cermeño, P., Falkowski, P.G., Romero, O.E., Schaller, M.F., Vallina, S.M. (2015) Continental erosion and the Cenozoic rise of marine diatoms. *PNAS* 112(14), 4239-4244.

5.7.12. Chen, B. and Hong Y. (2012) Testing for the Markov property in time series. *Econometric Theory* 28, 130-178.

5.7.13. Chow, P.-L. and Khasminskii, R.Z. (2011) Method of Lyapunov functions for analysis of absorption and explosion in Markov chains. *Problems of Information Transmission* 47(3), 232-250.

5.7.14. Cordeiro, J.D., Kharoufeh, J.P., Oxley, M.E. (2019) On the ergodicity of a class of level-dependent quasi-birth-and-death processes. *Advances in Applied Probability* 51(4), 1109-1128.

5.7.15. Clark, T.G., Bradburn, M.J., Love, S.B., Altman, D.G. (2003) Survival analysis part I: basic concepts and first analyses. *British Journal of Cancer* 89, 232-238.

5.7.16. Daley, D. J. and Vere-Jones, D. (2003) Basic properties of the Poisson process. In: *An Introduction to the Theory of Point Processes: Volume I: Elementary Theory and Methods, 2nd edition*, Springer, New York: 19-40.

5.7.17. Dawkins, P. Differential Equations: Laplace Transforms, Accessed on 31 December 2021, https://tutorial.math.lamar.edu/classes/de/StepFunctions.aspx.

5.7.18. Diaz, L.F.H., Harmon, L.J., Sugawara, M.T.C., Miller, E.T., Pennell, M.W. (2019) Macroevolutionary diversification rates show time dependency. *PNAS* 116(15), 7403-7408.

5.7.19. Domowitz, I. and El-Gamal, M.A. (2001) A consistent nonparametric test of ergodicity for time series with applications. *Journal of Econometrics* 102, 365-398.

5.7.20. Dunbar, S.R. (1988) A branching random evolution and a nonlinear hyperbolic equation. *SIAM J. Appl. Math.* 48(6), 1510-1526.

5.7.21. Flower, R.J. (1993) Diatom preservation: experiments and observations on dissolution and breakage in modern and fossil material. *Hydrobiologia* 269/270, 473-484.

5.7.22. Fraass, A.J., Kelly, D.C., Peters, S.E. (2015) Macroevolutionary history of the planktic foraminifera. *Annu. Rev. Earth Planet. Sci.* 43, 139-166.

5.7.23. Frigola, A., Prange, M., Schulz, M. (2018) Boundary conditions for the Middle Miocene Climate Transition (MMCT v1.0). *Geosci. Model Dev.* 11, 1607-1626.

5.7.24. Georgescu, P. and Zhang, H. (2013) A Lyapunov functional for a SIRI model with nonlinear incidence of infection and relapse. *Applied Mathematics and Computation* 219, 8496-8507.

5.7.25.   Hansen, J., Sato, M., Russell, G., Kharecha, P. (2013) Climate sensitivity, sea level and atmospheric carbon dioxide. *Phil. Trans. R. Soc.* A 371: 20120294.

5.7.26.   Harvey, B.P., Agostini, S., Kon, K., Wada, S., Hall-Spencer, J.M. (2019) Diatoms dominate and alter marine food-webs when CO2 rises. *Diversity* 11, 242.

5.7.27.   Hayashi, T., Krebs, W.N., Saito-Kato, M., Tanimura, Y. (2018) The turnover of continental planktonic diatoms near the middle/late Miocene boundary and their Cenozoic evolution. *PLoS ONE* 13(6): e0198003.

5.7.28.   Houben, A.J.P., van Mourik, C.A., Montanari, A., Coccioni, R., Brinkhuis, H. (2012) The Eocene-Oligocene transition: changes in sea level, temperature or both? *Palaeogeography, Palaeoclimatology, Palaeoecology* 335-336, 75-83.

5.7.29.   Jonkers, L., Hillebrand, H., Kucera M. (2019) Global change drives modern plankton communities away from the pre-industrial state. *Nature* 570, 372-375.

5.7.30.   Katz, M.E., Finkel, Z.V., Grzebyk, D., Knoll, A.H., Falkowski, P.G. (2004) Evolutionary trajectories and biogeochemical impacts of marine eukaryotic phytoplankton. *Annu. Rev. Ecol. Evol. Syst.* 35, 523-556.

5.7.31.   Katz, M. E., Fennel, K., Falkowski, P. G. (2007) Geochemical and biological consequences of phytoplankton evolution. In: *Evolution of Primary Production in the Sea*, Falkowski, P. and Knoll, A. (eds.), Elsevier, Amsterdam: 405–430.

5.7.32.   Kociolek, J.P. and Williams, D.M. (2015) How to define a diatom genus? Notes on the creation and recognition of taxa, and a call for revisionary studies of diatoms. *Acta Bot. Croat.* 74(2), 195-210.

5.7.33.   Lai, C. (2013) Issues concerning constructions of discrete lifetime models. *Quality Technology & Quantitative Management* 10(2), 251-262.

5.7.34.   Last, G. and Penrose, M. (2017) *Lectures on the Poisson Process*, Cambridge University Press, Cambridge, United Kingdom.

5.7.35.   Lazarus, D., Barron, J., Renaudie, J., Diver, P., Türke, A. (2014) Cenozoic planktonic marine diatom diversity and correlation to climate change. *PLoS ONE* 9, e84857.

5.7.36.   Lewitus, E., Bittner, L., Malviya, S., Bowler, C., Morlon, H. (2018) Clade-specific diversification dynamics of marine diatoms since the Jurassic. *Nature Ecology & Evolution* 2, 1715-1723.

5.7.37.   Lowery, C.M., Bown, P.R., Fraass, A.J., Hull, P.M. (2020) Ecological response of plankton to environmental change: thresholds for extinction. *Annual Reviews Earth Planet. Sci.* 48, 16.1-16.27.

5.7.38.   Lucantoni, D.M. (1991) New results on the single server queue with a batch Markovian arrival process. *Stochastic Models* 7(1), 1-46.

5.7.39.   Malviya, S., Scalco, E., Audic, S., Vincent, F., Veluchamy, A., Poulain, J., Wincker, P., Iudicone, D., deVargas, C., Bittner, L., Zingone, A., Bowler, C. (2016) Insights into global diatom distribution and diversity in the world's ocean. *PNAS* 113(11), E1516-E1525.

5.7.40.   Mauer, B.A. and McGill, B.J. (2011) Measurement of species diversity. In: *Biological Diversity: Frontiers in Measurment and Assessment*, Magurran, A.E. and McGill, B.J. (eds.), Oxford University Press, Oxford, U.K.: 55-65.

5.7.41.   Mayhew, P.J., Jenkins, G.B., Benton, T.G. (2008) A long-term association between global temperature and biodiversity, origination and extinction in the fossil record. *Proc. R.Soc. B* 275, 47-53.

5.7.42.   Methner, K., Campani, M., Fiebig, J., Löffler, N., Kempf, O., Mulch, A. (2020) Middle Miocene long-term continental temperature change in and out of pace with marine climate records. *Scientific Reports* 10:7989.

5.7.43.   Miller, H. and Orloff, J. Step and Delta Functions 18.031. Accessed on 29 December 2021, https://math.mit.edu/~stoopn/18.031/stepanddelta.pdf.

5.7.44.   Miller, K.G., Wright, J.D., Katz, M.E., Wade, B.S., Browning, J.V., Cramer, B.S., Rosenthal, Y. (2009) Climate threshold at the Eocene-Oligocene transition: Antarctic ice sheet influence on ocean circulation. In: *The Late Eocene Earth—Hothouse, Icehouse, and Impacts: Geological Society of America Special Paper 452*, Koeberl, C. and Montanari, A. (eds.), The Geological Society of America, Boulder, Colorado: 169-178.

5.7.45.   Miller, K.G., Kominz, M.A., Broning, J.V., Wright, J.D., Mountain, G.S., Katz, M.E., Sugarman, P.J., Cramer, B.S., Christie-Blick, N., Pekar, S.F. (2005) The Phanerozoic record of global sea-level change. *Science* 310, 1293-1298.

5.7.46.   Nair, N.U., Sankaran, P.G., Balakrishnan, N. (2018) *Reliability Modelling and Analysis in Discrete Time*, Academic Press, Cambridge, Massachusetts.

5.7.47.   Nakov, T., Beaulieu, J.M., Alverson, A.J. (2018). Accelerated diversification is related to life history and locomotion in a hyperdiverse lineage of microbial eukaryotes (Diatoms, Bacillariophyta) *New Phytologist* 219, 462-473.

5.7.48.   Nakov, T., Beaulieu, J.M., Alverson, A.J. (2019) Diatoms diversify and turnover faster in freshwater than marine environments. *Evolution* 73(12), 2497-2511.

5.7.49.   Nee, S. (2006) Birth-death models in macroevolution. *Annu. Rev. Ecol. Evol. Syst.* 37, 1-17.

5.7.50.   NIST/SEMATECH e-Handbook of Statistical Methods, Accessed on: 26 October 2021, https://doi.org/10.18434/M32189.

5.7.51.   Oreshkina, T.V. and Radionova, E.P. (2014) Diatom record of the Paleocene-Eocene thermal maximum in marine paleobasins of central Russia, Transuralia and adjacent regions. *Nova Hedwigia, Beiheft* 143, 307-336.

5.7.52.   Pappas, J.L. (2016) Multivariate complexity analysis of 3D surface form and function of centric diatoms at the Eocene-Oligocene transition. *Marine Micropaleontology* 122, 67-86.

5.7.53.   Park, J.Y. and Whang, Y.-J. (2005) A test of the martingale hypothesis. *Studies in Nonlinear Dynamics & Econometrics* 9(2), 1163-1192.

5.7.54.   Pimiento, C., Griffin, J.N., Clements, C.F., Silvestro, D., Varela, S., Uhen, M.D., Jaramillo, C. (2017) The Pliocene marine megafauna extinction and its impact on functional diversity. *Nature Ecology & Evolution* 1, 1100-1106.

5.7.55.   Pinseel, E., Janssesn, S.B., Verleyen, E., Vanormelingen, P., Kohler, T.J., Biersma, E.M., Sabbe, K., Van de Vijer, B., Vyverman, W. (2020) Global radiation in a rare biosphere soil diatom. *Nature Communications* 11:2382.

5.7.56.   Pinseel, E., Van de Vijver, B., Wolfe, A.P., Harper, M., Antoniades, D., Ashworth, A.C., Ector, L., Lewis, A.R., Perren, B., Hodgson, D.A., Sabbe, K., Verleyen, E., Vyverman, W. (2021) Extinction of austral diatoms in response to large-scale climate dynamics in Antarctica. *Sci. Adv.* 7:eabh3233.

5.7.57.   Pound, M.J. and Salzmann, U. (2017) Heterogeneity in global vegetation and terrestrial climate change during the late Eocene to early Oligocene transition. *Scientific Reports* 7:43386.

5.7.58.   Racki, G. (1999) Silica-secreting biota and mass extinctions: survival patterns and processes. *Palaeogeography, Palaeoclimatology, Palaeoecology* 154, 107-132.

5.7.59.   Rasmussen, D.A. and Stadler, T. (2019) Coupling adaptive molecular evolution to phylodynamics using fitness-dependent birth-death models. *eLife* 2019;8:e45562.

5.7.60.   Raup, D.M. (1985) Mathematical models of cladogenesis. *Paleobiology* 11(1), 42-52.

5.7.61.   Renaudie, J. (2016) Quantifying the Cenozoic marine diatom deposition history: links to the C and Si cycles. *Biogeosciences* 13, 6003-6014.

5.7.62.   Renaudie, J., Drews, E.-L., Böhne, S. (2018) The Paleocene record of marine diatoms in deep-sea sediments. *Foss. Rec.* 21, 183-205.

5.7.63. Rich, J.T., Neely, J.G., Paniello, R.C., Woelker, C.C.J., Nussenbaum, B., Wang, E.W. (2010) A practical guide to understanding Kaplan-Meier curves. *Otolaryngol Head Neck Surg.* 143(3), 331-336.

5.7.64. Rominger, A.J., Fuentes, M.A., Marquet, P.A. (2019) Nonequilibrium evolution of volatility in origination and extinction explains fat-tailed fluctuations in Phanerozoic biodiversity. *Scientific Advances* 5:eaat0122.

5.7.65. Roy, D. and Gupta, R.P. (1992) Classifications of discrete lives. *Microelectron. Reliab.* 32(10), 1459-1473.

5.7.66. Sancetta, C. (1999) Diatoms and marine paleoceanography In: *The Diatoms: Applications for the Environmental and Earth Sciences*, Stoermer, E.F., and Smol, J.P. (eds.), Cambridge University Press, Cambridge, U.K.: 374–386.

5.7.67. Sethi, D., Butler, T.O., Shuhaili, F., Vaidyanathan, S. (2020) Diatoms for carbon sequestration and bio-based manufacturing. *Biology* 9, 217.

5.7.68. Sigwart, J.D., Bennett, K.D., Edie, S.M., Mander, L., Okamura, B., Padian, K., Wheeler, Q., Winston, J.E., Yeung, N.W. (2019) Measuring biodiversity and extinction—present and past. *Integrative and Comparative Biology* 58(6), 1111-1117.

5.7.69. Sims, P.A., Mann, D.G., Medlin, L.K. (2006) Evolution of the diatoms: insights from fossil, biological and molecular data. *Phycologia* 45(4), 361-402.

5.7.70. Sluijs, A., Bowen, G.J., Brinkhuis, H., Lourens, L.J., Thomas, E. (2007) The Palaeocene-Eocene Thermal Maximum super greenhouse: biotic and geochemical signatures, age models and mechanisms of global change. In: *Deep-Time Perspectives on Climate Change: Marrying the Signal from Computer Models and Biological Proxies*, Williams, M., Haywood, A.M., Gregory, F.J., Schmidt, D.N. (eds.), The Micropalaeontological Society, Special Publications, The Geological Society, London: 323-349.

5.7.71. Sluijs, A., Bijl, P.K., Schouten, S., Röhl, U., Reichart, G.-J., Brinkhuis, H. (2011) Southern Ocean warming, sea level and hydrological change during the Paleocene-Eocene thermal maximum. *Clim. Past* 7, 47-61.

5.7.72. Spencer-Cervato, C. (1999) The Cenozoic deep sea microfossilrecord: explorations of the DSDP/ODP sample set using the Neptune database. *Palaeontol. Electronica* 2, 270.

5.7.73. Stel, V.S., Dekker, F.W., Tripepi, G., Zoccali, C., Jager, K.J. (2011) Survival analysis I: the Kaplan-Meier method. *Nephron. Clin. Pract.* 119, c83-c88.

5.7.74. Veillette, M.S. and Taqqu, M.S. (2011) A technique for computing the PDFs and CDFs of nonnegative infinitely divisible random variables. *J. Appl. Prob.* 48, 217-237.

5.7.75. Wang, S., Chen, A., Fang, J., Pacala, S.W. (2013) Speciation rates decline through time in individual-based models of speciation and extinction. *The American Naturalist* 182(3), E83-E93.

5.7.76. Weinberger, H.F. (1965) *A First Course in Partial Differential Equations with Complex Variables and Transform Methods*. Dover Publications, Inc., New York, New York, USA.

5.7.77. Winter, D., Sjunneskog, C., Scherer, R., Maffioli, P., Riesselman, C., Harwood, D. (2012) Pliocene-Pleistocene diatom biostratigraphy of nearshore Antarctica from the AND-1B drillcore, McMurdo Sound. *Global and Planetary Change* 96-97, 59-74.

5.7.78. Yakovlev, A. (2018) Energy current and computing. *Phil. Trans. R. Soc. A* 376: 20170449.

5.7.79. Zhu, L. and Hu, H. (2015) A stochastic SIR epidemic model with density dependent birth rate. *Advances in Difference Equations* 2015:330.

6

Diatom Food Web Dynamics and Hydrodynamics Influences in the Arctic Ocean

Abstract

Diatom food web dynamics are linked to the hydrodynamics of a given ocean system. Diatom biomass was measured as lipid content reflecting food quality source for consumers and used to simulate the dynamics of bloom formation during upwelling and changes in ice conditions on a seasonal basis in the Arctic Ocean. A Lattice Boltzmann model was used to incorporate macroscopic hydrodynamic variables and microscopic diatom taxon variables in the same simulation. *Chaetoceros, Fossula, Fragilariopsis, Haslea, Melosira, Navicula, Nitzschia, Pseudo-nitzschia,* and *Thalassiosira* represented diatom food web primary producer biomass of the Arctic Ocean. Changes were found to occur over time with increasing diatom blooms as a result of increased upwelling, while ice formation has diminished. On a seasonal basis, bloom formation in winter increased, then decreased in spring, and increased in summer, reflecting the changes in the timing of upwelling in the Arctic Ocean. With less ice formation, changes in diatom blooms have implications for changes in the Arctic food web dynamics as well as the climate regime.

Keywords: Diatom bloom, Lattice Boltzmann model, Arctic Ocean, biomass, ice formation, upwelling, advection, primary producers

6.1 Introduction

Diatoms are one of the most important primary producers in the world's oceans, being a key influence in the carbon cycle and sequestration as well as the silica cycle, especially throughout the Cenozoic (e.g., [6.12.30., 6.12.84., 6.12.143.]). Diatoms account for 40% of marine primary productivity (e.g., [6.12.59.]), 40% of marine primary production (e.g., [6.12.135.]) and 25 % of marine biomass (e.g., [6.12.76.]). Diatoms inhabit all areas of the oceans in a myriad of habitats, including extreme environments such as sea ice (e.g., [6.12.36.]). At any given time, diatom production and productivity may dictate the speed and direction of food web dynamics in the oceans (e.g., [6.12.135.]), especially in the high latitudes (e.g., [6.12.84.]).

Diatoms as primary producers are valuable as a food source for marine consumers. As a potential source for nutritional supplements (e.g., [6.12.14., 6.12.98.]) and biofuels for industrial production (e.g., [6.12.14., 6.12.80., 6.12.98.]), diatoms contain lipids that account for their value outside of their environment (e.g., [6.12.14., 6.12.98.]). As storage vehicles, diatoms are traceable via biomarkers (e.g., [6.12.108.]) over the course of geologic history (e.g., [6.12.123.]), indicating their high productivity over time and high lipid content as a percent of total biomass (e.g., [6.12.80., 6.12.108.]). Biomarkers have been used to detect diatom diversity during the Cenozoic (e.g., [6.12.121.]).

Sea ice is a determinant of primary productivity in the Arctic (e.g., [6.12.3., 6.12.41.]) and Antarctic (e.g., [6.12.105.]) Oceans. In marine food webs, sympagic diatoms constitute an important component of energy transfer and carbon fluctuation in the high latitudes (e.g., [6.12.68., 6.12.93.]). Diatom blooms may occur in response to warming temperatures, increased nutrients and sea ice melting (e.g., [6.12.8.]), under sea ice advection (e.g., [6.12.8., 6.12.63.]) or as a result of subglacial stratification and lack of convective mixing (e.g., [6.12.4.]). Changes in the amount of sea ice may be documented from sediment analyses with respect to paleoenvironments (e.g., [6.12.11.]). Sympagic diatom biomass or production may be tracked over geologic time as one way to understand available energy transfer in marine food webs (e.g., [6.12.93.]). Diatom productivity, fueled by nutrient availability and influenced by oceanic hydrodynamics, increases with respect to upwelling events (e.g., [6.12.145., 6.12.149.]), affecting marine food webs.

6.2   Purposes of this Study

Diatoms are an important food source for grazing zooplankton in sea ice laden waters, and specifically in the Arctic Ocean. Such waters may experience diatom blooms with respect to upwelling. Food web dynamics in the Arctic Ocean concerning changing diatom biomass, sea ice formation, upwelling, and blooms are simulated to produce different outcomes. Diatom lipid content is used as a proxy for biomass in simulation. With changes in the amount of sea ice present in the Arctic Ocean, simulations concerning the role of diatoms are of great importance in characterizing environmental conditions and the implications this has concerning climate change.

6.3   Background on Arctic Ocean Diatoms

6.3.1   Diatoms and their Relation to Sea Ice

In sea ice laden waters, diatoms are found in a variety of habitats and according to the degree of ice present [6.12.129., 6.12.130., 6.12.154.]. The Arctic Ocean receives an influx of biomass from the Pacific and Atlantic Oceans, thereby influencing marine food webs in terms of primary production in the northern hemisphere (e.g., [6.12.147.]). Arctic diatom genera include: *Nitzschia, Pseudo-nitzschia* and *Ceratoneis* found in the winter under sea ice; *Fragilariopsis* and *Fossula* found in spring melt ponds (e.g., [6.12.37.]); *Nitzschia, Navicula* and *Fossula* found in sea ice during the summer (e.g., [6.12.81., 6.12.107.]); *Thalassiosira* found in sea ice in spring (e.g., [6.12.81.]) or in marginal ice zones between spring and summer (e.g., [6.12.37.]); *Chaetoceros* found in the open water during the summer (e.g., [6.12.37.]). Dominant sympagic species include *Nitzschia frigida* (e.g., [6.12.12., 6.12.37.]) and *Melosira arctica* (e.g., [6.12.12., 6.12.15., 6.12.51.]). *Chaetoceros neogracilis* is dominant in pelagic waters of the Arctic (e.g., [6.12.37.]). Seasonality is related to the dominant genera present in the Arctic (e.g., [6.12.81., 6.12.141.]). Diatoms may grow in sea ice, under sea ice, or during sea ice melt (e.g., [6.12.5., 6.12.7., 6.12.90.]).

In the Barents Sea, sympagic diatoms induce the growth of pelagic diatoms, while pelagic and benthic diatoms are the sources of sympagic diatoms [6.12.12.]. These reciprocal influences are most evident from spring to summer in the Arctic [6.12.12., 6.12.65.]. Diatom sea ice species succession generally occurs with a transition from centrics such as *Porosira* and *Thalassiosira* in the spring to pennates such as some species of *Fragilariopsis*, *Navicula*, *Nitzschia*, and *Pseudo-nitzschia* in the summer (e.g., [6.12.65.]), although overlap in dominance does occur between centrics and pennates throughout the seasons (e.g., [6.12.81.]). *Fragilariopsis* and *Fossula* generally bloom in the spring, closely followed by *Thalassiosira* and *Chaetoceros* [6.12.72.] extending into the summer [6.12.1.]. *Chaetoceros* and *Melosira* may occur as vegetative cells as well as resting spores, extending their presence from spring into summer [6.12.72.]. *Chaetoceros* has produced resting spores in the North Water polynya as a result of nutrient depletion [6.12.16.]. Sea ice diatoms are seeds for inducing increases in biomass (e.g., [6.12.65.]), possibly inducing blooms (e.g., [6.12.4.]). The amount of sympagic diatoms released from the ice influences the magnitude of diatom biomass and the extent of a bloom [6.12.12.]. Bloom times are influenced by the degree of break up by the ice via the released diatoms [6.12.12.].

In the Arctic Ocean, there is an influx of freshwater which influences the degree of sea ice present (e.g., [6.12.3., 6.12.116., 6.12.152., 6.12.153.]). This influx induces an increase in diatom biomass with their high growth rate and are the dominant component of blooms [6.12.10.]. As sea ice changes occur, diatom blooms may consist of *Fossula arctica*, *Fragilariopsis cylindrus* and *Fragilariopsis oceanica* then transition to *Thalassiosira* spp. [6.12.72., 6.12.130.] and *Chaetoceros* [6.12.72.], while other pennate diatoms follow.

The degree of Arctic sea ice present induces changes in diatom biomass, production and productivity (e.g., [6.12.3.]) as well as the formation of blooms, whether under the ice or at the sea ice edge (e.g., [6.12.4., 6.12.6.]). For the Arctic Ocean, wind-driven events influence the degree of mixing in the water column with a lessening of the amount of sea ice feeding such events [6.12.3., 6.12.4.]. Advective currents [6.12.147.] may induce changes in Arctic diatom flora on a large-scale basis [6.12.3., 6.12.4.]. Advection may be representative of changes in seasons with respect to temperature, vertical mixing, and the status of ice formation (e.g., [6.12.12.]) with lateral current movements (e.g., [6.12.160.]). Convection may induce changes in the Arctic diatom flora as well. Sea ice formation is affected by convective currents being weakened by freshwater inputs, for example, and vertical convective currents affect the degree of vertical mixing or stratification present (e.g., [6.12.96.]). As less sea ice is formed, winds become more effective at driving advection of surface currents, inducing Ekman transport and vertical mixing (e.g., [6.12.106.]). More ice formation may induce weakened deep water convective circulation and more stratification (e.g., [6.12.96.]).

6.3.2 Sea Ice, Upwelling and Diatom Productivity

Within zones of sea ice or at the edge of sea ice, upwelling events occur (e.g., [6.12.25.]) via geostrophic (zonal and meridional) flow (e.g., [6.12.105.]). Upwelling occurs along the coastal ocean when warm water surface currents move offshore via Ekman transport, and these currents are then replaced with cold, deep nutrient rich waters that flow toward shore as a result of wind-driven air currents that are parallel to the shore (e.g., [6.12.18., 6.12.27.]). Surface winds are perpendicular to the water currents because of the Coriolis effect, inducing the upwelling event (Figure 6.1). Advection may characterize the movement of shallow

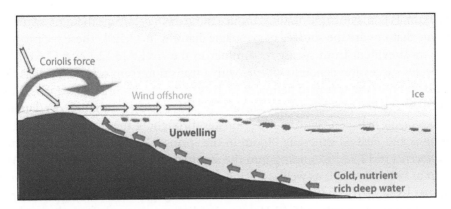

Figure 6.1 Diagram of coastal upwelling in sea ice laden waters such as the Arctic Ocean. Green masses are ice and open water diatoms.

waters as a result of upwelling [6.12.83.]. Downwelling occurs via Ekman pumping as surface waters are driven toward the shore with nutrient rich waters being cycled away from the shore into cold, deep waters (e.g., [6.12.85.]).

In sea ice laden waters, upwelling may enhance ice diatom production via increased vertical mixing, producing increased open water areas (e.g., [6.12.74.]). Diatom production may increase in polynyas in the Arctic, being influenced by upwelling (e.g., [6.12.136.]). In the Bering Sea, Walker circulation induced sea ice expansion at the Mid-Pleistocene Transition [6.12.130.], which in turn, played a role in deep water upwelling (e.g., [6.12.154.]) characteristic of waters in the Arctic [6.12.1.]. In the Cretaceous, the absence of sea ice in the Arctic meant that temperate taxa were a component of the primary producers and upwelling was a contributing factor in nutrient delivery [6.12.34.]. In modern times, Arctic sea ice has diminished and may be affecting upwelling, which in turn, has affected primary productivity and biomass (e.g., [6.12.3.]). Pelagic taxa such as *Thalassionema nitzschiodes* have become dominant as a result of upwelling, replacing taxa such as *Chaetoceros mitra* [6.12.1.]. Freshwater and brackish water diatom species are present in the Arctic (e.g., [6.12.1.]), and melting sea ice may contribute to an increased influx of such taxa (e.g., [6.12.154.]). With melting sea ice, the influx of temperate diatom species may be increasing as well [6.12.3.]. Seasonal sea ice induces phytoplankton blooms [6.12.130.] in which diatoms are the major component (e.g., [6.12.10., 6.12.129.]). Diatoms contribute the highest biovolume to Arctic Ocean blooms [6.12.5.].

In the high latitudes, diatoms account for the major component of primary production, phytoplankton biomass and primary productivity (e.g., [6.12.70.]). Diatom productivity in sea ice laden waters may induce diatoms to be dominant in blooms (e.g., [6.12.7., 6.12.70.]). Although blooms may be associated with seasonality (e.g., [6.12.37.]), upwelling events may induce a bloom (e.g., [6.12.90., 6.12.101.]), and upwelling may be seasonal as well (e.g., [6.12.157., 6.12.158.]). Changes in the amount and characteristics of sea ice and the degree of melting may be influential in productivity as well as bloom formation [6.12.4., 6.12.6., 6.12.7., 6.12.127.]. The amount of sea ice and the resultant diatom presence may be linked to the hydrodynamics of vertical mixing and upwelling (e.g., [6.12.78.]).

6.3.3 Diatom Lipid Content as a Proxy for Biomass

During the Holocene, diatom productivity was higher than it is in modern times with respect to upwelling, as indicated by measurement using diatom lipid content via biomarkers of sea ice such as IP_{25} (e.g., [6.12.11., 6.12.23., 6.12.107., 6.12.149., 6.12.155.]), which is a highly branched, sesterterpenoid in the class of acyclic isoprenoids [6.12.112.]. Upwelling present in sea ice laden waters may be associated with diatom productivity from which a diatom bloom may occur (e.g., [6.12.4., 6.12.145.]). Productivity may be parsed as a rate of production or biomass over time, and biomass with respect to a biomarker may be obtained via a proxy such as lipid content (e.g., [6.12.108.]). Measurement of biomass via lipid content is related to the proxy of lipid content representing sea ice in the Arctic (e.g., [6.12.22.]) and Antarctic (e.g., [6.12.73.]).

Diatoms possess lipids which are composed of triacylglycerol (TAG), galactolipids (GL), phospholipids (PL), sterols (ST), hydrocarbons (HC), and free fatty acids (FFA) [6.12.64.]. Along with carbohydrates, lipids are directly involved in being converted to the energy diatoms need in order to grow and survive, and diatom lipid content is essential in actively maintaining populations whether in marine or freshwater settings (e.g., [6.12.122.]). Under nutrient limitation, diatoms have been documented to increase their lipid production [6.12.45., 6.12.64., 6.12.111., 6.12.124.], and this phenomenon may be viewed as an indicator of environmental adaptation via unicell evolution (e.g., [6.12.31.]). Diatoms have genes that induce lipid biosynthesis [6.12.31.], and a phylogenetic signal may exist to indicate potential species selection concerning lipid content (e.g., [6.12.45., 6.12.53.]).

Diatom biomass is an important component of any marine food web (e.g., [6.12.56.]). Lipid content is an indicator of the food quality of diatoms for primary consumers (e.g., [6.12.19., 6.12.20., 6.12.64., 6.12.92-6.12.94, 6.12.151.]) and are identifiable as biomarkers such as IP_{25} (e.g., [6.12.20., 6.12.91.-6.12.94.]). Diatom lipid content provides a proxy for biomass (e.g., [6.12.108.]) or primary production available as quality food to primary consumers (e.g., [6.12.20.]). Such consumers may be present in the benthic (e.g., [6.12.19., 6.12.91.]) or pelagic (e.g., [6.12.19.]) environment.

Lipid content as a percentage of biomass for nine Arctic Ocean dominant diatoms has been compiled and reported in Table 6.1. A range of values have been reported covering many species and conditions. Nutrient conditions dictate the degree of lipid accumulation in diatoms (e.g., [6.12.24., 6.12.45.]), so a range of values for each diatom genus covering minima and maxima over all habitats were listed.

6.3.4 Diatom Biomass and the Hydrodynamics of Upwelling

Lipid content is related to nutrient depleted conditions and is not a factor in diatom buoyancy [6.12.124.]. Reynolds numbers of 0.002-0.2 have been calculated for diatoms [6.12.124., 6.12.125.], and downwelling has not been found to induce rapid sinking of diatoms, given the quantity of deposits and oozes containing diatoms [6.12.125.]. As a result, diatom transport proceeds according to the hydrodynamics of the water, and upwelling may be used as one mode of diatom movement at a macroscopic level.

The density of diatoms via biomass as percent lipid content may accumulate over time, and the velocity at which accumulation occurs may be dictated by hydrodynamics in relation to upwelling and the amount of ice present. Upwelling is generally a cylindrical flow

Table 6.1 Percent lipid biomass for nine dominant Arctic diatom genera.

Genus	Lipids (% dry weight)	References
Chaetoceros	7.63 - 14.86; 30.93 (summer) - 33.71 (winter); 32.77 - 37.38	[6.12.39.]; [6.12.32.]; [6.12.45.]
Fossula	0.84 - 11.43*	[6.12.155.]
Fragilariopsis	0.5; 0.84 - 11.43*	[6.12.81., 6.12.91.]; [6.12.155.]
Haslea	30.5 - 37.3	[6.12.88.]
Melosira	14.51 - 27.56; 32.84 (summer) - 33.18 (winter);	[6.12.45.]; [6.12.32.]
Navicula	20.76 - 40.12; 40.61 (summer) - 42.09 (winter); 24 - 51	[6.12.45.]; [6.12.32.]; [6.12.53.]
Nitzschia	22.73 - 40.12; 26 - 47	[6.12.45.]; [6.12.53.]
Pseudo-nitzschia	3.6 - 6.3; 13.5	[6.12.26., 6.12.35.]; [6.12.39.]
Thalassiosira	7.95 - 38.84; 19.42 - 24.00	[6.12.39.]; [6.12.45.]

*Values are a combination of *Fossula/Fragilariopsis*.

from deep water to surface water toward the shoreline, then dissipation of the water at the surface (e.g., [6.12.18.]). Hydrodynamics is often elucidated using Navier-Stokes equations (e.g., [6.12.117.]). However, a given system of Navier-Stokes equations is often not solvable or very complicated in order to represent the system being studied. Simulating a hydrodynamic system in a more accessible format is enabled by the lattice Boltzmann model (LBM) (e.g., [6.12.57., 6.12.58., 6.12.119.]). For the LBM, a one-dimensional (1D), two-dimensional (2D), three-dimensional (3D), or n-dimensional lattice may be used (e.g., [6.12.33., 6.12.104.]).

This model is obtained via the Chapman-Enskog expansion for a lattice gas automata model (e.g., [6.12.47., 6.12.57.]) by application of the Bhatnagar-Gross-Krook (BGK) collision operator where relaxation time is used (e.g., [6.12.13., 6.12.58.]). Because of this, an equivalence between Navier-Stokes equations and the LBM kinetic representation of dynamics in an incompressible fluid is attainable (e.g., [6.12.46., 6.12.47., 6.12.57., 6.12.58., 6.12.117.]). Both macroscopic (mass, velocity and energy in the form of temperature) and microscopic interactions are utilized in the LBM (e.g., [6.12.57., 6.12.58.], but the microscopic interactions enable equivalency to the incompressible Navier-Stokes equations at low Mach numbers (e.g., [6.12.57.]), where the ratio of flow velocity to the speed of sound is the Mach number (e.g., [6.12.47.]).

6.4   Lattice Boltzmann Model

For a 2D square lattice, e.g., a D2Q9 lattice, the center node or intersection is a particle defined as a diatom genus at rest (Figure 6.2a). The same applies to a 3D lattice such as the D3Q19 lattice (Figure 6.2b) in which the z-direction is a projection of the 2D lattice (e.g., [6.12.58.]). Other diatom genera in the lattice occurs at each lattice node or intersection. Particle shape does not affect outcome (e.g., [6.12.54.]), so diatom valve shape is not a factor. Movement from one node to any adjacent node in the lattice occurs as a result of speed changes of each genus (Figure 6.2). All numbered lattice nodes, and therefore all diatom genera except the rest node, are moving, thereby inducing interactions. This movement occurs among genera at a microscopic level, while hydrodynamics of oceanic water occurs at a macroscopic level.

The LBM is being used to simulate the hydrodynamics of water flow inducing the creation of a diatom bloom in sea ice laden waters. Movement in the lattice is represented as collision and streaming [6.12.13.]. Densities of genera will change slightly via these movements, and there are associated velocities of these movements. For example, in a 2D lattice, nine velocities are associated with each of nine distribution functions. Interactions among genera are induced by collisions and streaming, and these interactions dictate how genus distribution functions interact with each other. Hydrodynamically, the result is the space and time evolutions of the genera and their distributions on a macroscopic level in the marine environment via microscopic interactions of the genera. From this, equilibrium distribution functions are obtained, representing given outcomes. Different boundary conditions will produce different outcomes. For example, melting ice which may change nutrient concentrations may be used as boundary conditions to induce upwelling (e.g., [6.12.4.].

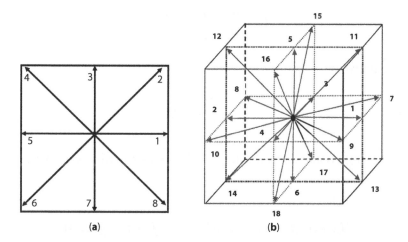

Figure 6.2 Examples of lattices for LBM: a, 2D as D2Q9; b, 3D as D3Q19. The particle at rest is 0 in the center of each lattice.

The lattice Boltzmann equation is a solution to the continuous Boltzmann equation in which the discrete outcome is obtained for use in the LBM [6.12.58.]. The Boltzmann equation with a basis in BGK via relaxation time to equilibrium is

$$\frac{\partial f}{\partial t} + c \cdot \nabla f + g \cdot \nabla_c f = -\frac{1}{\tau}(f - g) \tag{6.1}$$

where $f \equiv (x, t)$ is the velocity distribution function of lattice node x, t is the time, c is the microscopic velocity, ∇_c is the gradient operator in velocity or momentum space, τ is the relaxation time of the collision to local equilibrium, and g is the Maxwell-Boltzmann distribution given as

$$g \equiv \frac{\rho}{(2\pi RT)^{\frac{D}{2}}} \exp\left(-\frac{(c-u)^2}{2RT}\right) \tag{6.2}$$

where ρ, u and T are the macroscopic properties of density (of mass), velocity (momentum) and temperature (energy), R is the ideal gas constant, and D is dimension of the space [6.12.58., 6.12.60., 6.12.117., 6.12.119.]. The macroscopic properties are the moments of the velocity distribution function, f [6.12.58.].

For lipid density representing the diatoms, the Maxwell-Boltzmann distribution is the local equilibrium distribution function $g = f^{(eq)}$. The vector density along any direction with respect to the Maxwell-Boltzmann distribution may represent speeds of lipid density change. Streaming alternates with collisions and are one f_i from one discrete density, velocity and time to another f_{i+1} in the ith direction at the cth speed which occurs as links between lattice nodes (e.g., [6.12.119.]). When more than one genus arrives at a given node via streaming, a collision occurs at that node.

For discrete time, the collision equation becomes

$$\Omega = \frac{df}{dt} + \frac{1}{\tau}f = \frac{1}{\tau}g \tag{6.3}$$

where the time derivative

$$\frac{d}{dt} \equiv \frac{\partial}{\partial t} + c \cdot \nabla \tag{6.4}$$

travels in the direction obtained from c [6.12.58.]. For a small time-step, δ, the time evolution of f is

$$f(x + c\delta_t, t + \delta_t) - f(x, t) = -\frac{\delta_t}{\zeta}[f(x, t) - g(x, t)] \tag{6.5}$$

where $\tau \equiv \frac{\zeta}{\delta_t}$, which is the relaxation time in δ_t time steps for each collision time ζ (e.g., [6.12.58.]).

For an incompressible flow, conservation of mass (via density, ρ), momentum (velocity as u) and temperature (T) are obtained as integration in momentum space via quadrature (e.g., [6.12.47., 6.12.58.]). The discretized macroscopic ρ, u and T are given as

$$\rho = \sum_{\alpha} f_{\alpha} = \sum_{\alpha} g_{\alpha} \tag{6.6}$$

where ρ induces

$$\rho u = \sum_{\alpha} c_{\alpha} f_{\alpha} = \sum_{\alpha} c_{\alpha} g_{\alpha} \tag{6.7}$$

and

$$\rho \varepsilon = \frac{1}{2} \sum_{\alpha} (c_{\alpha i} - u)^2 f_{\alpha i} = \frac{1}{2} \sum_{\alpha} (c_{\alpha i} - u)^2 g_{\alpha i} \tag{6.8}$$

for the αth speed with $\alpha = 1, 2, \ldots, n$ in the ith direction, and $\varepsilon = \dfrac{D_0}{2} RT$ [6.12.58., 6.12.119.]. Small velocity expansion approaching the incompressible limit is obtained via low Mach number approximation [6.12.47.] of Eq. (6.2) which yields

$$f^{(eq)} = \frac{\rho}{(2\pi RT)^{\frac{D}{2}}} \exp\left(-\frac{c^2}{2RT}\right) \times \left[1 + \frac{(c \cdot u)}{RT} + \frac{(c \cdot u)^2}{2(RT)^2} - \frac{u^2}{2RT}\right] \tag{6.9}$$

forming the local equilibrium function, $f^{(eq)}$, up to $O(u^2)$ [6.12.57.], and $RT = c_{sound}^2 = \dfrac{c^2}{3}$ [6.12.58.]. Discretization of the c-space as a phase space is necessary in order to obtain a lattice structure (e.g., [6.12.47., 6.12.58.]) so that via quadrature, the discretized equilibrium function and lattice structure are obtained [6.12.58.].

6.5 Lattice Boltzmann Model and Hydrodynamics

6.5.1 Upwelling and Buoyancy

Generally, the Arctic Ocean consists of a polar mixed surface layer separated from an Atlantic layer of warm water at 200 to 500 m by the halocline in cold waters with a cold deep water layer below this, extending to beyond 2500 m [6.12.62.]. Upwelling may occur in the upper Atlantic layer within a 200m range (e.g., [6.12.148.]). In a multi-scale generalized setting, upwelling is advective in terms of surface currents, while oceanic currents occur in a convective fashion, driving large-scale deep water movement so that circulation is a double cell (e.g., [6.12.43.]).

During upwelling, flow is a rotation from deep to surface waters, moving toward the coastline. Water density is influenced by the amount of ice present, affecting temperature gradients, while influx of freshwater affects the salinity gradients (e.g., [6.12.25.]). Streaming and collisions of diatom genera occur within the rotation of water in this environment.

Upwelling occurs with respect to vertical buoyancy and is influenced by the influx of freshwater (e.g., [6.12.134.]) or melting ice (e.g., [6.12.4.] and is present during unstable rotating density (or gravity) currents [6.12.38.]. Upwelling [6.12.28.] as vertical buoyancy flux is given as

$$\Phi^+_{vertical} = \int_{\mathscr{v}} \rho u_{vertical}\, dV \tag{6.10}$$

where $\mathscr{v}$ is the flow domain, the vertical velocity is $u_{vertical} > 0$, and $V = u^2$ [6.12.38.]. The rotation occurs cylindrically (e.g., [6.12.66.]) whereby eventually, the upwelled current flow reaches an approximate geostrophic equilibrium [6.12.38.]. Convective cylindrical large-scale flow influences buoyancy and upwelling as advective smaller-scale flow. In the Arctic Ocean, buoyancy inputs affect the amount of sea ice present via advection, with ice thickness dictating the amount of convection at the surface (e.g., [6.12.29.]).

On a 2D lattice in shallow waters, streaming is given as

$$\Xi_{\alpha i} = f_{\alpha i}(x + c\delta_t, t + \delta_t) \tag{6.11}$$

with δ_t time steps, and collision occurs locally as

$$\Omega_{\alpha i}(x,t) = -\frac{1}{\tau}[f_{\alpha i}(x,t) - f_{\alpha i}^{(eq)}(x,t)] \tag{6.12}$$

[6.12.2., 6.12.119.]. The equation for circular cylindrical flow is given as

$$f_{\alpha i}(x + c\delta_t, t + \delta_t) - f(x,t) = -\frac{1}{\tau}[f_{\alpha i}(x,t) - f_{\alpha i}^{(eq)}(x,t)] + W_{\alpha i}\frac{\delta_t}{(c/\sqrt{3})^2}c_{\alpha i}F_i(x,t) \tag{6.13}$$

where the weight coefficient for each velocity is

$$W_{\alpha i} = \begin{cases} \dfrac{4}{9} & \alpha = 0 \\[2mm] \dfrac{1}{9} & \alpha = 1,3,5,7 \\[2mm] \dfrac{1}{36} & \alpha = 2,4,6,8 \end{cases} \tag{6.14}$$

[6.12.83., 6.12.97., 6.12.100., 6.12.118.] in the ith direction for α-speeds, and $F_i = -gr \, h_{sea \, water} \dfrac{\partial z_b}{\partial x_i} + \dfrac{sea \, water_i}{\rho_{sea \, water}} - \dfrac{bed_i}{\rho_{sea \, water}}$ is the force term with $h_{sea \, water}$ as water depth, $gr = 9.81 \, m/sec^2$ as gravity, z_b is bed elevation, $sea \, water_i$ is the sea water sheer stress, b_i is the bed sheer stress, and $\rho_{sea \, water}$ is sea water density [6.12.83.] which is 1024 kg/m³ for the Arctic Ocean [6.12.120.]. Reduced gravity is represented in F [6.12.113.]. The weight coefficient is the inverse of the relaxation rate $\tau = \dfrac{1}{W_{\alpha i}}$ [6.12.131.].

Advection in shallow waters from upwelling [6.12.83.] may induce an increase in biomass in sea ice laden waters such as the Arctic Ocean [6.12.106., 6.12.147.] and potentially induce early seasonal blooms (e.g., [6.12.133., 6.12.136., 6.12.137.]). The amount of ice present in the Arctic Ocean affects upwelling (e.g., [6.12.134.]) as does the advection of buoyancy [6.12.114.]. For dominant diatom genera in the Arctic Ocean, their movements of streaming and collisions occur as upwelling occurs. Streaming is advective [6.12.119.], inducing advective-diffusive collisions [6.12.83.]. The cylindrical flow equation including streaming and collisions is

$$\Xi_{\alpha i} - f_i(\mathbf{x}, t) = \Omega_{\alpha i} + W_{\alpha i} \frac{\delta_t}{(c/\sqrt{3})^2} c_{\alpha i} F_i(\mathbf{x}, t) \tag{6.15}$$

so that the probability of finding a particular diatom genus at a given node at t with velocity $c_{\alpha i}$ in the ith direction is $f_i(\mathbf{x}, t)$ at local equilibrium. Using this result and modifying Eq. (6.9),

$$f^{(eq)} = W_{\alpha i} \rho \left[1 + \frac{3(\mathbf{c}_i \cdot \mathbf{u})}{c^2} + \frac{9(\mathbf{c}_i \cdot \mathbf{u})^2}{2c^4} - \frac{3(\mathbf{u} \cdot \mathbf{u})}{2c^2} \right] \tag{6.16}$$

is the local equilibrium function [6.12.58., 6.12.97.] for each diatom genus.

6.5.2 Collisions and Streaming Densities of Diatom Genera during Upwelling

Diatom genera biomass represented by percent lipid content and the accumulated amount of biomass occurs with respect to collisions and streaming during upwelling. Collision densities are obtained for each diatom genus at the ijth node as

$$\rho_{ij}(\mathbf{x}, t) = \rho_{ij} - \lambda_{viscosity} \Delta t (\rho_{ij} - \rho_{ij}^{(eq)}) \tag{6.17}$$

where $\lambda_{viscosity}$ is a relaxation or decay coefficient with respect to viscosity (with $\lim \lambda \to 0 =$ no collisions), and

$$\rho_{ij}(\mathbf{x} + c_{ij} \Delta t, t + \Delta t) = \rho_{ij}(\mathbf{x}, t) \tag{6.18}$$

are the streaming densities [6.12.114.]. The relaxation coefficient is obtained as Δx or the spacing between two lattice points [6.12.113.]. For collisions and streaming,

$$\rho_{ij}(\mathbf{x} + c_{ij}\Delta t, t + \Delta t) = \rho_{ij}(\mathbf{x}, t) - \lambda_{viscosity}\Delta t(\rho_{ij}(\mathbf{x}, t) - \rho_{ij}^{(eq)}(\mathbf{x}, t)) \qquad (6.19)$$

(after [6.12.113., 6.12.114.]) is the total diatom genera density movement dynamics.

Diatom movement is affected by advection during upwelling. From Eq. (6.1), advection of buoyancy [6.12.114.] is given as

$$A_{Advection} = \frac{\partial}{\partial t} + c_{ij} \cdot \nabla \qquad (6.20)$$

in the ijth horizontal and vertical directions for each diatom genus, and

$$\frac{\partial \rho}{\partial t} + \nabla \cdot \mathbf{u} = 0 \qquad (6.21)$$

is the continuity equation at equilibrium, where $\mathbf{u}(\mathbf{x}, t) \equiv \sum_{ij} c_{ij}\rho_{ij}(\mathbf{x}, t)$ (after [6.12.114.]) with $\mathbf{u} \cdot \nabla\Phi_z^+$ with respect to upwelling (after [6.12.38., 6.12.114.]) and buoyancy.

6.5.3 Buoyancy and Ice

Buoyancy via Eqs. (6.7) and (6.10) with respect to the degree of ice formation or melting has a bearing on diatom genera. Sea ice measurement includes ice above and below the water line, and snow cover must be measured as well. At hydrostatic equilibrium, the relation between buoyancy and sea ice thickness is given as

$$H_{thickness} = \frac{h_{freeboard} \cdot \rho_{sea\,water} - h_{snow} \cdot (\rho_{sea\,water} - \rho_{snow})}{\rho_{sea\,water} - \rho_{ice}} \qquad (6.22)$$

where $h_{freeboard} = \dfrac{h_{snow}}{H_{thickness}}$ which is the total amount of ice and snow above the water line, h_{snow} is the amount of snow, and the other variables are densities of sea water, snow and ice [6.12.120.]. At equilibrium, sea water density is

$$\rho_{sea\,water} = \frac{\rho_{snow}\,h_{snow}}{h_{snow} - h_{snow}\big/H_{thickness}} \qquad (6.23)$$

and buoyancy becomes

$$\Phi_{vertical}^+ = \int_v \frac{\rho_{snow}\,h_{snow}}{h_{snow} - h_{snow}\big/H_{thickness}} u_{vertical}\,dV \qquad (6.24)$$

with $u_{vertical} > 0$ during upwelling with respect to sea ice, snow and thickness. Total buoyancy dynamics with regard to collisions and streaming for diatom genera is given as

$$\Phi_i^+(x+c_i\Delta t, t+\Delta t) = \Phi_i^+(x,t) - \lambda_{diffusivity}\Delta t(\Phi_i^+(x,t) - \Phi_i^{+(eq)}(x,t)) \qquad (6.25)$$

where $\lambda_{diffusivity}$ is a relaxation coefficient with respect to diffusivity [6.12.114.]. For velocity, density and buoyancy dynamics, outcomes from LBM simulation are geared toward achieving local equilibrium.

6.5.4 Upwelling and the Splitting of the Cylindrical Rotation of Currents

Generally, upwelling is describable as a process in which buoyancy drives flow via density changes in the water column inducing more flow horizontally (e.g., [6.12.97.]). With upwelling, vertical advective and convective currents are broken, and the smoothness in the overall cylindrical rotation buckles and may split (e.g., [6.12.38., 6.12.115.]). That buckling then splitting is obtained as the ratio of the Coriolis force to inertial forces when an apparent geostrophic equilibrium state occurs [6.12.38.]. The ratio is

$$C = \frac{\omega r_0}{u_\Phi} \qquad (6.26)$$

where ω is the rotational angular velocity, r_0 is the radius of the cylindrical lock, the buoyancy velocity is $u_\Phi = \sqrt{gr_0 h_{sea\,water}}$ with $gr_0 = gr\dfrac{\rho_1 - \rho_0}{\rho_0}$ as reduced gravity, and $h_{sea\,water}$ is water depth or lock height [6.12.38.]. The lock refers to a lock-release rotating system in which water is moving from the bottom up when physical changes in the water environment induce such movement [6.12.38.]. Instability with respect to gravity currents as the change in buoyancy via density gradients affects shallow waters (e.g., [6.12.97.]). For LBM simulation, $r_0 = h_{sea\,water}$ [6.12.38.].

Unsteady flow may be indicative of upwelling and may be indicated via

$$\frac{1}{2C} \approx \frac{1.8}{m} \approx \frac{r - r_0}{r_0} \qquad (6.27)$$

where C is related to m, the number of buckles, where m is an integer value [6.12.38.]. Splitting of the buckles means unstable currents represented by $m \geq 2$ for upwelling, and stable currents represented by $m = 1$ for no upwelling [6.12.38.].

6.6 Lattice Boltzmann Model: Diatom Bloom Density, Sea Ice and Upwelling

Diatom growth rate may double on a daily basis, sometimes exceeding phytoplankton biomass doubling [6.12.48.]. Phytoplankton growth rates, including diatoms, are controlled by zooplankton grazing and nutrient recycling in the euphotic zone [6.12.75.] characteristic of upwelling. For simulation, upwelling is used as the process to produce a diatom bloom in various sea ice conditions. Excessive diatom growth and production of a bloom is describable as an increase in biomass via percentage of lipid content at a high rate (velocity) of diatom cell division so that the growth rate is

$$\frac{d(\mu_{growth})}{dt} = \frac{\ln N_2 - \ln N_1}{t_2 - t_1} \tag{6.28}$$

where N is the cell concentration [6.12.37.] or diatom biomass [6.12.10.].

Diatoms in sea ice laden waters undergo a boom-and-bust cycle (e.g., [6.12.10., 6.12.49.]). The rate of change in biomass equation is

$$\frac{d(\mu_{biomass})}{dt} = \frac{\mu_{t_2} - \mu_{t_1}}{t_2 - t_1} \tag{6.29}$$

with the boom part of the cycle occurring at the maximum peak biomass, μ_{max}. To produce a bloom means a very high rate of biomass increase in diatoms. This increase may be determined for a fixed time or a time interval via advection. For a given time interval, the rate of diatom bloom density is

$$\frac{d(\rho_{Diatom_i})}{dt} = (\mu_i \cdot \rho_i) - (\xi_i \cdot \rho_i) - \nabla \cdot \mathbf{J_i} - \nabla \cdot \rho_i c_i \tag{6.30}$$

where $\mu_i \cdot \rho_i$ is the rate of biomass production, $\xi_i \cdot \rho_i$ is the rate of mortality, $\mathbf{J_i} = \rho_i \mu_i$ as the flux in diatoms relative to water movement, and $\nabla \cdot \rho_i c_i$ is an advection term [6.12.40.].

From season to season, the mortality rate of may be small and offset in simulation, considering that resting spores are a large component of diatoms in the Arctic Ocean which provide seeds to reestablish taxa during favorable conditions (e.g., [6.12.42.]). Additionally, the overall mortality rate from grazing loss may be approximately equal to the diatom growth rate (e.g., [6.12.9.]), inducing equilibrium conditions in an Arctic food web so that only growth rate needs to be considered. To induce a bloom, the rate of diatom density change may be rewritten as

$$\frac{d(\rho_{Diatom_i})}{dt} = (\mu_i \cdot \rho_i)_{rate\ of\ production} - \min[(\xi_i \cdot \rho_i) - \nabla \cdot \mathbf{J_i}] - G_{ice\ flux} \tag{6.31}$$

where $G_{ice\ flux} = \dfrac{dIce_{thickness}}{dt}$ and is the rate of change between ice thickness and open water [6.12.12.]. If $G_{ice\ flux}$ is positive, then sea ice is increasing, and if negative, then sea ice is melting [6.12.12.]. The modified rate of diatom bloom density equation is

$$\frac{d(\rho_{Diatom_i})}{dt} = (\mu_i \cdot \rho_i)_{rate\ of\ production} - G_{ice\ flux} \tag{6.32}$$

where positive $G_{ice\ flux}$ decreases bloom density, and negative $G_{ice\ flux}$ increases bloom density. That is, ice formation induces less open water and a reduction in diatom bloom density, while ice melting induces seeding and an increase in open water and an increase in diatom bloom density. In this way, the amount of sea ice cover is linked to diatom bloom dynamics and may be inferred on a seasonal basis (e.g., [6.12.156.]).

6.7   Lattice Boltzmann Model: Specifications for Simulation

Arctic Ocean general and ice habitat conditions with respect to season as well as bloom identifier genera are specified in Table 6.2. Cylindrical rotation of the system occurs with two or more revolutions to maintain conservation of mass and momentum and reach approximate geostrophic equilibrium (e.g., [6.12.38.]). Varying time steps and one relaxation time are used to reach outcomes up to bloom with respect to seasonality, change in diatom genera per season, upwelling, and ice conditions. A summary of Arctic Ocean scenarios is given in Table 6.3, and a generalized composite diagram based on Tables 6.1-6.3 was devised and presented as Figure 6.3.

Reynolds numbers are used as indicators of flow characteristics for each LBM simulation. For cylindrical flow, Reynolds numbers (Re) of 20 for steady flows and 100 for unsteady flows have been used [6.12.131., 6.12.159.], with $Re = 100$ as the initial indicator of upwelling. To induce vortex splitting, $Re > 100$ [6.12.89.] is used for upwelling with respect to ocean state. Reynolds numbers are related to density as $Re = \dfrac{\rho u h_{sea\ water}}{v_{dynamic}}$ [6.12.109.] where $v_{dynamic}$ is the dynamic viscosity of water motion. Lower Reynolds numbers mean that density is lower, favoring ice formation conditions (e.g., [6.12.2.]). To obtain Reynolds numbers for bloom conditions, diatom bloom density is used in relation to ocean state, and specifically with regard to season, ice density values are used.

The values for different ocean states of ice formation indicate seasonal changes in the Arctic Ocean in terms of buoyancy, upwelling and bloom dynamics. Arctic sea ice density ranges from 675 to 954 kg/m³ [6.12.61.], resulting in an average of 814.5 kg/m³ per year. Monthly variation in Arctic sea ice density ranges from is 832 to 916 kg/m³, covering August to February [6.12.61.], and from this, 905, 895, 835, and 880 kg/m³ are the average densities for winter, spring, summer, and autumn, respectively. Estimates for Arctic sea water density are 1024 to 1026 kg/m³ [6.12.12., 6.12.120.]. Melting sea ice is mediated by sea ice thickness, and in turn, this affects salinity concentrations (e.g., [6.12.12.]). However, melted sea ice density is in the range between that for sea ice and sea water at approximately 1000 kg/m³ (e.g., [6.12.95.]). Mean ice thickness for the Arctic Ocean historically has been 2.4 m in spring, 3.3 m in summer, 3.0 m in autumn, and 2.8 m in winter [6.12.17.].

Table 6.2 Characteristics of Arctic diatom genera for LBM simulations.

Genus	General habitat	References: general habitat	Season	Ice habitat	References: season and ice habitat	References: upwelling	References: bloom and season
Chaetoceros (vegetative cells and resting spores)	pelagic	[6.12.45.]	spring; spring into summer	in sea ice; under sea ice	[6.12.4.]; [6.12.37.]; [6.12.50.]; [6.12.72.]	[6.12.1., 6.12.129.]	[6.12.4.] (spring); [6.12.50, 6.12.72.] (spring into summer)
Fossula	sympagic	[6.12.129., 6.12.154.]	spring; spring into summer	melting ice, under sea ice	[6.12.37.]; [6.12.72., 6.12.154.]	[6.12.155.]	[6.12.4.] (spring), [6.12.72.] (spring); [6.12.141.] (spring); [6.12.155.] (summer)
Fragilariopsis	sympagic	[6.12.129., 6.12.154.]	into spring; spring into summer	melting ice, under sea ice	[6.12.4.]; [6.12.37.]; [6.12.65.]; [6.12.72., 6.12.154.]	[6.12.155.]	[6.12.4.] (winter into spring); [6.12.72.] (spring); [6.12.141.] (spring); [6.12.37.] (spring into summer)
Haslea	sympagic	[6.12.128.]	spring	in sea ice	[6.12.81.]	[6.12.110.]	[6.12.23.] (spring)
Melosira (vegetative cells and resting spores)	pelagic	[6.12.45.]	spring into summer	in sea ice, under sea ice	[6.12.4., 6.12.72.]; [6.12.15.]	[6.12.4.]	[6.12.72.] (spring into summer)
Navicula	benthic	[6.12.45., 6.12.129., 6.12.154.]	summer	in sea ice	[6.12.65.], 6.12.81.]	[6.12.140.]	[6.12.4.] (summer)

(Continued)

Table 6.2 Characteristics of Arctic diatom genera for LBM simulations. (*Continued*)

Genus	General habitat	References: general habitat	Season	Ice habitat	References: season and ice habitat	References: upwelling	References: bloom and season
Nitzschia	benthic	[6.12.45.]	winter; spring; summer	in sea ice; under sea ice	[6.12.37.]; [6.12.1.]; [6.12.65., 6.12.72., 6.12.81.]	[6.12.140.]	[6.12.4.] (winter); [6.12.37.] (summer)
Pseudo-nitzschia	benthic	[6.12.26., 6.12.35.]	winter; spring into summer	in sea ice; under sea ice	[6.12.103.]; [6.12.99.]; [6.12.72.]	[6.12.79.]	[6.12.4.] (winter); [6.12.72.] (spring into summer)
Thalassiosira (vegetative cells and resting spores)	pelagic	[6.12.45., 6.12.129., 6.12.154.]	spring; spring into summer	in sea ice; marginal sea ice	[6.12.4.]; [6.12.72., 6.12.81, 6.12.65.]; [6.12.37., 6.12.154.]	[6.12.1.]	[6.12.4.] (spring); [6.12.72.] (spring into summer)

Table 6.3 Arctic Ocean summary characteristics used in LBM simulation.

Ice season	Ice amount	Dominant diatoms	Bloom month	Upwelling season
spring	mostly ice and some melting ice cover	ice and melting ice taxa	early May	spring[a]
spring	less ice and more melting ice cover	ice taxa	middle May	spring[a]
summer	more open water (67%)[c] than ice (33%)[c] or melting ice cover[b]	pelagic taxa	middle June	summer[b]
autumn[c]	sea ice minimum; thinner ice[c]	pelagic taxa	weak bloom[b]	autumn[a]
winter	mostly ice cover; thicker ice and sea ice maximum[c]	ice and melting ice taxa	November, December[d]	winter[c, e]

[a] [6.12.140.], [b] [6.12.28.], [c] [6.12.134.], [d] [6.12.158.], [e] [6.12.85.]

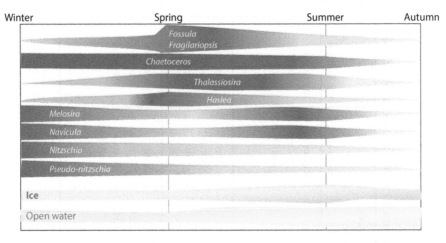

Figure 6.3 Generalized composite diagram of diatom genera seasonal presence, ice thickness and open water in the Arctic Ocean.

6.7.1   Overview of 2D LBM with Respect to Diatom Genera Lattice Nodes

For the D2Q9 square lattice, a diatom genus changes position via a collision with another diatom genus as $\mathbf{x},t$ after streaming as $\mathbf{x} + c\Delta t$, indicating genus horizontal movement to a neighboring lattice node. The lattice spacings are Δx and Δy in the x and y directions. For the nine diatom genera, $c = \dfrac{\Delta x}{\Delta t} = \dfrac{\Delta y}{\Delta t}$ are the nine velocities given as

$$c_{\alpha i 0} = (0,0) \qquad\qquad\qquad\qquad\qquad \alpha = 0$$

$$c_{\alpha i Odd} = \left(\cos\left[\frac{\pi}{4}(\alpha-1)\right], \sin\left[\frac{\pi}{4}(\alpha-1)\right] \right) \qquad \alpha = 1,3,5,7 \qquad (6.33)$$

$$c_{\alpha i Even} = \sqrt{2}\left(\cos\left[\frac{\pi}{4}(\alpha-1)\right], \sin\left[\frac{\pi}{4}(\alpha-1)\right] \right) \qquad \alpha = 2,4,6,8$$

[6.12.60., 6.12.83.] for the rest node, c_0, and eight moving diatom genera. The positions with respect to velocity on the lattice for the moving genera are $c_1 = (c,0)$, $c_2 = (c,c)$, $c_3 = (0,c)$, $c_4 = (-c,c)$, $c_5 = (-c,0)$, $c_6 = (-c,-c)$, $c_7 = (0,-c)$, $c_8 = (c,-c)$, where the odd subscripted velocities are movement in the x, y direction, and the even subscripted velocities are movement in the diagonal direction [6.12.113.]. That is, the directional velocity ijth vectors are

$$c_\alpha = \{(0,0), (1,0), (1,1), (0,1), (-1,1), (-1,0), (-1,-1), (0,-1), (1,-1)\} \quad (6.34)$$

from Eq. (33).

The velocities with horizontal, i, and vertical, j, components, and with respect to gravity, gr, have local equilibrium functions given in terms of ρ, u and c as

$$f_\alpha^{(eq)} = \begin{cases} \rho - \dfrac{5gr\rho^2}{6c^2} - \dfrac{2\rho}{3c^2}u_i u_i & \alpha = 0 \\[3mm] \dfrac{gr\rho^2}{6c^2} + \dfrac{\rho}{3c^2}c_{\alpha i}u_i + \dfrac{\rho}{2c^2}c_{\alpha i}c_{\alpha j}u_i u_j - \dfrac{\rho}{6c^2}u_i u_j & \alpha = 1,3,5,7 \\[3mm] \dfrac{gr\rho^2}{24c^2} + \dfrac{\rho}{12c^2}c_{\alpha i}u_i + \dfrac{\rho}{4c^2}c_{\alpha i}c_{\alpha j}u_i u_j - \dfrac{\rho}{24c^2}u_i u_j & \alpha = 2,4,6,8 \end{cases} \quad (6.35)$$

[6.12.83., 6.12.113.], and when rewritten with respect to Eqs. (6.14), (6.16) and (6.33) are more succinctly stated as

$$f_\alpha^{(eq)} = \begin{cases} \dfrac{4}{9}\rho\left(1 - \dfrac{3}{2}\mathbf{u}^2\right) \\[3mm] \dfrac{\rho}{9}\left[1 + 3(c_{\alpha i Odd}\cdot\mathbf{u}) + \dfrac{9}{2}(c_{\alpha i Odd}\cdot\mathbf{u})^2 - \dfrac{3}{2}\mathbf{u}^2\right] \\[3mm] \dfrac{\rho}{36}\left[1 + 3(c_{\alpha i Even}\cdot\mathbf{u}) + \dfrac{9}{2}(c_{\alpha i Even}\cdot\mathbf{u})^2 - \dfrac{3}{2}\mathbf{u}^2\right] \end{cases} \quad (6.36)$$

[6.12.60.]. Using Eq. (6.6),

$$\rho = \sum_{\alpha} f_{\alpha} = \frac{1}{u_i} \sum_{\alpha} c_{\alpha i} f_{\alpha} \tag{6.37}$$

and

$$u_i = \frac{1}{\rho} \sum_{\alpha} c_{\alpha i} f_{\alpha} \tag{6.38}$$

conserving mass and momentum [6.12.60., 6.12.83., 6.12.113.] in the ith coordinate direction, with the same applied to the jth coordinate direction. Summarizing the local equilibrium functions from Eqs. (6.13-6.15) and with respect to ρ and u via Eqs. (6.33-6.38) gives

$$f_{\alpha i}^{(eq)}(u,\rho) = W_{\alpha i}\rho(x,t)\left[1 + 3c_{\alpha i} \cdot u(x,t) + \frac{9}{2}(c_{\alpha i} \cdot u(x,t))^2 - \frac{3}{2}u(x,t)^2\right] \tag{6.39}$$

[6.12.131.] which reiterates Eq. (6.16).

For each diatom genus at each lattice node, ρ and u may be rewritten as

$$\rho(x,t) = \sum_i \rho_i(x,t) \tag{6.40}$$

and

$$u(x,t) = \sum_i c_i \rho_i(x,t) \tag{6.41}$$

where u is given in terms of ρ [6.12.114.]. Computation of the local equilibrium functions induce collision and buoyancy equations given as

$$\rho - \lambda_{viscosity}\Delta t(\rho - \rho^{(eq)}) = 0 \tag{6.42}$$

and

$$\Phi_{vertical}^+ - \lambda_{diffusivity}\Delta t(\Phi_{vertical}^+ - \Phi^{(eq)}) = 0 \tag{6.43}$$

with

$$u\Phi_{vertical}^+ = \sum_i c_i \Phi^{(eq)} \tag{6.44}$$

where c_i are buoyancy velocities [6.12.114.] so that u is given in terms of $\Phi_{vertical}^+$.

6.7.2   Buoyancy, Upwelling and Diatom Blooms in LBM via ρ and u

Using a D2Q9 lattice, a cross-section of flow is delineated via LBM. Rotation occurs around the cross-section and is controlled by the velocity, u, of the water in terms of upwelling. For LBM simulation, reduced gravity, shallow water dynamics and no slip or no stress boundary conditions may be applied (e.g., [6.12.60., 6.12.113., 6.12.114.]). This implies that for a cylindrical rotation, the curved boundaries are multi-reflection [6.12.131.] or bounce-back [6.12.100.]. For no slip conditions, $\Phi^+_{vertical} = 0$ for vertical velocity and buoyancy at the ocean bottom, and for no stress conditions at the cylinder wall and surface, $u = 0$ [6.12.114.]. At the no slip boundaries, incompressible conditions prevail, and viscosity is a very small number at the scale of the lattice, inducing the same conditions of stability as using a single relaxation time (e.g., [6.12.104.]). Relaxation time is related to viscosity as

$$\tau = \frac{6v+1}{2} \tag{6.45}$$

(e.g., [6.12.100., 6.12.159.]) where v is viscosity at which local equilibrium is achieved.

Buoyancy and the relation to amount of sea ice present is indicated by Eq. (6.44). To incorporate buoyancy into the LBM simulation

$$\Phi^+_{vertical} = -gr\frac{\rho_{taxon} - \rho_{ocean\ state}}{\rho_{taxon}} \tag{6.46}$$

where $\rho_{ocean\ state}$ is density of sea ice, melting ice or open water, and with reduced gravity, gr, and very small genus density values, the relation between buoyancy and ocean state density becomes

$$\rho_{ocean\ state} \approx \frac{\Phi^+_{vertical}}{gr} \tag{6.47}$$

[6.12.150.].

From Eq. (6.30), $\mathbf{J}_i = \rho_i u_i$ is the flux in diatoms relative to water movement in which water movement is dictated by buoyancy and upwelling via Eq. (6.10). From Eq. (6.44), u is related to vertical buoyancy via $u_\Phi = \sqrt{gr_0 h_{sea\ water}}$ from Eq. (6.26). For LBM simulation, ρ and u are the means by which buoyancy and upwelling are included in LBM simulation. An upwelling depth of $h_{sea\ water}$ = 150 m is used for the Arctic Ocean [6.12.25.]. Ice, melting ice and open water are the conditions that affect buoyancy and flow with regard to position and direction of diatom genera during upwelling.

By taking into account ice formation conditions, diatom bloom density induced via upwelling is obtained via Eq. (6.32). The rate of biomass production is the net growth rate [6.12.40.], and by using the advection term in Eq. (6.32), the net change in diatom bloom density is

$$d\rho_{Diatom_i} = (\mu_i \cdot \rho_i)_{rate\ of\ production} - \nabla \cdot \rho_i c_i \tag{6.48}$$

where biomass production in terms of ice status may be influenced by the advection of biomass in the water with respect to bloom formation. As a result,

$$\rho_{Diatom_i} = \frac{\mu_{max_i}}{100} \sum \rho_{ocean\ state} \tag{6.49}$$

where maximum diatom biomass as the percent lipid content and the sum of the ocean state densities influences bloom density.

6.8   Methods

Descriptive characteristics used for LBM simulations are based on Tables 6.1-6.3 and Figure 6.3. Numerical values reported in the specifications section are used in simulations, and a height of 150 m is used as well. LBM simulations of Arctic Ocean water flow were conducted as follows:

1. general examples of steady flow at $Re = 20$ and upwelling at $Re = 100$;
2. ocean state of ice, melting ice or open water densities for steady flow and upwelling;
3. all diatoms as bloom density with respect to ice, melting ice and open waters for steady flow and upwelling;
4. seasonal diatom bloom density during ice conditions for steady flow and upwelling.

General density values for sea ice, melting ice and open water as specified above were used as well as seasonal ice density values averaged from Ji *et al.* [6.12.61.]. Diatom bloom density values were calculated using Eq. (6.49) and sea ice density according to season. No slip boundary conditions were applied to the eastern coast (including continental slope and shelf, Figure 6.1) which is the western boundary, bottom bed, and eastern waters (as the eastern boundary), while the surface waters remained open, subject to wind. Changes in u reflect horizontal flow. All simulation results were depicted as diagrams. Additional diagrams were devised for upwelling with depth, with composites of those diagrams used to illustrate ocean state changes and diatom bloom density changes with respect to ocean state and season.

6.9   Results

LBM simulation results were depicted in diagrams of flow. Simulations were conducted for $h = 150$ m for steady and unsteady flow. For vortex splitting $h = 250$ m. Macroscopic velocities were $u = 1$, 10 or 50 for steady flow, upwelling and vortex splitting, respectively. To illustrate the methodology generally, steady and unsteady flow was depicted for Reynolds numbers of 20 and 100, respectively (Figure 6.4), depicting local equilibrium conditions.

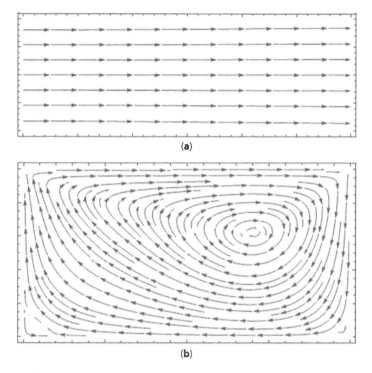

Figure 6.4 Flows in LBM: a, steady flow at Re = 20; b, unsteady flow at Re = 100.

To reiterate, average sea ice density = 815 kg/m³, melting ice density = 1000 kg/m³, and average open water density = 1025 kg/m³ as the values used in calculations. For Arctic sea ice, melting ice and open water, differences in steady flow were slight but detectable (Figure 6.5, column 1). Flow under each condition was generally from east to west boundaries, with the most dispersed flow for sea ice, followed by melting ice, and finally by open water (Figure 6.5). Upwelling was depicted for each ocean state, with similar patterns in which a tight vortex was formed toward the surface offshore (Figure 6.5, column 2). The tighter the vortex, the higher the upwelling. Profiles for sea ice and open waters were more similar to each other than to melting ice (Figure 6.5, column 2). At depth, upwelling with regard to ocean state illustrated that vortex splitting for melting ice and open waters were more similar to each other than to sea ice (Figure 6.6a, b, c). Downwelling profiles were very similar, although melting ice and open waters had tighter vortices than sea ice (Figure 6.6a, b, c). A composite profile was constructed for upwelling and ocean state reiterating upwelling and downwelling patterns (Figure 6.6d).

For all diatom genera, diatom bloom densities for the Arctic Ocean were calculated as 2291 kg/m³ for sea ice, 2811 kg/m³ for melting ice, and 2881 kg/m³ for open waters using maximum percent lipid content from Table 6.1. For steady flow, an east to west boundary pattern mostly summarizes conditions (Figure 6.7, column 1). As with ocean state conditions, upwelling was a similar pattern, except the vortex was generally more open (Figure 6.7, column 2). A secondary vortex at depth was just beginning to form in extended upwelling profiles (Figure 6.8a, b, c). Sea ice has a more vertical flow at the bottom than melting ice or open waters, and secondary vortices are more formed for sea ice and open waters than

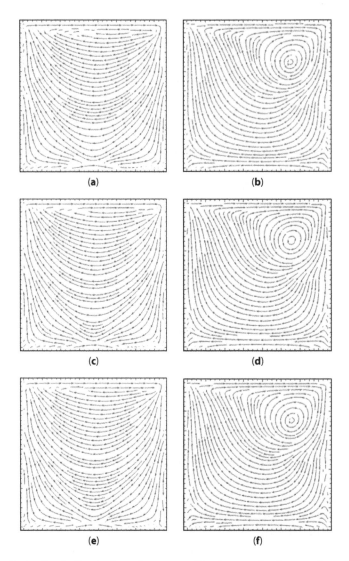

Figure 6.5 Arctic Ocean. Steady flow in first column, upwelling in second column: a, b, sea ice; c, d, melting ice; e, f, open waters.

melting ice although flow pattern for melting ice and open waters is more similar to each other than to sea ice (Figure 6.8a, b, c). A composite diagram illustrates sea ice being more different than melting ice or open waters during upwelling (Figure 6.8d).

Seasonal diatom bloom density with respect to sea ice in the Arctic Ocean was simulated in terms of changing seasons, referring to Table 6.2 and according to Table 6.3. From Table 6.1, genera used for winter were *Chaetoceros*, *Melosira*, *Navicula*, *Nitzschia*, and *Pseudo-nitzschia*. For spring, *Chaetoceros*, *Fossula*, *Fragilariopsis*, *Haslea*, and *Thalassiosira* values from Table 6.1 were used. Summer values from Table 6.1 were used for *Chaetoceros*, *Melosira*, *Navicula*, and *Thalassiosira*. Winter bloom density = 307 kg/m³, spring bloom density = 244 kg/m³, and summer bloom density = 299 kg/m³.

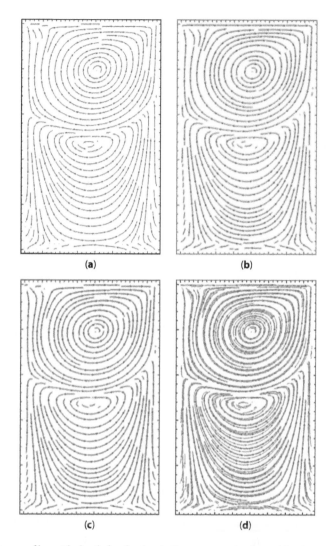

Figure 6.6 Upwelling profiles with depth for the Arctic Ocean: a, sea ice; b, melting ice; c, open water; d, superimposed ocean state profiles of sea ice (blue), melting ice (aqua) and open water (purple). The top is the surface, and the secondary vortex at the bottom is downwelling as a result of vortex splitting.

Steady flow was generally from east to west boundaries for each season with surface flow indicating a broader cyclonic pattern in winter, followed by spring and then summer (Figure 6.9, column 1). At depth, flow from east to west boundaries split for the summer diatom bloom in contrast to the winter and spring blooms (Figure 6.9, column 1). Upwelling was more similar between winter and spring where the vortex was slightly better formed and tighter in contrast to summer (Figure 6.9, column 2). Extended profiles of seasonal diatom bloom densities indicated that summer had the broadest upwelling vortex with spring having a tighter vortex, and winter having the tightest vortex for upwelling, although winter and spring were very similar (Figure 6.10a, b, c). Downwelling was better formed for summer and next for spring than it was for winter (Figure 6.10a, b, c). A composite diagram indicated the overall better formation of upwelling and diatom bloom density for winter

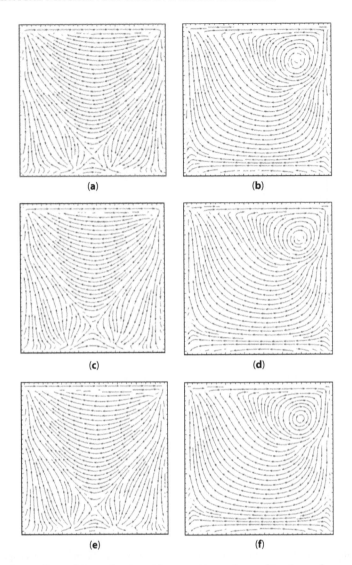

Figure 6.7 Arctic Ocean diatom bloom density with respect to sea ice, melting ice and open waters. Steady flow in first column, upwelling in second column: a, b, sea ice bloom density; c, d, melting ice bloom density; e, f, open water bloom density.

and spring, followed by summer, with downwelling best formed in summer, then spring, then winter (Figure 6.10d).

Seasonal diatom changes with sea ice and during upwelling conditions in the Arctic Ocean indicated that as ice density declined from winter to summer, percent lipid content declined as well, but bloom density dipped in the spring in contrast to winter and summer (Figure 6.11). From LBMs considering upwelling, diatom bloom density was highest in winter, very closely followed by spring, and the smallest bloom occurred in the summer (Figures 6.9, column 2 and 6.10). From LBMs, the spring diatom bloom increase may have been affected by potentially less ice present in addition to upwelling (Figures 6.9, column 2 and 6.10). A schematic was constructed to summarize and illustrate the

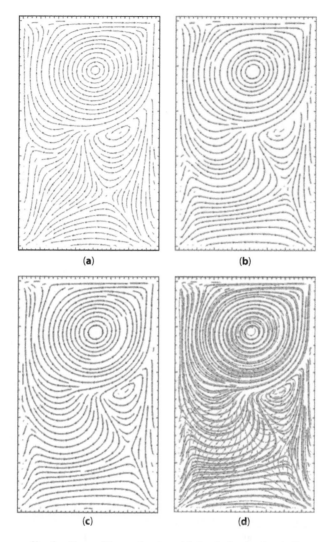

Figure 6.8 Upwelling profiles for diatom bloom density with depth for the Arctic Ocean: a, sea ice; b, melting ice; c, open water; d, superimposed ocean state profiles of sea ice (blue), melting ice (aqua) and open water (purple) in terms of diatom bloom density. The top is the surface, and the secondary vortex at the bottom is downwelling as a result of vortex splitting.

potential changes in the Arctic Ocean with regard to seasonal diatom blooms, sea ice and upwelling (Figure 6.12).

Seasonal diatom bloom density dynamics for the Arctic Ocean may be illustrated as a comparison between measured density as energy content and available density during upwelling, both of which are based on percent lipid content. Contrasting the outcomes from bloom energy with regard to an Arctic Ocean food web and LBMs ranked for blooms as available food, differences may be indicative of Arctic Ocean changing conditions concerning food web dynamics as well as ice conditions (Figure 6.13).

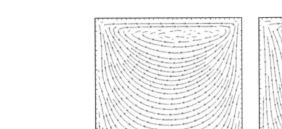

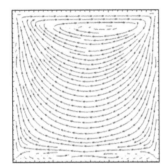
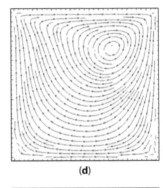
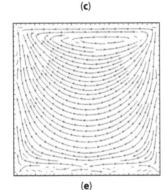

Figure 6.9 Arctic Ocean seasonal diatom bloom density with respect to sea ice. Steady flow in first column, upwelling in second column: a, b, winter bloom density; c, d, spring bloom density; e, f, summer bloom density.

6.10 Discussion

LBMs provided a way to combine a multitude of conditions in the Arctic Ocean to visualize seasonal and ocean state changes with regard to diatom bloom dynamics. Each scenario depicted via LBM enabled the association of specific diatom genera, their occurrence seasonally with respect to bloom density, and ocean state when upwelling occurs. Using percent lipid content of diatoms enabled inclusion of food quality sources with regard to conditions in the Arctic Ocean. In this regard, the usual biomass measurement was altered, but was useful in relation to trophic level dynamics for diatom blooms. Results were instructive in showing differences among ocean states and seasonal diatom bloom

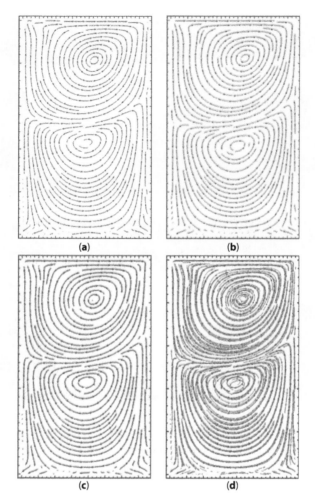

Figure 6.10 Upwelling profiles with depth for Arctic Ocean seasonal diatom bloom density with respect to sea ice: a, winter; b, spring; c, summer; d, superimposed upwelling profiles for winter (blue), spring (green) and summer (orange) diatom bloom densities. The top is the surface, and the secondary vortex at the bottom is downwelling as a result of vortex splitting.

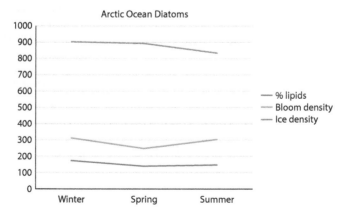

Figure 6.11 Seasonal diatom changes with respect to sea ice in the Arctic Ocean.

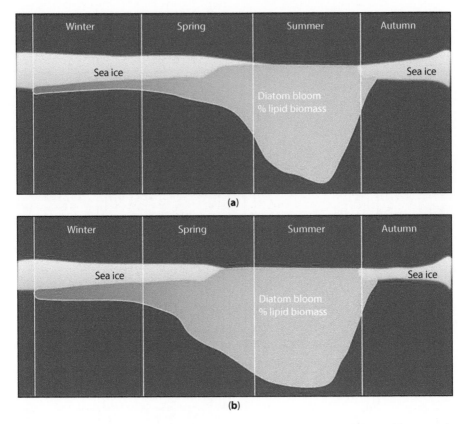

Figure 6.12 Schematic illustration of potential Arctic Ocean changes in seasonal diatom blooms with respect to: a, more sea ice, less upwelling; b, less sea ice, more upwelling.

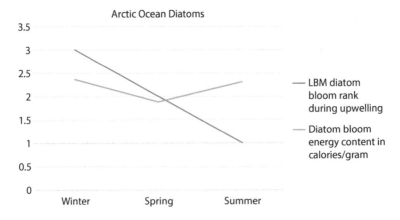

Figure 6.13 Relative comparison of seasonal diatom bloom energy available as food and LBMs ranked outcome with respect to upwelling.

density profiles for the Arctic Ocean and changes in ice regimes with the potential alteration of Arctic Ocean food webs.

The timing of upwelling and downwelling as it relates to bloom formation may be used as indicators of a shifting Arctic climate regime (e.g., [6.12.138.]). For example, there may be a lag effect for blooms to occur from upwelling during the summer so that the size of the bloom reduces as autumn approaches (e.g., [6.12.140.]), then reemerges in the winter and into the spring with reduced ice formation or thinning (e.g., [6.12.142.]). A shifting climate regime concerning ice formation and the effects this may have on food webs in the Arctic (e.g., [6.12.86.]) are a potential target of study using LBMs. A finer tuning of data could be used to compare past and present conditions in the Arctic Ocean as a guide for constructing possible future scenarios as Arctic ice diminishes over time (e.g., [6.12.3., 6.12.133.]). The LBM approach to modeling such a situation may be useful in this regard.

Other additions to the scenarios used could be instituted. For example, explicit changes in ice thickness may be incorporated into LBM simulations. The reduction in Arctic ice on a seasonal, annual, decadal, or longer time span (e.g., [6.12.71.]) could be included in LBM simulation and used in relation to climate change. Additionally, estimates for Arctic snow density are 320 kg/m^3 [6.12.120.] and could be used in simulations along with ice formation. Snow cover varies monthly with no snow in August, accumulation occurring in September to November, tapering off in December to January, and increasing during February to May [6.12.144.]. Using such modeling may be instructive in understanding changes in the Arctic Ocean with regard to photosynthetic capacity of diatoms or other primary producers and the influence this has on bloom or food web dynamics.

6.10.1   Arctic Diatom Food Web Dynamics: Other Potential Outcomes

Presently, diatoms are the major marine primary producer as the dominant silica-shelled unicell and carbon cycle contributor in terms of the biological pump as well as carbon production and sequestration (e.g., [6.12.30., 6.12.84., 6.12.143.]). As drivers of marine food webs, diatom dominance may be an indicator of their evolutionary trajectory throughout the Cenozoic. Lipid content has been shown to be an indicator of food value in diatoms phylogenetically [6.12.45., 6.12.53.]. On an evolutionary level, LBMs may enable studies of broader diatom food web dynamics on a geological time scale. Changes in food webs may be simulated on any time scale using LBM.

Additionally, lipid content may be indicative of ecological consequences in marine food webs (e.g., [6.12.21.]) in the Arctic where lipid-driven food webs are important [6.12.44.]. For example, domoic acid from *Pseudo-nitzschia* in the Arctic (e.g., [6.12.79.]) is accumulated by *Calanus* zooplankton grazers in their digestive systems (e.g., [6.12.87.]), which may produce negative effects in the trophic structure of the food web.

There is the potential for detrimental effects at higher trophic levels of the Arctic food web (e.g., [6.12.77.]) from diatom blooms. Excessive diatom growth may inhibit oxygen formation in pelagic waters potentially inducing hypoxic conditions, or during an upwelling event as increasing nutrients induce continued diatom production (e.g., [6.12.132.]), hypoxia may occur in coastal waters (e.g., [6.12.52., 6.12.132.]) or anoxia in deeper waters (e.g., [6.12.40.]). For these reasons and despite the chemical dynamics occurring in the water, the maximum density of the diatom bloom could be used to induce a change in the trophic structure of the food web (e.g., [6.12.126.]). Harmful algal blooms may be induced by increased bloom density of diatoms as yet another potential source of marine food web changes in the Arctic (e.g., [6.12.142.]).

LBM was used to simulate increasing upwelling inducing diatom bloom formation as an effect on Arctic food web dynamics. While a bloom may be viewed as an abundance of food for zooplankton grazers, another potential deleterious effect on the marine food web may occur as well (e.g., [6.12.12.]). Frequency of diatom blooms may be an indicator of marine food web anomalies [6.12.102.] such as trophic cascades (e.g., [6.12.139.]) by affecting higher trophic levels spatially and temporally, resulting from diminishing Arctic sea ice (e.g., [6.12.133.]) and human-induced factors (e.g., [6.12.142.]). Arctic food web changes may occur with increased algal mat formation (e.g., [6.12.55.]). As blooms increasingly occur, there may be an adaptive effect by sea ice diatoms to become more prevalent (e.g., [6.12.65.]).

Marine food web activity at the primary producer level may be indicative of the spatial and temporal arrangement of benthic and pelagic diatoms (e.g., [6.12.67.]). Diatom bloom events may be indicative of anomalous food web activity or signaling a change in the density of available high-quality food at the primary producer level. The frequency of diatom blooms may be understood with respect to the frequency of sea ice cover and upwelling events, whether over geologic time or during the present with regard to climate change (e.g., [6.12.133.]). There is a close relation between ice-algal and pelagic food webs so that anomalies in one food web may affect the other (e.g., [6.12.68.]). As upwelling events become more evident, increased bloom formation may induce the timing of seasonal, annual or longer-term time span changes.

6.10.2   Diatom Blooms: Influences over Time and Space

The relation between seasonal bloom and carbon export may be studied using LBM. Ice conditions and diatom blooms with respect to climate conditions may be simulated to understand changes in amount of ice and its condition on a seasonal basis [6.12.146.]. Extending such studies over long periods of time or on a comparative basis between recent Arctic Ocean conditions and those from the Cenozoic may be interesting to map the changes in climate over geologic time.

LBM analyses may be instituted in a multitude of ways. An age model could be incorporated into the LBM (e.g., [6.12.83.]) so that, for example, diatom genera from the Pleistocene to Pliocene are included. Pelagic genera from *Neodenticula* and *Coscinodiscus* were dominant in the Pliocene with lesser contributions to productivity from *Kisseleviella* and *Stephanopyxis* in the neritic zone or the benthic diatom *Cocconeis* [6.12.130.]. In the Pleistocene, the dominant contributors to productivity were from *Neodenticula* and *Actinocyclus*, followed by upwelling species from *Chaetoceros* and ice-dwelling *Fragilariopsis* [6.12.129.]. While some of these genera echo those used in the current LBM, additional analyses with respect to paleo-reconstructions (e.g., [6.12.23.]) may be used in LBM simulations of upwelling or fluctuations in ice presence [6.12.129., 6.12.130.] as well as changes in productivity or bloom formation [6.12.22.]. The spatial distribution of benthic and pelagic diatoms during upwelling time intervals may be simulated using LBM (e.g., [6.12.83.]) with respect to changes in ice cover conditions over geologic time.

Diatom loss rates were not explicitly considered but may be included in simulation more directly by using sinking rates (e.g., [6.12.40.]) or by ice entrapment from melting and refreezing of ice (e.g., [6.12.12.]). Influence of dinoflagellates as competitors, growth of additional diatoms or other algal taxa, duration of upwelling, position of upwelling to the coastal area, other processes or currents regimes, and other factors may be included or substituted in the simulation as well. Nutrient concentration changes with respect to changes in salinity and temperature in terms of the degree of melting ice may be included to simulate the effects on diatoms (e.g., [6.12.12.]).

Factors at a cellular level or life cycle stage may be considered for inclusion in LBM analyses. For example, cell growth and adaptive resistance to environmental conditions (e.g., [6.12.82.]), or the persistence of diatom resting spores or cells as aquatic seeds to produce resilient taxa (e.g., [6.12.42.]) may be used on both an ecological and evolutionary level in LBM studies.

6.11 Summary and Future Research

Because of its versatility in applications, LBM and its various forms may provide a variety of ways to analyze the relation among food web trophic levels and hydrodynamics. LBM is amenable to including thermodynamics as well as dynamical systems so that heat transfer (e.g., [6.12.119.]) or chaotic dynamics (e.g., [6.12.69.]), respectively, may be simulated. Diatom movement concerning chemical and physical parameters as well as biotic factors may be simulated with the LBM to study spatial and temporal distributions. Of importance, analyses concerning silica limitation effects on diatoms or carbon cycling and sequestration via diatoms and their effects on climate could be simulated via LBM. The halocline, pycnocline and thermocline may be incorporated into oceanic LBM studies. While the current study was confined to the Arctic and sea ice coverage, other boundaries may be specified for particular topics. LBM utility is evident with regard to potential studies in ecological, evolutionary, environmental, and climate studies as well as food web dynamics studies.

LBM can be used with finite element, finite differences, finite volume modeling or other methods. Lagrangian and Eulerian-based LBMs have been used (e.g., [6.12.83.]) as well as entropy-based LBM (e.g., [6.12.119., 6.12.131.]). LBM applicability with respect to different mathematical regimes is evident. In terms of hydrodynamics, applications of LBM using turbulent flows (i.e., high Reynolds numbers) or mixed laminar and turbulent flows (e.g., [6.12.119.]) may be instituted. Surface vorticities may be included in LBM to study shallow water interactions (e.g., [6.12.100.]) concerning the Arctic Ocean (e.g., [6.12.134.]), or upwelling (e.g., [6.12.38.]). Multiple relaxation times may be used (e.g., [6.12.131.]). Many avenues of study are available using LBM simulation.

6.12 References

6.12.1.   Agafonova, E., Polyakova, Y., Novichkova, Y. (2020) The diatom response to Postglacial environments in the White Sea, the European Arctic. *Marine Micropaleontology* 161, 101927.

6.12.2.   Andreas, E.L., Paulson, C.A., Williams, R.M., Lindsay, R.W., Businger, J.A. (1979) The turbulent heat flux from Arctic leads. *Boundary-Layer Meteorology* 17(1), 57-91.

6.12.3.   Ardyna, M. and Arrigo, K.R. (2021) Phytoplankton dynamics in a changing Arctic Ocean. *Nature Climate Change* 10, 892-903.

6.12.4.   Ardyna, M., Mundy, C.J., Mayot, N., Matthes, L.C., Oziel, L., Horvat, C., Leu, E., Assmy, P., Hill, V., Matrai, P.A., Gale, M., Melnikov, I.A., Arrigo, K.R. (2020) Under-ice phytoplankton blooms: shedding light on the "invisible" part of Arctic primary production. *Front. Mar. Sci.* 7:608032.

6.12.5.   Ardyna, M., Mundy, C.J., Mills, M.M., Oziel, L., Grondin, P.-L., Lacour, L., *et al.* (2020) Environmental drivers of under-ice phytoplankton bloom dynamics in the Arctic Ocean. *Elem Sci Anth* 8, 30

6.12.6.   Arrigo, K.R., Perovich, D.K., Pickart, R.S., Brown, Z.W., van Dijken, G.L., *et al.* (2012) Massive phytoplankton blooms under Arctic sea ice. *Science* 336(6087), 1408.

6.12.7.   Arrigo, K.R., Perovich, D.K., Pickart, R.S., Brown, Z.W., van Dijken, G.L., *et al.* (2014) Phytoplankton blooms beneath the sea ice in Chukchi Sea. *Deep Sea Res. Pt. 2* 105, 1-16.

6.12.8.   Assmy, P., Fernández-Méndez, M., Duarte, P., Meyer, A., Randelhoff, A., Mundy, C.J., *et al.* (2017) Leads in Arctic pack ice enable early phytoplankton blooms below snow-covered sea ice. *Scientific Reports* 7, 40850.

6.12.9.   Barber, R.T. and Hiscock, M.R. (2006) A rising tide lifts all phytoplankton: growth response of other phytoplankton taxa in diatom-dominated blooms. *Global Biogeochemical Cycles* 20, GB4S03.

6.12.10.   Behrenfeld, M.J., Hu, Y., O'Malley, R.T.O., Boss, E.S., Hostetler, C.A., *et al.* (2016) Annual boom-bust cycles of polar phytoplankton biomass revealed by space-based lidar. *Nature Geoscience* 10, 118-124.

6.12.11.   Belt, S.T., Massé, G., Rowland, S.J., Poulin, M., Michel, C., LeBlanc, B. (2007) A novel chemical fossil of palaeo sea ice: IP25. *Organic Geochemistry* 38, 16-27.

6.12.12.   Benkort, D., Daewel, U., Heath, M., Schrum, C. (2020) On the role of biogeochemical coupling between sympagic and pelagic ecosystem compartments for primary and secondary production in the Barents Sea. *Front. Environ. Sci.* 8, 548013.

6.12.13.   Bhatnagar, P.L., Gross, E.P., Krook, M. (1954) A model for collision processes in gases. I. Small amplitude processes in charged and neutral one-component systems. *Physical Review* 94(3), 511-525.

6.12.14.   Bhattacharjya, R., Marella, T.K., Tiwari, A., Saxena, A., Singh, P.K., Mishra, B. (2020) Bioprospecting of marine diatoms *Thalassiosira, Skeletonema* and *Chaetoceros* for lipids and other value-added products. *Bioresource Technology* 318, 124073.

6.12.15.   Boetius, A., Albrecht, S., Bakker, K., Bienhold, C., Felden, J., Fernández-Méndez, M., *et al.* (2013) Export of algal biomass from the melting Arctic sea ice. *Science* 339(6126), 1430-1432.

6.12.16.   Booth, B.C., Larouche, P., Bélanger, S., Klein, B., Amiel, D., Mei, Z.P. (2002) Dynamics of *Chaetoceros socialis* blooms in the North Water. *Deep-Sea Research II* 49, 5003-5025.

6.12.17.   Bourke, R.H. and Garrett, R.P. (1987) Sea ice thickness distribution in the Arctic Ocean. *Cold Regions Science and Technology* 13, 259-280.

6.12.18.   Brink, K.H. (1983) The near-surface dynamics of coastal upwelling. *Prog. Oceanog.* 12, 223-257.

6.12.19.   Brown, T.A. and Belt, S.T. (2012) Closely linked sea ice—pelagic coupling in the Amundsen Gulf revealed by the sea ice diatom biomarker IP25. *Journal of Plankton Research* 34(8), 647-654.

6.12.20.   Brown, T.A. and Belt, S.T. (2012) Identification of the sea ice diatom biomarker IP25 in Arctic benthic macrofauna: direct evidence for a sea ice diatom diet in Arctic heterotrophs. *Polar Biology* 35, 131-137.

6.12.21.   Brown, T.A., Belt, S.T., Ferguson, S.H., Yurkowski, D.J., Davison, N.J., Barnett, J.E.F., Jepson, P.D. (2013) Identification of the sea ice diatom biomarker IP25 and related lipids in marine mammals: a potential method for investigating regional variations in dietary sources within higher trophic level marine systems. *J. Exper. Mar. Biol. Ecol.* 441, 99-104.

6.12.22.   Brown, T.A., Belt, S.T., Philippe, B., Mundy, C.J., Massé, G., Poulin, M., Gosselin, M. (2011) Temporal and vertical variations of lipid biomarkers during a bottom ice diatom bloom in the Canadian Beaufort Sea: further evidence for the use of the IP25 biomarker as a proxy for spring Arctic sea ice. *Polar Biology* 34, 1857-1868.

6.12.23.   Brown, T.A., Belt, S.T., Tatarek, A., Mundy, C.J. (2014) Source identification of the Arctic sea ice proxy IP25. *Nature Communications* 5:4197.

6.12.24.   Brown, T.A., Rad-Menéndez, C., Ray, J.L., Skaar, K.S., Thomas, N., Ruiz-Gonzalez, C., Leu, E. (2020) Influence of nutrient availability on Arctic sea ice diatom HBI lipid synthesis. *Organic Geochemistry* 141, 103977.

6.12.25.   Buckley, J.R., Gammelsrød, T., Johannessen, J.A., Johannessen, O.M., Røed, L.P. (1979) Upwelling: oceanic structure at the edge of the Arctic ice pack in winter. *Science* 203, 165-167.

6.12.26.   Budge, S.M. and Parrish, C.C. (1999) Lipid class and fatty acid composition of *Pseudo-nitzschia* multiseries and *Pseudo-nitzschia pungens* and effects of lipolytic enzyme deactivation. *Phytochemistry* 52, 561-566.

6.12.27.   Carmack, E. and Chapman, D.C. (2003) Wind-driven shelf/basin exchange on an Arctic shelf: the joint roles of ice cover extent and shelf-break bathymetry. *Geophysical Research Letters* 30(14), 1778.

6.12.28.   Carmack, E., Barber, D., Christensen, J., Macdonald, R., Budels, B., Sakshaug, E. (2006) Climate variability and physical forcing of the food webs and carbon budget on panarctic shelves. *Progress in Oceanography* 71, 145-181.

6.12.29.   Carmack, E., Polyakov, I., Padman, L., Fer, I., Hunke, E., Hutchings, J., *et al.* (2015) Toward quantifying the increasing role of oceanic heat in sea ice loss in the new Arctic. Bulletin of the *American Meteorological Society* 96(12), 2079-2105.

6.12.30.   Cermeño, P., Falkowski, P.G., Romero, O.E., Schaller, M.F., Vallina, S.M. (2015) Continental erosion and the Cenozoic rise of marine diatoms. *PNAS* 112(14), 4239-4244

6.12.31.   Chan, C.X., Baglivi, F.L., Jenkins, C.E., Bhattachrya, D. (2013) Foreign gene recruitment to the fatty acid biosynthesis pathway in diatoms. *Mobile Genetic Elements* 3, e27313.

6.12.32.   Chen, Y.-C. (2012) The biomass and total lipid content and composition of twelve species of marine diatoms cultured under various environments. *Food Chemistry* 131, 211-219.

6.12.33.   Chen, S., Wang, Z., Shan, X., Doolen, G.D. (1992) Lattice Boltzmann computational fluid dynamics in three dimensions. *Journal of Statistical Physics* 68(3/4), 379-400.

6.12.34.   Chin, K., Bloch, J., Sweet, A., Tweet, J., Eberle, J., Cumbaa, S., Witkowski, J., Harwood, D. (2008) Life in a temperate Polar sea: a unique taphonomic window on the structure of Late Cretaceous Arctic marine ecosystem. *Proc. R. Soc. B* 275, 2675-2685.

6.12.35.   Cointet, E., Wielgosz-Collin, G., Méléder, V., Gonçalves, O. (2019) Lipids in benthic diatoms: a new suitable screening procedure. *Algal Research* 39, 101425.

6.12.36.   Costa, R.R., Mendes, C.R.B., Ferreira, A., Tavano, V.M., Dotto, T.S., Secchi, E.R. (2018) Large diatom bloom off the Antarctic Peninsula during cool conditions associated with the 2015/2016 El Niño. *Communications Earth & Environment* 2:252.

6.12.37.   Croteau, D., Lacour, T., Schiffrine, N., Morin, P.-I., Forget, M.-H., Bruyant, F., Ferland, J., Lafond, A., Campbell, D.A., Tremblay, J.-É., Babin, M., Lavaud, J. (2022) Shifts in growth light optima among diatom species support their succession during the spring bloom in the Arctic. *Journal of Ecology* 110, 1356-1375.

6.12.38.   Dai, A. and Wu, C.-S. (2018) High-resolution simulations of unstable cylindrical gravity currents undergoing wandering and splitting motions in a rotating system. *Physics of Fluids* 30, 026601.

6.12.39.   d'Ippolito, G., Sardo, A., Paris, D., Vella, F.M., Adelfi, M.G., Botte, P., Callo, C., Fontana, A. (2015) Potential of lipid metabolism in marine diatoms for biofuel production. *Biotechnology for Biofuels* 8, 28.

6.12.40.   Donaghay, P.L. and Osborn, T.R. (1997) Toward a theory of biological-physical control of harmful algal bloom dynamics and impacts. *Limnol. Oceanogr.* 42(5, part 2), 1283-1296.

6.12.41.   Dupont, F. (2012) Impact of sea-ice biology on overall primary production in a biophysical model of the pan-Arctic Ocean. *Journal of Geophysical Research* 117, C00D17.

6.12.42.   Ellegaard, M. and Ribeiro, S. (2018) The long-term persistence of phytoplankton resting stages in aquatic 'seed banks.' *Biol. Rev.* 93, 166-183.

6.12.43.   Endoh, M. (1977) Double-celled circulation in coastal upwelling. *Journal of the Oceanographical Society of Japan* 33, 30-37.

6.12.44.   Falk-Petersen, S., Pavlov, V., Berge, J., Cottier, F., Kovacs, K.M., Lydersen, C. (2015) At the rainbow's end: high productivity fueled by winter upselling along an Arctic shelf. *Polar Biology* 38, 5-11.

6.12.45.   Fields, F.J. and Kociolek, J.P. (2015) An evolutionary perspective on selecting high-lipid-content diatoms (Bacillariophyta). *J Appl Phycol* 27, 2209-2220.

6.12.46.   Frisch, U. (1991) Relaton between the lattice Boltzmann equation and the Navier-Stokes equations. *Physica D* 47, 231-232.

6.12.47.   Frisch, U., d'Humières, D., Hasslacher, B., Lallemand, P., Pomeau, Y., Bivet, J.-P. (1987) Lattice gas hydrodynamics in two and three dimensions. *Complex Systems* 1, 649-707.

6.12.48.   Furnas, M. (1990) In situ growth rates of marine phytoplankton: approaches to measurement, community and species growth rates. *Journal of Plankton Research* 12, 1117-1151.

6.12.49.   Gilbertson, R. Langan, E, Mock, T. (2022) Diatoms and their microbiomes in complex and changing Polar oceans. *Front. Microbiol.* 13, 786764.

6.12.50.   Gogorev, R.M. and Samsonov, N.I. (2016) The genus *Chaetoceros* (Bacillariophyta [sic]) in Arctic and Antarctic. *Novosti Sist. Nizsh. Rast.* 50, 56-111.

6.12.51.   Gosselin, M., Levasseur, M., Wheeler, P.A., Horner, R.A., Booth, B.C. (1997) New measurements of phytoplankton and ice algal production in the Arctic Ocean. *Deep. Res. Part II Top. Stud. Oceanogr.* 44, 1623-1644.

6.12.52.   Grantham, B.A., Chan, F., Nielsen, K.J., Fox, D.S., Barth, J.A., Huyer, A., Lubchenco, J., Menge, B.A. (2004) Upwelling-driven nearshore hypoxia signals ecosystem and oceanographic changes in the northeast Pacific. *Nature* 429, 749-754.

6.12.53.   Griffiths, M.J. and Harrison, S.T.L. (2009) Lipid productivity as a key characteristic for choosing algal species for biodiesel production. *J Appl Phycol* 21, 493-507.

6.12.54.   Guan, Y., Guadarrama-Lara, R., Jia, X., Zhang, K., Wen, D. (2017) Lattice Boltzmann simulation of flow past a spherical/non-spherical particle in a wide range of Reynolds number. *Advanced Powder Technology* 28(6), 1486-1494.

6.12.55.   Hancke, K., Dristiansen, S., Lund-Hansen, L.C. (2022) Highly productive ice algal mats in Arctic melt ponds: primary production and carbon turnover. *Front. Mar. Sci.* 9, 841720.

6.12.56.   Harvey, B.P., Agostini, S., Kon, K., Wada, S., Hall-Spencer, J.M. (2019) Diatoms dominate and alter marine food-webs when CO2 rises. *Diversity* 11, 242.

6.12.57.   He, X. and Luo, L.-S. (1997) Lattice Boltzmann model for the incompressible Navier-Stokes equation. *Journal of Statistical Physics* 88(3/4), 927-944.

6.12.58. He, X. and Luo, L-S. (1997) Theory of the lattice Boltzmann method: from the Boltzmann equation to the lattice Boltzmann equation. *Physical Review E* 56(6), 6811-6817.

6.12.59. Hildebrand, M., Davis, A.K., Smith, S.R., Traller, J.C., Abbriano, R. (2012) The place of diatoms in the biofuels industry. *Biofuels* 3(2), 221-240.

6.12.60. Inamuro, T., Yoshino, M., Ogino, F. (1995) A non-slip boundary condition for lattice Boltzmann simulations. *Physics in Fluids* 7, 2928.

6.12.61. Ji, Q., Li, B., Pang, X., Zhao, X., Lei, R. (2021) Arctic sea ice density observation and its impact on sea ice thickness retrieval from CryoSat-2. *Cold Regions Science and Technology* 181, 103177.

6.12.62. Jones, E.P. (2001) Circulation in the Arctic Ocean. *Polar Research* 20(2), 139-146.

6.12.63. Johnsen, G., Norli, M., Moline, M., Robbins, I., von Quillfeldt, C., Sørensen, K., Cottier, F., Berge, J. (2018) The advective origin of an under-ice spring bloom in the Arctic Ocean using multiple observational platforms. *Polar Biology* 41, 1197-1216.

6.12.64. Jónasdóttir, S.H. (2019) Fatty acid profiles and production in marine phytoplankton. *Marine Drugs* 17, 151.

6.12.65. Kauko, H.M., Olsen, L.M, Duarte, P., Peeken, I., Granskog, M.A., Johnsen, G., et al. (2018) Algal colonization of young Arctic sea ice in spring. *Front. Mar. Sci.* 5, 199.

6.12.66. Kemper, J., Riebesell, U., Graf, K. (2022) Numerical flow modeling of artificial ocean upwelling. *Front. Mar. Sci.* 8, 804875.

6.12.67. Koch, C.W., Cooper, L.W., Lalande, C., Brown, T.A., Frey, K.E., Grebmeier, J.M. (2020) Seasonal and latitudinal variations in sea ice algae deposition in the Northern Bering and Chukchi Seas determined by algal biomarkers. *PLoS ONE* 15(4), e0231178.

6.12.68. Kohlbach, D., Graeve, M., Lange, B.A., David, C., Peeken, I., Flores, H. (2016) The importance of ice algae-produced carbon in the central Arctic Ocean ecosystem: food web relationships revealed by lipid and stable isotope analyses. *Limnol. Oceanogr.* 61, 2027-2044.

6.12.69. Koshel, K.V. and Prants, S.V. (2006) Chaotic advection in the ocean. *Physics – Uspekhi* 11, 1151-1178.

6.12.70. Krause, J.W. and Lomas, M.W. (2020) Understanding diatoms' past and future biogeochemical role in high-latitude seas. *Geophysical Research Letters* 47, e2019GL085602.

6.12.71. Kwok, R., Cunningham, G.F., Wensnahan, M., Rigor, I., Zwally, H.J., Yi, D. (2009). Thinning and volume loss of the Arctic Ocean sea ice cover: 2003-2008. *Journal of Geophysical Research* 114, C07005.

6.12.72. Lafond, A., Leblanc, K., Quéguiner, B., Moriceau, B., Leynaert, A., Cornet, V., Legras, J., Ras, J., et al. (2019) Late spring bloom development of pelagic diatoms in Baffin Bay. *Elem Sci Anth* 7, 44.

6.12.73. Lamping, N., Müller, J., Hefter, J., Mollenhauer, G., Haas, C., Shi, Z., Vorrath, M.-E., Lohmann, G., Hillenbrand, C.-D. (2021) Evaluation of lipid biomarkers as proxies for sea ice and ocean temperatures along the Antarctic continental margin. *Clim. Past* 17, 2305-2326.

6.12.74. Lannuzel, D., Tedesco, L., van Leeuwe, M., Campbell, K., Flores, H., et al. (2020) The future of Arctic sea-ice biogeochemistry and ice-associated ecosystems. *Nature Climate Change* 10, 983-992.

6.12.75. Laws, E.A. (2013) Evaluation of In Situ phytoplankton growth rates: a synthesis of data from varied approaches. *Annual Review of Marine Science* 5, 247-268.

6.12.76. Leblanc, K., Aristegui, J., Armand, L., Assmy, P., Beker, B., Bode, A., Breton, E., et al. (2012) A global diatom database – abundance, biovolume and biomass in the world ocean. *Earth Syst. Sci. Data* 4, 149-165.

6.12.77. Lefebvre, K.A., Quakenbush, L., Frame, E., Huntingon, K.B., Sheffield, G., Stimmelmayr, R., *et al.* (2016) Prevalence of algal toxins in Alaskan marine mammals foraging in a changing arctic and subarctic environment. *Harmful Algae* 55, 13-24.

6.12.78. Le Fouest, V., Zakardjian, B., Saucier, F.J., Starr, M. (2005) Seasonal versus synoptic variability in planktonic production in a high-latitude marginal sea: the Gulf of St. Lawrence (Canada). *Journal of Geophysical Research* 110, C09012.

6.12.79. Lelong, A., Hégaret, H., Soudant, P., Bates, S.S. (2012) *Pseudo-nitzschia* (Bacillariophyceae) species, domoic acid and amnesic shellfish poisoning: revisiting previous paradigms. *Phycologia* 51(2), 168-216.

6.12.80. Levitan, O., Dinamarca, J., Hochman, G., Falkowski, P.G. (2014) Diatoms: a fossil fuel of the future. *Trends in Biotechnology* 32(3), 117-124.

6.12.81. Limoges, A., Massé, G., Weckström, K., Poulin, M., Ellegaard, M., Heikkilä, M., Geilfus, N.-Z., Sejr, M.K., Rysgaard, S., Ribeiro, S. (2018) Spring succession and vertical export of diatoms and IP25 in a seasonally ice-covered high Arctic fjord. *Front. Earth Sci.* 6, 226.

6.12.82. Lindström, H.J.G. and Friedman, R. (2020) Inferring time-dependent population growth rates in cell cultures undergoing adaptation. *BMC Bioinformatics* 21, 583.

6.12.83. Liu, H., Ding, Y., Wang, H., Zhang, J. (2015) Lattice Boltzmann method for the age concentration equation in shallow water. *Journal of Computational Physics* 299, 613-629.

6.12.84. Malviya, S., Scalco, E., Audic, S., Vincent, F., Veluchamy, A., Poulain, J., Wincker, P., Iudicone, D., deVargas, C., Bittner, L., Zingone, A., Bowler, C. (2016) Insights into global diatom distribution and diversity in the world's ocean. *PNAS* 113(11), E1516-E1525.

6.12.85. Meneghello, G., Marshall, J., Timmermans, M.-L., Scott, J. (2018) Observations of seasonal upwelling and downwelling in the Beaufort Sea mediated by sea ice. *Journal of Physical Oceanography* 48(4), 795-805.

6.12.86. MERCINA (Marine Ecosystem Responses to Climate in the North Atlantic) Working Group (Greene, C.H. *et al.*) (2012) Recent Arctic climate change and its remote forcing of northwest Atlantic shelf ecosystems. *Oceanography* 25(3), 208-213.

6.12.87. Miesner, A.K., Lundholm, N., Krock, B., Nielsen, T.G. (2016) The effect of Pseudo-nitzschia seriata on grazing and fecundity of *Calanus finmarchicus* and *Calanus glacialis*. *Journal of Plankton Research* 38(3), 564-574.

6.12.88. Mimouni, V., Ulmann, L., Tremblin, G., Robert, J.-M. (2003) Desaturation of linoleic acid in the marine diatom *Haslea ostrearia* Simonsen (Bacillariophyceae). *Cryptogamie, Algol.* 24(3), 269-276.

6.12.89. Mokhasi, P. (2019) Building a lattice Boltzmann-based wind tunnel with the Wolfram language, notebook archive, https://notebookarchive.org/2019-11-3pvcd07.

6.12.90. Mundy, C.J., Gosselin, M., Ehn, J., Gratton, Y., Rossnagel, A., Barber, D.G., Martin, J., *et al.* (2009) Contribution of under-ice primary production to an ice-edge upwelling phytoplankton bloom in the Canadian Beaufort Sea. *Geophysical Research Letters* 36, L17601.

6.12.91. Nichols, P.D., Palmisano, A.C., Smith, G.A., White, D.C. (1986) Lipids of the Antarctic sea ice diatom *Nitzschia cylindrus*. *Phytochemistry* 25(7), 1649-1653.

6.12.92. Nichols, P.D., Palmisano, A.C., Volkman, J.K., Smith, G.A., White, D.C. (1988) Occurrence of an isoprenoid C25 diunasaturated [sic] alkene and high neutral lipid content in Antractic [sic] sea-ice diatom communities. *Journal of Phycology* 24(1), 90-96.

6.12.93. Nichols, P.D., Palmisano, A.C., Rayner, M.S., Smith, G.A., White, D.C. (1989) Changes in the lipid composition of Antarctic sea-ice diatom communities during a spring bloom: an indication of community physiological status. *Antarctic Science* 1(2), 133-140.

6.12.94. Nichols, P.D., Palmisano, A.C., Rayner, M.S., Smith, G.A., White, D.C. (1990) Occurrence of novel C30 sterols in Antarctic sea-ice diatom communities during a spring bloom. *Organic Geochemistry* 15(5), 503-508.

6.12.95. Noerdlinger, P.D. and Brower, K.R. (2007) The melting of floating ice raises the ocean level. *Geophys. J. Int.* 170, 145-150.

6.12.96. Nummelin, A., Ilicak, M., Li, C., Smedsrud, L.H. (2016) Consequences of future increased Arctic runoff on Arctic Ocean stratification, circulation, and sea ice cover. *J. Geophys. Res. Oceans* 121, 617-637.

6.12.97. Ottolenghi, L., Prestininzi, P., Montessori, A., Adduce, C., La Rocca, M. (2018) Lattice Boltzmann simulations of gravity currents. *European Journal of Mechanics/B Fluids* 67, 125-136.

6.12.98. Parrish, C.C., deFreitas, A.S.W., Bodennec, G., Macpherson, E.J., Ackman, R.G. (1991) Lipid composition of the toxic marine diatom, *Nitzschia pungens*. *Phytochemistry* 30(1), 113-116.

6.12.99. Percopo, I., Ruggiero, M.V., Balzano, S., Gourvil, P., Lundholm, N., Siano, R., *et al.* (2016) *Pseudo-nitzschia arctica* sp. nov., a new cold-water cryptic *Pseudo-nitzschia* species within the *P. pseudodelicatissima* complex. *J. Phycol.* 52(2), 184-199.

6.12.100. Perumal, D.A., Kumar, G.V.S., Dass, A.K. (2014) Lattice Boltzmann simulation of flow over a circular cylinder at moderate Reynolds numbers. *Thermal Science* 18(4), 1235-1246.

6.12.101. Pitcher, G.C., Figueiras, F.G., Hickey, B.M., Moita, M.T. (2010) The physical oceanography of upwelling systems and the development of harmful algal blooms. *Prog Oceanogr.* 85(1-2), 5-32.

6.12.102. Post, E., Bhatt, U.S., Bitz, C.M., Brodie, J.F., Fulton, T.L., Hebblewhite, M., Kerby, J., Kutz, S.J., Stirling, I., Walker, D.A. (2013) Ecological consequences of sea-ice decline. *Science* 341, 519-524.

6.12.103. Poulin, M., Daugbjerg, N., Gradinger, R., Ilyash, L., Ratkova, T., von Quillfeldt, C. (2011) The pan-Arctic biodiversity of marine pelagic and sea-ice unicellular eukaryotes: a first-attempt assessment. *Mar. Biodiv.* 41, 13-28.

6.12.104. Qian, Y.H., d'Hunières, D., Lallemand, P. (1992) Lattice BGK models for Navier-Stokes equation. *Europhysics Letters* 17(6), 479-484.

6.12.105. Ramadhan, A., Marshall, J., Meneghello, G., Illari, L., Speer, K. (2022) Observations of upwelling and downwelling around Antarctica mediated by sea ice. *Frontiers in Marine Science* 9:864808.

6.12.106. Randelhoff, A. and Sundfjord, A. (2018) Short commentary on marine productivity at Arctic shelf breaks: upwelling, advection and vertical mixing. *Ocean Sci.* 14, 293-300.

6.12.107. Ren, J., Chen, J., Bai, Y., Sicre, M.-A., Yao, Z., Lin, L., *et al.* (2020) Diatom composition and fluxes over the Northwind Ridge, western Arctic Ocean: impacks of marine surface circulation and sea ice distribution. *Progress in Oceanography* 186, 102377.

6.12.108. Reuss, N. and Poulsen, L.K. (2002) Evaluation of fatty acids as biomarkers for a natural plankton community. Afield study of a spring bloom and a post-bloom period off West Greenland. *Marine Biology* 141, 423-434.

6.12.109. Reynolds, O. (1883) An experimental investigation of the circumstances which determine whether the motion of water shall be direct or sinuous, and of the law of resistance n parallel channels. *Phil. Trans. Roy. Soc.* 174, 935-982.

6.12.110. Ribeiro, S., Limoges, A., Massé, G., Johansen, K.L., Colgan, W., Weckström, K., *et al.* (2021) Vulnerability of the North Water ecosystem to climate change. *Nature Communications* 12, 4475.

6.12.111. Richardson, B., Orcutt, D.M., Schwertner, A.H., Martinez, C.L., Wickline, H.E. (1969) Effects of nitrogen limitation on the growth and composition of unicellular algae in continuous culture. *Applied Microbiology* 18(2), 245-250.

6.12.112. Robson, J.N. and Rowland, S.J. (1986) Identification of novel widely distributed sedimentary acyclic sesterterpenoids. *Nature* 324, 561-563.

6.12.113. Salmon, R. (1999) The lattice Boltzmann method as a basis for ocean circulation modeling. *Journal of Marine Research* 57, 503-535.

6.12.114. Salmon, R. (1999b) Lattice Boltzmann solutions of the three-dimensional planetary geostrophic equations. *Journal of Marine Research* 57, 847-884.

6.12.115. Saunders, P.M. (1973) The instability of a baroclinic vortex. *Journal of Physical Oceanography* 3(1), 61-65.

6.12.116. Serreze, M.C., Barrett, A.P., Slater, A.G., Woodgate, R.A., Aagaard, K., Lammers, R.B., Steele, M., Moritz, R., Meredith, M., Lee, C.M. (2006) The large-scale freshwater cycle of the Arctic. *Journal of Geophysical Research* 111, C11010.

6.12.117. Shan, X., Yuan, X.-F., Chen, H. (2006) Kinetic theory representation of hydrodynamics: a way beyond the Navier-Stokes equation. *J. Fluid Mech.* 550, 413-441.

6.12.118. Sharma, K.V., Straka, R., Tavares, F.W. (2017) New cascaded thermal lattice Boltzmann method for simulation of advection-diffusion and convective heat transfer. *International Journal of Thermal Sciences* 118, 259-277.

6.12.119. Sharma, K.V., Straka, R., Tavares, F.W. (2020) Current status of Lattice Boltzmann Methods applied to aerodynamic, aeroacoustic, and thermal flows. *Progress in Aerospace Sciences* 115, 100616.

6.12.120. Shen, X., Ke, C.-Q., Wang, Q., Zhang, J., Shi, L., Zhang, X. (2021) Assessment of Arctic sea ice thickness estimates from ICESat-2 using IceBird airborne measurements. *IEEE Transaction on Geoscience and Remote Sensing* 59(5), 3764-3775.

6.12.121. Shiine, H., Suzuki, N., Motoyama, I., Hasegawa, S., Gladenkov, A.Y., Gladenkov, Yu.B., Ogasawara, K. (2008) Diatom biomarkers during the Eocene/Oligocene transition in the Il'pinskii Peninsula, Kamchatka, Russia. *Palaeogeography, Palaeoclimatology, Palaeoecology* 264, 1-10.

6.12.122. Sicko-Goad, L. and Andresen, N.A. (1991) Effect of growth and light/dark cycles on diatom lipid content and composition. *Journal of Phycology* 27, 710-718.

6.12.123. Simoneit, B.R.T. (2004) Biomarkers (molecular fossils) as geochemical indicators of life. *Advances in Space Research* 33, 1255-1261.

6.12.124. Smayda, T.J. (1970) The suspension and sinking of phytoplankton in the sea. Oceanog. *Marine Biol. Ann. Rev.* 8, 353-414.

6.12.125. Smayda, T.J. (1971) Normal and accelerated sinking of phytoplankton in the sea. *Marine Geol.* 11, 105-122.

6.12.126. Smayda, T.J. (2003) What is a bloom? A commentary. *Limnol. Oceangr.* 42(5, part 2), 1132-1136.

6.12.127. Spall, M.A., Pickart, R.S., Brugler, E.T., Moore, G.W.K., Thomas, L., Arrigo, K.R. (2014) Role of shelfbreak upwelling in the formation of a massive under-ice bloom in the Chukchi Sea. *Deep-Sea Research II* 105, 17-29.

6.12.128. Stoynova, V., Shanahan, T.M., Hughen, K.A., de Vernal, A. (2013) Insights into Circum-Arctic sea ice variability from molecular geochemistry. *Quaternary Science Reviews* 79, 63-73.

6.12.129. Stroynowski, Z., Abrantes, F., Bruno, E. (2017) The response of the Bering Sea gateway during the Mid-Pleistocene Transition. *Palaeogeogr. Palaeoclimatol. Palaeoecol.* 485, 974-985.

6.12.130. Stroynowski, Z., Ravelo, A.C., Andreasen, D. (2015) A Pliocene to Recent history of the Bering Sea at site U1340A, IODP expedition 323. *Paleoceanography* 30, 1641-1656.

6.12.131. Strniša, F., Urbic, T., Plazl, I. (2020) A lattice Boltzmann study of 2D steady and unsteady flows around a confined cylinder. *Journal of the Brazilian Society of Mechanical Sciences and Engineering* 42, 103.

6.12.132. Sun, J.-Z., Wang, T., Huang, R., Yi, Z., Zhang, D., *et al.* (2022) Enhancement of diatom growth and phytoplankton productivity with reduced O2 availability is moderated by rising CO2. *Communications Biology* 5:54.

6.12.133. Tedesco, L., Vichi, M., Scoccimarro, E. (2019) Sea-ice algal phenology in a warmer Arctic. *Sci. Adv.* 5, eaav4830.

6.12.134. Timmermans, M.-L. and Marshall, J. (2020) Understanding Arctic Ocean circulation: a erview of ocean dynamics in a changing climate. *Journal of Geophysical Research: Oceans* 125, e2018JC014378.

6.12.135. Tréguer, P., Bowler, C., Moriceau, B., Dutkiewicz, S., Gehlen, M., Aumont, O., Bittner, L., Dugdale, R., Finkel, Z., Iudicone, D., Jahn, O., Guidi, L., Lasbleiz, M., Leblanc, K., Levy, M., Pondaven, P. (2017) Influence of diatom diversity on the ocean biological carbon pump. *Nature Geoscience* 11, 27-37.

6.12.136. Tremblay, J.-E., Gratton, Y., Fauchot, J., Price, N.M. (2002) Climatic and oceanic forcing of new, net, and diatom production in the North Water. *Deep-Sea Research II* 49, 4927-4946.

6.12.137. Tremblay, J.-E., Michel, C., Hobson, K.A., Gosselin, M., Price, N.M. (2006) Bloom dynamics in early opening waters of the Arctic Ocean. *Limnol. Oceanogr.* 51(2), 900, 912.

6.12.138. Tremblay, J.-É., Bélanger, S., Barber, D.G., Asplin, M., Martin, J., Darnis, G., Fortier, L., Gratton, Y., Link, H., Archambault, P., Sallon, A., Michel, C., Williams, W.J., Philippe, B., Gosselin, M. (2011) Climate forcing multiplies biological productivity in the coastal Arctic Ocean. *Geophysical Research Letters* 38, L18604.

6.12.139. Trombetta, T., Vidussi, F., Roques, C., Scotti, M., Mostajir, B. (2020) Marine microbial food web networks during phytoplankton bloom and non-bloom periods: warming favors smaller organism interactions and intensifies trophic cascade. *Front. Microbiol.* 11, 502336.

6.12.140. Vonnahme, T.R., Persson, E., Dietrich, U., Hejdukova, E., Dybwad, C., Elster, J., Chierici, M., Gradinger, R. (2020) Subglacial upwelling in winter/spring increases under-ice primary production. *The Cryosphere Discussions*, 1-45. https://doi.org/10.5194/tc-2020-326

6.12.141. von Quillfeldt, C.H. (2000) Common diatom species in Arctic spring-blooms: their distribution and abundance. *Botanica Marina* 43(6), 499-516.

6.12.142. Walsh, J.J., Dieterle, D.A., Chen, F.R., Lenes, J.M., Maslowski, W., Cassano, J.J., Whitledge, T.E., *et al.* (2011) Trophic cascades and future harmful algal blooms within ice-free Arctic Seas north of Bering Strait: a simulation analysis. *Progress in Oceanography* 91, 312-343.

6.12.143. Wang, X.-W., Liang, J.-R., Luo, C.-S., Chen, C.-P., Gao, Y.-H. (2014) Biomass, total lipid production, and fatty acid composition of the marine diatom *Chaetoceros muelleri* in response to different CO_2 levels. *Bioresource Technology* 161, 124-130.

6.12.144. Warren, S.G., Rigor, I.G., Untersteiner, N., Radionov, V.F., Bryazgin, N.N., Aleksandrov, Y.I. (1999) Snow depth on Arctic sea ice. *Journal of Climate* 12(6), 1814-1829.

6.12.145. Wasiłowska, A., Tatur, A., Rzepecki, M. (2022) Massive diatom bloom initiated by high winter sea ice in Admiralty Bay (King George Island, South Shetlands) in relation to nutrient concentrations in the water column during the 2009/2010 summer. *Journal of Marine Systems* 226, 103667.

6.12.146. Wassmann, P., and Reigstad, M. (2011) Future Arctic Ocean seasonal ice zones and implications for pelagic-benthic coupling. *Oceanography* 24(3), 220-231.

6.12.147. Wassmann, P., Kosobokova, K.N., Slagstad, D., Drinkwater, K.F., Hopcroft, R.R., Moore, S.E., Ellingsen, I., Nelson, R.J., Carmack, E., Popova, E., Berge, J. (2015) The contiguous domains of Arctic Ocean advection: trails of life and death. *Progress in Oceanography* 139, 42-65.

6.12.148. Williams, W.J., Carmack, E.C., Shimada, K., Melling, H., Aagaard, K., Macdonald, R.W., Ingram, R.G. (2006) Joint effects of wind and ice motion in forcing upwelling in Mackenzie Trough, Beaufort Sea. *Continental Shelf Research* 26, 2352-2366.

6.12.149. Willmott, V., Rampen, S.W., Domack, E., Canals, M., Sinninghe Damsté, J.S., Schouten, S. (2010) Holocene changes in *Proboscia* diatom productivity in shelf waters of the north-western Antarctic Peninsula. *Antarctic Science* 22(1), 3-10.

6.12.150. Wirth, A. (2015) A guided tour through buoyancy driven flows and mixing. Master. Buoyancy driven flows and mixing, France. 68 pp. cel-01134112v5.

6.12.151. Wold, A., Darnis, G., Søreide, J.E., Leu, E., Philippe, B., Fortier, L., *et al.* (2011) Life strategy and diet of Calanus glacialis during winter-spring transition in Amundsen Gulf, south-eastern Beaufort Sea. *Polar Biology* 34, 1929-1946.

6.12.152. Woodgate, R.A., Weingartner, T., Lindsay, R. (2010) The 2007 Bering Strait oceanic heat flux and anomalous Arctic sea-ice retreat. *Geophysical Research Letters* 37, L01602.

6.12.153. Woodgate, R.A., Weingartner, T., Lindsay, R. (2012) Observed increases in Bering Strait oceanic fluxes from the Pacific to the Arctic from 2001-2011 and their impacts on the Arctic Ocean water column. *Geophysical Research Letters* 39, L24603.

6.12.154. Worne, S., Stroynowski, Z., Kender, S., Swann, G.E.A. (2021) Sea-ice response to climate change in the Bering Sea during the Mid-Pleistocene Transition. *Quaternary Science Reviews* 259, 106918.

6.12.155. Xiao, X., Fahl, K., Müller, J., Stein, R. (2015) Sea-ice distribution in the modern Arctic Ocean: biomarker records from trans-Arctic Ocean surface sediments. *Geochimica et Cosmochimica Acta* 155, 16-29.

6.12.156. Yan, D., Yoshida, K., Nishioka, J., Ito, M., Toyota, T., Suzuki, K. (2020) Response to sea ice melt indicates high seeding potential of the ice diatom *Thalassiosira* to spring phytoplankton blooms: a laboratory study on an ice algal community from the Sea of Okhotsk. *Frontiers in Marine Science* 7, doi: 10.3389/fmars.2020.00613

6.12.157. Yang, J. (2006) The seasonal variability of the Arctic Ocean Ekman transport and its role in mixed layer heat and salt fluxes. *Journal of Climate* 19, 5366-5387.

6.12.158. Yang, J. (2009) Seasonal and interannual variability of downwelling in the Beaufort Sea. *Journal of Geophysical Research* 114, C00A14.

6.12.159. Yu, D., Mei, R., Luo, L.-S., Shyy, W. (2003) Viscous flow computations with the method of lattice Boltzmann equation. *Progress in Aerospace Sciences* 39, 329-3.

6.12.160. Zhang, J., Rothrock, D., Steele, M. (2000) Recent changes in Arctic sea ice: the interplay between ice dynamics and thermodynamics. *Journal of Climate* 13(7), 3099-3114.

Part III
GENERAL AND SPECIAL FUNCTIONS IN DIATOM MACROEVOLUTIONARY SPACES

Part III

7

Diatom Clade Biogeography: Climate Influences, Phenotypic Integration and Endemism

Abstract

Diatom biogeography is important in understanding the distribution of taxa with respect to phylogenetics. A data set of freshwater diatom genera per clade was used to determine the distribution of those diatom clades worldwide. Shortest tour for all genera as well as shortest tours per clade were analyzed heuristically via the Traveling Salesman Problem (TSP). Naviculales, Cymbellales, Fragilariales, Thalassiosiroids, and Achnanthidiales were determined to be the most widely distributed clades from continent to continent. Bacillariales, Rhizosoleniales and Thalassiophycidales had no endemics and occurred on more than one continent. The remaining clades included endemics as well as a presence of genera on more than one continent. When characterized using the Köppen climate classification system, most clades had a large presence on North America, Australasia and Asia with a lesser presence on South America, Africa, Europe, and Antarctica. Freshwater diatom distribution was associated with climate. Endemics were mostly present in the Naviculales and on Asia. Distance decay of composite distance gradation from cosmopolitanism to endemism was demonstrated for all genera with respect to ecological similarity. The phenotypic novelty of a biraphid trait as a result of phenotypic integration was associated with endemic genera.

Keywords: Clade, endemism, dispersal, Traveling Salesman Problem, biogeography-based optimization, phylogeography, phenotypic integration, phenotypic novelty.

7.1 Introduction

Biogeography of organisms is an important facet of evolutionary history on the distribution of organisms spatially and temporally. Diatoms, a major contributor to the carbon (e.g., [7.7.4.]) and silica cycles (e.g., [7.7.92.]) as well as being the basis of marine and freshwater food webs, have a worldwide distribution that spans, at least, from the Jurassic to the present [7.7.77.]. Traditionally, island biogeography, as espoused by the MacArthur-Wilson hypothesis, gives principles of organismal distribution as the relation between speciation and extinction [7.7.46.]. Typically, the notion of an "island" is applied to restricted geographic areas in order to analyze particular organisms with respect to certain aspects of their biological existence such as ecological or reproductive considerations. Such applications may include restrictions via land barriers or other natural features, as vicariance is a commonly invoked reason for geographic isolation (e.g., [7.7.53.]). Temporal change from extinct to extant taxa, their phylogenetic evolution and their biogeographic range, or the combination of similar topographic regions for larger scale spatial biogeographic analyses

Janice L. Pappas. *Mathematical Macroevolution in Diatom Research*, (243–276) © 2023 Scrivener Publishing LLC

or invoking continental drift via plate tectonics allude to the tempo, mode and distributional evidence of biogeographical patterns [7.7.82.].

For diatoms, biogeographic patterns have some similarities to macro-organism distributions, although the patterns for microorganisms may be weak and may exhibit differences from macro-organisms with respect to ecological considerations [7.7.79.]. Diatom distribution worldwide may be influenced by latitudinal gradients [7.7.59.] in some instances with respect to climate, but not necessarily in all cases [7.7.79.]. Environmental influences have been documented for freshwater diatoms with respect to their distribution on a local [7.7.85.] or regional scale (e.g., [7.7.6.]), but climate considerations are important at a global scale.

For microorganisms, biogeographic patterns emerge as a result of the mechanisms of selection, drift, dispersal, and mutation [7.7.23.]. Of these processes, dispersal is a mechanism of diatom patterns (e.g., [7.7.29., 7.7.42.]) as a result of water dynamics, airborne transit (e.g., [7.7.25.]) or transport by adhering to other organisms [7.7.28.], and even via humans (e.g., [7.7.85.]). Dispersal biogeography is important at short and longer time scales with respect to evolution [7.7.53.]. Despite this, some diatoms are found on specific biologic or physical substrate [7.7.24., 7.7.75.], restricting their occurrence to circumscribed areas. Because of their ecological status, many diatom species have been documented to occur in preferred habitats, having narrow environmental tolerances with respect to light availability, temperature, pH, nutrient availability (e.g., phosphate or nitrate), micronutrient availability (e.g., iron), among other parameters. As such, selection may play a role in the ecological status of diatoms, restricting their ability to disperse on a global basis [7.7.23.]. Dispersal may be random as well when coupled with drift and mutation [7.7.23.], inducing speciation as a result [7.7.15.].

Diatom dispersal biogeography may be passive, when considering plankton, or active when considering that diatoms may be transported by birds or other vectors (e.g., [7.7.85.]). From an initial site, diatom dispersal may occur in patterns similar to those found in microbial taxa [7.7.23.]. One ubiquitous dispersal pattern in microbial biogeography is describable as distance decay [7.7.56.]. That is, the largest number of a specific group of taxa disperse such that the destination of dispersed agents declines in numbers with respect to the length of the distance from the original source. However, when considering passive dispersal and the scale of the diatom community, distance decay may not be found because environmental conditions may be influential in biogeographic patterns to a greater degree [7.7.49.]. The degree to which environmental and ecological considerations relate to diatom biogeographical patterns may be discernable with respect to geographic and classification or taxonomic scales of analysis. As one example, diatom biogeographical patterns may be determined on a global scale via inclusion of environmental and ecological conditions that are matched to a worldwide geographic analysis.

When considering clades, biogeographic distribution may depend on lineage divergence times, and as a result, disjunct distributions occur in contrast to vicariance as the primary influence [7.7.15.]. For diatoms, clade biogeography may occur via dispersal or other mechanisms, and endemism as well as disjunct groups may result (e.g., [7.7.31.]). For microorganisms, geographic barriers may be one way which induces reproductive isolation, which in turn, may induce endemism (e.g., [7.7.5.]). At specific ecological regimes, diatoms may become endemic, especially concerning landlocked lakes or other isolated bodies of freshwater (e.g., [7.7.34.]). Diatoms have been documented to exhibit endemism (e.g., [7.7.31.,

7.7.88.]), as endemism has been established for microorganisms as a result of molecular analyses (e.g., [7.7.5., 7.7.26., 7.7.87.]).

While most diatoms are said to be cosmopolitan (e.g., [7.7.31.]), this designation is dependent on taxonomic identification. As a widely known problem, many diatom genera are bins for species that may only superficially resemble the designated type for a given genus so that much taxonomic work is necessary. Regardless, genera may be a useful classification tool because biogeographical patterns may be discerned based on genus diversity [7.7.19.]. Importantly, conscription of genera to clades is also a fruitful way of understanding the relation between diatoms and their biogeographic patterns concerning the degree to which endemism or cosmopolitanism exists. Biogeographic patterns of clades and their genera encompass inferences about lineages with respect to endemism or cosmopolitanism.

Endemism for diatoms may be a time-sensitive phenomenon, occurring at relatively small to large isolated geographic areas around the globe. Endemic diatoms may also be time-sensitive with regard to the distance of potential, viable transport, and whether taxa find empty niches that are conducive to adaptation in a new area. Because of viability, the shortest distance traveled in the least amount of time may cover the extent of transport that identifies some diatoms as endemics. Emigration and immigration of taxa may have a selective cost in terms of adaptation to a new location [7.7.13., 7.7.81.], and diatoms that are transported by vectors or as a result of large-scale hydrologic events may exhibit similar movement as a "stepping-stone" phenomenon [7.7.13., 7.7.53.]. The result may provide a means of establishment or expansion of their range, potentially as endemics, disjuncts or over many geographic venues. Diatom dispersal of a given taxon, even if it occurs partially, as an inefficient mechanism of transport, or at a short distance, may induce endemism, depending on the ecological or environmental conditions present [7.7.53.].

Dispersal may be induced or limited by ecological or environmental stress as well as life cycle stage of an organism [7.7.81.] or via gene flow and genetic isolates as well as reproductive mode [7.7.42.]. Diatoms undergo size reduction during vegetative reproduction [7.7.47., 7.7.66.-7.7.67.], and the dispersal of the asexual phase of the diatom life cycle, given sufficient time, may produce descendants that are genetically different from one another [7.7.7.], exhibiting adaptation to different ecological niches ([7.7.42.]), and potentially, becoming geographically isolated [7.7.62.]. Asexual reproduction, phenotypic plasticity, exponential growth, and transport via organismal and other vectors has produced the geographical distribution, ecological adaptation, physiological differences, and genetic variability over various time scales evidenced by diatom worldwide existence as seen today [7.7.2., 7.7.42.]. While finding diatoms to be widespread and ubiquitous is true, it is just as true that genomic isolates of some diatoms exist in restricted habitats and ecological regimes as geographically isolated entities as a result of the necessity of sexual reproduction induced as MacDonald-Pfitzer limits are attained [7.7.42.].

7.1.1   Biogeography and Climate

Climate plays an important role in diatom biogeography. Dispersal according to climate may be understood on a worldwide basis by using the Köppen classification system [7.7.10., 7.7.39.-7.7.41., 7.7.43., 7.7.63.]. A composite profile of climate types based on the system categories for each continent are given as five major categories: tropical, dry, temperate/mild, snow/continental/cold, and polar (Table 7.1). Köppen's system has undergone a number of modifications with the inclusion of subcategories, but essentially consists of worldwide temperature and precipitation data defining climates with respect to vegetation on land and in the oceans [7.7.10., 7.7.39.-7.7.41., 7.7.43., 7.7.63.]. This system is also based on latitudinal gradients not only according to temperature and precipitation, but also vegetation and seasonal patterns [7.7.10.].

7.1.2   Mapping Biogeographic Patterns

Biogeographic patterns may be closely linked to migration, immigration, emigration, or other descriptors of organismal movement temporally and spatially. Without knowing the origination site of a given organism, biogeographic patterns are descriptive of relative spatial movement between circumscribed areas for restricted periods of time. Such patterns may reveal mechanisms describing movement in which particular processes may be at work, either randomly or directionally with regard to selection. One such mechanism that may be explanatory with regard to microorganisms as well as macro-organisms is dispersal [7.7.13.].

Dispersal of genera or their clades produces a pattern that may have at least three manifestations. One is a directed pattern in which there is a clear start and end; that is, dispersal may be goal-oriented. There may be multiple starts and multiple ends, but the directionality of these subpatterns would be evident. The other pattern than may emerge is cyclical. There is no definitive start or end, but there is a pattern of connectedness among taxa or clades. Rather than a directed goal, reduction in the number of clade members may occur, but dispersion may mean a return to the original site of the given clade. In this case, the cycle or circuit represents the dispersion of genera or clades without explicit reference to an end goal. The third case involves random dispersal. No pattern is evident among taxa and clades, and therefore there is no pattern for describing directionality and end goal or cycles or circuits among taxa and clades.

Because freshwater diatoms cannot be identified as migratory per se, distributional patterns may be viewed as the most efficient way that diatoms inhabit continents biogeographically. For diatoms, dispersal biogeography may be characterized as cyclical patterns because starts and ends of dispersion are not necessarily known. Cyclical patterns are the most efficient way that diatom dispersal occurs worldwide, and such patterns are used as a general basis to depict and analyze biogeographical patterns.

7.1.3   Biogeography as an Optimization Problem

Biogeographic distribution of taxa is a result of geologic and evolutionary processes. As continental drift continues, biogeography may be documented over geologic time. Biogeography may be documented on a global or local scale as well. Evolutionary processes and outcomes occurring are evident at various temporal and spatial scales biogeographically.

Biogeography-based optimization (BBO) is a suite of metaheuristic evolutionary algorithms used to assess fitness functions with respect to biogeographic distribution [7.7.76.]. Phylogeography entails BBO such that evolutionary considerations are included in biogeographic pattern determination via statistical techniques associated with molecular biology (e.g., [7.7.9.]). For microorganisms in which validated phylogenetic trees have yet to be attained, GIS methods are not applicable, genetic data is scarce, or molecular and morphological data are incomplete, and as a result, BBO and phylogeography are not currently used. This is the case with diatoms. Utilization of other techniques and methodologies to aid in the inclusion of phylogenetic information (to the extent that it is available) and evidence of particular evolutionary mechanisms may be applied in biogeographical pattern analysis.

Analyzing biogeography as an optimization problem may be viewed as a combinatorial problem as well. Using combinatorial optimization techniques and phylogenetic information, biogeographic distribution may be determined as a macroevolutionary pattern. Diatom biogeography entails phylogenetic considerations as well as geographical setting. At a macroevolutionary scale, diatom biogeography may be analyzed with regard to clades on a continental basis as a combinatorial optimization problem.

7.1.4   Biogeographic Pattern and Spatial Rate of Change

Distance decay has been identified as a prevalent pattern in dispersal biogeography (e.g., [7.7.56.]), especially concerning microorganism dispersal [7.7.23., 7.7.56.]. In other studies, freshwater phytoplankton undergoing passive dispersal have not been found to exhibit distance decay within a circumscribed biogeographic region (e.g., [7.7.49.]). Temporal and spatial considerations with regard to scale of the biogeographic area analyzed are key to determining whether distance decay applies.

Similarity with respect to distance is usually measured to determine distance decay (e.g., [7.7.80.]). Less similar taxa occur with the increase in distance from the source location so that distance decay is evident (e.g., [7.7.80.]). However, similarity is usually assessed with respect to ecological conditions on the scale of community composition in which geographical temporal and spatial considerations are viewed [7.7.56.]. Colonization [7.7.28., 7.7.52.-7.7.53.] or invasion [7.7.60.] may be a consideration with regard to assessing biogeographic pattern even though colonization or invasion may be a localized adaptive result rather than a cause of dispersal [7.7.28.].

Table 7.1 Köppen climate classification category (with additions from the Holdridge system) per continent with respect to lines of latitude as demarcations. 1 = present, 0 = absent. Sum of the climate conditions per continent are given.

	North America	Europe	South America	Asia	Australasia	Africa	Antarctica
Tropical	0	0	1	1	1	1	0
Dry/Desert	1	0	1	1	1	1	0
Temperate/Mild	1	1	1	1	1	1	0
Snow/ Continental/ Cold	1	1	0	1	0	0	0
Polar	1	1	1	1	0	0	1
SUM	4	3	4	5	3	3	1

From the Köppen climate classification system, subcategories have been delineated to characterize specific temperature, humidity and precipitation regimes that coincide with seasonality [7.7.8.]. These subcategories are descriptive of circumscribed areas of each continent [7.7.3.]. Descriptors common among regions of continents are used as ecological categories and proxies of similarity in genera, exhibiting similar ecological tolerances and responses to their particular environment. Cosmopolitan genera are widespread, and therefore do not need to disperse, representing shortest distance of transport. Endemic genera have the narrowest ranges and would need to be transported farthest to be re-established, and therefore require the longest distance for dispersal. Widely distributed and disjunct genera represent the gradient in between cosmopolitan and endemic genera. This scale is used as a distance indicator to be measured with respect to ecological similarity of all genera.

7.1.5 Biogeography, Phenotypic Integration and Phenotypic Novelty

Changes in biological morphologies are phenotypes as a result of the interactions of genotypes with the environment (e.g., [7.7.86.]). Morphological variation from mutation, environmental influences or both may produce n-variants. Those morphotypes that result from genetic changes record phenotypic variation and plasticity (e.g., [7.7.86.]). Selection acts on the phenotype via mutation and lineage sorting (e.g., [7.7.20.]). Characteristics of the phenotype may be informative about biogeographic patterns with respect to diversification (e.g., [7.7.19.]). Phenotypic traits and their response to the environment may be compiled and used in analyses to understand biogeographic distributions of organisms [7.7.19.].

One such characteristic is phenotypic novelty (e.g., [7.7.17.]) which refers to a "key innovation" (e.g., [7.7.44.]) with regard to a body plan (e.g., [7.7.64., 7.7.70.]), a new non-homologous structure not present in an ancestor, or a new character as a result of variation over time and distinct from any character state found in an ancestor (e.g., [7.7.64.]). A phenotypic novelty is considered to be a discontinuous character from the ancestral state [7.7.64.] and occurs at a macroevolutionary level [7.7.64., 7.7.70.].

Novelty also implies that a character that appears for the first time cannot be adaptive as a result of selection [7.7.64., 7.7.70.] because the novel structure is not a result of a trait that currently exists [7.7.64.]. Not all evolution is adaptive [7.7.22., 7.7.36.-7.7.38., 7.7.64.], but a novel character may become functional, then at that point, adaptation may result, or the novelty is viewed as a constraint [7.7.68.]. Novel characters may appear as a result of developmental changes (e.g., [7.7.69.]) as well as variational changes so that a novelty may appear relatively abruptly in contrast to incremental variational change over time [7.7.64.].

Diatom biogeography analysis may include fossil as well as extant taxa. In the Paleocene, some diatoms evolved the phenotypic innovation of the raphe system [7.7.61., 7.7.77., 7.7.91.]. A diatom phenotypic novelty is the raphe (Figure 7.1). This morphological structure is a slit in the sternum that may have siliceous thickening as central nodules (Figure 7.1). At its first appearance during evolution, the raphe occurred in some extant taxa and was absent in other related fossil taxa [7.7.77.]. The raphe appearing on both valves of a diatom—that is, the diatom cell is biraphid—enables those diatom vegetative cells to be motile [7.7.11., 7.7.75., 7.7.77.]. The raphe has appeared and disappeared multiple times during diatom evolutionary history [7.7.33., 7.7.35.], but for those taxa which are biraphid,

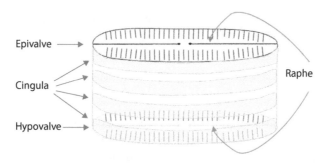

Figure 7.1 Diagram of a diatom vegetative cell with a raphe.

motility is a functional ability and example of adaptive radiation at a clade level [7.7.24., 7.7.35., 7.7.51.].

In the case of diatoms, a morphological character such as the raphe, is a phenotypic trait as a novelty, enabling diatoms to be motile by an alteration in their body plan (e.g., [7.7.75.]). Tracing a phenotypic novelty on a worldwide basis may provide insight into possible biogeographic patterns with respect to evolutionary processes such as dispersal. The role of a phenotypic novelty that is characteristic of a clade representing evolutionary change with respect to biogeographic distribution may be understood via phenotypic integration [7.7.68.].

Phenotypic integration as evidenced by a phenotypic novelty may be related to endemism because this characteristic of organisms alludes to adaptation or constraint with respect to specific environmental conditions [7.7.27.]. The limited distribution of endemics, in contrast to widely distributed cosmopolitan taxa, speaks to the lack of dispersion over large spatial and temporal scales [7.7.30.]. The biraphid trait in diatoms, while characteristic of many genera within clades, may be compared to endemism occurrences as an indicator of morphological evolution. On a worldwide scale spatially, and implicitly, on a geologic time scale, the relation between biraphidism and endemism may be mapped.

7.2   Purposes of this Study

To understand freshwater diatoms as they are biogeographically distributed worldwide, developing a pictorial map of diatom genera and clades is one of the goals of this study. As the major mode of distribution, dispersal with respect to biogeographic patterns on a continental scale is analyzed in terms of the Köppen climate classification system. The relation between climate and geographic distribution of freshwater diatoms will be established.

Biogeographic analysis is used to assess diatom clade distribution worldwide, and especially, where endemism occurs. Using clades, biogeographic patterns are elucidated worldwide and with respect to climate and ecological variables. A narrow range of phenotypic variability may be characteristic of endemic taxa as well as a narrow range of ecological conditions. The relation between endemism and phenotypic integration via a novel phenotypic trait is developed. That is, the adaptive ability of diatom endemics that have the phenotypic novelty trait of being biraphid is investigated as to whether such genera are more phenotypically integrated endemics than genera with the phenotypic traits of being monoraphid or araphid endemics.

7.3 Methods

Data for analyses are obtained from Tables 1 and 2 from Kociolek [7.7.31.] as given in the publication and available online supplementary data spreadsheet files. One set of data consist of freshwater diatom genera and their worldwide distributions as the categories of endemic, disjunct, widely distributed, or cosmopolitan from Table 1 [7.7.31.]. Another set of data consist of freshwater diatom genera and their incidence in each continent from Table 2 [7.7.31.]. Clade membership of genera is used according to Kociolek [7.7.31.] and Theriot *et al.* [7.7.83.].

For this study, a data matrix was constructed with row labels of clades and their genera listed in the order given in Table 1 from Kociolek [7.7.31.], and each column heading for the incidence of occurrence for each genus was ordered as the continents of Asia, North America, Europe, South America, Africa, Australasia, and Antarctica. The order of the continents was determined by the Shannon diversity index value for each continent (Figure 7.2), with the highest diversity occurring in the first column of continents decreasing in value to the lowest value in the seventh column of continents.

Three items are worth noting about the data tables from Kociolek [7.7.31.]. In Table 2 [7.7.31.], *Fideliacyclus* is incorrectly listed as occupying North and South America. *Fideliacyclus* occurs only on North America [7.7.78.]. The listed continental descriptors in Table 2 [7.7.31.] for *Stauronella* and *Stauroforma* have been switched; *Stauronella* occurs in Europe and Asia, while *Stauroforma* is cosmopolitan. Also, in Table 2 [7.7.31.], *Terpsinoë*

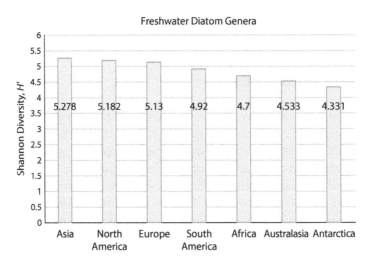

Figure 7.2 Shannon diversity for 249 freshwater diatom genera. Diversity values are superimposed on each continent's column.

should have an "X" mark in columns for Asia [7.7.55., 7.7.84.] and Africa [7.7.57.] as this genus is cosmopolitan, not disjunct. These items were corrected in the data sets used for analyses in this study.

7.3.1   Freshwater Diatom Dispersal Biogeography and the Traveling Salesman Problem

The overall distribution of freshwater diatom genera from continent to continent is to be determined via the most efficient cyclical path. This path as a closed loop is a solution to the shortest tour problem [7.7.90.] and is analyzed using the Traveling Salesman Problem (TSP, [7.7.71.]) which is a combinatorial optimization technique. For all genera, TSP is used to find shortest tours. For each clade, TSP is used to determine the shortest tour of genera from continent to continent. The magnitude of the presence of genera per clade occurring on each continent is obtained from matrix plots per clade.

To start, a graph is used as the basis of the tour. The graph may be a complete graph and may contain one or more Hamiltonian circuits (tours) [7.7.90.]. Although many solutions may be present because TSP is NP-hard, the solution to the TSP is finding Hamiltonian circuits that minimize the cost of traversing each edge in each graph as the most efficient, shortest total distance of a tour. As the goal, a given solution to the TSP is considered to be an optimal tour of all vertices. A balance between computational efficiency and optimality must be observed to obtain an optimal solution while minimizing cost (e.g., [7.7.73., 7.7.93.]).

To determine solutions to TSP and find shortest tours, heuristic and metaheuristic algorithms are used. Brute force means finding every solution to the problem explicitly and choosing the optimal outcome. As this is time and efficiency prohibitive, nearest neighbor, branch and bound (e.g., [7.7.73.]), [7.7.12.], and the greedy algorithms are examples of heuristics (e.g., [7.7.50., 7.7.73.]) employed to search for near-optimal solutions. Once such solutions are found, *opt*, tabu-search, Lin-Kernighan, and genetic algorithms as well as simulated annealing with regard to selecting, removing and reconnecting edges at places in the tour path are used to improve tour outcomes [7.7.73.]. The addition of such methods to the initial approach is an example of a metaheuristic application [7.7.50.]. TSP is essentially a method of sorting in which the minimum number of steps (tour points) are taken to get from start to end in the shortest amount of time, and that brute force, heuristics and metaheuristics may be used in finding solutions to TSP.

TSP may be generally described as a graph theory problem (e.g., [7.7.18.]). Vertices are tour stops and represented by clades or genera, and edges are representations of the connections among the clades or genera. Let $TSP = \{(Graph, f, C): K_n = (V, E)$ for n-nodes that are

V vertices with E edges where function $f: V \times V$ and $Cost = \begin{cases} C_{ij} = C_{ji} \text{ if symmetric} \\ C_{ij} \neq C_{ji} \text{ if asymmetric} \end{cases}$. If $\forall\ i$,

j, k, and $C_{ij} \leq C_{ik} + C_{kj}$, then C satisfies the triangle inequality [7.7.93.]. A distance matrix is associated with E so that the cost of a tour is minimized (e.g., [7.7.73.]).

A symmetric shortest tour may be summarized as $\sum_{j=1}^{n} x_{ij} = n$-vertices, where $x_{ij} \geq 0$, $i = 1, 2, \ldots, n$, $i \neq j$, and $x_{ij} \equiv x_{ji}$ [7.7.14.]. Minimization of the distance between the ith, jth vertex and the $i+1$th, $j+1$th vertex is represented by the matrix $D(x) = \sum_{i>j} d_{ij} x_{ij}$, where all distances are non-negative. The generalized paradigm to be solved is to minimize $f(x)$ subject to $Ax \leq b$ [7.7.14.]. That is, find the minimum number of m-shortest edges between n-vertices of a linear system as an optimal solution.

TSP may be broadly described as a combinatorial optimization technique following a generalized schema that includes constraints to obtain the desired outcome [7.7.18., 7.7.93.]. For $x_{ij} = \begin{cases} 1, \text{if tour edge } (i, j) \text{ is used} \\ 0, \text{otherwise} \end{cases}$, we want to minimize $f(x) = \sum_{i} \sum_{j} x_{ij} c_{ij}$, where c_{ij} is the cost of traveling on tour (i, j) (i.e., distance), $\sum_{i}^{n} x_{ij} = 1$ with $i = 1, 2, \ldots, n$ and $j = 2, 3, \ldots, n$, $\forall V$, $\sum_{j}^{n} x_{ij} = 1$ with $j = 1, 2, \ldots, n$ and $i = 2, 3, \ldots, n$, $\forall V$, meaning that n-vertices, V, are constrained to be visited only once without retracing, and $n = x_{i_1, j_1} + x_{i_2, j_2} + \ldots + x_{i_n, j_n}$, where $i_2 = j_1, i_3 = j_2, \ldots, i_1 = j_n$ constrains the tour to end at the tour's starting point, completing the circuit [7.7.18., 7.7.93.]. To eliminate the formation of subtours or disconnections, $\sum_{i \in S}^{n} \sum_{j \in S}^{n} x_{ij} \leq |S| - 1, \forall S \subset V$, and $S \neq \emptyset$ [7.7.65., 7.7.93.]. Constraints ensure that the resultant tour satisfies all criteria for the given problem to be solved and achieves an optimal tour as the outcome [7.7.18., 7.7.65., 7.7.93.].

7.3.2   Freshwater Diatom Biogeographic Patterns with Respect to Climate

The magnitude of freshwater diatom clade presence per continent may be influenced by climate. Each continent is classifiable with respect to climate conditions using the Köppen climate classification system [7.7.10., 7.7.40.]. Assignment of a classification category to each continent is determined with respect to lines of latitude used as demarcations (Table 7.1).

For each clade, total number of genera per continent is tabulated and paired with the sum of the climate classification for the matching continent. For TSP, this matrix is used to obtain shortest tours for each clade. Magnitude of the contribution of each continent to each tour stop is obtained via the matrix plot of the input data for each clade.

From the input data used with respect to the Köppen climate classification system, relative magnitude of the TSP tour stops is plotted for each continent for each clade. Relative distribution is plotted on a worldwide basis in terms of magnitude of the tour stops of the shortest tour given by size differences and color coding of vertices and edges for each tour for each clade.

7.3.3   Diatom Biogeographic Patterns and Distance Decay

The relation between organismal similarities and spatial distance has been used in biogeographic assessments via distance decay modeling (e.g., [7.7.58.]). Distance decay may be calculated as $I = A d^{-c}$, where I is the interaction effect between taxonomic entities, A is a constant, d is distance between geographic areas, and c is a coefficient [7.7.56.].

The variables used to determine whether distance decay applies have been modified to suit the data used in this study. The Köppen climate classification system categories and subcategories based on the environmental and ecological variables of temperature, precipitation, humidity, and vegetation were used to determine ecological similarity. Sum of the categories and subcategories for each genus was calculated with respect to the continents on which the genus is found. Sum of the number of continents occupied by each genus is used as a proxy for distance. The more continents that are occupied, the less distance of dispersion needed to become established elsewhere. The fewer continents that are occupied, the more distance of dispersion needed to become established elsewhere. Biogeographical distribution patterns of cosmopolitan, widely distributed, disjunct, and endemic were used as composite distance categories representing continental occupation as the capacity to disperse. A plot of ecological similarity versus composite distance was used to determine the trend as ecological similarity changes from cosmopolitan to widely distributed to disjunct to endemic status. Rate of change is determined as the slope of a trendline.

7.3.4   Endemism and Continental Area

The relation between endemics in each clade and area of each continent is given as

$$\sum_{\substack{j=1 \\ i=1}}^{n,m} \sum_{k}^{q} \frac{\%Endemics_{j_i} / Clade_j}{Continental\ Area_k} \tag{7.1}$$

where the ith endemic genus of the jth clade is summed to obtain n-endemics/clade over each continental area, k. From this, a plot of endemic genus "richness" [7.7.30.] versus continental area for all clades is used to determine the accumulation of endemics per continent. Continental area in kilometers is obtained from the National Geographic Family Reference Atlas of the World [7.7.54.]. Although continent area may be equivocal, the same general value is used consistently in all calculations.

7.3.5   Endemism and Dispersal Distance

Endemics in each clade with respect to the difference in distances between continents from shortest distance tours is given as

$$\sum_{\substack{j=1 \\ i=1}}^{n,m} \sum_{k=1}^{q} \frac{\%Endemics_{j_i} / Clade_j}{Continent_{k+1} - Continent_k} \tag{7.2}$$

where the ith endemic genus of the jth clade is summed to obtain n-endemics/clade over each continent, k. Distance between two continents is used to rank the continents with respect to number of endemic genera per continent, and the values are plotted. Distance between continents is obtained from an Internet search with the keywords: "Distance between [Name of Continent] and [Name of Another Continent]. While the value depends on the site selected withing a given continent, the same ballpark figure is used consistently in all calculations.

7.3.6   Endemism, Phenotypic Integration and Phenotypic Novelty: The Raphe

As an evolutionary consideration, phenotypic integration means that complex phenotypes require multiple traits with respect to functional capability [7.7.68.]. In this regard, a phenotypic novelty identifies an evolutionary event in which structural change in an organism induces or enables a new functional behavior. One such novel structural feature is the synapomorphy of the raphe in diatoms (e.g., [7.7.35.]). The raphid structural system enables locomotion in diatoms (e.g., [7.7.16., 7.7.72.]). As a structural system, the raphid system may have one or two fissures and may be present as deep slits in the diatom valve, and each section of the raphe may be separated by a siliceous central nodule [7.7.32., 7.7.75.]. Having raphes on both valves is structurally a more phenotypically integrated feature than having a raphe on only one valve or not having a raphe at all. Having a raphe on one valve is structurally more phenotypically integrated than not having a raphe on either valve. Although taxa may have had two raphes and lost both or one of them throughout evolutionary history, the current groupings of biraphid taxa reflect phylogenetic clades [7.7.35.].

The relation between endemics and phenotypic integration of the novelty trait—the raphid, and specifically, the biraphid character—is analyzed. Two trait assessments are made. First, clades having the araphid or raphid traits were binned in groups for comparison. Second, clades with the biraphid trait with genera, exhibiting motility in their vegetative state, were compared with araphid and monoraphid clades grouped together, representing traits of non-motility. In this treatment, monoraphid clades were considered to be non-motile because of their propensity to be attached and not necessarily engaging in active locomotion (e.g., [7.7.89.] on *Achnanthes*). For both comparisons, trait variation categories among clades are determined and expressed as % variance. Two trait plots are constructed: one is %variance of raphid versus %variance of araphid trait; the other is %variance of biraphid clades versus % variance of araphids and monoraphid traits. Trait plots represent phenotypic integration.

Results from trait plots of phenotypic integration were plotted versus biogeographical distribution category, after Hermant *et al.* [7.7.27.], Figure 1. Percentages of the number of genera that are widely distributed, disjunct or endemic were binned to represent biogeographical distribution. Cosmopolitan taxa were not included because of the taxonomic uncertainty associated with some of the genera per clade. That is, genus membership in some clades may be *sensu lato* rather than *sensu stricto*. Phenotypic integration is used as the gradient from araphid to monoraphid to biraphid traits. The plot is used to determine the relation between the biraphid trait and endemics.

7.4   Results

A total of 249 genera from 14 clades of freshwater diatoms was used in biogeographic analyses. A stacked bar graph of all genera per continent for each clade is illustrated in Figure 7.3. The Naviculales dominate all continents with the number of genera, while the Coscinodiscales, Biddulphiales, Rhizosoleniales, Thalassiophycidales, and Rhopalodiales

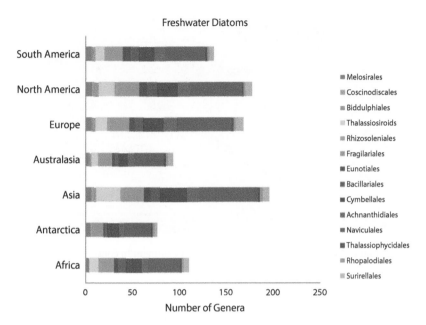

Figure 7.3 Number of genera per continent for each freshwater diatom clade.

have the fewest number of genera per continent (Figure 7.3). Thalassiosiroids are prominent on Asia, North America and to a lesser degree present on South America, Europe, Africa, and Australasia (Figure 7.3). Genera from this clade are for the most part absent on Antarctica. Fragilariales are approximately evenly distributed among all continents, as is the case for Cymballes. Achnanthidiales is noteworthy on Asia, Europe, North and South America with a lesser presence on Australasia, Africa and Antarctica (Figure 7.3).

A TSP tour of 249 genera was constructed for continents worldwide using Shannon diversity as a guide for order of continents as a constraint (Figure 7.4). Tour length was 240.326. Distribution of freshwater genera occurs from Asia to North America to Europe, spanning the northern hemisphere (Figure 7.4). From Europe, dispersal occurs to South America then to Africa to Australasia, and finally to Antarctica, covering the southern hemisphere (Figure 7.4). Antarctica back to Asia completes the tour and the dispersal of freshwater diatom genera (Figure 7.4).

TSP was applied to the 14 clades to produce a shortest tour with respect to continents (Figure 7.5). Input data were the sum of genera per clade in the order of Melosirales, Coscinodiscales, Biddulphiales, Thalassiosiroids, Rhizosoleniales, Fragilariales, Eunotiales, Bacillariales, Cybellales, Achnanthidiales, Naviculales, Thalassiophycidales, Rhopalodiales, and Surirellales for continents in the order of Asia, North America, Europe, South America, Africa, Australasia, and Antarctica. Tour length was 74.5498. Naviculales, Cymbellales, Fragilariales, Thalassiosiroids, Melosirales, and Achnanthidiales were determined to be the most widely distributed clades from continent to continent. Coscinodiscales, Rhizosoleniales, Rhopalodiales and Thalassiophycidales were the least distributed in contrast to other clades (Figure 7.5).

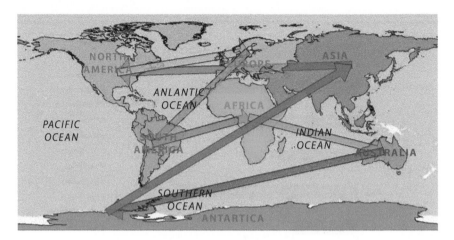

Figure 7.4 Shortest tour of 249 freshwater diatom genera from continent to continent. Tour length = 240.326.

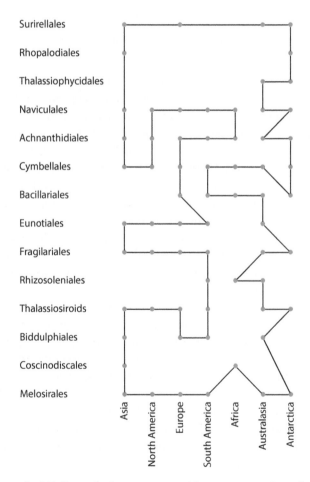

Figure 7.5 Shortest tours for 249 diatom freshwater genera with respect to continent for each clade. Tour length = 74.5498.

7.4.1   Clade Shortest Tours from Continent to Continent with Respect to Climate

For all 249 genera, a TSP tour was devised for continents worldwide with respect to Köppen climate classification categories (Figure 7.6). Tour length was 240.301. The pattern depicts freshwater genera dispersing across the northern hemisphere from North America to Asia, then from Asia across the southern oceans to Antarctica (Figure 7.6). From there, dispersal occurs in the southern hemisphere from Australasia to Africa to South America (Figure 7.6). Crossing from southern to northern hemisphere, freshwater genera are found in Europe, then disperse to North America completing the cyclical tour (Figure 7.6).

TSP was applied to genera for each clade with respect to the Köppen climate classification system (Table 7.1). Tour lengths for each clade are reported in Table 7.2. Shortest tours ranged from around 8.4 for Thalassiophycidales to 69.9 for Naviculales (Table 7.2). Matrix plots were used to interpret magnitude of each continent on each clade tour (Figure 7.7).

Clades with the largest number of genera, including Thalassiosiroids, Naviculales and Cymbellales, have more "linear" tour plots than smaller clades including Melosirales, Fragilariales, Achnanthidiales, Eunotiales, Bacillariales, and Surirellales (Figure 7.7). The clades with the smallest number of genera such as Biddulphiales, Coscinodiscales, Rhizosoleniales, Thalassiophycidales, and Rhopalodiales have a more "triangular" tour (Figure 7.7).

Matrix plots of Rhizosoleniales, Thalassiophycidales and Rhopalodiales do not have large magnitudes of influence from any continent with respect to the categories given in the Köppen climate classification system (Figure 7.7). Melosirales, Biddulphiales, Achnanthidiales, and Bacillariales have large influences with respect to Köppen categories for North America (Figure 7.7).

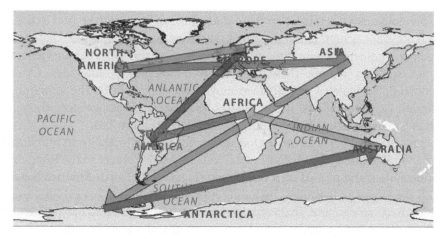

Figure 7.6 Shortest tour of 249 freshwater diatom genera from continent to continent worldwide with respect to climate. Tour length = 240.301.

Table 7.2 Shortest tour lengths of clades over all continents.

Clade	Tour length
Melosirales	12.0552
Coscinodiscales	10.6700
Biddulphiales	13.9508
Thalassiosoids	60.9161
Rhizosoleniales	10.8482
Fragilariales	23.1179
Eunotiales	18.2913
Bacillariales	18.7241
Cymbellales	39.5116
Achnanthidiales	20.9754
Naviculales	69.7472
Thalassiophycidales	8.3592
Rhopalodiales	8.5373
Surirellales	15.6832

7.4.2   Magnitude of Clade Tour Stops from Continent to Continent with Respect to Climate

From the previous analyses, tour order from clade shortest tours with respect to the Köppen climate classification system and matrix plot data were combined to plot clade tours with magnitude per continent on world maps (Figure 7.8). Shortest tours of continents for each clade depict similarity among Thalssiosiroids, Rhizosoleniales, Eunotiales, Cymbellales, Achnanthidiales, Naviculales, and Thalassiophycidales (Figure 7.8). The remaining clades of Melosirales, Coscinodiscales, Biddulphiales, Fragilariales, Bacillariales, Rhopalodiales, and Surirellales have unique shortest tours of continents (Figure 7.8).

Relative magnitude of clade distribution worldwide was extracted from individual clade TSP tours and magnitude of each continental tour stop with respect to the Köppen climate classification system and plotted on a world map (Figure 7.9). North America has the most clades with the largest magnitudes, including Melosirales, Biddulphiales, Rhizosoleniales, Fragilariales, Bacillariales, and Thalassiophycidales, while Australasia has the second most clades with the largest magnitudes including Coscinodiscales and Naviculales (Figure 7.9). Eunotiales, Achnanthidiales and Surirellales have the largest magnitudes in Asia, while Thalassiosiroids in Antarctica, Cymbellales in Africa, and Rhopalodiales in Europe have the largest clade magnitudes (Figure 7.9). All of the continents have varying degrees of magnitude of all clades with the exception of Africa and Antarctica, both of which are missing

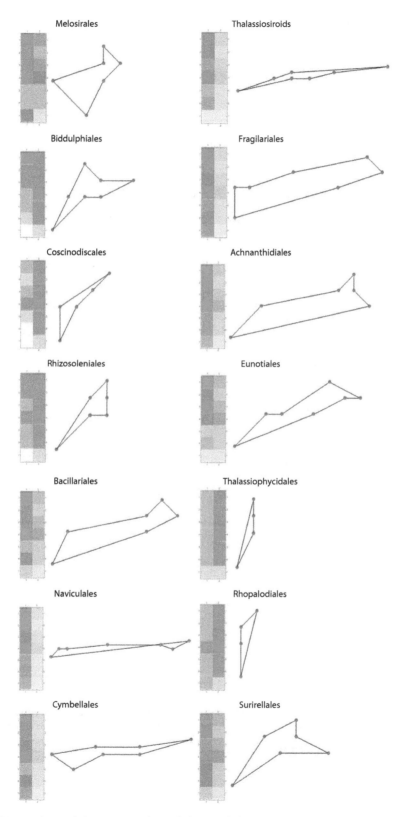

Figure 7.7 Matrix plots and shortest tours for each diatom clade with respect to climate. Matrix plots show magnitude of tour stops for each continent with respect to the Köppen climate classification system.

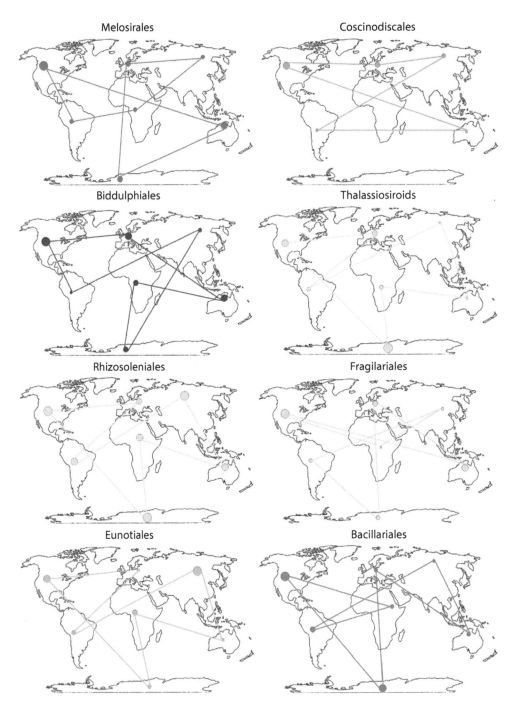

Figure 7.8 Shortest tour of each diatom clade with respect to the Köppen climate classification system plotted from continent to continent where the size of the dots represents magnitude of genera per continent. Each clade is color coded.   (*Continued*)

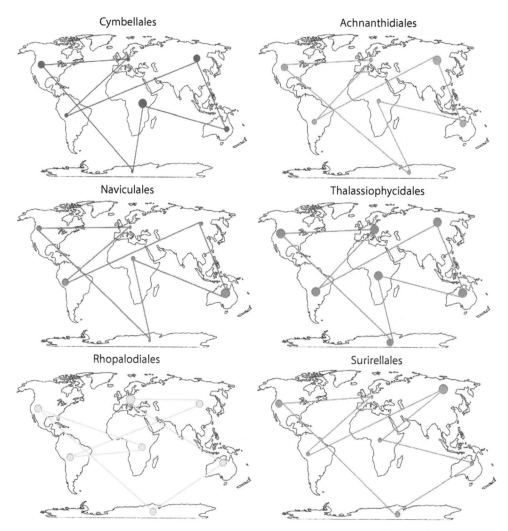

Figure 7.8 (Continued) Shortest tour of each diatom clade with respect to the Köppen climate classification system plotted from continent to continent where the size of the dots represents magnitude of genera per continent. Each clade is color coded.

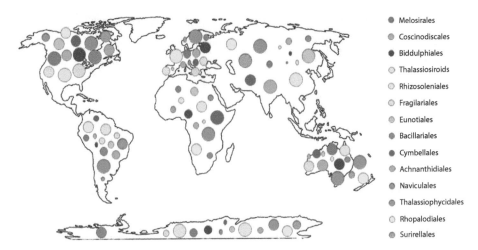

Figure 7.9 Relative diatom clade distribution worldwide. Size of dot signifies magnitude of each clade tour stop on each continent with respect to the Köppen climate classification system.

Coscinodiscales (Figure 7.9). The continents with the smallest magnitudes of clades are Europe, South America, and Antarctica (Figure 7.9).

7.4.3   Ecological Similarity, Biogeographical Distribution and Distance Decay

Ecological similarity was determined for genera per clade via categories and subcategories from the Köppen climate classification system and compiled in Table 7.3. Continental gradients of combined ecological characterizations were summarized in Figure 7.10. A plot of ecological similarity versus composite distance based on biogeographical distribution patterns is given in Figure 7.11. A least-squares best-fit curve was $y_{Ecological\ Similarity} = -3.5719x_{Distance} + 29.255$ with $R^2 = 0.9696$. Rate of change was -3.5719, signifying a decay rather than growth profile. Change in similarity with respect to distance on a continent-to-continent basis indicated distance decay (Figure 7.12).

7.4.4   Biogeographical Distribution of Freshwater Diatom Genera

TSP was applied to the 14 clades to produce a shortest tour with respect to biogeographical distribution categories (Figure 7.13). Input data were the sum of genera per clade in the order listed above, and biogeographical distribution bins were endemic, disjunct, widely distributed, and cosmopolitan. Tour length was 45.8863. Bacillariales, Rhizosoleniales and Thalassiophycidales were not found as endemics on any continent, while most clades included endemics as well as a presence of genera on more than one continent (Figure 7.13). Clades with genera occurring in all biogeographical distribution categories were Naviculales, Cymbellales, Eunotiales, Fragilariales, Thalassiosiroids, and Melosirales (Figure 7.13).

Table 7.3 Abbreviated temperature, humidity, precipitation, and vegetation consolidated subcategories of the Köppen climate classification system. Subcategories reflect ecological conditions of each continent worldwide.

Ecological conditions	Köppen categories/subcategories	North America	Europe	South America	Asia	Africa	Australasia	Antarctica
temperature/precipitation	tropical/monsoon	0	0	1	0	0	0	0
temperature/vegetation	tropical/rainforest	0	0	0	1	1	0	0
temperature/vegetation	tropical/savanna	0	0	0	0	1	1	0
humidity/precipitation	dry/semi-arid	1	0	0	1	1	1	0
humidity/precipitation	dry/arid	1	0	1	1	1	1	0
temperature/vegetation	continental/boreal	1	1	0	1	0	0	0
temperature/humidity	temperate/dry winter	0	0	1	0	0	1	0
temperature/humidity	temperate/Mediterranean-humid	1	1	0	1	1	0	0
temperature/vegetation	tundra	1	0	0	0	0	0	0
temperature/vegetation	polar	0	0	0	0	0	0	1
	SUM	5	2	3	5	5	4	1

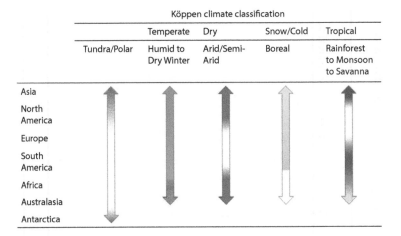

Figure 7.10 Continental gradients of Köppen climate classification categories and subcategories consisting of temperature, precipitation, humidity, and vegetation as composite ecological characterization.

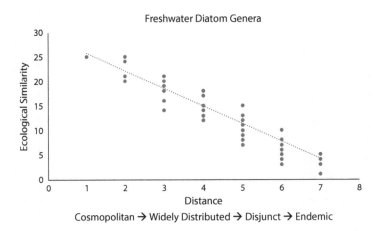

Figure 7.11 Ecological similarity based on Köppen classification subcategories (Table 7.3) of temperature, precipitation, humidity and vegetation gradients worldwide. Least-squares regression produced a best-fit curve of $y_{Ecological\ Similarity} = -3.5719x_{Distance} + 29.255$. $R^2 = 0.9696$.

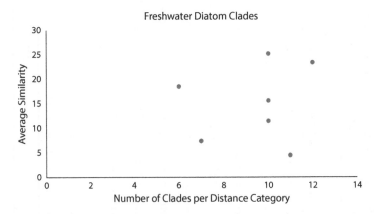

Figure 7.12 Average similarity per clade versus number of clades in each distance category.

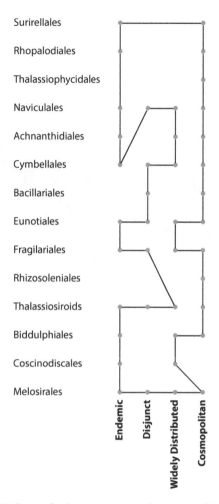

Figure 7.13 Shortest tour for 249 diatom freshwater genera with respect to biogeographical distribution category for each clade. Tour length = 45.8863.

7.4.5   Endemics in each Clade and on each Continent

The Naviculales have the largest number of endemic genera per clade (Figure 7.14). In descending order, endemic genera occur in Thalassiosiroids, Cymbellales, Eunotiales, Fragilariales, Melosirales, Coscinodiscales, and Surirellales (Figure 7.14). Those clades with very few endemic genera are Rhopalodiales, Achnanthidiales and Biddulphiales (Figure 7.14). Rhizosoleniales, Thalassiophycidales and Bacillariales have no endemic genera (Figure 7.14).

The number of endemic genera per continent indicates that Africa has the least number of endemics, while Asia has the most (Figure 7.15). Australasia, Antarctica and Europe have approximately the same number of endemic genera, albeit very few, while North and South America have the same number of endemic genera (Figure 7.15).

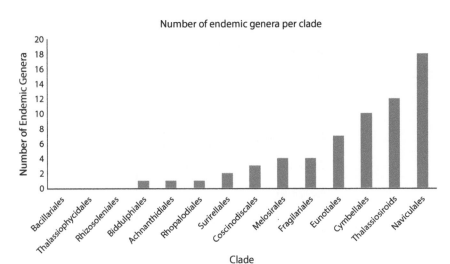

Figure 7.14 Number of endemic genera in each clade. No endemics occur in Bacillariales, Thalssiophycidales and Rhizosoleniales, while Naviculales has the most endemics.

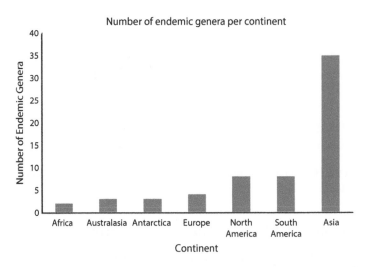

Figure 7.15 Number of endemic genera on each continent. Africa has the fewest number of endemics, while Asia has the most.

7.4.6   Phenotypic Integration and Relation to Geographic Distribution

Araphid versus raphid traits (Figure 7.16, top of right column) and araphid-monoraphid versus biraphid traits were calculated as %variances and plotted to represent phenotypic integration (Figure 7.16, bottom of right column). Results from phenotypic integration were plotted versus the geographic distribution of binned %widely distributed to %disjunct to %endemic genera (Figure 7.16, left column). Biraphid trait characterizes endemic genera to a greater degree in contrast to araphid and monoraphid traits (Figure 7.16, left column). Biraphid clades have more endemics than araphid and monoraphid clades. The biraphid

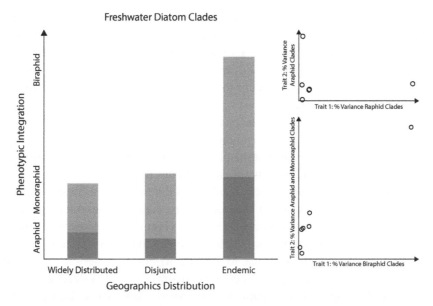

Figure 7.16 Phenotypic integration plotted versus geographic distribution of freshwater diatom clades as % widely distributed, disjunct and endemic genera; blue column represents araphid to monoraphid trait, and orange column represents biraphid trait (left column). % Variance of clades plotted as raphid versus araphid traits (top of right column) and biraphid versus araphid and monoraphid traits (bottom of right column) representing phenotypic integration.

trait clades are more highly phenotypically integrated than clades with monoraphid or araphid traits. The biraphid trait is a phenotypic novelty that signifies an evolutionary change in diatom clades that enabled motility.

7.5 Discussion

Freshwater diatom biogeography was treated as a combinatorial optimization problem using TSP. No assumptions were made about origination or end points of distribution of genera or clades so that no timeline was implied. Shortest tours produced cyclical biogeographic patterns on a worldwide continent-to-continent basis such that the minimal distance among tour points was achieved. Justified order of the clades and continents in TSP analysis enabled the construction of a single shortest tour as the optimal solution.

For the application of different rules at the outset, TSP analysis could be used to produce a better solution. Two rules that would be very useful would be the actual order of diatom clades with respect to most recent common ancestor grading to the ancestor or gradation from the most ancient to most recent clade with respect to geologic time. In this way, more definitive statements about biogeographic patterns and the processes that influence them could be made on a macroevolutionary level. For the temporal aspect of biogeographic distribution, scheduling could be used in conjunction with TSP (e.g., [7.7.1.]). If TSP is viewed as a "planning" problem, then scheduling may be viewed as a way to define tour stops and examine biogeography as a step-wise process.

Ensemble or hybrid algorithms may be used to enhance or improve solutions to optimization approaches in biogeographic pattern detection (e.g., [7.7.45.]). With the multitude of approaches and methodologies using TSP as well as the many optimization approaches available, best possible outcomes in assessing biogeographic patterns and processes are mostly dependent on the specific problem to be solved, available data, and state of knowledge about the organism of interest.

Overall, shortest tour of the continents for all genera reflected Shannon diversity per continent (Figures 7.2, 7.4 and 7.6). Northern hemisphere continents with high Shannon diversity were connected as were southern hemisphere continents with lower Shannon diversity. Shortest distance between continents as transport routes is a reasonable expectation, given constraints of biotic and abiotic transport modes and diversity of freshwater diatoms.

From shortest tour with respect to the continents, clades represented by most of the continents showed the greatest amount of dispersal (Figure 7.5). However, dispersal was not detected to be linear. Outcomes from the tour showed that Melosirales and Fragilariales dispersed from Asia to North America to Europe to South America, but because the clade is not present in Africa, dispersal between Australasia and Antarctica was isolated. For Thalassiosiroids, dispersal was from Asia to North America to Europe, but occurrence in South America was isolated as was the dispersal from Australasia to Antarctica. Cymbellales, Achnanthidiales and Naviculales occurred in all continents with different dispersal regimes. Cymbellales dispersed between Asia and North America and among South America, Africa and Australasia. Isolated occurrences were evident in Europe and Antarctica. Achnanthidiales dispersed among Europe, South America and Africa as well as between Australasia and Antarctica with isolated occurrences in Asia and North America. From North America to Europe to South America to Africa, Naviculales dispersed as was the case between Australasia and Antarctica. Occurrence in Asia was the only isolated continent for Naviculales (Figure 7.5).

Dispersal with respect to biogeographical distribution categories of endemic, disjunct, widely distributed, and cosmopolitan showed that shortest tours for clades were non-linear except for Melosirales (Figure 7.13). For clades that occurred in all categories, dispersal was the same for Eunotiales and Fragilariales where endemic and disjunct occurred together as did widely distributed and cosmopolitan members of each clade. Endemics, disjuncts and widely distributed members of Thalassiosiroids occurred together but apart from cosmopolitan members. Cymbellales and Naviculales had disjunct and widely distributed members together with endemics and cosmopolitan members occurring separately. Biddulphiales had widely distributed and cosmopolitan members occurring together with endemics as a separate occurrence. Surirellales had endemics connected to cosmopolitan members. All other clades had separate occurrences of their members with respect to biogeographical distribution categories (Figure 7.13).

7.5.1   Freshwater Diatom Clade Dispersal and Climate

A TSP map of freshwater diatom dispersal worldwide is a useful tool in looking at co-occurrences of clades. Adding Köppen climate classification categories to the analysis enables the viewing of dispersal with respect to constraints on clades and the effect that has on incidence of occurrence.

Taking into account climate categories, North America is the most important continent for all clade occurrences in contrast to Asia (Figure 7.4), despite Asia having the highest diversity (Figure 7.2). Australasia is the next important continent with respect to climate and all clade occurrences, followed by Asia (Figure 7.4). Europe, Africa, South America, and Antarctica have approximately the same magnitude of clade occurrences, but with different clades being most important. For Europe the important clades with respect to climate are Rhopalodiales and Thalassiophycidales, for Africa the most important clades are Cymbellales and Thalssiophycidales, for South America the most important clades are Thalassiophycidales, Naviculales, Rhopalodiales, and Rhizosoleniales, and for Antarctica, Thalassiosiroids and Rhizosoleniales are the most important clades (Figure 7.4).

Reasons for diatom dispersal with respect to climate vary. Latitudinal gradients for freshwater diatom clades may affect bipolar latitude distributions (e.g., [7.7.59., 7.7.88.]) with respect to climate. Location of endemics from freshwater diatom clades have been found in different basins according to latitude [7.7.59.]. Climate change may affect diatom biogeographic distribution (e.g., [7.7.74.]) regardless of latitude.

7.5.2   Distance Decay as the Pattern of Dispersal in Freshwater Diatom Biogeography

Distance decay was demonstrated in terms of generalized biogeographical distribution characterizations of endemics, disjuncts, widely distributed, and cosmopolitan for freshwater diatom genera. Ecological similarity based on Köppen climate classification categories and subcategories measured against biogeographical distribution pattern indicated a negative slope (Figure 7.11). However, a better understanding about the movements of diatom genera and better tracking of diatom presence with respect to a time sequence would be necessary to determine actual distance decay as the biogeographical pattern for freshwater diatoms. The similarity change in distance on a continent to continent basis is minimal, if at all (Figure 7.12) as expected [7.7.56.].

7.5.3   Phenotypic Integration and Diatom Biogeography

Phenotypic integration has been hypothesized to play a role in morphological evolution, potentially influencing directionality [7.7.21.]. Despite variation, phenotypic integration patterns may be conserved across large clades [7.7.21.]. For diatoms, the biraphid trait conferring motility represents the most phenotypically integrated clade members that are endemics (Figure 7.16a and b).

The biraphid trait is a functionally relevant phenotypic trait shared across clades. Phenotypic integration is influenced by ecological conditions, and adaptation to specialized environments may induce endemism [7.7.27.]. Comparing diatom lineages, endemics are largely found in biraphid genera (Figure 7.14). Despite taxonomic uncertainties surrounding diatom genera, the clades encompassing those genera and their environmental or ecological characteristics are distinguishable as endemics from disjuncts and widely distributed genera (Figure 7.16b). Cosmopolitan genera embody the taxonomic uncertainties and need additional study before this biogeographical distribution category can be definitively assigned. More endemics, disjuncts or widely distributed genera may be dissected from those clades containing broadly-conceived genera bins. Diversity may increase via splitting some of the current bins into countable new genera designations.

Phenotypic integration via a phenotypic novelty trait may be useful in determining gradients of change in ecological and environmental conditions on a local scale and climate gradients on a global scale. Such gradients may be useful in determining the stability or fragility of a given freshwater lake or river with respect to similar bodies of water or within a given region or continent for conservation purposes. Assessment of evolutionary changes on a longer time scale may be inferred by studying phenotypic integration of novel diatom morphological characters with respect to climate change as proxy indicators of such assessment.

7.6   Summary and Future Research

Diatom biogeography analyzed at the clade level enabled understanding of the distribution of freshwater diatoms via a dispersal mechanism and with regard to climate. The phenotypic novelty of the biraphid trait was found to be associated with endemism in freshwater diatom genera. Using TSP was one way to invoke a combinatorial optimization schema to produce biogeographic distributional patterns for analysis. Genera and clades were used in lieu of species data, showing that biogeographic analyses may be elucidated on a macroevolutionary level.

To help complete the picture, a study on extinct versus extant freshwater taxa distributions, change in continental position over geologic time, and the change in clade distribution is necessary. TSP and scheduling (e.g., [7.7.48.]) may be one way to address this. Other optimization techniques or mutation rates, drift, or other evolutionary considerations need to be included in analyses. More information regarding development, life cycle stages and morphological as well as molecular representation of taxa needs to be incorporated into diatom biogeographical analyses. Phenotypic integration of other traits may be of interest in diatom biogeography.

The world tour of freshwater diatom genera was mirrored by the world tour with respect to climate variables (Figures 7.4 and 7.6). The link between freshwater biodiversity and climate has been well established, and climate change has been determined to be an emerging threat (e.g., [7.7.74.]). Biogeographic assessment is important in understanding the relation between freshwater biodiversity and climate change on a worldwide basis and in conservation efforts. Research efforts in this vein are of the utmost importance, and additional biogeographic analyses of freshwater diatoms cannot start too soon.

7.7   References

7.7.1.   Bagchi, T.P., Gupta, J.N.D., Sriskandarajah, C. (2006) A review of TSP based approaches for flowshop scheduling. *European Journal of Operational Research* 169, 816-854.

7.7.2.   Baker, H.G. (1955) Self-compatibility and establishment after "long-distance" dispersal. *Evolution* 9, 347-349.

7.7.3. Beck, H.E., Zimmermann, N.E., McVicar, T.R., Vergopolan, N., Berg, A., Wood, E.F. (2018) Data descriptor: present and future Köppen-Geiger climate classification maps at 1-km resolution. *Nature|Scientific Data* 5, 180214.

7.7.4. Benoiston, A.-S., Ibarbalz, F.M., Bittner, L. Guidi, L., Jahn, O., Dutkiewicz, S., Bowler, C. (2017) The evolution of diatoms and their biogeochemical functions. *Phil. Trans. R. Soc. B* 372: 20160397.

7.7.5. Boenigk, J., Pfandl, K., Garstecki, T., Harms, H., Novarino, G., Chatzinotas, A. (2006) Evidence for geographic isolation and signs of endemism within a protistan morphospecies. *Applied and Environmental Microbiology* 72(8), 5159-5164.

7.7.6. Bouchard, G., Gajewski, K., Hamilton, P.B. (2004) Freshwater diatom biogeography in the Canadian Arctic Archipelago. *Journal of Biogeography* 31(12), 1955-1973.

7.7.7. Casteleyn, G., Leliaert, F., Backeljau, T., Debeer, A.-E., Kotaki, Y., Rhodes, L., Lundholm, N., Sabbe, K., Vyverman, W. (2010) Limits to gene flow in a cosmopolitan marine planktonic diatom. *Proc. Nat. Acad. Sci.* 107(29)12952-12957.

7.7.8. Chan, D. and Wu, Q. (2015) Significant anthropogenic-induced changes of climate classes since 1950. *Scientific Reports* 5:13487.

7.7.9. Chan, L.M., Brown, J.L., Yoder, A.D. (2011) Integrating statistical genetic and geospatial methods brings new power to phylogeography. *Molecular Phylogenetics and Evolution* 59, 523-537.

7.7.10. Chen, D. and Chen, H.W. (2013) Using the Köppen classification to quantify climate variation and change: an example for 1901-2010. *Environmental Development* 6, 69-79.

7.7.11. Chen, L., Weng, D., Du, C., Wang, J., Cao, S. (2019) Contribution of frustules and mucilage trails to the mobility of diatom Navicula sp. *Nature|Scientific Reports* 9, 7342.

7.7.12. Christofides, N. (1976) Worst-case analysis of a new heuristic for the traveling salesman problem. Report 388, Graduate School of Industrial Administration, Carnegie Mellon University, Pittsburg, Pennsylvania, USA.

7.7.13 . Crisp, M.D., Trewick, S.A., Cook, L.G. (2011) Hypothesis testing in biogeography. *Trends in Ecology and Evolution* 26(2), 66-72.

7.7.14. Danzig, G., Fulkerson, R., Johnson, S. (1954) Solution of a large-scale traveling-salesman problem. Technical Report P-510, RAND Corporation, Santa Monica, California, USA.

7.7.15. deQueiroz, A. (2005) The resurrection of oceanic dispersal in historical biogeography. *Trends in Ecology and Evolution* 20(2), 68-73.

7.7.16. Edgar, L.A. and Pickett-Heaps, J.D. (1983) The mechanism of diatom locomotion. I. An ultrastructural study of the motility apparatus. *Proc. R. Soc. Lond. B* 218, 331-343.

7.7.17. Erwin, D.H. (2015) Novelty and innovation in the history of life. *Current Biology* 25, R930-R940.

7.7.18. Filip, E. and Otakar, M. (2011) The travelling salesman problem and its application in logistic practice. *WSEAS Transactions on Business and Economics* 8(4), 163-173.

7.7.19. Fragoso, G.M., Poulton, A.J., Yashayaev, I.M., Head, E.J.H., Johnsen, G., Purdie, D.A. (2018) Diatom biogeography from the Labrador Sea revealed through a trait-based approach. *Frontiers in Marine Science* 5, Article 297 DOI 10.3389/fmars.2018.00297.

7.7.20. Godhe, A. and Rynearson, T. (2017) The role of intraspecific variation in the ecological and evolutionary success of diatoms in changing environments. *Phil Trans. R. Soc. B* 372, 20160399.

7.7.21. Goswami, A., Smaers, J.B., Soligo, C., Polly, P.D. (2014) The macroevolutionary consequences of phenotypic integration: from development to deep time. *Phil. Trans. R. Soc. B* 369, 20130254.

7.7.22. Gould, S.J. and Lewontin, R.C. (1979) The spandrels of san marco and the panglssian paradigm: a critique of the adaptionist programme. *Proceedings of the Royal Society of London B: Biological Sciences* 205, 581-598.

7.7.23. Hanson, C.A., Fuhrman, J.A., Horner-Devine, M.C., Martiny, J.B.H. (2012) Beyond biogeographic patterns: processes shaping the microbial landscape. *Nature Reviews|Microbiology* 10, 497-506.

7.7.24. Harper, M. (1969) Movement and migration of diatoms on sand grains. *British Phycological Journal* 4(1), 97-103.

7.7.25. Harper, M. and McKay, R.M. (2010) Diatoms as markers of atmospheric transport. In *The Diatoms: Applications for the Environmental and Earth Sciences, 2nd edition*, Smol, J.P. and Stoermer, E.F. (eds.), Cambridge University Press, U.K.: 552-559.

7.7.26. Hedlund, B.P. and Staley, J.T. (2004) Microbial endemism and biogeography. In *Microbial Diversity and Bioprospecting*, Bull, A.T. (ed.), ASM Press, Washington, D.C.

7.7.27. Hermant, M., Prinzing, A., Vernon, P., Convey, P., Hennion, F. (2013) Endemic species have highly integrated phenotypes, environmental distributions and phenotype-environment relationships. *Journal of Biogeography* 40, 1583-1594.

7.7.28. Incagnone, G., Marrone, F., Barone, R., Robba, L., Naselli-Flores, L. (2015) How do freshwater organisms cross the "dry ocean"? A review on passive dispersal and colonization processes with a special focus on temporary ponds. *Hydrobiologia* DOI 10.1007/s10750-014-2110-3.

7.7.29. Keck, F., Franc, A., Kahlert, M. (2018) Disentangling the processes driving the biogeography of freshwater diatoms: a multiscale approach. *Journal of Biogeography* 45(7), 1582, 1592.

7.7.30. Kier, G., Kreft, H., Lee, T.M., Jetz, W., Ibisch, P.L., Nowicki, C., Mutke, J., Barthlott, W. (2009) A global assessment of endemism and species richness across island and mainland regions. *Proceedings of the National Academy of Sciences* 106(23), 9322-9327.

7.7.31. Kociolek, J.P. (2018) A worldwide listing and biogeography of freshwater diatom genera: a phylogenetic perspective. *Diatom Research* 33(4), 509-534.

7.7.32. Kociolek, J.P., Spaulding, S.A., Lowe, R.L. (2015) Bacillariophyceae: the raphid diatoms, Chapter 16. In *Freshwater Algae of North America: Ecology and Classification 2nd edition*, J.D., Sheath, R.G., Kociolek, J.P. (eds.), Academic Press, San Diego, California: 709-772.

7.7.33. Kociolek, J.P., Stepanek, J.G., Lowe, R.L., Johansen, J.R., Sherwood, A.R. (2013) Molecular data show the enigmatic cave-swelling diatom *Diprora* (Bacillariophyceae) to be a raphid diatom. *Eur. J. Phycol.* 48(4), 474-484.

7.7.34. Kociolek, J.P., Kopalova, K., Hamsher, S.E., Kohler, T.J., Van de Vijver, B., Convey, P., McKnight, D.M. (2017) Freshwater diatom biogeography and the genus *Luticola*: an extreme case of endemism in Antarctica. *Polar Biology* DOI 10.1007/s00300-017-2090-7.

7.7.35. Kociolek, J.P., Williams, D.M., Stepanek, J., Liu, Q., Liu, Y., You, Q., Karthick, B., Kulikovskiy, M. (2019) Rampant homoplasy and adaptive radiation in pennate diatoms. *Plant Ecology and Evolution* 152(2), 131-141.

7.7.36. Koonin, E.V. (2009) Evolution of genome architecture. *Int J Biochem Cell Biol* 41(2), 298-306.

7.7.37. Koonin, E.V. (2009) Survey and summary: Darwinian evolution in the light of genomics. *Nucleic Acids Research* 37(4), 1011-1034.

7.7.38. Koonin, E.V. (2016) Splendor and misery of adaptation, or the importance of neutral null for understanding evolution. *BMC Biology* 14:114.

7.7.39.   Köppen, W. (1884) Die Wärmezonen der Erde, nach der Dauer der heissen, gemäs-
sigten und kalten Zeit und nach der Wirkung der Wärmezonen auf die organische Welt
betrachtet. *Meteorol. Z.* 1, 215–226.

7.7.40.   Köppen, W. (1900) Versuch einer Klassifikation der Klimate,vorzugsweise nach ihren
Beziehungen zur Pflanzenwelt. *GeographischeZeitschrift* 6, 657–679.

7.7.41.   Köppen, W. (2011) The thermal zones of the Earth according to the duration of hot, mod-
erate and cold periods and to the impace of heat on the organic world. *Meteorologische
Zeitschrift* 20(3), 351-360. [translation of Köppen, W. (1884) by E. Volken and S.
Brönnimann]

7.7.42.   Koester, J.A., Berthiaume, C.T., Hiranuma, N., Parker, M.S., Iverson, V., Morales, R.,
Ruzzo, W.L., Armbrust, E.V. (2018) Sexual ancestors generated an obligate asexual and
globally dispersed clone within the model diatom species *Thalassiosira pseudonana.
Nature|Scientific Reports* 9, 10492.

7.7.43.   Kottek, M., Grieser, J., Beck, C., Rudolf, B., Rubel, F. (2006) World map of the Köppen–
Geiger climate classification updated. *Meteorologische Zeitschrift* 15(3), 259–263.

7.7.44.   Love, A.C. (2003) Evolutionary morphology, innovation, and the synthesis of evolutionary and developmental biology. *Biology and Philosophy* 18, 309-345.

7.7.45.   Ma, H., Simon, D., Fei, M., Shu, X., Chen, Z. (2014) Hybrid biogeography-based evolutionary algorithms. *Engineering Applications of Artificial Intelligence* 30, 213-224.

7.7.46.   MacArthur, R.M. and Wilson, E.O. (1967) *The Theory of Island Biogeography.* Princeton
University Press, New Jersey, USA.

7.7.47.   MacDonald, J.D. (1869) On the structure of the diatomaceous frustule, and its genetic
cycle. *Annals Magazine Natural History Series* 4(3), 1-8.

7.7.48.   Mahi, M., Baykan, Ö.K., Kodaz, H. (2015) A new hybrid method based on particle
swarm optimization, ant colony optimization and 3-opt algorithms for the Traveling
Salesman Problem. *Applied Soft Computing* 30, 484-490.

7.7.49.   Mazaris, A.D., Moustaka-Gouni, M., Michaloudi, E., Bobori, D.C. (2010) Biogeographical
patterns of freshwater micro- and macroorganisms: a comparison between phytoplankton, zooplankton and fish in the eastern Mediterranean. *Journal of Biogeography* DOI
10.1111/j.1365-2699.2010.02294.x.

7.7.50.   Milwel, M.M. (2016) Traveling salesman problem mathematical description. https://
www.researchgate.net/publication/322132334_Travelling_Salesman_Problem_
Mathematical_Description, accessed on 21 September 2020.

7.7.51.   Nakov, T., Beaulieu, J.M., Alverson, A.J. (2018) Accelerated diversification is related to
life history and locomotion in hyperdiverse lineage of mirobial eukaryotes (Diatoms,
Bacillariophyta). *New Phytologist* 219, 462-473.

7.7.52.   Naselli-Flores, L., Termine, R., Barone, R. (2016) Phytoplankton colonization patterns.
Is species richness depending on distance among freshwaters and on their connectivity?
Hydrobiologia 764, 103-113.

7.7.53.   Nathan, R. (2013) Dispersal biogeography. In *Encyclopedia of Biodiversity, 2nd edition,*
Volume 2, Academic Press, Waltham, Massachusetts: 539-561.

7.7.54.   National Geographic Family Reference Atlas of the World (2006) Washington, DC:
National Geographic Society (U.S.).

7.7.55.   Navarro, J.N. and Lobban, C.S. (2009) Freshwater and marine diatoms from the western
pacific islands of Yap and Guam, with notes on some diatoms in damselfish territories.
Diatom Research 24, 123-157.

7.7.56.   Nekola, J.C. and White, P.S. (1999) The distance decay of similarity in biogeography and
ecology. *Journal of Biogeography* 26, 867-878.

7.7.57. N'Guessan, K.R., Aboua, B.R.D., Tison-Rosebery, J., Ouattara, A., Kouamelan, E.P. (2018) Biodiversity and ecology of epilithic diatoms in the Agnéby River, Ivory Coast. *African Journal of Aquatic Science* 43(2), 131-140.

7.7.58. Okubo, A. and Levin, S.A. (1989) A theoretical framework for data analysis of wind dispersal of seeds and pollen. *Ecology* 70, 329–338.

7.7.59. Oyanedel, P., Vega-Retter, C., Scott, S., Hinojosa, L.F., Ramos-Jiliberto, R. (2008) Finding patterns of distribution for freshwater phytoplankton, zooplankton and fish, by means of parsimony analysis of endemicity. *Revista Chilena de Historia Natural* 81, 185-203.

7.7.60. Padisák, J., Vasas, G., Borics, G. (2016) Phycogeography of freshwater phytoplankton: traditional knowledge and new molecular tools. *Hydrobiologia* 764, 3-27.

7.7.61. Pantocsek, J. (1889) Beiträge zur Kenntniss der fossilen Bacillarien Ungarns. II Brackwasser Bacillarien. Julíus Platzko, Nagy-Tapolcsány, Hungary.

7.7.62. Patarnello, T., Volckaert, E., Castilho, R. (2007) Pillars of Hercules: is the Atlantic Mediterranean transition a phylogeographical break? *Molecular Ecology* 16, 4426-4444.

7.7.63. Peel, M.C., Finlayson, B.L., McMahon, T.A. (2007) Updated world map of the Köppen-Geiger climate classification. *Hydrol. Earth Syst. Sci.* 11, 1633-1644.

7.7.64. Peterson, T. and Müller, G.B. (2016) Phenotypic novelty in evo-devo: the distinction between continuous and discontinuous variation and its importance in evolutionary history. *Evolutionary Biology* 43, 314-335.

7.7.65. Pferschy, U. and Staněk, R. (2017) Generating subtour elimination constraints for the TSP from pure integer solutions. *Central European Journal of Operations Research* 25, 231-260.

7.7.66. Pfitzer, E. (1871) Untersuchungen über Bau und Entwicklung der Bacillariaceen (Diatomaceen) [Studies on construction and development of Bacillariaceae (Diatomaceae)] [German]. In: *Botanische Abhandlungen aus dem Gebiete der Morphologie und Physiologie [Botanical Treatises in the Field of Morphology and Physiology]*. J.L.E.R. von Hanstein, (ed.) Adolph Marcus, Bonn, Germany: [i]-vi, 1-189, 186 pls.

7.7.67. Pfitzer, E. (1882) Die Bacillariaceen (Diatomaceae) [Bacillariaceae (Diatomaceae)] [German]. In: *Encyklopaedie der Naturwissenschaften. I. Abteilung. I. Thiel: Handbuch der Botanik*. A. Schenk, (ed.) Verlag von Eduard Trewendt, Breslau. 2: 403-445.

7.7.68. Pigliucci, M. (2003) Phenotypic integration: studying the ecology and evolution of complex phenotypes. *Ecology Letters* 6, 265-272.

7.7.69. Pigliucci, M. (2008) What, if anything, is an evolutionary novelty? *Philosophy of Science* 75, 887-898.

7.7.70. Pigliucci, M. and Müller, G.B. (eds.) (2010) *Evolution: The Extended Synthesis*, MIT Press, Cambridge, United Kingdom.

7.7.71. Platzman, L.K. and Bartholdi, J.J. (1989) Spacefilling curves and the planar ravelling salesman problem. *Journal of the ACM* 36(4), DOI 10.1145/76359.76361.

7.7.72. Poulsen, N.C., Spector, I., Spurck, T.P., Schultz, T.F., Wetherbee, R. (1999) Diatom gliding is the result of an actin-myosin motility system. *Cytoskeleton* 44(1), 23-33.

7.7.73. Rego, C., Gamboa, D., Glover, F., Osterman, C. (2011) Traveling salesman problem heuristics: leading methods, implementations and latest advances. *European Journal of Operational Research* 211, 427-441.

7.7.74. Reid, A.J., Carlson, A.K., Creed, I.F., Eliason, E.J., Gell, P.A., Johnson, T.J., Kidd, K.A., MacCormack, T.J., Olden, J.D., Ormerod, S.J., Smol, J.P., Taylor, W.W., Tockner, K., Vermair, J.C., Dudgeon, D., Cooke, S.J. (2019) Emerging threats and persistent conservation challenges for freshwater biodiversity. *Biol. Rev.* 94, 849-873.

7.7.75.   Round, F.E., Crawford, R.M. and Mann, D.G. (1990) *The Diatoms, Biology & Morphology of the Genera.* Cambridge University Press, Cambridge, U.K.

7.7.76.   Simon, D. (2008) Biogeography-based optimization. *IEEE Transactions on Evolutionary Computation* 17(6), 702-713.

7.7.77.   Sims, P.A., Mann, D.G., Medlin, L.K. (2006) Evolution of the diatoms: insights from fossil, biological and molecular data. *Phycologia* 45(4), 361-402.

7.7.78.   Siver, P.A., Wolfe, A.P., Edlund, M.B. (2016) *Fideliacyclus wombatiensis* gen. et sp. nov. – a Paleocene non-marine centric diatom from northern Canada with complex frustule architecture. *Diatom Research* DOI 10.1080/0269249X.2016.1256351.

7.7.79.   Soininen, J. and Teittinen, A. (2019) Fifteen important questions in the spatial ecology of diatoms. *Freshwater Biology* 64, 2071-2083.

7.7.80.   Soininen, J., McDonald, R., Hillebrand, H. (2007) The distance decay of similarity in ecological communities. *Ecogeography* 30, 3-12.

7.7.81.   Spear, S.F., Balkenhol, N., Fortin, M.-J., McRae, B.H., Scribner, K. (2010) Use of resistance surfaces for landscape genetic studies: considerations for parameterization and analysis. *Molecular Ecology* 19, 3576-3591.

7.7.82.   Steinbauer, M.J., Schweiger, A.H., Irl, S.D.H. (2016) Biogeography, patterns in. In *Encyclopedia of Evolutionary Biology, Volume 1,* Kliman, R.M. (ed.), Academic Press, Oxford: 221-230.

7.7.83.   Theriot, E.C., Ashworth, M.P., Nakov, T., Ruck, E., Jansen, R.K. (2015) Dissecting signal and noise in diatom chloroplast protein encoding genes with phylogenetic information profiling. *Molecular Phylogenetics and Evolution* 89, 28-36.

7.7.84.   Tuji, A. (2018) A new freshwater diatom, *Terpsinoe muninensis* sp. nov., from the Ogasawara Islands, Japan. *Mem. Natl. Mus. Nat. Sci.,* Tokyo 52, 5-15.

7.7.85.   Vanormelingen, P., Verleyen, E., Vyverman, W. (2007) The diversity and distribution of diatoms: from cosmopolitanism to narrow endemism. *Biodiversity and Conservation* 17, 393-405.

7.7.86.   Via, S. and Lande, R. (1985) Genotype-environment interaction and the evolution of phenotypic plasticity. *Evolution* 39(3), 505-522.

7.7.87.   Vyverman, W., Verleyen, E., Wilmotte, A., Hodgson, D.A., Willems, A., Peeters, K., Van de Vijver, B., De Wever, A., Leliaert, F., Sabbe, K. (2010) Evidence for widespread endemism among Antarctic micro-organisms. *Polar Science* 4, 103-113.

7.7.88.   Vyverman, W., Verleyen E., Sabbe, K., Vanhoutte, K., Sterken, M., Hodgson, D.A., Mann, D.G., Juggins, S., Van de Vijver, B., Jones, V., Flower, R., Roberts, D., Chepurnov, V.A., Kilroy, C., Vanormelingen, P., De Wever, A. (2007) Historical processes constrains patterns in global diatom diversity. *Ecology* 88, 1924-1931.

7.7.89.   Wang, Y., Lu, J., Mollet, J.-C., Gretz, M.R., Hoagland, K.D. (1997) Extracellular matrix assembly in diatoms (Bacillariophyceae). *Plant Physiology* 113, 1071-1080.

7.7.90.   Weisstein, E.W. (2002) *CRC Concise Encyclopedia of Mathematics, 2nd ed.,* Chapman & Hall/CRC, London, United Kingdom.

7.7.91.   Witt, O.N. (1886) Ueber den Polierschiefer von Archangelsk-Kurojedowo im Gouv. Simbirsk. *Verhandlungen der Russisch-Kaiserlichen Mineralogischen Gesellschaft zu St. Petersbur,* series 2. 22, 137-177.

7.7.92.   Yool, A. and Tyrrell, T. (2003) Role of diatoms in regulating the ocean's silicon cycle. *Global Biogeochemical Cycles* 17(4), 1103, 14-1 – 14-21.

7.7.93.   Yousif, M. and Al-Khateeb, B. (2018) A novel metaheuristic algorithm for multiple traveling salesman problem. *Journal of Advanced Research in Dynamical & Control Systems* 10, 13-Special Issue, 2113-2122.

8

Cell Division Timing and Mode of the Diatom Life Cycle

Abstract

The diatom life cycle features the unique occurrence of size diminution during vegetative reproduction. This stage of development as well as the other stages of the diatom life cycle make for a complicated life history of this unicell. The timing of the longest phase of the diatom life cycle—vegetative reproduction—and reaching the MacDonald-Pfitzer limit induce sexual reproduction, followed by auxosporulation and production of an initial cell which restores size to the new vegetative cell. The Mackey-Glass equation was modified to represent each stage of the diatom life cycle. Switches between each stage were modeled according to the Mackey-Glass equation as well so that switch timing was instantaneous and spontaneous. Overall, timing of the end of each stage of the diatom life cycle induced a chaotic outcome as a signal to switch to the next stage. Stable periodic outcomes were indicative of each life cycle stage. Using a Mackey-Glass system enabled the introduction of subtle changes in life cycle stages that may provide the means to model individual life cycles to represent different diatom taxa.

Keywords: Mitosis, meiosis, Mackey-Glass equation, size diminution, auxosporulation, initial cell, cell division, feedback loop

8.1 Introduction

The diatom life cycle is one of the most complicated compared to other unicells. The diatom life cycle is diplontic (e.g., [8.9.19., 8.9.25., 8.9.30., 8.9.41., 8.9.60., 8.9.88.]) and has features of animal and plant cells (e.g., [8.9.28.]). One unique feature is size diminution or reduction during vegetative reproduction [8.9.67., 8.9.80., 8.9.81.] in which cytokinesis is coupled to mitosis (e.g., [8.9.55.]). Not all diatoms exhibit this feature (e.g., [8.9.60.]). Once an approximately 30% reduction has occurred, diatoms form auxospores after undergoing sexual reproduction in order to restore their size (e.g., [8.9.5., 8.9.22., 8.9.59., 8.9.67., 8.9.80., 8.9.81., 8.9.88.]). However, some diatoms restore their size asexually (e.g., [8.9.60.]). Various modes of achieving size restoration have been modeled to describe this facet of the diatom life cycle (e.g., [8.9.115.]).

Another feature of the diatom life cycle occurs in response to the environment. Diatoms are capable of forming resting spores during times of environmental stress and is an evolved survival strategy (e.g., [8.9.92.]). Resting spores may remain as such for years to millennia [8.9.92.]. As an evolutionary model for existence, diatoms have a multi-step approach to ensuring that longevity is sustainable.

Size reduction during vegetative reproduction may not be adequately modeled by the MacDonald-Pfitzer schema (e.g., [8.9.115.]). The assumption is that size reduction proceeds in a constant stepwise fashion. However, size reduction may happen gradually or abruptly [8.9.25.]. The timing of each cell division may be linear or non-linear. Regardless, the general overview of the MacDonald-Pfitzer rule as size reduction during vegetative cell division is still useful as an initial guidance for that stage of the diatom life cycle.

8.1.1   Evolution of Diatom Cell Division Dynamics

Cell division induces evolutionary implications for diatoms over the course of their biological history. Diatoms originated via secondary endosymbiosis with regard to progenitor red and green algal plastids (e.g., [8.9.73.]). The wide range of environmental conditions under which diatoms have adapted and survived have been determined to be a result of novel regulatory mechanisms concerning their ability to integrate signaling networks with cell cycle dynamics (e.g., [8.9.53.]). Specifically, nutrient sensing via cyclin genes (e.g., [8.9.53.]) and blue light modulation [8.9.54.] are directly involved in regulating diatom cell division. The onset of cell division is induced by light modulation via cyclin genes and the cell cycle checkpoints of DNA replication and mitosis as well [8.9.53., 8.9.54.]. Diatom cell signaling induces growth and reproduction that may influence grazing capacity in the oceans, and diatom responses to nutrient limiting conditions via signaling may induce a reduction in growth and reproduction (e.g., [8.9.23., 8.9.107.]).

8.1.2   Diatom Life Cycle as a Dynamical System

The diatom life cycle may be represented as a dynamical system consisting of a four-stage process:

> Stage 1: asexual (vegetative) reproduction and size diminution via mitosis.
> Stage 2: sexual reproduction via meiosis I and II.
> Stage 3: auxospore formation.
> Stage 4: size restoration via initial cell formation.

A generalized diatom life cycle schematic is given in Figure 8.1. The diatom survival strategy of resting spore formation [8.9.92.] may be construed as another life cycle stage, occurring potentially during mitosis [8.9.60.] or after auxosporulation [8.9.25., 8.9.59., 8.9.90.]. Resting spores are a source of regenerated vegetative cell reproduction [8.9.92.]. For the four-stage life cycle, timing of each stage is definable according to results from natural and laboratory analyses in which actual time intervals have been observed and measured in hours, days, weeks or years (e.g., [8.9.5., 8.9.59.]).

Stage 1 is defined by starting with a vegetative cell, then dividing so that size reduction occurs (Figure 8.2a). The end of this stage culminates in reduction at the 30% level. Stage 2 is the production of gametes that lead to sexual reproduction (Figure 8.2c). Between Stages 1 and 2, a switch occurs (Figure 8.2b). Fertilization produces a zygote, and auxospore formation and expansion constitutes Stage 3 (Figure 8.2e). Between Stages 2 and 3, another switch occurs (Figure 8.2d). An initial cell is produced from auxosporulation, defining Stage 4 (Figure 8.2g). Between Stages 3 and 4, the third switch is defined for the life cycle

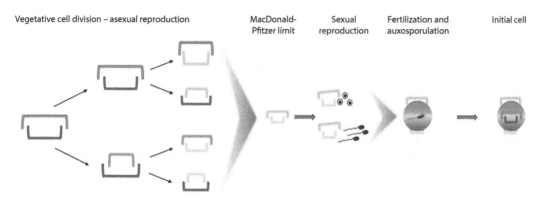

Figure 8.1 The diatom life cycle.

Vegetative cell division – asexual reproduction and MacDonald-Pfitzer limit		Sexual reproduction		Fertilization and auxosporulation		Initial cell	
n-cell divisions: weeks to years [1]	switch 1	hours to weeks [1]	switch 2	hours to days [1]	switch 3	hours to days [1]	switch 4
a		b	c	d	e	f	g	h

[1] Kaczmarska and Ehrman 2021

Figure 8.2 Timing for each diatom life cycle stage.

(Figure 8.2f). Finally, the initial cell becomes the first vegetative cell via a switch between Stages 4 and 1 (Figure 8.2h), and the life cycle is complete (Figures 8.1 and 8.2).

The diatom life cycle is a dynamical system via each life cycle stage being dependent on timing and the interval of duration for each stage. However, linkage to circadian oscillators as regulators of timing has not been established, with signaling via cyclin genes being implicated in diatom life cycle regulation (e.g., [8.9.53.]). Timing and each stage's time interval vary over multiple time scales so that as a dynamical system, there is a dependence on time delays (e.g., [8.9.12.]) in the diatom life cycle.

8.1.3   Diatom Cell Division and Growth Rate

For unicells, cell division induces a growth rate (e.g., [8.9.69., 8.9.85., 8.9.111.]). How often cells divide is a frequency and may be given as

$$Cell_{frequency} = \frac{\left(n + \dfrac{n}{2}\right)}{m + \left(n + \dfrac{n}{2}\right)} \tag{8.1}$$

where m is the number of non-dividing cells, and n is the number of divided cells [8.9.85., 8.9.111.]. The growth rate for the maximum cell division frequency is

$$\mu_{growth} = \frac{1}{\ln(1 + Cell_{\max frequency})} \tag{8.2}$$

[8.9.111.]. The growth rate equation may be rewritten as

$$\mu_{growth} Cell_{frequency} = \frac{\ln(1 + Cell_{frequency})}{\ln 2} \left(\frac{1}{t} \right) \tag{8.3}$$

where t is time [8.9.85.].

For diatoms, mitosis is cellular organelle division and distribution during cell separation, producing new diatom cells [8.9.64., 8.9.79., 8.9.102.]. Mitosis, as $Cell_{number}$ is a function of t and may be represented as

$$Cell_{number}(t) = \int_{t}^{t+\tau} dN(t) = N(t+\tau) - N(t) \tag{8.4}$$

where the number of cells that complete cell separation equals N, the number of cells from time t to τ as the time interval in between interphase and cell separation or cytokinesis, from preprophase to telophase [8.9.28., 8.9.69.]. The number of cells in mitosis is

$$Cell_{mitosis}(t) = \frac{N(t+\tau)}{N(t)} - 1 \tag{8.5}$$

with the maximum number of cells in mitosis per day as

$$Cell_{mitosis}(t_{growth\ day}) = N(t_0)(1 + Cell_{max\ mitosis}) \tag{8.6}$$

which may be used to estimate the total number of cells undergoing mitosis [8.9.69.]. The specific growth rate during mitosis per day is

$$\mu_{mitosis} = \frac{1}{t_{growth\ day}} \ln(1 + Cell_{max\ mitosis}) = \frac{1}{t_{growth\ day}} \ln \left(\frac{N(t+\tau)}{N(t)} \right) \tag{8.7}$$

[8.9.69.], although actual growth rates may be less than or greater than results from Eq. (8.6). Examples of growth rates via mitosis for *Ditylum brightwelli* and *Biddulphia mobiliensis* [8.9.99.] as well as *Chaetoceros armatum* [8.9.65.] may be represented by $\mu_{mitosis} = Cell_{max\ mitosis}$ [8.9.69.], although this may not necessarily represent all diatoms.

8.1.4   Diatom Cell Size Diminution during Mitosis

Diatom cell division may occur with a reduction in the size of subsequent cells during mitosis in the diatom life cycle. Size diminution may be represented by a decay rate while the mitotic cell division rate is a growth rate with respect to number of cells produced. That is, the diatom cell decays in size, while the number of cells produced during mitosis grows.

Size diminution is closely related to cell volume and is reciprocally related to growth rate [8.9.70.]. The relation between size diminution and mitotic rate may be given as

$$V_{size\ diminution} = \frac{1}{\mu_{mitosis}}$$

(8.8)

so that the size diminution rate per day is

$$V_{size\ diminution} = \frac{t_{size\ diminution}}{\ln(1 + Cell_{max\ mitosis})}$$

(8.9)

where the maximum mitotic rate in days is inversely proportional to size diminution rate. The fraction of time that mitosis occurs during cell number growth when there is constant cell division is

$$\left(\frac{t_{mitosis}}{\tau}\right)\ln 2 = \ln(1 + Cell_{mitosis})$$

(8.10)

[8.9.26.]. As a result, the fraction of time that size diminution occurs during mitosis is

$$t_{size\ diminution} = V_{size\ diminution}\left[\left(\frac{\tau}{t_{mitosis}}\right)\ln 2\right]$$

(8.11)

so that

$$\mu_{mitosis} = \frac{1}{t_{size\ diminution}\left[\left(\frac{t_{mitosis}}{\tau}\right)\ln 2\right]} = \frac{1}{V_{size\ diminution}}$$

(8.12)

reiterating that the mitotic rate is inversely proportional to the size diminution rate. From Eq. (8.12), the average rate of size diminution per generation is generally a negative exponential function (e.g., [8.9.1.]).

8.1.5   Diatom Cell Division during Meiosis

Once the MacDonald-Pfitzer threshold is reached, the diatom vegetative cell is induced to produce haploid cells that become either oogonia (eggs) or spermatogonia (sperm) via meiosis [8.9.34., 8.9.59.]. After first dividing during meiosis I, meiosis II occurs as replication of the haploid cells [8.9.59.]. In terms of size, egg cells are larger in size than sperm cells [8.9.61.].

As a two-step process, diatom meiosis involves growth in cell numbers with half the chromosomes, then a further growth in cell numbers with the full number of chromosomes [8.9.59.]. During meiosis I, the base pairs—adenine-thymine (AT) and guanine-cytosine (GC)—content of DNA from the mother cell is reallocated to the daughter cells [8.9.101.]. The conversion from one set of base pairs to the other occurs as a rate per generation [8.9.101.]. The meiotic I rate of the changes in base pairs is

$$\lambda_{meiosis\ I} = \frac{1}{q + w} \ln\left(\frac{p_{initial} - p_{final}}{p_{CG\ transient} - p_{final}} \right) \tag{8.13}$$

where q and w are rates of generational change for AT and CG base pairs, respectively, and the fraction of generations of GC, $p_{CG\ transient}$, that change from initial to final equilibrium during the reallocation of base pairs with respect to AT as $(1 - p_{CG\ transient})$ are changes in generations (after [8.9.101.]). With changing generations, there is the potential for a time delay so that the meiotic I rate becomes

$$\lambda_{meiosis\ I} = \frac{1}{t_{(AT+CG)}} \ln\left(\frac{N(t+\tau)}{N(t)} \right) \tag{8.14}$$

where the fraction of generations changing is $p(\cdot) = \exp\left[\ln\frac{N_t}{N_0} / t \right]$ as obtained via the spontaneous mutation rate (e.g., [8.9.62.]), and the number of generations of AT and CG changes from t to $t + \tau$.

For meiosis II, haploid cell replication occurs so that twice the number of daughter cells are produced with reference to meiosis I (e.g., [8.9.59.]). Numbers of cells replicating are given as

$$\lambda_{meiosis\ II} = \left[\frac{2}{t_{(AT+CG)}} \ln\left(\frac{N(t+\tau)}{N(t)} \right) \right] \tag{8.15}$$

which is the meiotic II rate estimated to enable formation of the diploid zygote.

8.1.6   Diatom Cell Division after Meiosis

After meiosis, the zygote expands via auxosporulation [8.9.59.]. Generally, the auxospore undergoes isodiametric and anisodiametric expansion in centric diatoms (e.g., [8.9.56., 8.9.59.]) and elongation in a tubular fashion in pennate diatoms (e.g., [8.9.59., 8.9.61., 8.9.77.]). Auxospore expansion occurs in hours to days [8.9.59.] as a sequential process (e.g., [8.9.61.]). Overall, auxospore expansion is represented as

$$\kappa_{auxospore} = \frac{1}{t - \tau} \ln\left(\frac{N_{auxospore}(t+\tau)}{N_{auxospore}(t)} \right) \tag{8.16}$$

where the time from zygote to expanded auxospore does not undergo cell division.

From the mature auxospore, an initial cell is produced over hours to days and may resemble a misshapen vegetative cell [8.9.59., 8.9.94.]. Initial cells may develop vegetatively as well (e.g., [8.9.95.]). Within the auxospore, the initial cell undergoes divisions, resulting in the breakage of the auxospore wall (e.g., [8.9.96.]). The size of the initial cell may be as large or larger than the auxospore from which it came [8.9.59., 8.9.61., 8.9.95.] and may be as large or larger than the first vegetative cell resulting from the initial cell [8.9.59., 8.9.95., 8.9.96.]. The production of the initial cell is represented as

$$\zeta_{initial} = \frac{1}{t-\tau} \ln\left(\frac{N_{initial}(t+\tau)}{N_{initial}(t)} \right) \tag{8.17}$$

and from the initial cell, the vegetative part of the diatom life cycle resumes (e.g., [8.9.88., 8.9.96.]).

8.2   Purposes of this Study

The evolution of various developmental pathways induces the success of organisms in the biological world. Diatoms have evolved with a unique and complicated life cycle that has enabled their proliferation throughout the Cenozoic to be a dominant influence in the oceans as primary producers (e.g., [8.9.48.]), in the silica and carbon cycles (e.g., [8.9.11.]), in carbon sequestration (e.g., [8.9.97.]), and as environmental and climate indicators (e.g., [8.9.72.]). The timing and length of life cycle stages occur as cell size changes, and this forms the basis of generating a generalized model. The unique life cycle enables the capability of diatoms to exist in a myriad of conditions which attests to their ability to adapt, survive and thrive throughout millions of years.

8.3   Background on the Diatom Cell Cycle

8.3.1   Diatom Life Cycle Timing: Stages

The four-staged diatom life cycle is definable as an initial value problem. Each life cycle stage (Stages 1-4) may have different initial values. Between stages, switches (next section) may have different initial values from each other and from stages. Specification of initial values is necessary to properly initiate each diatom life cycle stage.

To model the four-staged diatom life cycle, each stage must be characterized by change in cell size with respect to change in growth rate in number of cells. That is, each stage has its own cell number growth rate and geometric characteristics (Figures 8.1 and 8.2). Cell number growth rate increase or decrease may include time delays in the life cycle process within each stage. Switches occur and may or may not be delays between stages.

For Stage 1 (Figure 8.1), mitosis proceeds, resulting in n-cells in which size reduction may or may not occur. The more interesting case is the one for size reduction, so this will be designated as a qualifier for Stage 1. While the number of cells is variable, size reduction induces the accumulation of smaller and smaller cells until the MacDonald-Pfitzer limit is reached. Expansion of each mitotic cell increases in the girdle band region so that overall, there may not be a reduction in total cell volume [8.9.90.] for this stage. However, the rate of cell size change is a decay rate during size reduction. Time scale for Stage 1 is the longest of any stage in the diatom life cycle and constitutes the major phenotypic representation of the diatom cell (Figure 8.2a).

After reaching the MacDonald-Pfitzer limit, male and female cells are produced inducing sexual reproduction as given for Stage 2 (Figure 8.1). The time interval is much shorter than it is for Stage 1 (Figure 8.2c). The female cell or egg may enlarge at a rate different from the male cell or sperm, inducing a cell size difference for the first part of Stage 2 of the diatom life cycle. The second part of Stage 2 is the replication of haploid gametes (Figure 8.2c).

At Stage 3, fertilization occurs producing a zygote which induces auxosporulation and expansion of the cell (Figure 8.1). Auxospores vary by shape where there are spherical, tubular and multipolar forms [8.9.59.]. Stage 3 is the shortest duration compared to Stages 1 and 2 (Figure 8.2e). The time scale for Stage 3 applies to Stage 4 as well (Figure 8.2g). Formation of an initial cell induces the largest cell volume as the restoration of cell size occurs (e.g., [8.9.90.]). Cell size for Stage 3 may be comparable to Stage 4 (e.g., [8.9.59.]). From the resultant initial cell, the more recognizable vegetative cell is produced to start the diatom life cycle process once again.

Diatom life cycle stages and switches may be representative of multiple feedback loops. As an example of a negative feedback loop, distribution of cellular organelles may lag behind each mitotically divided cell, and as a result, a delay is activated to enable each new cell time to acquire the proper cellular contents [8.9.37.]. Negative feedback loops induce oscillations, enabling stability and steady-state dynamics [8.9.36., 8.9.37.]. Positive feedback loop oscillations would cover processes involving activation or inactivation of cues or switches such as starting or stopping mitosis [8.9.35., 8.9.37.]. The periodicity of the oscillations from negative feedback loops is also a property of positive feedback loops [8.9.37.].

At equilibrium, perturbations may induce instability. However, stability may return multiple times resulting in sustained oscillations. This may induce evolution toward a limit cycle [8.9.45.]. Multiple states of equilibrium may induce bistability [8.9.36., 8.9.37., 8.9.45.]. Oscillations induce bifurcations which are the result of switches via "on-off" transitions inducing a bistable regime during, e.g., mitosis [8.9.36., 8.9.37.].

8.3.2   Diatom Life Cycle Timing: Switches

Between diatom life cycle stages, changes occur in which there is a switch from one stage to the next. Genetic toggle switches that are bistable (e.g., [8.9.40., 8.9.47.]) may be used as models for switches between life cycle stages. Switches represent a type of delay which may be distinct from the delays within each stage. Stability of each switch is necessary to ensure successful transition between life cycle stages. As a result, the geometric properties of such switches (e.g., [8.9.2.]) are necessarily utilized to induce the linking of successive stages together.

The generalized genetic toggle switch when applied to the diatom life cycle is

$$\frac{dx}{dt} = \frac{\alpha_1}{1+y^\beta} - \delta_1 x \tag{8.18}$$

and

$$\frac{dy}{dt} = \frac{\alpha_2}{1+x^\gamma} - \delta_2 y \tag{8.19}$$

where x and y represent the end and beginning of sequential life cycle stages, respectively α_1, represents "on" or activation at the beginning of a stage, and α_2 represents "off" or repression at the end of the same stage, β and γ represent mutual cooperativity between the end of one stage and the beginning of another one, with $\delta_1 x$ and $\delta_2 y$ representing decay rates (after [8.9.40., 8.9.47.]). For the diatom life cycle, switches between stages proceed in the forward direction so that as one stage ends ("off"), the next stage begins ("on"). That is, a switch exists in two exclusive states [8.9.9.] between life cycle stages. Switching in the forward direction induces stability [8.9.4., 8.9.52.] for each stage in the diatom life cycle. Switching between life cycle stages is bistable (e.g., [8.9.9., 8.9.66.]).

8.3.3 Diatom Life Cycle Timing: Cell Behavior

The Hill-Langmuir equation [8.9.50., 8.9.51., 8.9.63.] is applicable to cell division (e.g., [8.9.17.]) and is given as

$$\vartheta(t,N) = \frac{r}{1+N^\gamma} \tag{8.20}$$

where ϑ is a rate of cell division, and $\gamma > 0$ (e.g., [8.9.17.]). Cell division rate is a function of the number of cells and time and is comparable to a mitotic rate (e.g., [8.9.91.]).

Diatoms use quorum sensing to regulate behavior with regard to environmental cues externally and internally involving signaling (e.g., [8.9.114.]). Quorum sensing is the mode of cellular communication that controls cell density via gene regulation (e.g., [8.9.29.]). For example, timing of the switching from vegetative to sexual reproduction is induced via quorum sensing in *Thalassiosira weissflogii* [8.9.32.], i.e., quorum sensing has a role in diatom cell division (e.g., [8.9.114.]).

Quorum sensing and quenching may be used to model promoting and restricting cell division occurrences, respectively, in each diatom life cycle stage. Additionally, quorum sensing and quenching may be used to infer stability to achieve cell division success and induce or suppress diatom cell growth, depending on the presence of other algal or bacterial influences (e.g., [8.9.114.]). In spite of this, there is a regularity of cell behavior so that synchronicity dictates diatom cell division (e.g., [8.9.46.]).

Switching must occur optimally (e.g., [8.9.38., 8.9.112.]) between stages to ensure cell cycle progression. The cell cycle induces transitions in which both negative and positive feedback loops induce oscillatory behavior [8.9.37., 8.9.45.]. As a result, bistability is induced in that switch activation (on) or repression (off) is fast and complete [8.9.36., 8.9.37.]. Switches act as memory devices to activate or halt cell cycle activity including oscillations [8.9.27.].

8.4   Modeling the Diatom Life Cycle: Timing of Stages and Switches

8.4.1   Delay Differential Equations

During vegetative reproduction, diatoms may or may not undergo size reduction, and the rate of change from one division to the next is not necessarily constant. The timing may be non-linear, and to model this phenomenon, delay differential equations (DDEs) are suitable to use. Additionally, sexual reproduction inducing auxosporulation and initial cell formation as well as the possibility of resting spore formation and duration attest to the non-linear character of the entire diatom life cycle and life strategy.

Cell division across the spectrum of biological organisms may have an aspect of synchronicity that may be recoverable via DDEs [8.9.20.]. When synchronous, there is a specific reset state that may be defined in a sequence of stages (e.g., [8.9.106.]). When asynchronous, oscillations may occur (e.g., [8.9.78.]). In either case, timing and time interval duration may be represented by DDEs and may be applied to cell division systems.

Characteristics of DDEs include oscillatory behavior as periodic (e.g., [8.9.20.]) which is also characteristic of cellular processes including cell division via signaling (e.g., [8.9.29.]). Such cellular processes at a microscopic level induce macroscopic resultant life cycle stage outcomes from cell division. Oscillations in cellular behavior may be synchronous or asynchronous from collective interaction of multiple cellular processes [8.9.29.]. Whether synchronous and in phase or asynchronous and out of phase, oscillations result and are dependent on cell density [8.9.29.]. Quorum sensing may occur when oscillations are synchronous in which a resultant decrease in amplitude or damping occurs from the start to the end of a time evolution step [8.9.29.]. This dynamical process may be useful in characterizing a given diatom life cycle stage. An asynchronous regime might induce incoherent oscillations over time and possibly produce a chaotic outcome. DDEs are useful in modeling synchronous and asynchronous behavior (e.g., [8.9.78.]).

Oscillatory behavior may be dictated via bifurcation during cell division (e.g., [8.9.37.]). Bifurcation may induce stability (e.g., [8.9.43., 8.9.110.]) from "on" to "off" switches between diatom life cycle stages. Hopf bifurcation may be present to indicate the kind of stability present with respect to the kind of oscillations present [8.9.43., 8.9.78., 8.9.110.] and may correspond to the limit cycle [8.9.45.].

8.4.2   Solutions to DDEs

Within each DDE regime for each diatom life cycle stage, quorum sensing and quenching may be explicitly or implicitly represented. As a result, some life cycle stages may have delays with abrupt time changes, while other delays may produce longer timespan changes. Changes from very small to large time steps may induce non-solution of the DDEs resulting in stiffness (e.g., [8.9.98.]) which may occur at any stage. That is, the abruptness in changing between different time scales may be present as stiffness in DDEs.

To address stiffness, Runge-Kutta (R-K) methods have been used to enable solution to DDEs (e.g., [8.9.20., 8.9.74., 8.9.98.]). Implicit R-K methods enable coverage of widely different time steps versus explicit R-K methods that are more applicable to small time steps over a given time span. Implicit R-K may be very efficient when used only if there is stiffness present [8.9.98.]. For non-stiff DDEs, explicit R-K (e.g., [8.9.100.]) or Gragg smoothing (e.g., [8.9.98.]) is more efficient. For some DDE systems, switching between stiff and non-stiff detection induces efficiency in obtaining accurate solutions when time steps are small (e.g., [8.9.21.]).

In particular, solutions to stiff DDEs have been obtained via 4th-order R-K methods (e.g., [8.9.20.]), Hermite interpolation polynomials (e.g., [8.9.74.]), or 4th- to 7th-order Runge-Kutta-Fehlberg methods along with Hermite interpolation polynomials (e.g., [8.9.76.]). Linear multi-step methods have been used for stiff DDEs where there are constant delays [8.9.21.]. The backward Euler method (e.g., [8.9.98.]) as an implicit rule and the symmetrization of R-K methods for both implicit and explicit rules (e.g., [8.9.24.]) are extrapolation techniques enabling the switching between non-stiff and stiff solutions to DDEs. In any case, stiffness needs to be dealt with to produce a workable DDE system that induces stable solutions [8.9.98.].

8.4.3   Mackey-Glass System of DDEs

Many DDEs have been devised to represent a number of dynamical biological systems. The Mackey-Glass equation (M-G) [8.9.68.] has been used in modeling physiological phenomena such as respiration rate via Hill's equation [8.9.42.] or blood cell production rate, including changes from normal to abnormal cell division and is given as

$$\frac{dN}{dt} = \frac{rN_\tau}{1 + N_\tau^\gamma} - bN \tag{8.21}$$

where $\frac{dN}{dt}$ is the mature cell density rate, the first term on the right-hand side of the equation is total cell production rate, $N_\tau = N(t - \tau)$ is the time lag for mature cell production, and the second term on the right-hand side of the equation is the cell mortality rate [8.9.13., 8.9.68.]. The production rate includes a time delay, τ, so that a change in N depends on N_τ or N at time $(t - \tau)$. Discrete growth rate is $1 + N_\tau^\gamma$, while bN is also a decay rate. The exponent γ indicates the steepness of the stepwise growth curve as it increases over time. The parameters r, γ, and b may be obtained from empirical observations or data and used as initial conditions.

For the M-G, time lag is constant and a single delay. This provides a starting point for application to diatom cell division vegetatively or at other stages in the life cycle. The M-G is useful in modeling cell division changes within or between stages having bifurcation and periodic characteristics (e.g., [8.9.20.]) such as the diatom life cycle stages [8.9.39.] as a result of resetting or perturbation occurrences of the life cycle occurring as a result of environmental influences (e.g., [8.9.22.]). Depending on the life cycle process, modifications of the M-G including multiple delays [8.9.15.], positive and negative delays [8.9.16.], or variable delays [8.9.13.] may need to be instituted in modeling.

The M-G at small delays induces a Hopf bifurcation converging to a periodic solution, while at large delays, chaotic patterns may emerge (e.g., [8.9.87.]). Chaotic outcomes are prevalent in biological systems (e.g., [8.9.44., 8.9.103.]), although such outcomes may be indicative of pathological conditions or instabilities (e.g., [8.9.42.]) in a cell cycle system. Diatom colony growth follows a Lindenmayer (or L-) system, accounting for the self-similarity and fractal properties [8.9.46.] associated with cell division and growth during the life cycle. Such properties are associated with chaotic outcomes. However, the M-G applied to the diatom life cycle at each stage and switch would have initial conditions specified to induce periodic rather than chaotic outcomes so that stability is achieved.

8.4.4   Mackey-Glass System: Stage 1 of the Diatom Life Cycle

Stages of the diatom life cycle and the switches between stages have similar structure with respect to the M-G. The exception is with Stage 1 because of size diminution during mitosis (Figure 8.1) that may occur over weeks to years (Figure 8.2). Diatom mitosis is a growth process in terms of cell numbers, while size diminution is a decay process. Using Eqs. (8.8) and (8.12), Stage 1 in terms of the M-G may be given as

$$\frac{dN_{Stage1}}{dt} = a_{2delays}\frac{rN_\tau\ h(t)}{1+N_\tau^\gamma\ g(t)} - bN_{Stage1} \tag{8.22}$$

where $h(t)$ and $g(t)$ represent mitosis and size diminution, respectively. For two delays used that coincide with each other, $a_{2delays} = \gamma = 2$ (e.g., [8.9.16.]). Alternatively,

$$\frac{dN_{Stage1}}{dt} = a_{many\ delays}\frac{r(\mu_\tau\cdot v_\tau)}{1+\mu_\tau^\gamma} - bN_{Stage1} \tag{8.23}$$

may be used to represent many coincident delays in which size diminution occurs at each step of the cell division process during mitosis, as given in the numerator of the first term on the right-hand side of the equation. Size diminution induces a slowing of the growth rate in cell numbers during Stage 1 (e.g., [8.9.3.]), and may be specified for many coincident delays in terms of the MacDonald-Pfitzer limit.

Stage 1 ends when the MacDonald-Pfitzer limit is reached, inducing a chaotic rather than a stable periodic outcome. Obtaining a chaotic outcome is achieved by using the longest time delay, and coefficients for size diminution are included so that the Stage 1 M-G becomes

$$\frac{dN_{Stage1}}{dt} = \frac{r(\mu_\tau\cdot[a_{MacDonald-Pfizer}\cdot v_\tau])}{1+\mu_\tau^\gamma} - bN_{Stage1} \tag{8.24}$$

where $0.3 \leq a_{MacDonald - Pfizer} \leq 1.0$ includes size values ranging from the MacDonald-Pfizer limit (0.3 for 30% size reduction) to the first vegetative cell (1.0) produced from the initial cell.

8.4.5 Mackey-Glass System: Stage 2 of the Diatom Life Cycle

Stage 2 of the diatom life cycle is characterized by sexual reproduction (Figure 8.1). Meiosis I occurs via gametogenesis with the production of haploid gametes (e.g., [8.9.10.]) in which cytokinesis may occur [8.9.59., 8.9.71.]. Oogenesis and spermatogenesis often occur within hours (Figure 8.2) in which eggs and sperm are formed [8.9.59.]. Meiosis II occurs in which haploid gametes replicate. No cytokinesis occurs during meiosis II in centric diatoms [8.9.59., 8.9.61., 8.9.71., 8.9.96.].

For application of the M-G to Stage 2, the two-step meiotic process involves cell differentiation and the changing of base pairs during the first step and cell division for second step. For meiosis I,

$$\frac{dN_{Stage2_I}}{dt} = \frac{r_I(\lambda_\tau)}{1+(\lambda_\tau^\gamma)} - bN_{Stage2_I} \tag{8.25}$$

where λ is the meiotic rate. For meiosis II,

$$\frac{dN_{Stage2_{II}}}{dt} = \frac{r_{II}(\lambda_\tau)}{1+(\lambda_\tau^\gamma)} - bN_{Stage2_{II}} \tag{8.26}$$

where $r_{II} = 2r_I$ with twice as many cells resulting.

8.4.6 Mackey-Glass System: Stages 3 and 4 of the Diatom Life Cycle

For Stage 3, fertilization occurs, and an auxospore is formed in which cell expansion proceeds (e.g., [8.9.59.]). Cell division does not occur. Auxosporulation is represented as

$$\frac{dN_{Stage3}}{dt} = \frac{r(\kappa_\tau)}{1+(\kappa_\tau^\gamma)} - bN_{Stage3} \tag{8.27}$$

based on the auxospore sequential expansion rate to maturation.

Stage 4 is the production of an initial cell vegetatively or from the auxospore and is represented as

$$\frac{dN_{Stage4}}{dt} = \frac{r(\zeta_\tau)}{1+(\zeta_\tau^\gamma)} - bN_{Stage4} \tag{8.28}$$

based on the initial cell undergoing cell divisions before emerging from the auxospore.

8.4.7   Mackey-Glass System: The Diatom Life Cycle Switches

During the diatom life cycle, switching from one stage to the next is an event that has the same structure as the M-G. From Eqs. (8.18) and (8.19) as switching rates, switching time is exponentially driven in this bistable system (e.g., [8.9.18.]). Once the MacDonald-Pftizer limit is reached, a switch occurs ending Stage 1 and starting Stage 2 (Figure 8.2b). Cytokinesis finishes and mitosis ceases at the end of Stage 1, then Stage 2 begins with gamete production (Figure 8.1). The switching time may be modelled as

$$\tau_{switching} \simeq \frac{\mathcal{A}}{g_{production}\alpha} \exp\left[\frac{2}{\alpha(1+\delta)}\ln\frac{1}{\delta}\right] \tag{8.29}$$

where $\mathcal{A}$ is a coefficient, $g_{production}$ is the production rate of Stage 1, α is α_1 for "on" or α_2 for "off," and δ is δ_1 and δ_2 decay rates [8.9.18.]. Using symmetric parameters for α and δ, spontaneous switches for the average switching time is

$$\tau_{spontaneous\ switching} \cong \frac{\alpha_1\ g_{production}}{\alpha_2\ d^2_{degradation}} \tag{8.30}$$

where $d_{degradation}$ is the degradation rate [8.9.9., 8.9.66.] which is proportional to the decay rate [8.9.109.]. Spontaneous switching is bistable and provides suitable approximation of the switching time (e.g., [8.9.9.]). Switching rates between Stages 2 and 3, Stages 3 and 4, and Stages 4 and 1 occur in the same fashion. In modeling the diatom life cycle, spontaneous switching times between all stages occur via instantaneous rates. That is, switching occurs in an instant.

Instantaneous or instant switches that occur between life cycle stages may have the form of the M-G. The solution to the M-G for a switch is defined so that a bistable steady-state is achieved. Using Eqs. (8.18), (8.19) and (8.20) in view of Eqs. (8.28) and (8.29), the M-G of the "on" and "off" switches are given as

$$\frac{dN_{switch\ on}}{dt} = \frac{rN_{\tau\ on}}{1+N^{\beta}_{\tau\ off}} - bN_{switch\ on} \tag{8.30}$$

and

$$\frac{dN_{switch\ off}}{dt} = \frac{rN_{\tau\ off}}{1+N^{\gamma}_{\tau\ on}} - bN_{switch\ off} \tag{8.31}$$

but because instant switches are used, $a_1 \equiv a_2$. By using spontaneous switches, $\delta_1 \equiv \delta_2$. For $\beta = \frac{1}{\gamma}$, bistability is induced for large values of r [8.9.40.]. For t and τ with respect to instant switching,

$$t_{switch}(\tau_{switch}) = \begin{cases} t_{switch\,0} + (N_\tau - t_{switch\,0})\tau_{switch} & 0 \le \tau_{switch} \le 1 \\ N_\tau + (t_{switch\,f} - N_\tau)(\tau_{switch} - 1) & 1 \le \tau_{switch} \le 2 \end{cases} \tag{8.32}$$

where $t_{switch} \in [t_{switch\,0}, t_{switch\,f}]$ for $\sigma_{switching\ sequence} = [t_{switch\,0}, t_{switch\,f}]$ (after [8.9.112.]). For N_τ, $\frac{dN_\tau}{dt} = 0$, and $N_\tau(0) = t_{switch\,1}$ is the switching instant (after [8.9.112.]). Rewriting the M-G,

$$\frac{dN_{switch}(\tau)}{d\tau} = (N_\tau - t_{switch\,0})f_1 \tag{8.33}$$

for $\tau \in [0,1)$, and

$$\frac{dN_{switch}(\tau)}{d\tau} = (t_{switch\,f} - N_\tau)f_2 \tag{8.34}$$

for $\tau \in [1,2]$ with $\frac{dN_\tau}{dt} = 0$ for f (after [8.9.112.]). One result is that N_τ may be equal to N_t as an instant switch (e.g., [8.9.38.]). The spontaneous, instant switch between life cycle stages is

$$\frac{dN_{switch\,1}}{dt} = \frac{rN_\tau}{1 + N_\tau^\beta} - bN_{switch\,1} \tag{8.35}$$

where $\tau_{switch\,1}$ is evaluated with respect to the inequalities in Eq. (8.32). For the diatom life cycle, the spontaneous, instant switch is non-linear and occurs as one stage ends and the next stage begins.

8.5 Methods

Simulations if diatom life cycle stages and switches are based on the M-G. Using the M-G system requires testing with stiffness extrapolation methods. All stages are run in which all parameters are defined numerically, and τ, which is used to advance each stage with constant multiple time delays, is defined by the same values for each stage.

Time delays are instituted according to Figure 8.2. Stage 1 is represented by the largest time spread, followed by Stage 2, with the smallest time interval representing Stages 3 and 4 (Figure 8.2). Multiple constant time lags are simulated for each life cycle stage, although for Stages 2, 3 and 4, the time lags are scaled additionally with other parameters so that outcomes are representative of the time scale differences with respect to all stages.

Stage 1 is simulated according to Eq. (8.24). For the rest of the diatom life cycle, Eqs. (8.25) and (8.26) are used for Stage 2, Eq. (8.27) is used for Stage 3, and Eq. (8.28) is used for Stage 4. Values for r, y and b are held constant. Simulation outcomes for stages are depicted as feedback loops and time series plots according to each time delay instituted for simulation. At each stage, time delay plots are used to depict stability for the completion of the given stage as well as instability which is depicted as chaotic behavior that would be exhibited when the life cycle stage is extended beyond its actual time boundaries. That is, chaotic behavior signifies the time beyond the end of a given stage.

Between stages, spontaneous, instant switches are instituted according to Eq. (8.35) with respect to Eq. (8.32). Because the diatom life cycle is cyclical and a feedback loop, the same switch is used from Stage 4 to Stage 1. The switches used have a minimal impact on the life cycle while representing the transition from one stage to the next. A schematic of spontaneous, instant switches will be illustrated, and M-G results will be plotted.

All M-G results for the diatom life cycle are compiled as a matrix by assembling of the time series for each stage and presented diagrammatically. Amplitudes for each time delay are plotted to depict the timing of life cycle stages. A line of demarcation is made between stable periodic and unstable chaotic outcomes. Maximum amplitudes for each time delay that is stable periodic signifies diatom life cycle stage timing.

A composite plot is devised to illustrate maximum stable amplitudes for all stages of the diatom life cycle. From all results an ensemble plot is constructed to represent the timing of stages and switches of the diatom life cycle.

8.6   Results

Simulations for each diatom life cycle stage were depicted as plots in terms of feedback loops and time series oscillations. Time was set on the interval $t \in [300, 500]$ as transient outcomes were no longer present at $t = 300$, and resolved outcomes were indicative of the given stage at $t = 500$. Time delay values for each stage were $\tau = 5, 10, 15, 30,$ and 100, and size diminution coefficients, respectively, for these time delays were $a_{MacDonald-Pfizer} \cdot v_{\tau} = 1.0,$ 0.8, 0.6, 0.4, and 0.3 (Figure 8.3). The values of the other parameters were $r = 0.25$ for Stages 1, 2-meiosis I, and 4, $r = 0.5$ for Stage 2-meiosis II, $r = 1$ for Stage 3, $y = 10$ for all stages except Stage 4 with $y = 12$, and $b = 0.1$ for all stages except Stage 3 with $b = 0.2$.

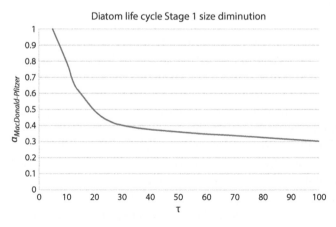

Figure 8.3 Diatom life cycle Stage 1 size diminution chart from maximum to minimum size with respect to time delays.

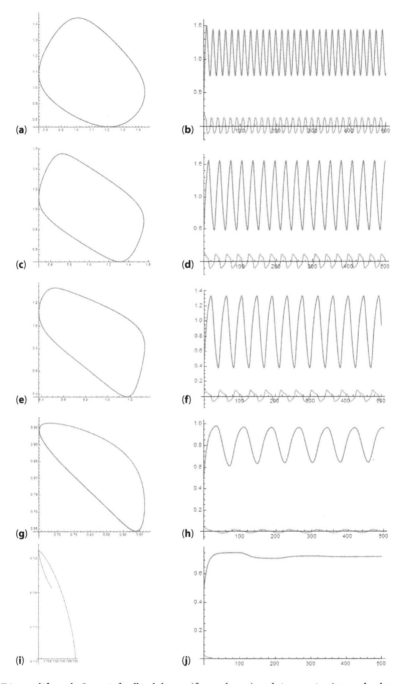

Figure 8.4 Diatom life cycle Stage 1 feedback loops (first column) and time series (second column). M-G system of mitosis with cell size diminution for time delays: a, b, $\tau = 5$; c, d, $\tau = 10$; e, f, $\tau = 15$; g, h, $\tau = 30$; i, j, $\tau = 100$. For all feedback loops, $r = 0.25$; $\gamma = 10$; $b = 0.1$. First (blue curves) and second derivative (orange curves) oscillations are plotted as times series.

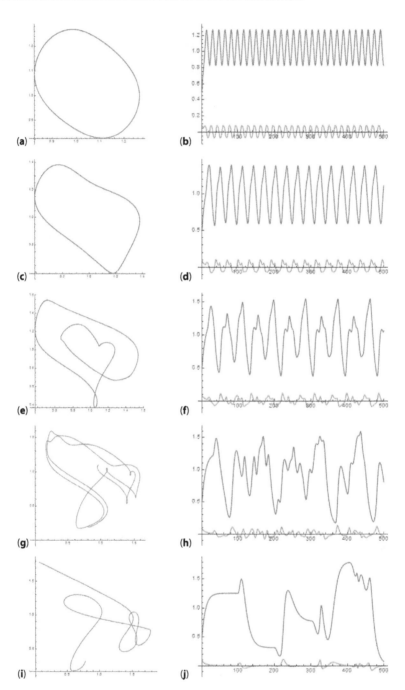

Figure 8.5 Diatom life cycle Stage 2, meiosis I feedback loops (first column) and time series (second column). M-G system of meiosis I where the time delays are: a, b, $\tau = 5$; c, d, $\tau = 10$; e, f, $\tau = 15$; g, h, $\tau = 30$; i, j, $\tau = 100$. For all feedback loops, $r = 0.25$; $\gamma = 10$; $b = 0.1$. First (blue curves) and second derivative (orange curves) oscillations are plotted as times series.

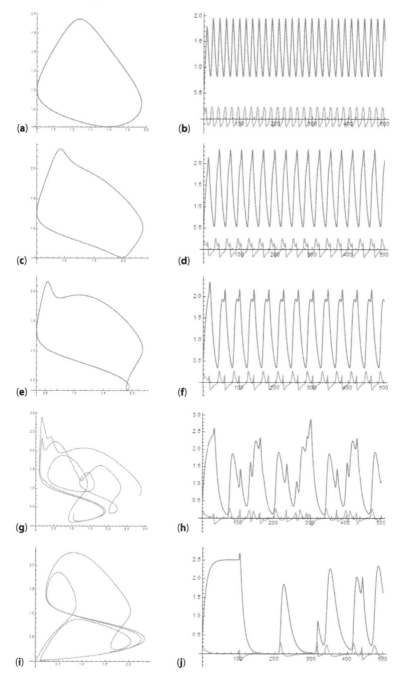

Figure 8.6 Diatom life cycle Stage 2, meiosis II feedback loops (first column) and time series (second column). M-G system of meiosis II where the time delays are: a, b, $\tau = 5$; c, d, $\tau = 10$; e, f, $\tau = 15$; g, h, $\tau = 30$; i, j, $\tau = 100$. For all feedback loops, $r = 0.5$; $\gamma = 10$; $b = 0.1$. First (blue curves) and second derivative (orange curves) oscillations are plotted as times series.

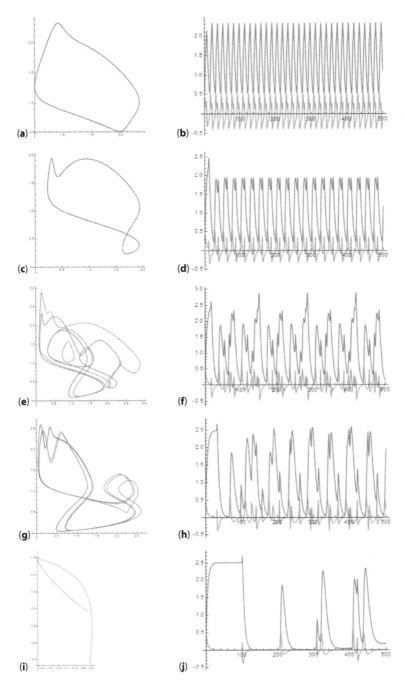

Figure 8.7 Diatom life cycle Stage 3 feedback loops (first column) and time series (second column). M-G system of auxosporulation where the time delays are: a, b, $\tau = 5$; c, d, $\tau = 10$; e, f, $\tau = 15$; g, h, $\tau = 30$; i, j, $\tau = 100$. For all feedback loops, $r = 1$; $y = 10$; $b = 0.2$. First (blue curves) and second derivative (orange curves) oscillations are plotted as times series.

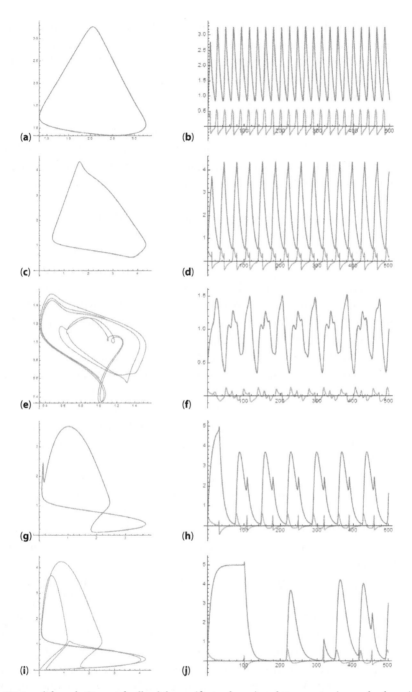

Figure 8.8 Diatom life cycle Stage 4 feedback loops (first column) and time series (second column). M-G system of initial cell production where time delays are: a, b, $\tau = 5$; c, d, $\tau = 10$; e, f, $\tau = 15$; g, h, $\tau = 30$; i, j, $\tau = 100$. For all feedback loops, $r = 0.25$; $y = 12$; $b = 0.1$. First (blue curves) and second derivative (orange curves) oscillations are plotted as times series.

Stage 1 at all τ values depicted stability except at $\tau = 100$ for the MacDonald-Pfitzer limit (Figure 8.4). For Stage 2 meiosis I and II, chaotic outcomes occurred for $\tau = 30$ and $\tau = 100$ (Figures 8.5 and 8.6) where the time scale is approximately 7 to 10 times shorter than the time scale for Stage 1 (Figure 8.4). For Stage 3 (Figure 8.7), $\tau = 15$, 30, and 100 induced chaotic outcomes as is the case for Stage 4 (Figure 8.8).

For Stage 1, the MacDonald-Pfitzer size diminution limit used in the M-G was at 30%. To further investigate the behavior of the M-G for $\tau = 100$ at the approximate MacDonald-Pfitzer limit, outcomes at 32, 35, 38, and 40% were analyzed. Chaotic, unstable outcomes were induced at 30, 32 and 35% (Figure 8.9a-f), while at 38 and 40%, stable periodicity was achieved (Figure 8.9g-j). The stability outcome is similar to that for $\tau = 30$ and the 40% limit (Figure 8.4g, h).

A schematic of spontaneous, instant switches patterned after the exclusive genetic switch (after [8.9.9., 8.9.18., 8.9.66.]) is given in Figure 8.10. Spontaneous, instant switches between life cycle stages were plotted as feedback loops and time series (Figure 8.11). With $\sigma_{switching\ sequence} = [t_{switch\ 0}, t_{switch\ f}]$, the rate of spontaneous, instant switching for $0 \leq \tau_{switch} \leq 2$ is less than the rate for each life cycle stage (Figure 8.12a) thereby preserving bistability (Figure 8.12b). The relation between $N_{\tau\ switch}$, as log a with respect to a_1 and a_2, and τ_{switch} depicts the time delay between the switch being $N_{\tau\ on}$ ("on") or a_1 and $N_{\tau\ off}$ ("off") or a_2 between life cycle stages (Figure 8.12).

M-G time series for each time delay for each diatom life cycle stage was assembled into a matrix (Figure 8.13). A red line of demarcation indicated stable periodic outcomes above the line and unstable chaotic outcomes below the line (Figure 8.13). Maximum stable periodic outcomes represented each life cycle stage with the comparative timing of each stage represented by the number of time delays exhibiting stability. Stage 1 was stable from $\tau = 5$ to 30, Stage 2 as 2-I and 2-II represented stability via $\tau = 5$ to 15, and Stages 3 and 4 exhibited stability for $\tau = 5$ to 10 (Figure 8.13).

Maximum amplitudes for each diatom life cycle stage were plotted with respect to time delays for stable periodic outcomes up to $\tau = 30$ (Figure 8.14). The curves for each stage illustrated the progression of the diatom life cycle, with Stage 1 at the bottom to Stage 4 at the top (Figure 8.14). The slopes for each stage got progressively larger. Stage 1 had a negative slope at -0.0045, followed by positive slopes for Stages 2-I, 2-II, 3, and 4 with 0.002, 0.01, 0.02, and 0.05, respectively (Figure 8.14).

An ensemble sequence of stages and switches to represent the timing of the generalized diatom life cycle was devised as a diagram (Figure 8.15). The ensemble diagram depicted the changes in time series maximum amplitudes associated with each life cycle stage. Switch timings were depicted as linear connectors in view of Eq. (8.29). For the switches, "on" is green and "off" is red with a change in maximum amplitude up to 0.2 (Figure 8.15). As the life cycle stages progressed over time, the overall trend was an increase in maximum amplitude (Figures 8.13 and 8.15). A negative slope occurred for Stage 1, while positive slopes occurred for Stages 2, 3 and 4 (Figure 8.15).

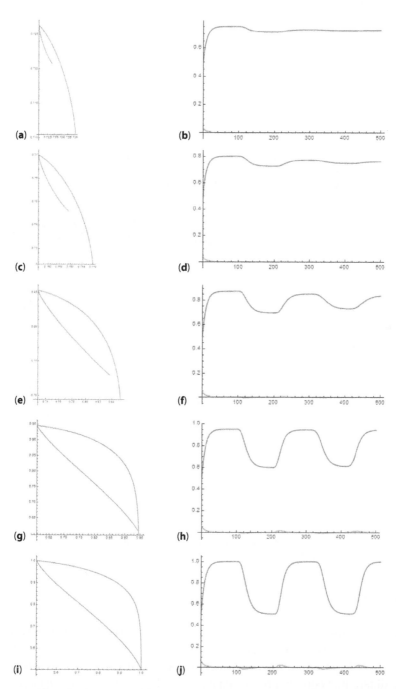

Figure 8.9 Diatom life cycle MacDonald-Pfitzer cell size limit during Stage 1. Feedback loops (first column) and time series (second column) are plotted using $\tau = 100$: a, b, 30%; c, d, 32%; e, f, 35%; g, h, 38%; i, j, 40%. First (blue curves) and second derivative (orange curves) oscillations are plotted in the times series plots. Rows 1, 2 and 3 depict chaotic instability, while rows 4 and 5 depict stable periodicity.

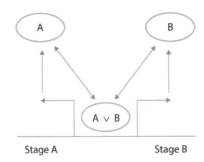

Figure 8.10 Schematic of spontaneous, instant switch from state A to state B.

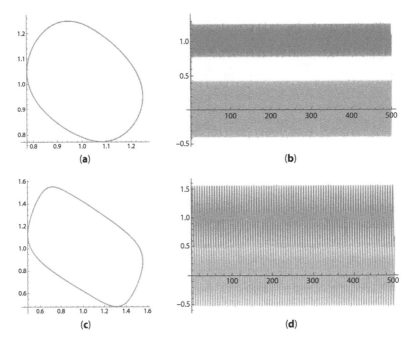

Figure 8.11 Spontaneous, instant switches used between stages of the diatom life cycle: a, b, $\tau = 1$; c, d, $\tau = 2$. Other parameters are $r = 1$, $b = 0.5$, and $\beta = 10$.

8.7   Discussion

The generalized schema for the diatom life cycle was suitably modeled using the M-G system. Each step of the way during the life cycle was dependent on changes in cell number growth (or cell density), cell size and timing. Each step was a variant of the M-G in this regard. Using such a delay differential equation model enabled overlapped timing and generations of cells as well as accounting for the fraction of cell deaths. The M-G system illustrated the institution of necessary control in the diatom life cycle so that chaos does not ensue.

Using the M-G, a negative slope for Stage 1 is reflective of size diminution as it occurs during mitosis over time (Figure 8.15) which may reflect a negative feedback loop (e.g., [8.9.52.]). Positive slopes for the remaining stages may reflect positive feedback loops (e.g., [8.9.45.]). Oscillatory behavior in stages may induce negative as well as positive feedback loops (e.g., [8.9.45.]), and the transition from one oscillatory time-delayed state to another

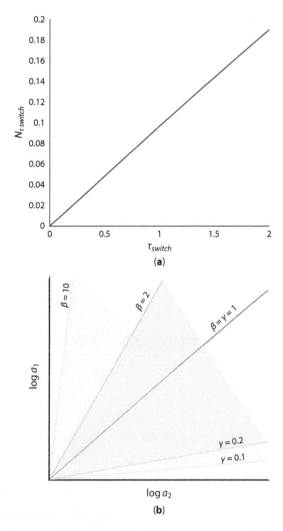

Figure 8.12 Spontaneous, instant switches: a, on-off switch from $0 \leq \tau_{switch} \leq 1$ to $1 \leq \tau_{switch} \leq 2$; b, bifurcation of on-off switch (central blue line) and bistability regions. For large values of r, $\beta = \dfrac{1}{\gamma}$.

one may be indicative of a mixture of positive and negative feedback loops (e.g., [8.9.45.]) in the diatom life cycle.

A normal diatom life cycle implies that the phenotype is preserved such that stability is present for stages as evidenced by the limit cycles (Figures 8.4-8.8). That is, life cycle stages are identifiable as stable periodic (e.g., [8.9.8.]). For M-G system simulations of abnormal diatom life cycle stages, chaotic outcomes may be an adaptive response to extreme environmental conditions, possibly inducing teratogenic phenotypes. Additionally, chaotic outcomes may be a response to mutagenic changes induced at the genetic level and expressed by the phenotype. That is, detecting potential chaotic outcomes may be used to infer possible adaptive dynamics evidenced by resultant diatom phenotypes. Stable oscillatory cell behavior may be indicative of stable cell phenotypes [8.9.104.].

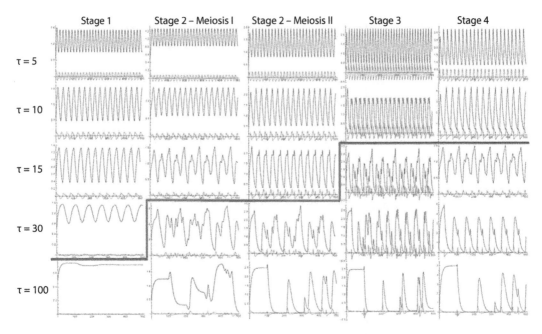

Figure 8.13 Diatom life cycle modeled via a M-G system. Red line divides upper matrix of stable periodicity from lower matrix of unstable chaoticity.

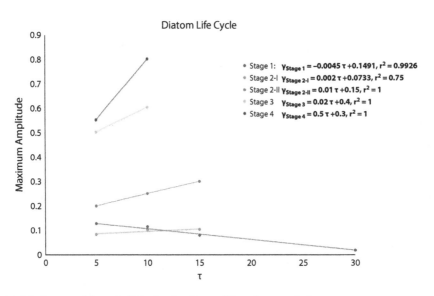

Figure 8.14 Maximum stable periodicity for the diatom life cycle stage timing.

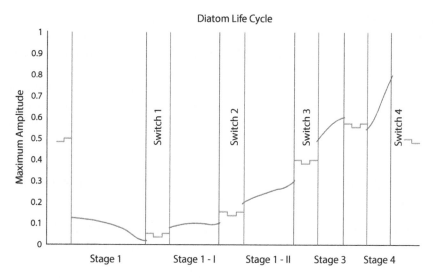

Figure 8.15 Diagram of stages (blue curves) and switches (gold) in terms of maximum amplitude over the diatom life cycle. End of a stage is indicated at a red line and start of a stage is indicated at a green line.

Stability and instability may be tested and confirmed via Lyapunov exponents (e.g., [8.9.104.]). Lyapunov spectra may be considered in a dynamical system such as the diatom life cycle so that positive Lyapunov exponents as an indication of chaotic outcomes may be associated with possible internal as well as external cell developmental anomalies.

8.7.1 Cell Size Control and the Diatom Cell Cycle Structure

Diatom cell size is controlled on a molecular level (e.g., [8.9.45., 8.9.55., 8.9.82.]) during the life cycle. The timing of size diminution constitutes a size control network dictated by activators and inhibitors during cell division [8.9.82.]. Evolution of a size control network is dictated by feedback mechanisms involving concentrations of activators and inhibitors that affect the duration of the cell cycle [8.9.82.]. Feedback loops are related to the occurrence of steady-state conditions (e.g., [8.9.45.]) and are indicative of cell size volume with respect to stability of growth [8.9.82.]. In terms of growth, cell cycle timing controls cell size so that mother and daughter diatom cells have similar volumes over the course of a cell cycle [8.9.82.] as a mode of inducing stability.

Over longer time spans, a generation of diatom cells is produced. A generation is considered to be a group of cells at approximately the same size before and after cell division [8.9.49.]. Diatom cell size restoration after diminution that occurs during mitosis occurs on a very short time frame (e.g., [8.9.59.]), and after over 100 generations, cell size restoration occurs as a sawtooth patterned oscillatory time series [8.9.49.] both experimentally [8.9.7.] and as a simulation [8.9.49.].

As the number of generations increases, the length of time between continuous cell size diminution and restoration increases, then size restoration occurs rapidly [8.9.49.]. Size diminution may occur over years, while size restoration may occur over hours to days (e.g., [8.9.59.]). Overall, the average cell size of diatoms is more broadly affected by environmental conditions (e.g., [8.9.53., 8.9.54]) as dictated by the rate of population growth during different nutrient regimes at different times of the year and grazing pressures (e.g., [8.9.22., 8.9.49.]). Auxospore formation via sexual reproduction (or vegetatively) contributes but a small fraction of the cells to the total population of diatoms over generations [8.9.49.]. However, the diplontic life cycle is evidence of the evolutionary success of diatoms (e.g., [8.9.55.]). In all, diatom cell size frequency during the life cycle is controlled over multiple time scales that exhibits a distribution ranging from unimodality (e.g., [8.9.49.]) to bimodality (e.g., [8.9.93.]).

8.7.2 Potential Alterations to the Mackey-Glass System when Applied to the Diatom Life Cycle

The M-G was useful in devising a schema to relate cell number growth, cell size and timing during the diatom life cycle. One potential alteration to the M-G is the fractional M-G (e.g., [8.9.31.]) as a way to better fine tune particular stages in the life cycle.

Along this line, another alteration may include more specific information that is modeled. Using the M-G, Stage 1 is obtained from Eq. (8.24). However, cell densities from cell division means that half the resultant cells are the same size and the other half are reduced in size. To allow for this outcome, Stage 1 using the M-G system from Eq. (8.23) may be modified as

$$\frac{dN_{Stage1}}{dt} = \left(a_{n-delays} \left[\frac{r(\mu_\tau \cdot v_\tau)}{1 + \mu_\tau^\gamma} \right] + \left[\frac{r(\mu_\tau)}{1 + \mu_\tau^\gamma} \right] \right) - bN_{Stage1} \qquad (8.36)$$

where the first term on the right-hand side of the equation represents size diminution, and the second term after the addition operator represents size preservation. Stage 1 ends when the MacDonald-Pfitzer limit is reached, inducing a chaotic rather than a stable periodic outcome. Obtaining a chaotic outcome is achieved by using the longest time delay, and coefficients for size diminution are included so that from Eqs. (8.23) and (8.24), the Stage 1 M-G becomes

$$\frac{dN_{Stage1}}{dt} = \left(\frac{r\left(\mu_\tau \cdot [a_{MacDonald-Pfizer} \cdot v_\tau] \right)}{1 + \mu_\tau^\gamma} + \frac{r(\mu_\tau)}{1 + \mu_\tau^\gamma} \right) - bN_{Stage1} \qquad (8.37)$$

where $0.3 \leq a_{MacDonald-Pfizer} \leq 1.0$ would be the range to cover size diminution. Other variations may be instituted with regard to particular diatom taxa to capture differences among taxon-specific diatom life cycles (e.g., [8.9.19.]).

A possible alternative to the bistable genetic toggle switch is the λ-genetic switch in which "on" and "off" are dualities in a single switch [8.9.33.]. Switch timing is key to devising such a switch via a Markov process (e.g., [8.9.9.]), including a hidden Markov model (e.g., [8.9.33.]). Presuming that the diatom life cycle is a Markov process would require justification for each stage and switch in terms of stochasticity. Then, Markov methods may be applicable to the diatom life cycle. The M-G could be instituted using Markovian switches connecting life cycle stages into a single schema.

8.7.3 Mackey-Glass Systems: Utility and Applications

Forward genetics (e.g., [8.9.89.]) used in conjunction with M-G modeling may be applicable to the diatom cell and life cycle. Associating genetic steps with the possible time delays in a M-G system may be useful in understanding diatom cell dynamics with respect to molecular identifiers or triggers of cellular events during life cycle stages or switches.

Relation between phenotypic plasticity and instantaneous phenotypic switching that depends on the stress indicator present may be modeled using the M-G with the time delay instituted to achieve the switch [8.9.105.]. Phenotypic switching may be useful in understanding morphological outcomes in diatom vegetative reproduction.

Determining chaotic versus periodic time series in a M-G system is useful in characterizing potential size or morphological outcomes. Chaoticity may be modeled as outcomes from high mutation rates (e.g., [8.9.62.]), or high cell enlargement rates inducing aberrant forms (e.g., [8.9.95.]). Detecting abruptness in cell size diminution [8.9.95.] may be analyzed using a M-G system in which changes in time scales result in chaotic outcomes.

The Levenberg-Marquardt algorithm may be used to minimize the error function associated with the M-G time series as to when a chaotic solution will occur [8.9.113.]. As a method of non-linear least squares curve fitting, Levenberg-Marquardt has been found to be efficient when compared to gradient descent or Gauss-Newton algorithms [8.9.113.] and covers a wide range of circumstances that may be found via simulation using the M-G. Levenberg-Marquardt may enable the fine-tuning of outcomes from using a M-G system.

8.7.4 Potential Additional Analyses of Results from Mackey-Glass Systems

More information may be gleaned from M-G results of the diatom life cycle. For stable periodic outcomes, decomposition of each time series as an eigensystem may be performed (e.g., [8.9.75.]). The M-G time series enables further analyses using Fourier spectral analysis (e.g., [8.9.75.]) or the Hilbert-Huang transform (e.g., [8.9.86.]) to study properties of each diatom life cycle stage. Such methods may aid in the classification of differences in life cycle stages for specific diatom taxa (e.g., [8.9.19.]) beyond whether a diatom is a centric or pennate. Using the M-G and additional analyses, there is the potential to study the association of differences in life cycle stages at different time scales (e.g., [8.9.108.]) with respect to phylogenetic inferences (e.g., [8.9.61.]).

8.8 Summary and Future Research

Using the M-G opens the door to many avenues of studying the life cycle or other biological processes for diatoms. Methods for finding error rates concerning M-G time series or other statistical ways of verifying results may be instituted to improve outcomes (e.g., [8.9.113.]).

The M-G, and more broadly, DDEs enables modeling stability and instability in dynamic processes that diatoms may exhibit.

The characteristics of the time lag used in modeling may take many forms. Stepwise cell number growth may be represented via neutral DDEs [8.9.20., 8.9.84.]. Hyperbolic (e.g., [8.9.20., 8.9.115.]) or stochastic properties (e.g., [8.9.20.]) with respect to time delay may be incorporated into DDEs as well. In each case, particular DDEs may be applicable to particular stages of the diatom life cycle. Alternatively, one type of DDE may be modified for each diatom life cycle stage as was done using the M-G system. More detailed systems of DDEs may be devised to model the timing of the sequential stages or switches of the diatom life cycle in terms of other factors such as molecular or genetic (e.g., [8.9.104.]) to ecological or environmental conditions (e.g., [8.9.105.]).

Other DDEs may be applicable in modeling processes such as the diatom life cycle stages. The Ikeda equation has been used to study bistable optical rings in which light intensity induces changing resonance states [8.9.57., 8.9.58.]. Expressed as $\frac{dN}{dt} = \mu \sin(N(t-\tau)) - \alpha N(t)$ where μ, τ and α are parameters with initial conditions $N(t \leq 0) = 0.1$, the Ikeda equation represents multi-stable periodic states resulting from successive bifurcations with a non-linear time lag [8.9.57., 8.9.106.]. Another possibility is Nicholson's blowflies equation (e.g., [8.9.14., 8.9.15.]) given as $\frac{dN}{dt} = [rN(h(t))e^{-aN(g(t))} - bN]$. These are but two of the many DDEs that could be used to model various aspects of the diatom life cycle such as particular stages or possibly being associated to particular genetic, molecular or cellular processes such as mitosis, cytokinesis or meiosis.

More generally, forward DDEs may be applicable (e.g., [8.9.84.]) to biological processes such as genetic diversity and molecular evolution (e.g., [8.9.62.]). Distributed delay differential or integro-differential equations (e.g., [8.9.14., 8.9.16.]) may provide modeling regimes of biological processes as well. DDEs may be used in conjunction with neural networks (e.g., [8.9.6.]) or with respect to supervised or unsupervised learning (e.g., [8.9.83.]). As a basic equation that is applicable to many biological systems, DDEs may be a starting point to induce more intricate modeling of biological processes, as illustrated by the application of the M-G to the diatom life cycle in this study.

8.9 References

8.9.1. Amato, A., Orsini, L., D'Alelio, D., Montresor, M. (2005) Life cycle, size reduction patterns, and ultrastructure of the pennate planktonic diatom *Pseudo-nitzschia delicatissima* (Bacillariophyceae) *J. Phycol.* 41, 542-556.

8.9.2. An, Q., Beretta, E., Kuang, Y., Wang, C., Wang, H. (2019) Geometric stability switch criteria in delay differential equations with two delays and delay dependent parameters. *J. Differential Equations* 266, 7073-7100.

8.9.3. Ando, Y. and Katano, T. (2018) Effect of cell size on growth rates of two *Skeletonema* species and their competitive interactions. *J. Phycol.* 54(6), 850-859.

8.9.4. Andrecut, M. and Kauffman, S.A. (2006) Noise in genetic toggle switch models. *Journal of Integrative Bioinformatics* 3(1), 63-77.

8.9.5.   Annunziata, R., Mele, B.H., Marotta, P., Volpe, M., Entrambasaguas, L., Mager, S., Stec, K., d'Alcalà, M.R., Sanges, R., *et al.* (2022) Trade-off between sex and growth in diatoms: molecular mechanisms and demographic implications. *Sci. Adv.* 8, eabj9466.

8.9.6.   Anumasa, S. and Srijith, P.K. (2020) Delay differential neural networks. Retrieved December 21, 2022: https://arxiv.org/pdf/2012.06800.pdf

8.9.7.   Armbrust, E.V. and Chisholm, S.W. (1992) Patterns of cell size change in a marine centric diatom: variability evolving from clonal isolates. *J. Phycol.* 28(2), 146-156.

8.9.8.   Bartha, F.A., Krisztin, T., Vígh, A. (2021) Stable periodic orbits for the Mackey-Glass equation. *J. Diff. Eqns.* 296, 15-49.

8.9.9.   Barzel, B. and Biham, O. (2008) Calculation of switching times in the genetic toggle switch and other bistable systems. *Phys. Rev. E* 78, 041919.

8.9.10.   Basu, S., Patil, S., Mapleson, D., Russo, M.T., Vitale, L., Fevola, C., Maumus, F., Casotti, R., *et al.* (2017) Finding a partner in the ocean: molecular and evolutionary bases of the response to sexual cues in a planktonic diatom. *New Phytologist* 215, 140-156.

8.9.11.   Benoiston, A.-S., Ibarbalz, F.M., Bittner, L., Guidi, L., Jahn, O., Dutkiewicz, S., Bowler, C. (2017) The evolution of diatoms and their biogeochemical functions. *Phil. Trans. R. Soc. B* 372: 20160397.

8.9.12.   Beuter, A., Bélair, J., Labrie, C. (1993) Feedback and delays in neurological diseases: a modeling study using dynamical systems. *Bull. Math. Biol.* 55(3), 525-541.

8.9.13.   Berezansky, L. and Braverman, E. (2006) Mackey-Glass equation with variable coefficients. *Computers and Mathematics with Applications* 51, 1-16.

8.9.14.   Berezansky, L. and Braverman, E. (2016) Boundedness and persistence of delay differential equations with mixed nonlinearity. *Appl. Math. Comput.* 279, 154-169.

8.9.15.   Berezansky, L. and Braverman, E. (2017) A note on stability of Mackey-Glass equations with two delays. *Journal of Mathematical Analysis and Applications* 450, 1208-1228.

8.9.16.   Berezansky, L. and Braverman, E. (2019) On stability of delay equations with positive and negative coefficients and applications. *Journal of Analysis and its Applications* 38(2), 157-189.

8.9.17.   Berezansky, L., Braverman, E., Idels, L. (2005) Delay differential equations with Hill's type growth rate and linear harvesting. *Computers Math. Appl.* 49, 549-563.

8.9.18.   Biancalani, T. and Assaf, M. (2015) Genetic toggle switch in the absence of cooperative binding: exact results. *Phys. Rev. Lett.* 115(2), 208101.

8.9.19.   Bilcke, G., Ferrante, M.I., Montresor, M., De Decker, S., De Veylder, L., Vyverman, W. (2022) Life cycle regulation. In: *The Molecular Life of Diatoms*, Falciatore, A. and Mock, T. (eds.), Springer Nature Switzerland AG, Switzerland: 205-228.

8.9.20.   Bocharov, G.A. and Rihan, F.A. (2000) Numerical modelling in biosciences using delay differential equations. *J. Computational Appl. Math.* 125, 183-199.

8.9.21.   Bocharov, G.A., Marchuk, G.L., Romanyukha, A.A. (1996) Numerical solutions by LMMs of stiff delay differential systems modelling an immune response. *Numer. Math.* 73, 131-148.

8.9.22.   Bowler, C., De Martino, A., Falciatore, A. (2010) Diatom cell division in an environmental context. *Current Opinion in Plant Biology* 13(6), 623-630.

8.9.23.   Brownlee, C. (2008) Diatom signalling: deadly messages. *Curr. Biol.* 18(12), R518-R519.

8.9.24.   Chan, R.P.K. and Gorgey, A. (2013) Active and passive symmetrization of Runge-Kutta Gauss methods. *Appl. Numer. Math.* 67, 64-77.

8.9.25.   Chepurnov, V.Q., Mann, D.G., Sabbe, K., Vyverman, W. (2004) Experimental studies on sexual reproduction in diatoms. *Int. Rev. Cytol.* 237, 91-154.

8.9.26.   Chung, K.-T., Nilson, E.H., Case, M.J., Marr, A.G., Hungate, R.E. (1973) Estimation of growth rate from mitotic index. *Appl. Microbiol.* 25(5), 778-780.

8.9.27.   Dalchau, N., Szép, G., Hernansaiz-Ballesteros, R., Barnes, C.P., Cardelli, L., Phillips, A., Csikász-Nagy, A. (2018) Computing with biological switches and clocks. *Natural Computing* 17, 761-779.

8.9.28.   De Martino, A., Amato, A., Bowler, C. (2009) Mitosis in diatoms: rediscovering an old model for cell division. *Bioessays* 31, 874-884.

8.9.29.   De Monte, S., d'Ovidio, F., Danø, S., Sørensen, P.G. (2007) Dynamical quorum sensing: population density encoded in cellular dynamics. *PNAS* 104(47), 18377-18381.

8.9.30.   Edlund, M.B. and Stoermer, E.F. (1997) Ecological, evolutionary, and systematic significance of diatom life histories. *Journal of Phycology* 33, 897–918.

8.9.31.   El-Sayed, A.M.A., Salman, S.M., Elabd, N.A. (2016) On a fractional-order delay Mackey-Glass equation. *Adv. in Difference Eqns.* 2016, 137.

8.9.32.   Falciatore, A. and Bowler, C. (2002) Revealing the molecular secrets of marine diatoms. *Annu. Rev. Plant Biol.* 53, 109-130.

8.9.33.   Fang, X., Liu, Q., Bohrer, C., Hensel, Z., Han, W., Wang, J., Xiao, J. (2018) Cell fate potentials and switching kinetics uncovered in a classic bistable genetic switch. *Nat. Comm.* 9, 2787.

8.9.34.   Ferrante, M.I., Entrambasaguas, L., Johansson, M., Töpel, M., Kremp, A., Montresor, M., Godhe, A. (2019) Exploring molecular signs of sex in the marine diatom *Skeletonema marinoi*. *Genes* 10, 494.

8.9.35.   Ferrell, Jr., J.E. (2002) Self-perpetuating states in signal transduction: positive feedback, double-negative feedback and bistability. *Curr. Opin. Cell Biol.* 14(2), 140–148.

8.9.36.   Ferrell, Jr., J.E. (2008) Feedback regulation of opposing enzymes generates robust, all-or-non bistable responses. *Curr. Biol.* 18(6), R244-R245.

8.9.37.   Ferrell, Jr., J.E. (2013) Feedback loops and reciprocal regulation: recurring motifs in the systems biology of the cell cycle. *Curr. Opin. Cell Biol.* 25(6), 676-686.

8.9.38.   Fu, J. and Li, H. (2022) Dynamic optimization of state-dependent switched systems with free switching sequences. *Automatica* 148, 110747.

8.9.39.   Fuhrmann-Lieker, T., Kubetschek, N., Ziebarth, J., Klassen, R., Seiler, W. (2021) Is the diatom sex clock a clock? *J. R. Soc. Interface* 18, 20210146.

8.9.40.   Gardner, T.S., Cantor, C.R., Collins, J.J. (2000) Construction of a genetic toggle switch in Escherichia coli. *Nature* 403, 339-342.

8.9.41.   Geitler, L. (1935) Reproduction and life history in diatoms. *The Botanical Review* 1(5), 149-161.

8.9.42.   Glass, L. and Mackey, M.C. (1979) Pathological conditions resulting from instabilities in physiological control systems. *Ann. New York Acad. Sci.* 316(1), 214-235.

8.9.43.   Glass, D.S., Jin, Z., Riedel-Kruse, H. (2021) Nonlinear delay differential equations and their application to modeling biological network motifs. *Nat. Commun.* 12(1), 1788-1806.

8.9.44.   Goldberger, A.L. (1991) Is the normal heartbeat chaotic or homeostatic? *News Physiol. Sci.* 6, 87-91.

8.9.45.   Goldbeter, A. (2002) Computational approaches to cellular rhythms. *Nature* 420(6912), 238-245.

8.9.46.   Harbich, T. (2021) On the size sequence of diatoms in clonal chains. In: *Diatom Morphogenesis [DIMO, Volume in the series: Diatoms: Biology & Applications, series editors: Richard Gordon & Joseph Seckbach].* V. Annenkov, J. Seckbach and R. Gordon, (eds.) Wiley-Scrivener, Beverly, MA, USA: 69-92.

8.9.47.   Harrison, J. and Yeung, E. (2021) Stability analysis of parameter varying genetic toggle switches using Koopman operators. *Mathematics* 9, 3133.

8.9.48.   Harvey, B.P., Agostini, S., Kon, K., Wada, S., Hall-Spencer, J.M. (2019) Diatoms domi-nate and alter marine food-webs when CO2 rises. *Diversity* 11, 242.

8.9.49.   Hense, I. and Beckmann, A. (2015) A theoretical investigation of the diatom cell size reduction-restitution cycle. *Ecological Modelling* 317, 66-82.

8.9.50.   Hill, A.V., (1910) The possible effects of the aggregation of the molecules of haemoglobin on its dissociation curves. *J. Physiol.* 40, iv-vii.

8.9.51.   Hill, A.V. (1913) XLVII. The combinations of haemoglobin with oxygen and with carbon monoxide. *Biochem. J.* 7(5), 471-480.

8.9.52.   Hillenbrand, P., Fritz, G., Gerland, U. (2013) Biological signal processing with a genetic toggle switch. *PLoS ONE* 8(7), e68345.

8.9.53.   Huysman, M.J.J., Martens, C., Vandepoele, K., Gillard, J., Rayko, E., Heijde, M., Bowler, C., Inzé, D., Van de Peer, Y., De Veylder, L., Vyverman, W. (2010) Genome-wide analysis of the diatom cell cycle unveils a novel type of cyclins involved in environmental signal-ing. *Genome Biology* 11, R17.

8.9.54.   Huysman, M.J.J., Fortunato, A.E., Matthijs, M., Costa, B.S., Vanderhaeghen, R., Van den Daele, H., Sachse, M., Inzé, D., Bowler, C., Kroth, P.G., Wilhelm, C., Falciatore, A., Vyverman, W., De Veylder, L. (2013) AUREOCHROME1a-mediated induction of the diatom-specific cyclin dsCYC2 controls the onset of cell division in diatoms (Phaeodactylum tricornutum). *The Plant Cell* 25, 215-228.

8.9.55.   Huysman, M.J.J., Vyverman, W., De Veylder, L. (2014) Molecular regulation of the dia-tom cell cycle. *J. Exper. Bot.* 65(10), 2573-2584.

8.9.56.   Idei, M., Osada, K., Sato, S., Toyoda, K., Nagumo, T., Mann, D.G. (2012) Gametogenesis and auxospore development in *Actinocyclus* (Bacillariophyta). *PLoS ONE* 7(8), e41890.

8.9.57.   Ikeda, K. and Matsumoto, K. (1987) High-dimensional chaotic behavior in systems with time-delayed feedback. *Physica* 29D, 223-235.

8.9.58.   Ikeda, K., Daido, H., Akimoto, O. (1980) Optical turbulence: chaotic behavior of trans-mitted light from a ring cavity. *Physical Review Letters* 45(9), 709-712.

8.9.59.   Kaczmarska, I. and Ehrman, J.M. (2021) Enlarge or die! An auxospore perspective on diatom diversification. *Organisms Diversity & Evolution* 21, 1-23.

8.9.60.   Kaczmarska, I., Poulíčková, A., Sato, S., Edlund, M. B., Idei, M., Watanabe, T., Mann, D. G. (2013) Proposals for a terminology for diatom sexual reproduction, auxospores and resting stages. *Diatom Research* 28(3), 263–294.

8.9.61.   Kaczmarska, I., Ehrman, J.M., Ashworth, M.P. (2022) Sexual reproduction and auxo-spore development in the diatom *Biddulphia biddulphiana*. *PLoS ONE* 17(9), e0272778.

8.9.62.   Krasovec, M., Sanchez-Brosseau, S., Piganeau, G. (2019) First estimation of the sponta-neous mutation rate in diatoms. *Genome Biol. Evol.* 11(7), 1829-1837.

8.9.63.   Langmuir, I. (1918) The adsorption of gases on plane surfaces of glass, mica and plati-num. *J. Amer. Chem. Soc.* 40(9), 1361-1403.

8.9.64.   Lauterborn, R. (1896) Untersuchungen uber Bau, Kernteilung und Bewegung der Diatomeen. W. Engelmann, Leipzig.

8.9.65.   Lewin, J.C. and Rao, V.N. (1975) Blooms of surfzone diatoms along the coast of the Olympic Peninsula, Washington. 6. Daily periodicity phenomena associated with *Chaetoceros armatum* in its natural habitat. *J. Phycol.* 11, 330-338.

8.9.66.   Loinger, A., Lipshtat, A., Balaban, N.Q., Biham, O. (2007) Stochastic simulations of genetic switch systems. *Phys. Rev. E* 75, 021904.

8.9.67.   MacDonald, J.D. (1869) On the structure of the diatomaceous frustule, and its genetic cycle. *Annals Magazine Natural History Series* 4(3), 1-8.

8.9.68. Mackey, M.C. and Glass, L. (1977) Oscillation and chaos in physiological control systems. *Science* 197(4300), 287-289.

8.9.69. McDuff, R.E. and Chisholm, S.W. (1982) The calculation of in situ growth rates of phytoplankton populations from fractions of cell undergoing mitosis: a clarification. *Limnol. Oceanogr.* 27, 783-788.

8.9.70. Mizuno, M. (1991) Influence of cell volume on the growth and size reduction of marine and estuaringe diatoms. *J. Phycol.* 27(4), 473-478.

8.9.71. Mizuno, M. (2006) Evolution of meiotic patterns of oogenesis and spermatogenesis in centric diatoms. *Phycol. Res.* 54(1), 57-64.

8.9.72. Morin, S. Gómez, N., Tornés, E., Licursi, M., Rosebery, J. (2016) Benthic diatom monitoring and assessment of freshwater environments: standard methods and future challenges. In: *Aquatic Biofilms – Ecology, Water Quality and Wastewater Treatment*, Romaní, A.M., Guasch, H., Balaguer, M.D., (eds.), Caister Academic Press, Norfolk, U.K.: 111-124.

8.9.73. Moustafa, A., Beszteri, B., Maier, U.G., Bowler, C., Valentin, K., Bhattacharya, D. (2009) Genomic footprints of a cryptic plastid endosymbiosis in diatoms. *Science* 324, 1724-1726.

8.9.74. Neves, K.W. (1975) Automatic integration of functional differential equations, an approach. *ACM Trans. Math. Soft.* 1, 357-368.

8.9.75. Niang, O., Deléchelle, É., Lemoine, J. (2010) A spectral approach for sifting process in empirical mode decomposition. *IEEE Trans. Signal Process.* 58(11), 5612-5623.

8.9.76. Oberle, H.J. and Pesch, H.J. (1981) Numerical treatment of delay differential equations by Hermite interpolation. *Numer. Math.* 37, 235-255.

8.9.77. Patil, S., Moeys, S., von Dassow, P., Huysman, M.J.J., Mapleson, D., De Veylder, L., Sanges, R., *et al.* (2015) Identification of the meiotic toolkit in diatoms and exploration of meiosis-specific SP011 and RAD51 homologs in the sexual species *Pseudo-nitzschia multistriata* and *Seminavis robusta*. *BMC Genomics* 16, 930

8.9.78. Pender, J., Rand, R.H., Wesson, E. (2017) Ques with choice via delay differential equations. *International Journal of Bifurcation and Chaos* 27(4), 1730016.

8.9.79. Pickett-Heaps, J.D., Schmid, A.M., Tippit, D.H. (1984) Cell division in diatoms. *Protoplasma* 120, 132-154.

8.9.80. Pfitzer, E. (1869) Über den bau and zellteilung der diatomeen. *Botanische Zeitung* 27, 774-776

8.9.81. Pfitzer, E. (1871) Untersuchungen über Bau und Entwicklung der Bacillariaceen (Diatomaceen) [Studies on construction and development of Bacillariaceae (Diatomaceae)] [German]. In: *Botanische Abhandlungen aus dem Gebiete der Morphologie und Physiologie [Botanical Treatises in the Field of Morphology and Physiology]*. J.L.E.R. von Hanstein, (ed.) Adolph Marcus, Bonn, Germany: [i]-vi, 1-189, 186 pls.

8.9.82. Proulx-Giraldeau, F, Skotheim, J.M., François, P. (2022) Evolution of cell size control is canalized towards adders or sizers by cell cycle structure and selective pressures. *eLife* 11:e79919.

8.9.83. Rackauckas, C., Ma, Y., Martensen, J., Warner, C., Zubov, K., Supekar, R., Skinner, D., Ramadhan, A., Edelman, A. (2021) Universal differential equations for scientific machine learning. Retrieved on December 21, 2022: https://arxiv.org/pdf/2001.04385.pdf

8.9.84. Richard, J.-P. (2003) Time-delay systems: an overview of some recent advances and open problems. *Automatica* 39, 1667-1694.

8.9.85. Rivkin, R.B. (1986) Radioisotopic method for measuring cell division rates of individual species of diatoms from natural populations. *Applied and Environmental Microbiology* 51(4), 769-775.

8.9.86. Rodríguez, R., Bila, J., Mexicano, A., Cervantes, S., Ponce, R., Mghien, N.B. (2014) Hilbert-Huang transform and neural networks for electrocardiogram modeling and prediction. *2014 10th International Conference on Natural Computation (ICNC)* 2014, 561-567.

8.9.87. Röst, G. and Wu, J. (2007) Domain-decomposition method for the global dynamics of delay differential equations with unimodal feedback. *Proc. R. Soc. A* 463, 2655-2669.

8.9.88. Round, F.E., Crawford, R.M. and Mann, D.G. (1990) *The Diatoms, Biology & Morphology of the Genera*. Cambridge University Press, Cambridge, U.K.

8.9.89. Saade, A. and Bowler, C. (2009) Molecular tools for discovering the secrets of diatoms. *Bioscience* 59, 757-765.

8.9.90. Sabater, S. (2009) Diatoms. In: *Encyclopedia of Inland Waters*, Likens, G.E. (ed.), Elsevier, Amsterdam: 149-156.

8.9.91. Salmina, K., Bljko, A., Inashkina, I., Staniak, K., Dudkowska, M., Podlesniy, P., Rumnieks, F., *et al.* (2020) "Mitotic slippage" and extranuclear DNA in cancer chemoresistance: a focus on telomeres. *Int. J. Mol. Sci.* 21, 2779.

8.9.92. Sanyal, A., Larsson, J., van Wirdum, F., Andrén, T., Moros, M., Lönn, M., Andrén, E. (2021) Not dead yet: diatom resting spores can survive in nature for several millennia. *Am. J. Bot.* 109, 67-82.

8.9.93. Sarno, D., Zingone, A., Montresor, M. (2010) A massive and simultaneous sex event of two *Pseudo-nitzschia* species. *Deep-Sea Research II* 57, 248-255.

8.9.94. Sato, S., Nagumo, T., Tanaka, J. (2004) Auxospore formation and the morphology of the initial cell of the marine araphid diatom *Gephyria media* (Bacillariophyceae). *J. Phycol.* 40, 684-691.

8.9.95. Sato, S., Mann, D.G., Nagumo, T., Tanaka, J., Tadano, T., Medlin, L.K. (2008) Auxospore fine structure and variation in modes of cell size changes in *Grammatophora marina* (Bacillariophyta). *Phycologia* 47(1), 12-27.

8.9.96. Schmid, A.M.M. and Crawford, R.M., (2001) *Ellerbeckia arenaria* (Bacillariophyceae): formation of auxospores and initial cells. *Eur. J. Phycol.* 36(4), 307–320.

8.9.97. Sethi, D., Butler, T.O., Shuhaili, F., Vaidyanathan, S. (2020) Diatoms for carbon sequestration and bio-based manufacturing. *Biology* 9, 217.

8.9.98. Shampine, L.F. and Gear, C.W. (1979) A user's view of solving stiff ordinary differential equations. *SIAM Review* 21(1), 1-17.

8.9.99. Smayda, T.J. (1975) Phased cell division in natural populations of the marine diatom *Ditylum brightwellii* and the potential significance of diel phytoplankton in the sea. *Deep-Sea Research* 22, 151-165.

8.9.100. Sofroniou, M. and Spaletta, G. (2004) Construction of explicit Runge-Kutta pairs with stiffness detection. *Math. Comp. Model.* 40, 1157-1169.

8.9.101. Sueoka, N. (1962) On the genetic basis of variation and heterogeneity of DNA base composition. *Proc. Natl. Acad. Sci. USA* 48, 582-592.

8.9.102. Tippit, D.H. and Pickett-Heaps, J.D. (1977) Mitosis in the pennate diatom *Surirella ovalis*. *The Journal of Cell Biology* 73, 705-727.

8.9.103. Toker, D., Sommer, F.T., D'Esposito, M. (2020) A simple method for detecting chaos in nature. *Commun. Biol.* 3, 11-23.

8.9.104. Uthamacumaran, A. and Zenil, H. (2022) A review of mathematical and computational methods in cancer dynamics. *Front. Oncol.* 12, 850731.

8.9.105. Utz, M., Jeschke, J.M., Loeschcke, V., Gabriel, W. (2014) Phenotypic plasticity with instantaneous by delayed switches. *J. Theor. Biol.* 340, 60-72.

8.9.106. Valli, D., Muthuswamy, B., Banerjee, S., Ariffin, M.R.K., Wahab, A.W.A., Ganesan, K., Subramaniam, C.K., Kurths, J. (2014) Synchronization in coupled Ikeda delay systems. *Eur. Phys. J. Special Topics* 223, 1465-1479.

8.9.107. Vardi, A., Bidle, K., Kwityn, C., Thompson, S.M., Callow, J.A., Callow, M.E., Falkowski, P., Bowler, C. (2008) A diatom gene regulating nitric-oxide signaling and susceptibility to diatom-derived aldehydes. *Curr. Biol.* 18 (12), 895-899.

8.9.108. von Dassow, P. and Montresor, M. (2011) Unveiling the mysteries of phytoplankton life cycles: patterns and opportunities behind complexity. *J. Plankton Res.* 33(1), 3-12.

8.9.109. Watson, P.Y. and Fedor, M.J. (2009) Determination of intracellular RNA folding rates using self-cleaving RNAs. *Methods in Enzymology* 468, 259-286.

8.9.110. Wei, J. and Fan, D. (2007) Hopf bifurcation analysis in a Mackey-Glass system. *Internation. J. Bifurcation Chaos* 17(6), 2149-2157.

8.9.111. Weiler, C.S. and Eppley, R.W. (1979) Temporal pattern of division in the genus Ceratium and its application to the determination of growth rate. *J. Exp. Mar. Biol. Ecol.* 39, 1-24.

8.9.112. Xu, X. and Antsaklis, P.J. (2002) An approach to switched systems optimal control based on parameterization of the switching instants. *IFAC Proceedings Volumes* 35(1), 365-370.

8.9.113. Zhao, J., Li, Y., Yu, Z., Zhang, X. (2014) Levenberg-Marquardt algorithm for Mackey-Glass chaotic time series prediction. *Discrete Dynamics in Nature and Society* 2014, 193758.

8.9.114. Zhou, J., Lyu, Y., Richlen, M.L., Anderson, D.M., Cai, Z. (2016) Quorum sensing is a language of chemical signals and plays an ecological role in algal-bacterial interactions. *Critical Reviews in Plant Sciences* 35(2), 81-105.

8.9.115. Ziebarth, J., Seiler, W.M., Fuhrmann-Lieker, T. (2023) Size-resolved modeling of diatom populations: old findings and new insights. In: *The Mathematical Biology of Diatoms [DMTH, Volume in the series: Diatoms: Biology & Applications, series editors: Richard Gordon & Joseph Seckbach].* J.L. Pappas, (ed.) Wiley-Scrivener, Beverly, MA, USA.

9

Diatom Morphospaces, Tree Spaces and Lineage Crown Groups

Abstract

Diatom morphospaces have been based on morphological attributes such as length or shape which are analyzed using statistical or deterministic methodologies. Such morphospaces have been used to analyze taxonomic or ecological properties of related diatom taxa. The idea of occupied versus unoccupied morphospace has not been adequately addressed in either diatom research or applications of other organisms in either a theoretical or empirical morphological context. To remedy this, an initial diatom morphospace was constructed of genera representing their diverse array of morphologies. A multivariate technique was used to obtain morphospaces of the generalized groups of taxa labeled as: all genera, all centric diatoms, and all pennate diatoms. Four taxonomic (structural) groups—radial centrics, polar centrics, araphid pennates, and raphid pennates—consist of nine lineage crown groups which were used as boundary conditions for morphospace construction. Network morphospaces were devised for each lineage crown group using morphological data based on multiple shape attributes. The resultant morphospace encompassed nine submorphospaces each of which contained a network of genera. Occupied morphospace was the sum of the network submorphospaces, and unoccupied morphospace was the complement. Diatom morphospace with lineage-based boundaries for submorphospaces are interpretable via evolutionary inferences.

Keywords: Unoccupied morphospace, tree space, metric space, topological space, vector space, lineage crown groups, networks, novelty

9.1 Introduction

The biological analysis of morphology quantitatively has necessitated the usage of geometric constructs that enable the study of the commonality of the attributes being studied. Many biologists looked to D'Arcy Thompson's [9.13.108.] treatise *On Growth and Form* as guidance for the principles that determine the geometric relation among measured morphological attributes within a spatial setting. Other avenues of measuring growth and form ensued inducing the necessity to compile such quantities in a cohesive picture. Biological morphospace analyses provided a way to depict developmental, ecological, evolutionary, or taxonomic morphological features such that an understanding of the relational nature of biological organisms emerged.

Biological morphospaces have taken many forms, each having specific characteristics. Theoretical morphospaces are used to illustrate the relation among modeled phenotype attributes and/or use arbitrary axes to depict a relation among phenotypes (e.g., [9.13.71., 9.13.94.]) or are not morphospaces per se but a tabulation of phenotype attributes

Janice L. Pappas. *Mathematical Macroevolution in Diatom Research*, (313–354) © 2023 Scrivener Publishing LLC

(e.g., [9.13.107.]). Empirical morphospaces are used to illustrate the relation among pheno-types based on measurements (e.g., [9.13.42., 9.13.71., 9.13.103.]). In the theoretical case, the morphospace is a geometric structure, but the relation among phenotypes is not com-mensurate with appropriate scaling among variables or necessarily produce a picture of independent variables [9.13.74.]. With the empirical case, the morphospace is a geomet-ric structure exhibiting the univariate or multivariate statistical properties of the relation among phenotypes, but need not be deterministic. That is, if a new input datum is used, the entire analysis needs to be redone, and the outcome of the previous analysis need not have anything to do with the subsequent one.

Statistical morphospaces apply only to the data they represent, are not generalizable to overall phenotype relations among n-organisms, and additional input data may be obscured by already defined groups, not leading to any clearer outcome of the analysis. Statistical mor-phospaces may be n-dimensional, but they usually undergo dimension reduction to depict only two dimensions (possibly three) for interpretation purposes (e.g., [9.13.58., 9.13.74.]). Additionally, missing data induces possible misinterpretation of outcomes inducing poten-tially biased results (e.g., [9.13.36.]). Deterministic morphospaces are geometric structures that depict subgroup phenotypes related by defined variables that represent specific group phenotype descriptors. However, such morphospaces may suffer from the same problems that statistical morphospaces do, thereby obscuring accurate interpretation of phenotype descriptors. In all cases, morphospace analysis proceeds by being based on closely related organisms and is not amenable to input of distantly-related lineages [9.13.10.].

All morphospaces are geometrical in structure (e.g., [9.13.74.]). Euclidean geometry forms the basis of morphospaces, regardless of the content of the input data. Euclidean mor-phospaces enable the application of distance metrics except in affine morphospaces (e.g., [9.13.10.]). For real numbers, $\mathbb{R}^n$, a two-dimensional (2D) Euclidean space distance metric is

$$d(x, y) = \sqrt{(x_1 - y_1)^2 + (x_2 - y_2)^2 + \cdots + (x_n - y_n)^2} \tag{9.1}$$

with

$$\mathbf{x}^T \mathbf{y} = \langle \mathbf{x}, \mathbf{y} \rangle = \left\langle \begin{bmatrix} x_1 \\ x_2 \\ \vdots \\ x_n \end{bmatrix}, \begin{bmatrix} y_1 \\ y_2 \\ \vdots \\ y_n \end{bmatrix} \right\rangle = \sum_{i=1}^{n} x_i y_i = x_1 y_1 + x_2 y_2 + \cdots + x_n y_n \tag{9.2}$$

as the inner product or dot product of two vectors, $\mathbf{x}$ and $\mathbf{y}$. Via the Cauchy-Schwartz inequality, $|\langle x, y \rangle| \leq ||x|| \, ||y||$, the triangle inequality, $||x + y|| \leq ||x|| + ||y||$ follows, giving the matrix norm. This result induces $||x + y||^2 \leq ||x||^2 + 2\langle x, y \rangle + ||y||^2$ [9.13.113.]. For a vector, $\mathbf{x} = \begin{bmatrix} x_1 \\ x_2 \\ \vdots \\ x_n \end{bmatrix}$, the L^2-norm or vector length is $|\mathbf{x}|_2 = |\mathbf{x}| = \sqrt{x_1^2 + x_2^2 + \cdots + x_n^2}$ [9.13.56.].

Distance metrics are used to assess proximity of phenotypes in a morphospace analysis.

Euclidean space is a metric space [9.13.56.], and therefore a Euclidean morphospace is a metric space. Euclidean morphopaces are related to other metric spaces in which various distance metrics may be used. A metric space is a generalization of a Euclidean space in that a distance function may be determined and is calculable on that space [9.13.56.]. Morphospace distance metrics are usually Euclidean distances; however, other metrics may be used.

9.1.1   Euclidean Spaces are Subspaces of Hilbert and Banach Spaces

Related to Euclidean spaces are Hilbert and Banach spaces (e.g., [9.13.113.]). A Hilbert space is an inner product space given via a vector norm such as the L^2-norm and encompasses a Euclidean space. Define two continuous real functions as f and g. Then,

$$\left(\sum_{i=1}^{n} |f_i g_i|\right)^2 \leq \sum_{i=1}^{n} f_i^2 \sum_{i=1}^{n} g_i^2 \tag{9.3}$$

and

$$\langle f, g \rangle = \sum_{i=1}^{\infty} f_i g_i = \langle (f_1, f_2, \cdots), (g_1, g_2, \cdots) \rangle \tag{9.4}$$

define an n-dimensional Hilbert space [9.13.56.] which has the L^2-norm [9.13.113.].

A Hilbert space with an inner product norm is always a Banach space [9.13.113.]. A Banach space may be an n-dimensional space that has a Euclidean or L^2-norm [9.13.113.]. Other finite or infinite n-dimensional spaces are vector spaces as Banach spaces in which a supremum norm for a function, f, is defined as

$$\| f \| = \sup_{x \in \mathbb{R}} | f(x) |. \tag{9.5}$$

A Banach space with a supremum norm is not an inner product space, and therefore is not a Hilbert space [9.13.113.]. In any case, the norms for Euclidean, Hilbert and Banach spaces are vector norms [9.13.113.], which induce a metric vector morphospace. The extent to which metric spaces may be used encompasses the potential of using such spaces in a broader sense to aid in interpreting the relation among a given groups of phenotypes in morphospace.

9.1.2   From Geometrical to Topological Spaces as Mathematical Morphospaces

A metric space is a collection that consists of a set given as $S \subseteq T$ $(x \in S) \Longrightarrow (x \in T)$, where $S \cup T = \{x | x \in S \text{ or } x \in T\}$ and $S \cap T = \{x | x \in S \text{ and } x \in T\}$ for the union and intersection, respectively [9.13.56.], and a distance function used on that set may be used as a morphospace. A subset that is open in a metric space is given as $V \subseteq S$, for $x \in V$ provided that

$\delta > 0$ so that $Ball_\delta(x) \subseteq V \subseteq S$ with $Ball_\delta(x) = \{y | y \in S, d(x, y) < \delta\}$ [9.13.56.]. That is, the open ball, $Ball_\delta(x)$, has radius δ with respect to x for any $y \in S$ also in V, and V is an open set [9.13.56.]. For a collection of open sets, $\{V_a\}$, $a \in A$ in metric space, S, a are indices in set A, and $\cup_a V_a$ is the union of open sets; for a finite intersection of open sets, the result is also an open set [9.13.56.].

As a result, specified subsets of a collection of subsets that are definable as open sets are descriptive as the topology on a metric space provided that the union of open sets and finite intersection of sets is closed [9.13.56.]. From this, a topological space is a set that has a specified number of subsets which are open sets that belong to the topological space [9.13.56.]. That is, there is a set which consists of open subsets found in a topological space so that the set is also the topological space.

A more generalized way of categorizing morphospaces is at hand by defining a topological space. A metric space is a kind of topological space so that the generalization covers all potential morphospaces. More than one topology may be determined on a given set. Open sets are topological properties of topological spaces [9.13.56.] as are closed sets under finite intersection and arbitrary union or arbitrary intersection and finite union [9.13.113.].

A topological space is a metric space, and a vector norm, such as the L^2-norm, may be associated to a topological space, so that a topological vector space results. An n-dimensional Euclidean space is a vector space of $\mathbb{R}^n$ real numbers [9.13.113.] as well as a topological space, because operations on the elements of an n-dimensional Euclidean space are continuous and closed under vector addition and scalar multiplication [9.13.113.]. The elements of $\mathbb{R}^n$ are n-vectors [9.13.113.], indicating that an n-dimensional Euclidean space is a topological vector space.

A topological vector space enables the use of multiple spaces within a larger morphospace. Euclidean, Hilbert and Banach spaces are all topological vector spaces. The relation among n-spaces within a larger morphospace may be determined as a result of the properties of topological vector spaces. By using topological vector spaces as the basis of morphospace analysis, flexibility is provided with the congregation or incorporation of disparate data sets into a common morphospace.

9.2   Occupied and Unoccupied Morphospace

The impetus for the interest in morphospaces was piqued by studies involving interpretations as to what was meant by occupied versus unoccupied morphospace in phenotypic spaces (e.g., [9.13.64., 9.13.94., 9.13.96.]). Typically, a convex hull (e.g., [9.13.63.]) is drawn with regard to proximity of resultant taxon points in the morphospace, and these circumscribed groups constitute occupied morphospace. However, no accounting is ever given on what the empty space between points (i.e., vertices) consists of within a convex hull as the assumption is that the space within a convex hull represents morphological variation. This may not be the case, because point (vertex) proximity is relative to each pair of axes selected from n-dimensional axes in multivariate statistical morphospaces. That is, a morphospace is a model of m-points in n-dimensional space in which visualization is a non-trivial problem at four (4D) or more dimensions [9.13.101.].

Graphical drawing of a convex hull may lead to errors or misinterpretation in group determination from dimension to dimension [9.13.62., 9.13.101.]. Each point would need to be tested over n-dimensions to determine group membership fully defined by a given convex hull (e.g., [9.13.80.]). Redundancies may result, and because of high dimensionality, computational time may be a cost (e.g., [9.13.61., 9.13.115.]). Resorting to approximations may be insufficient regarding morphospace occupation. Additionally, calculation entails minimization of the connections between points (or edges between vertices) (e.g., [9.13.61., 9.13.115.]), which entails calculating geodesics (e.g., [9.13.73.]). All of this affects the interpretation of morphospace as to which parts are occupied or unoccupied.

Measures of disparity (e.g., [9.13.29., 9.13.30., 9.13.49.]) have been used to determine unoccupied morphospace, and they are statistical and dependent on the particular research question being raised, the specific group of organisms being studied, whether lineages are considered, sensitivity, sample size, and whether there is missing data (e.g., [9.13.25., 9.13.43.]). Different kinds of morphological similarity are measured differently with respect to various disparity measures (e.g., [9.13.43.]), and this affects measurement of morphospace occupation (e.g., [9.13.10.]).

One of the problems with determinations of occupied versus unoccupied morphospaces involves the approach. On the one hand, some studies have been based on filling ecological space with respect to lineage diversity to represent morphospace occupation [9.13.91.]. On the other hand, other studies show how canalization via developmental and life cycle processes induces morphological consolidation, confining morphologies to particular conglomerations in terms of morphospace occupation [9.13.91.]. Morphospace occupation may result from such development biases or the degree of incongruity between genotype and phenotype, or randomness rather than some adaptive evolutionary value (e.g., [9.13.22.]) or some other factor. Regardless, occupied morphospace is accounted for, but in any case, what constitutes unoccupied morphospace is referred to only obliquely with regard to all other possible or potential morphospaces. A consensus on the kind of measurement across all morphospace studies of occupied versus unoccupied morphospace has been elusive because of the varying circumstances and perspectives of morphospace studies.

In terms of morphospace occupation, all possible morphologies may be obtained via combinatorial methods (e.g., [9.13.69.]). Expanding the possible morphologies that could occupy morphospace, generated morphotypes via modeling along with measured phenotype traits of actual specimens comprise a combined data set that could cover n-possible morphologies in morphospace. Such combinatorial morphological data are also multivariate, so the resultant morphospace would be n-dimensional and statistical. Determination of unoccupied space still may be a problem to resolve without determining morphospace boundaries. In this vein, a finite list of phenotype trait descriptors may be used to provide boundaries for a morphospace so that degree of occupation may be determined for the particular traits. This outcome does not address the problem of multiple spaces within a greater unoccupied morphospace. However, the combinatorial approach would enable n-spaces to be subsumed within a larger unoccupied morphospace.

9.3   Purposes of this Study

Diatoms, which are eukaryotic unicells of evolutionary and ecological significance, are the microorganisms of study in this morphospace analysis. Mathematical morphospaces are devised with respect to diatom lineage crown groups, having a two-fold purpose. One is the construction of a lineage crown group-based morphospaces in which submorphospaces are related within the larger morphospace via a combinatorial metric. Within each submorpho-hospace, Euclidean metrics are used. The boundaries of each submorphospace are dictated by the lineage crown groups, which have been transformed into topological vector spaces, then as metric spaces. As a result, a constrained lineage-based diatom morphospace will result in a metric space.

The second purpose is to determine unoccupied diatom morphospace, given the open-ended understanding of what is meant by unoccupied space. Such a measure will need to be devised to be independent of the taxa considered or morphologies measured. Occupied morphospace will be determined via networks with respect to diatom lineage crown groups, which are close-ended entities. From this, unoccupied morphospace will be determined as a complementary value to occupied morphospace.

9.4   Morphospace Structure and Dynamics

9.4.1   Morphospace Networks and All Possible Morphologies

Morphospaces depict the relation among phenotypes where an evolutionary interpretation may be only implicitly or obliquely interpreted. To overcome this, matching phylogenies to morphospaces has been used as one way to add an evolutionary understanding to morpho-spaces (e.g., [9.13.84., 9.13.104.]), but such matchings do not include all possible variations implicit in a phylogeny or morphospace. Incorporating phylogenetic structures into mor-phospaces or combining the attributes of graph structures with morphological evolutionary data may provide a dynamic element to morphospace analysis. Such models of morpho-space dynamics include branching random walks (e.g., [9.13.91.]) and network morpho-spaces (e.g., [9.13.5., 9.13.17., 9.13.77., 9.13.98.]). Acyclic directed graphs form the basis of such morphospaces in order to exhibit hierarchical structure (e.g., [9.13.55.]), thereby enabling comparison with phylogenies.

The inclusion of all possible morphologies may involve using network morphospaces (e.g., [9.13.5.]). Such morphospaces involve determining structural traits that define each dimension, e.g., using a network adjacency matrix or spatial growth model to generate a network via optimization [9.13.5.]. Empirical networks may be embedded in a morpho-space with relevant structural traits. Subregions of the morphospace may be associated with particular parts of the network topology [9.13.5.]. Theoretical morphologies may be included as potential morphologies to extend the occupied range of morphospace, and a network morphospace may accommodate such an outcome (e.g., [9.13.17.]). Generating networks and utilizing morphospace metrics may involve a general framework such as mul-tidimensional scaling (MDS) that enables the visualization of networks and determinations of distances in morphospaces [9.13.55.].

9.4.2   Networks, Hierarchy and Morphospace

Complex networks (e.g., [9.13.1., 9.13.17.]) may address the problem of determining unoccupied morphospace. Clustered groups may be thought of as hubs that are connected by edges in space (e.g., [9.13.1.]). Hubs may take various shapes. One possibility is a pyramidal (or conical) structure (e.g., [9.13.17.]) that may represent the directionality of the morphologies, reflecting ordering or partial ordering within a hub, or a nesting entity of n-nestings. That is, directionality is implied as morphospace occupation would progress from the least to the largest number of entities in an acyclic tree graph within a hub and between hubs, resulting in a network morphospace (Figure 9.1). Pyramid structures are used in conceptualizations of community ecology (e.g., [9.13.109.]), demographics (e.g., [9.13.90.]) and satellite imagery (e.g., [9.13.110.]), and network morphospaces would qualify as potential pyramidal structures.

Hub connections determine the type of topology exhibited by the network. For a given distance between n-points in morphospace, the complex network is a tree topology (e.g., [9.13.72.]). For such a morphospace topology, the structure may induce the interpretation of the directions of morphological change as a hierarchy. For n-morphospace tree topologies, the degree of hierarchical structure may emerge to define the network morphospace. The hierarchical structure may be present over multiple scales and be representative of the network structure [9.13.13.] of the morphospace.

Within each hub in a morphospace, a network may exist in which edges are connectors of hub members. The net result may be a complex nested network in which the smallest groups are embedded in larger and larger spaces, depending on the extent of the largest network. That largest network may define the boundaries of the morphospace in which embedded network group clusters, n-hubs, exist. The space that is not connected may be indicative of unoccupied morphospace.

Spanning trees are connected acyclic networks and may be subgraphs of a larger network (e.g., [9.13.37.]). For a n-vertex network, the number of edges in a spanning tree is $n - 1$, and the number of spanning trees within a given network is determined via Cayley's formula [9.13.12.]. Minimum spanning trees (MSTs) are the minimal weighted paths of connected acyclic networks (e.g., [9.13.76.]) that also may exhibit shortest distance paths (e.g., [9.13.8.]). MSTs are but one kind of network that may be used to represent connected hub members of submorphospaces or morphospaces.

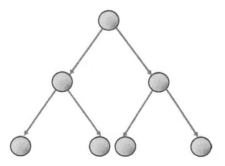

Figure 9.1 Pyramidal structure of an acyclic directed graph.

An embedded network is representative of embedded spaces, and the arrangement of the embedding may exhibit, either in part or wholly, a hierarchical topology [9.13.17.]. This hierarchy may induce a constraint in morphospace so that minimum to maximum points are partially ordered. The constrained hierarchical topology may be used to determine the characteristics of non-occupied network morphospace as to whether it represents possible morphologies or has no evolutionary interpretative basis [9.13.17.].

9.5   Phylogeny Structure and Phylogenetic Dynamics

9.5.1   Phylogenetic Trees and Mapped Traits

Phylogenetic trees are used to depict the relation among taxa implicitly giving the evolutionary significance of that relation. Crown groups define clades ([9.13.20.]). Phylogenies may be depicted as node-based or stem-based, i.e., branch-based trees (e.g., [9.13.20., 9.13.70.]). In both cases, phylogenetic trees are graphs of vertices and edges with no circuits, where the terminal nodes are labelled and known as leaves (e.g., [9.13.75.]). Unlike cladograms, phylogenetic trees depict unequal branch length. Phylogenetic tree shape is exhibited by its topology or branching arrangement as well as branch length (edge weight). Phylogenetic trees may be generated combinatorically or probabilistically (e.g., [9.13.55.]). A combinatorial-based tree results in relative distances between taxa defining evolutionary relations (e.g., [9.13.54.]), while a probabilistically-based tree results from likelihood or other probability-based methods [9.13.55.]. In either case, a phylogenetic tree depicts hierarchical relations, but not those relations resulting from processes such as hybridization, recombination, introgression, or horizontal gene transfer (e.g., [9.13.54.]). A phylogenetic tree is an incomplete representation of evolutionary history for a group of related taxa.

The input data for phylogenetic analyses may have a molecular or morphological basis, and a phylogenetic tree from one type of data may not be in accordance with a phylogenetic tree from the other type of data (e.g., [9.13.44.]). Mapping traits onto phylogenetic trees is not necessarily a one-to-one result. Variation exists in molecular and gene trees just as it does in morphological trees [9.13.44.]. Evolution acts on the phenotype resulting in morphological variation, while genetic mutations are not always expressed in morphologies (e.g., [9.13.97., 9.13.100.]). There is no reason to expect that morphological data will map perfectly to a gene, genomic or molecular-based phylogenetic tree (e.g., [9.13.100.]).

To overcome the shortfalls of phylogenetic trees, phylogenetic networks have been used (e.g., [9.13.54., 9.13.55.]). To overcome the shortfalls of mapping morphological data to gene or molecular trees, morphospace analysis must include allowances for variation that may be incorporated into phylogenetic trees. While there is no mention of "unoccupied lineages" in analyses of phylogenetic trees or networks, whether based on gene or morphological input, unoccupied morphospace is said to exist only as a result of morphospace analysis. There are constraints on phylogenies working to prevent a phenotype from existing or inducing the loss of a phenotype over geologic time [9.13.66.]. The close-ended phylogenetic tree and the open-ended morphospace need to be brought into line to make some headway on the relation between these two spaces.

9.5.2   Phylogenetic Trees and the Geometry of Tree Spaces

In phylogenetic analyses, often, multiple trees result that may not converge to a best possible tree; i.e., there are various ways to determine a consensus tree (e.g., [9.13.7., 9.13.48.]). The set of trees that result depict the space occupied by all possible trees for a given analysis. However, that space is not a vector space [9.13.81.]. In fact, that space is not a metric space. Instead, it is a statistical space.

Combining phylogenies and morphospaces requires a geometric transformation of the phylogenetic tree. Graph matching in which a morphospace is skeletonized and matched to a phylogenetic tree (e.g., [9.13.84.]) means using two different types of trees, one being Euclidean (the morphospace) and the other (phylogeny) being non-Euclidean. A geometric phylogenetic tree enables measurement via transformation into a metric space via a vector space. The result could potentially be a geometric morphospace embedded within a phylogenetic geometric tree space so that the resultant space depicts a lineage-based and potentially bounded occupied morphospace.

To enable the fitting of a morphospace within a phylogenetic space, a transformation of the phylogenetic space to a vector space is necessary (e.g., [9.13.9.]). Geometrically, the tree vector space must have a metric property associated to it whereby path lengths or geodesics are measurable [9.13.9.], with the n-dimensional nature of phylogenetic trees and their tree spaces being evident [9.13.60., 9.13.78.]. From this, the vector space [9.13.60.], geometrically, must be a metric space in which morphospace and tree space measurements may be obtained. More broadly, the topological properties of the phylogenetic vector space must be evident so that a topological vector space may be devised in which the morphospace resides.

The vector space of phylogenetic trees may be determined in various ways including via principal components analysis (PCA) (e.g., [9.13.81.]) or MDS [9.13.47., 9.13.59.], which have been used to devise morphospaces. While these techniques are statistical, they are also algebraically geometric, and outcomes are plotted in a geometric metric space in which points are vectors. Phylogenetic trees would be analyzed via distance metrics—e.g., Euclidean—with respect to vectors. Measurement would be pairwise distances between leaves, which are the arrowheads of the vectors (e.g., [9.13.75.]).

Other geometric spaces may be used as well. For example, phylogenetic trees may be analyzed as structures of flat Banach spaces (e.g., [9.13.46.]) so that martingales in such spaces may be used to induce a phylogenetic tree to be in a metric space [9.13.92.]. In this way, the metrics in phylogenetic tree spaces may be obtained via embedding a tree space, say, within a morphospace in which a common tree pattern for each space results in this combined space. Alternatively, a Hilbert projective metric space may be used in which the distance metric involves differences between maxima and minima from pairwise point measurements [9.13.75.]. Thus, a metric within the tree space and morphospace are obtained as a common metric, such as Euclidean distance (e.g., [9.13.34.]), i.e., the L^2-norm [9.13.59.], or some other linear or geodesic measure [9.13.9., 9.13.73.].

Metric tree spaces are topological tree spaces [9.13.2., 9.13.117.] that may be used as representation of n-phylogenetic trees. Networks of phylogenies may comprise the topological tree space. As within morphospaces, phylogenies in tree space may be described as subtrees of a larger phylogenetic tree. Within this tree space, a metric, such as the nearest neighbor interchange (NNI) enables measuring the proximity of phylogenies to one another or changes in a phylogenetic tree (e.g., [9.13.2., 9.13.95.]). The topology of the tree may change or remain the same when moving from one subtree to the next within the tree space [9.13.2.]. As with network morphospaces, phylogenetic networks may incorporate implicit and explicit subnetworks [9.13.54.] analogous to theoretical and empirical morphospaces, respectively.

With n-trees (or subtrees) in tree space, a metric may be determined within a combinatorial framework [9.13.117.], resulting in a metric tree space [9.13.2.]. Generating phylogenetic trees as a combinatorial outcome of n-trees may be depicted within an n-dimensional space [9.13.60.]. One way to express the combinatorial properties of metric tree spaces may be given via weighted tree edges (or paths) in the process of constructing n-phylogenetic trees [9.13.54., 9.13.78.]. As another example, Markov processes (e.g., [9.13.27., 9.13.28., 9.13.60.]) may be used with respect to combinatorial resultant n-phylogenetic tree spaces. In any case, where a product space of edges results [9.13.78.], a probabilistic measure (e.g., [9.13.33., 9.13.34.]) may be induced, and measurements within a combined metric tree space and morphospace would be probabilistically-based.

No matter which avenue is used to obtain a geometric phylogenetic tree, the geometry enables the embedding of a geometric phylogenetic tree within a geometric morphospace, or vice versa. The resultant space enables analysis of the combined phylogenetic tree and morphospace from a geometric basis.

9.6 Measuring Occupied Morphospace: Clustering Coefficients

Clusters as hubs may be numerically represented by clustering coefficients [9.13.57.]. For a network that is spatially growing, number of edges among vertices in each cluster, when connected via the average shortest path lengths from a vertex, v, that is positioned in the center of the cluster, may be used to calculate a clustering coefficient [9.13.57.]. For a single cluster as

$$C_v = \frac{|Edges_v|}{\binom{k_v}{2}_i} \tag{9.6}$$

where the absolute value of the number of actual edges (connections) is in the neighborhood of vertex, v is $Edges_v = \sum_i v_i$, and this value is divided by the number of possible edges [9.13.111.] with k_v neighbors—i.e., k_v choose 2—which is all the connections among the vertices in a single cluster. One result is that this may reflect n-possible morphologies to occupy morphospace.

For network morphospaces with non-spatial clustering, coefficients for bounded submorphospaces of MSTs may be calculated as

$$C_{Network\ Submorphospace} = \frac{|\ Edges_{Network\ Submorphospace}\ |}{\left(\dfrac{k_{Network\ Submorphospace\ Vertices}}{2}\right)} Area_{Submorphospace} \qquad (9.7)$$

where $\left(\dfrac{k_v}{2}\right) = \left(\dfrac{k_{Network\ Submorphospace\ Vertices}}{2}\right)$ and $Area_{Submorphospace} = (crown(y)_{Max} - crown(y)_{Min})$ $(crown(x)_{Max} - crown(x)_{Min})$, which is calculated from the boundary coordinate vertices of each crown group. The resultant numerical values are measures of occupied submorphospaces.

For n-submorphospaces, the total occupied morphospace is calculated as

$$Total_{Occupied\ Morphospace} = \sum_{i=1}^{n} C_{Network\ Submorphospace_i} \qquad (9.8)$$

where clustering coefficients for n-crown group bounded network clusters are summed, while unoccupied morphospace is calculated as

$$Total_{Unoccupied\ morphospace} = 1 - \sum_{i=1}^{n} C_{Network\ Submorphospace_i}. \qquad (9.9)$$

9.7   A Brief Background on Diatom Morphospaces

Theoretical and empirical morphospace analysis is a staple of works used to elucidate the relation among organisms with respect to morphometric, taxonomic, developmental, evolutionary, or ecological considerations. Separate morphospaces for qualitative and quantitative phenotype data are usually devised, because the ability to combine disparate kinds of data provide difficulties in terms of differences in units or scaling (e.g., [9.13.50.]). Or, if disparate data sets are combined, one kind of data is treated via conversion or becomes a constraint for another kind of data that is segregated in a separate data file. For diatoms, it is no different. Diatom morphospaces embody these issues, but have taken two general forms in which delineation of deterministic or statistical bases characterizes such morphospaces.

Deterministic diatom morphospaces have been devised with regard to ecology [9.13.82.-9.13.84.] as well as 3D surface morphology [9.13.86.]. Many more diatom studies have been conducted resulting in multivariate statistical morphospaces usually devised with PCA or discriminant analysis (DA) [9.13.87.]. A host of such studies chiefly were used as an aid in taxonomic decision making. Other morphospaces have been devised in terms of multiple taxa on a macroevolutionary level involving planktonic marine diatoms (e.g., [9.13.63.]), morphological and functional complexity (e.g., [9.13.85.]), or distinguishing between two different diatom morphospaces for *Gomphonema* and *Reimeria* (e.g., [9.13.31.]).

Input data for diatom morphospace studies varies from qualitative descriptors to measurements of morphological features. Most studies use two-dimensional (2D) outline data of shapes or landmark/pseudolandmarks to form the basis of diatom morphospace analysis [9.13.87.]. The resultant morphospaces may be interpretable from the species to population level. Because most studies embody statistical morphospaces, variance or distance metrics are used to measure morphological proximity [9.13.87.]. Interpretation of diatom morphologies or phenotypes in a morphospace, regardless of the kind of input data used, induces the result that the minimization of variance or distance is illustrated by proximity in low dimensional morphospace (e.g., [9.13.87.]).

Because diatoms exhibit specific developmental or life cycle attributes such as vegetative size reduction [9.13.68., 9.13.88.-9.13.89.] and ecological or habitat preferences because of their biological indicator status (e.g., [9.13.67.]), these factors may be used to induce a boundary for the morphospace as a whole or via boundaries of submorphospaces therein. Vegetative size reduction and ecological preferences may be used as defining characteristics of submorphospaces within a larger morphospace. Such submorphospaces may appear to be compressed, with morphological variants exhibiting proximity in similarity, or such variants may be nested in which, e.g., a size-reduced taxon sequence is embedded in a particular habitat or with regard to a specific substrate or in a chemically or physically circumscribed environment. Developmental and ecological morphospaces may be exhibited in a common morphospace with regard to temporal sequences (e.g., vegetative size reduction) and spatial distributions (e.g., temperature gradients).

Diatom submorphospaces may be measurable as clusters embedded within the larger morphospace. Proximity may be determined via networks in which clusters serve as connected hubs (e.g., [9.13.1.]). On a larger scale, the arrangement of submorphospaces may be nested or partially to totally ordered, and this arrangement may be present as a hierarchical structure within an n-dimensional morphospace. In this way, hierarchy induces a constraint on an n-dimensional morphospace.

The geometry of the submorphospaces may present as at least two general shapes. One is a pyramidal shape for each subspace, representing nestedness, and the other shape is a tree in which sequences are relationally structured and ordered [9.13.16.]. Pyramidal shape of submorphospaces as connected hubs in a network could be construed as a network morphospace. Constructing a diatom morphospace in this way may induce a hierarchical n-dimensional morphospace as a tree structure.

9.8 Mathematical Morphospaces in the Context of a Diatom Phylogeny

Usually, diatom morphospaces are constructed as a result of input representing numerical data (e.g., counted striate) or measurements (e.g., valve length) or geometry (e.g., shape outline), representing morphological attributes [9.13.87.]. Only the relation among morphological attributes is determined by the result of morphospace analysis (e.g., [9.13.87.]), and from this, inferences may be made with regard to taxonomic or other morphologically-based significance. Habitat, ecology (including functional dynamics), developmental, or evolutionary significance may be implicitly inferred.

A diatom phylogeny may be used as the explicit context and basis for a diatom mathematical morphospace analysis as well.

Actual morphologies may be supplemented by theoretical morphologies to represent the potential morphological variation of a given diatom taxon, as theoretical morphospaces enable including all possible morphologies (e.g., [9.13.5.]). This coupled with vegetative size reduction induces multiple-scaled morphological variations for the given diatom taxon, and theoretically-based as well as empirically-based morphospace analyses may be subsumed in one outcome for a diatom morphospace. Using multiple-scaled processes as the heart of morphospace dynamics enables relating diatom vegetative reproduction to taxonomic proximity to ecological context, and when considering diatom phylogenetic analyses, a holistic picture of implicitly interacting processes may emerge.

Diatom phylogenies on the whole have been incomplete. Although improvements have been made over the years, no one phylogenetic tree is representative of the distribution of diatoms throughout evolutionary history. However, advancements have been made such that a consensus on diatom lineage crown groups has emerged based on the structural gradation hypothesis [9.13.105.], and this result may be used to form the basis in which to evaluate a diatom morphospace. Crown groups depict the inference of diversity from the stem (ancient) group members to crown (modern) group members with respect to time (e.g., [9.13.11.]), and as such, may be used to induce an implicit evolutionary description of morphospace dynamics.

9.9   Methods

9.9.1   Input Data for Morphospace Analysis

A generalized diatom phylogeny was devised by Theriot *et al.* [9.13.105.] of lineage crown groups of structural grades of diatoms. In Theriot *et al.* [9.13.105.], Figure 1, four different structural groups were depicted consisting of radial centrics, polar centrics, araphids, and raphids. For each structural group, nine major clades were identified as: three for radial centrics; three for polar centrics; two for araphids; raphids as the only monophyletic structural group [9.13.105.]. A reconstructed version with color coding of the lineage crown groups is illustrated in Figure 9.2. Taxa for each lineage crown group was obtained from Theriot *et al.* [9.13.106.] supplementary Table.S1.Taxa. GenbankNumbers.

For diatom morphospace analysis, assignment of taxa was accomplished by using the Legendre coefficient bins listed in Chapter 1, Tables 1.1 and 1.5, as morphological traits of diatom shape. Table 1.1 was modified with the assignment of Legendre coefficient numbers to basic valve shapes (Table 9.1). Genera listed in Table.S1.Taxa.GenbankNumbers [9.13.106.] were used as input data for binning according to Table 9.1 and Chapter 1, Table 1.5. Three Legendre coefficient bins were assigned per genus to cover morphological shape variation for each genus. The order of coefficients was listed from smallest to largest per genus. The input data was expanded by coding for the diatom structural or taxonomic (1 to 4) and lineage crown groups (1 to 9) for each genus. A Legendre coefficient bin value matrix resulted in 128 genera rows by three columns of Legendre coefficient bin numbers, a column of taxonomic codes, and a column of lineage crown group codes, resulting in a 128

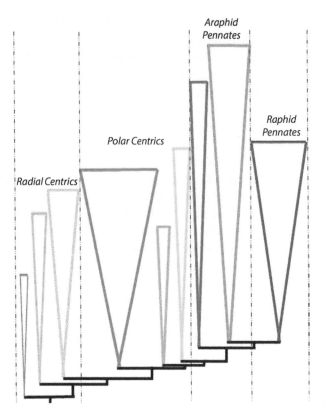

Figure 9.2 Redrawn structure of diatom lineage crown groups from [9.13.105.], Figure 1. Crown groups are color coded so that radial centrics are oranges, polar centrics are greens, araphid pennates are blues, and raphid pennates are purple.

by 5 matrix (Table 9.2). Assignment of each genus to the appropriate crown group was verified by consulting the ANSP Diatom New Taxon File [9.13.93.] and the Diatom of North America website [9.13.102.].

9.9.2   Diatom Lineage Crown Groups Embedded in a Metric Space

To obtain a diatom morphospace within which are bounded submorphospaces, the tree that is composed of lineage crown groups, which is a consolidation of lineages from the original phylogenetic tree [9.13.105.], needs to be embedded in a metric space inducing spatially measurable distances. To accomplish this, each lineage crown group may be viewed as a cluster or hub of networked genera that are n-hubs of the complete tree of lineage crown groups. With n-genera comprising a given lineage crown group, each genus within is a member of a partially ordered set of n-genera. The set is the network of subtrees (or subsets) within the lineage crown group.

   For the purpose of embedding a lineage crown group in a metric space, the phylogenetic tree topology or the branching pattern and branch lengths are only relevant to the extent that a pyramidal shape is depicted. Only the boundaries of a lineage crown group play a role in placement of submorphospaces within a metric morphospace.

Table 9.1 Basic valve shapes, descriptions and Legendre coefficient bins.*

Legendre coefficient number	Valve shape description	Valve shape
0	Square and rounded square forms.	
2	Circular and modified circular forms with 5 or greater edges.	
4	Triangular and some modified triangular forms.	
12	Hyperbolic triangular forms.	
14	Cruciform	
20	> 10 edges on forms – multiple undulations.	

*See Chapter 1, Tables 1.1 and 1.5 for more information.

A lineage crown group may be an open set or a closed set with finite intersection and arbitrary union, or arbitrary intersection and finite union so that a topological space results. A topological space is a metric space, and each lineage crown group is, therefore a metric space. If the shape of the lineage crown group is taken into consideration, then the equal length legs of the triangle are vectors, where the arrowhead is located at the points connecting the legs to the hypotenuse. These vectors signify that phylogenetic activity (e.g., ancestor-descendant speciation) emanates from the basal node of the lineage crown group to the terminal nodes. That is, the lineage crown group is a directional cyclic graph.

Planting all lineage crown metric spaces in the configuration that they present as a phylogenetic tree induces their fixed position in a larger metric space. That is, an isometry exists between the original lineage crown tree and the lineage crown tree metric space. A distance metric must be sensitive to and indicative of the way in which the connections among lineage crown groups was devised (Figure 9.2). In this vein, NNI distance is used, which is the minimum number of moves required to convert one tree into a second tree (e.g., [9.13.19.]); that is, the minimum number of moves necessary to transform one tree into another one is the NNI distance [9.13.65.]. The NNI distance is NP-complete

Table 9.2 Taxonomic, crown group and Legendre coefficients assigned to 128 genera from [9.13.105.], Table S1.

Genus	Taxonomic group	Crown group	Legendre 1	Legendre 2	Legendre 3
Acanthoceros	2	5	2	2	2
Achnanthes	4	9	2	4	6
Actinocyclus	1	3	2	2	2
Actinoptychus	1	3	2	2	2
Amphipentas	2	4	2	20	20
Amphipleura	4	9	6	14	14
Amphitetras	2	4	14	14	14
Amphora	4	9	4	4	6
Arcocellulus	2	4	2	2	2
Ardissonea	2	6	6	6	18
Asterionella	3	8	14	18	18
Asterionellopsis	3	7	14	14	18
Attheya	4	9	2	2	2
Aulacodiscus	1	3	2	2	2
Aulacoseira	1	2	2	2	2
Bacillaria	4	9	6	6	14
Bellerochea	2	4	2	2	2
Berkeleya	3	8	6	6	6
Biddulphia	2	4	8	16	20
Biddulphiopsis	2	6	8	16	20
Bleakeleya	3	7	14	18	18
Brockmanniella	2	4	6	6	6
Caloneis	4	9	2	2	2
Campylosira	2	4	4	4	4
Centronella	3	8	12	12	12
Cerataulina	2	5	2	2	2
Cerataulus	2	4	2	0	0

(*Continued*)

Table 9.2 Taxonomic, crown group and Legendre coefficients assigned to 128 genera from [9.13.105.], Table S1. (*Continued*)

Genus	Taxonomic group	Crown group	Legendre 1	Legendre 2	Legendre 3
Chaetoceros	2	5	2	2	2
Chrysanthemodiscus	2	6	2	2	2
Climaconeis	4	9	2	2	2
Cocconeis	4	9	2	2	2
Corethron	1	1	2	2	2
Coscinodiscus	1	3	2	2	2
Craticula	4	9	6	6	14
Ctenophora	3	8	14	18	18
Cyclophora	3	8	14	18	18
Cyclostephanos	1	3	2	2	2
Cyclotella	1	3	2	2	2
Cylindrotheca	4	9	6	6	6
Cymatopleura	4	9	6	2	6
Cymatosira	2	4	2	2	2
Dactylosolien	2	5	2	2	2
Delphineis	3	7	2	2	2
Denticula	3	8	6	6	14
Detonula	1	3	2	2	2
Diatoma	3	8	2	2	2
Dimeregramma	3	7	2	2	2
Diploneis	4	9	2	2	2
Ditylum	2	4	2	2	2
Ellerbeckia	1	2	2	2	2
Endictya	1	2	2	2	2
Eucampia	2	5	2	2	2
Eunotia	4	9	8	12	22
Eunotogramma	2	4	2	4	12

(*Continued*)

Table 9.2 Taxonomic, crown group and Legendre coefficients assigned to 128 genera from [9.13.105.], Table S1. (*Continued*)

Genus	Taxonomic group	Crown group	Legendre 1	Legendre 2	Legendre 3
Extubocellulus	2	4	2	2	2
Fallacia	4	9	2	2	2
Florella	4	9	2	2	2
Fragilariforma	3	8	8	14	16
Gomphonema	4	9	14	14	14
Grammatophora	3	8	6	16	16
Guinardia	2	4	2	2	2
Gyrosigma	4	9	14	14	14
Hemiaulus	2	5	2	2	2
Hyalodiscus	1	2	2	2	2
Hyalosira	3	8	14	14	14
Koernerella	3	7	2	2	2
Lampriscus	2	6	2	4	12
Lemnicola	4	9	2	2	2
Leptocylindrus	1	1	2	2	2
Leyanella	2	4	2	2	2
Licmophora	3	8	10	10	10
Lithodesmioides	2	4	20	20	20
Lithodesmium	2	4	4	4	4
Manguinea	4	9	6	6	6
Mastodiscus	2	4	2	2	2
Mastogloia	4	9	6	6	6
Melosira	1	2	2	2	2
Meuneira	4	9	6	6	6
Minidiscus	1	3	2	2	2
Minutocellus	2	4	2	2	6
Nanofrustulum	3	8	2	2	2

(Continued)

Table 9.2 Taxonomic, crown group and Legendre coefficients assigned to 128 genera from [9.13.105.], Table S1. (*Continued*)

Genus	Taxonomic group	Crown group	Legendre 1	Legendre 2	Legendre 3
Navicula	4	9	2	4	6
Neidium	4	9	4	6	6
Nitzschia	4	9	6	14	18
Odontella	2	4	2	2	2
Opephora	3	8	10	10	10
Palmerina	2	4	2	2	2
Papiliocellulus	2	4	2	2	2
Paralia	1	2	2	2	2
Perideraion	3	7	2	2	14
Phaeodactylum	4	9	6	18	6
Pinnularia	4	9	6	8	16
Placoneis	4	9	4	4	4
Plagiogramma	3	7	2	2	14
Plagiogrammopsis	2	4	6	6	6
Planktoniella	1	3	2	2	2
Pleurosira	2	4	2	2	2
Podocystis	3	8	10	10	10
Porosira	1	3	2	2	2
Proboscia	1	2	2	2	2
Psammoneis	3	7	2	2	2
Pseudosolenia	2	4	2	2	2
Pseudostaurosira	3	8	2	2	14
Rhabdonema	3	8	2	6	14
Rhaphoneis	3	7	4	4	4
Rhizosolenia	2	4	2	2	2
Roundia	1	3	2	2	2
Scoliopleura	4	9	6	6	6

(*Continued*)

Table 9.2 Taxonomic, crown group and Legendre coefficients assigned to 128 genera from [9.13.105.], Table S1. (*Continued*)

Genus	Taxonomic group	Crown group	Legendre 1	Legendre 2	Legendre 3
Stauroneis	4	9	6	6	14
Staurosirella	3	8	2	2	14
Stellarima	1	3	2	2	2
Stephanopyxis	1	2	2	2	2
Stictocyclus	1	3	2	2	2
Striatella	3	7	14	14	14
Surirella	4	9	2	10	14
Synedra	3	8	14	18	18
Synedropsis	3	8	14	14	14
Tabellaria	3	8	6	6	6
Tabularia	3	8	6	6	18
Terpsinoë	2	5	16	16	16
Tetracyclus	3	8	2	20	20
Thalassiosira	1	3	2	2	2
Toxarium	2	6	6	6	6
Triceratium	2	4	4	4	4
Trieres	2	4	2	2	2
Trigonium	2	6	4	4	4
Tryblionella	4	9	6	2	2
Urosolenia	2	5	2	2	2

[9.13.19.] as a combinatorial metric. However, in the case of lineage crown groups, tree matching is not required, because all crown groups are triangular shapes. NNI distance is used here in terms of the spatial position of each lineage crown group with respect to each other and the entire tree structure in the larger metric space. Coordinate pairs of vertices are used to plot each lineage crown group metric space locally in the larger global metric space.

The entire lineage crown group tree is depicted in 2D as a planar structure (Figure 9.2). The triangles lend themselves to being anchored at their vertices relative to one another in lineage crown group tree space. The capping of the "leaves" at the terminal vertices (nodes) enables relaxation of combinatorically positional outcomes of *n*-phylogenetic trees because

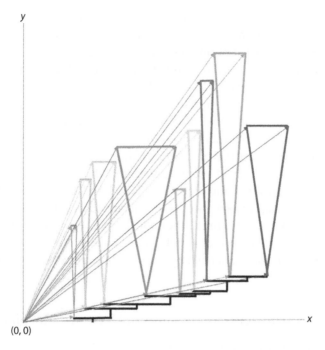

Figure 9.3 Diatom lineage crown groups with rays emanating from the origin to vertices of each crown group, signifying coordinate values in metric space. See Figure 9.2 legend for color labels.

of the low dimensionality. As a result, embedding a 2D lineage crown group tree in a metric space reduces the problem to a simpler outcome in creating a morphospace.

Each attached triangle from the linear crown group tree is anchored in metric space in the following way. Place the entire lineage crown group tree in an arbitrary position within a metric space defined by x- and y-axes. Draw three rays for each triangle—one at the basal vertex and two at the terminal vertices where all rays start at the origin, $(x_0, y_0) = \{0,0\}$ (Figure 9.3). From this, vertex point coordinates may be obtained for each lineage crown group in a Euclidean space even though the spatial relation among all lineage crown groups is obtained via NNI.

9.9.3   Diatom Submorphospaces Embedded in a Metric Morphospace

Triangular crown groups provide the boundaries for the nine submorphospaces. From the diatom input matrix of ordinal data (Table 9.2), non-metric MDS (NMDS) [9.13.45.] was used to ordinate genera from each lineage crown group. For each ordination, NMDS Euclidean coordinates were used to create planar metric trees [9.13.21.] as planar network graphs. Placement of the graphs is obtained via NNI [9.13.3.] so that submorphospaces were spatially in accordance with the lineage crown group tree. That is, Euclidean coordinates were used to plot each submorphospace, while NNI was used to plot those submorphospaces to be spatially correct with respect to the lineage crown tree.

In 2D, triangular structures were created via bracketing of the planar networks for the submorphospaces. Bracketing a planar tree (e.g., [9.13.17., 9.13.21.]) is used to create "crown groups" for submorphospaces. Because each submorphospace is a metric space via

NMDS, embedding in the larger morphospace induces a metric space. It should be noted that although the submorphospace boundaries are from the lineage crown groups, directionality is not implied by those boundaries, only relative spacing.

Each network submorphospace was plotted via NMDS Euclidean coordinates within the appropriate lineage crown group metric space. The connections between submorphospaces are a network and comprise the larger network morphospace. That is, once all nine NMDS ordinations were completed and networks devised, these planar network graphs were reassembled in their original configuration according to the lineage crown group branching and embedded within the larger metric morphospace.

To accomplish this reassembly of submorphospaces, the distance between crown group submorphospaces was measured and plotted via NNI [9.13.14., 9.13.15., 9.13.19., 9.13.32., 9.13.52.]. The diatom lineage crown groups were arranged as connected submorphospaces in the same configuration and spacing as they appear in Theriot *et al.* [9.13.105.], Figure 1. Using NNI as the metric ensured that a biologically meaningful [9.13.114.] morphospace was devised that reflected the spatial arrangement of lineage crown groups as phylogenetically based.

9.9.4   Clustering Coefficients as Measures of Occupied Morphospace

Clustering coefficients of submorphospace are related to the lineage network in the neighborhood of each crown group and calculated as measures of lineage crown group bounded occupied morphospace. That is, boundary conditions are the lineage crown groups as planar graphs. The sum of those submorphospaces is the total occupied morphospace (Eq. 9.8). The complement is total unoccupied morphospace (Eq. 9.9).

9.10   Results

Conventional diatom morphospaces were ordinated using NMDS to produce plots for all genera (Figure 9.4a), centric diatoms (Figure 9.4b) and pennate diatoms (Figure 9.4c). NMDS ordinations were produced for each taxonomic (structural) group from the diatom lineage crown tree for radial centrics (Figure 9.5a), polar centrics (Figure 9.5b), araphid pennates (Figure 9.5c), and raphid pennates (Figure 9.5d). NMDS ordinations were constructed for submorphospaces of crown groups 4 through 9 (Figure 9.6). STRESS and coefficients of determination were reported for each taxonomic group, crown group, all genera, all centric diatoms, and all pennate diatoms (Table 9.3). Crown groups 1 through 3 were not multivariate, but colinear.

Each NMDS ordination is depicted without networking for the purpose of embedding in crown group submorphospaces. Diatom lineage crown groups were embedded into a Euclidean metric space according to the Methods Sections 9.8.2 and 9.8.3. From Figure 9.3, Euclidean coordinates were obtained for the vertices of each crown group and identified in Euclidean metric space in which x- and y-axes were normalized to [0, 1] (Figure 9.7). Then, each NMDS submorphospace was embedded in each crown group morphospace via rescaling equations (Table 9.4 and Figure 9.8), and the crown group bounded submorphospaces were depicted in Figure 9.9.

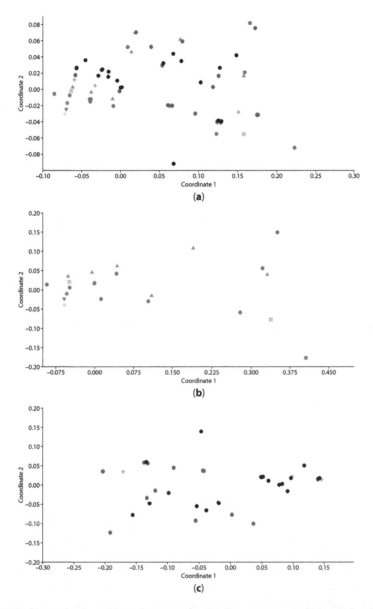

Figure 9.4 NMDS of Legendre shape bin assignments for 128 diatom genera in [9.13.105.] and [9.13.106.]: a, NMDS for all genera; b, NMDS for all centric diatoms; c, NMDS for all pennate diatoms.

From NMDS ordinations of crown group submorphospace, MSTs were obtained as network submorphospaces (Figure 9.10). Rescaled network submorphospaces from MSTs from bounded crown groups were depicted in Figure 9.11. The rescaled network submorphospaces were used to calculate occupied submorphospace and morphospace coefficients (Table 9.5). Total occupied morphospace according to Eq. 9.8 was 0.05627, and total unoccupied morphospace according to Eq. 9.9 was 0.94373.

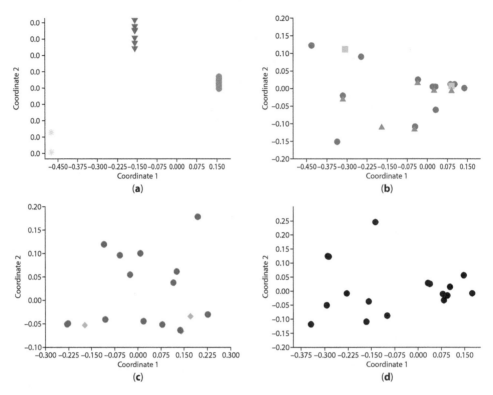

Figure 9.5 NMDS for each taxonomic group based on Legendre shape bin assignments. a, Radial centrics; b, polar centrics; c, araphid pennates; d, raphid pennates. Crown groups for each taxonomic group are color coded.

9.11 Discussion

Morphospace analysis begins with multivariate statistical spaces but is more complicated than this when considering a phylogenetic basis. The concept of unoccupied morphospace is not just the lack of apparent points in an eigenvector diagram. Measurement of occupied and unoccupied morphospace require a broader approach that has been enabled by considering the basic structure of graphs as networks, that networks are geometric and topological spaces, and that phylogenetic tree space may be used as a constraint on morphospace.

Many models exist of hierarchical networks, and the application of the concept of a network to morphospace analysis is useful in the determination of what is meant by boundaries with regard to occupied morphospace. For diatoms, using lineage crown groups as a network basis of morphospace analysis started with a distance-based multivariate technique, NMDS, then proceeded to transform the initial morphospace into network embedded submorphospaces via lineage crown groups. The lineage crown groups were previously converted to metric spaces to enable the embedding of each ordination's results. The resultant network lineage crown group morphospace is a metric vector morphospace, and therefore a topological vector space. Conceptually, the broad outline of topological spaces provided the rationale and impetus to convert lineage crown groups to geometric entities that could be used as constraints to morphospace. This lineage constrained morphospace explicitly depicts occupied morphospace boundaries and the connections among submorphospaces

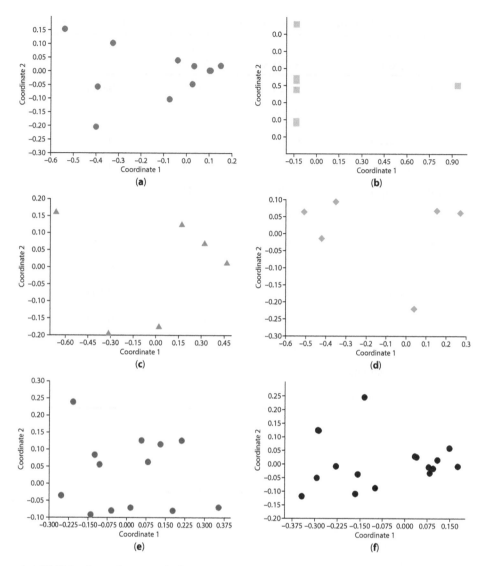

Figure 9.6 NMDS submorphospaces for lineage crown groups: polar centrics, crown groups 4 (a), 5 (b) and 6 (c); araphid pennates, crown groups 7 (d) and 8 (e); raphid pennates (f). Radial centrics, crown groups 1, 2 and 3 are not multivariate, but colinear.

geometrically. With this in mind, other distance-based multivariate techniques, including linear or non-linear PCA, may be used.

Morphospace analysis entails combinatorial possibilities. That is, calculating all possible morphologies is not a countable finite quantity. However, with all possible morphologies constrained by lineage crown groups as boundaries, a finite morphospace is calculable. Clustering coefficients of submorphospaces measured occupied morphospace so that given explicit boundary conditions—namely, lineage crown groups—unoccupied morphospace could be calculated as the complement of occupied morphospace. Because of the boundaries imposed on submorphospace structures, calculating occupied morphospace was relatively straight-forward in comparison to calculating metric tree spaces (e.g., [9.13.9., 9.13.34.]).

Table 9.3 NMDS ordination results of STRESS and coefficients of determination.

Lineage group	STRESS	Axis 1 R^2	Axis 2 R^2
Radial Centrics	0.424	0.9996	0.608
Polar Centrics	0.100	0.973	0.313
Araphid Pennates	0.062	0.885	0.057
Raphid Pennates	0.066	0.908	0.302
Crown Group 4	0.185	0.984	0.658
Crown Group 5	0.620	1	0
Crown Group 6	0.040	0.967	0.013
Crown Group 7	0.096	0.947	0.019
Crown Group 8	0.067	0.837	0.103
Crown Group 9	0.066	0.908	0.302
All Genera	0.047	0.951	0.173
All Centric diatoms	0.077	0.986	0.401
All Pennate diatoms	0.056	0.894	0.101

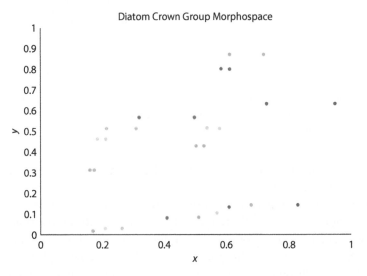

Figure 9.7 Crown group metric morphospace. Coordinates are plotted for each crown group's vertices where $x, y \in [0, 1]$. See Figure 9.2 legend for color labels.

Table 9.4 Table of rescaled NMDS equations for each submorphospace bounded by each crown group from which the new rescaled coordinates are plotted within each crown group submorphospace. The x-direction for crown groups 1, 2 and 3 are colinear so that rescaled NMDS equations are not applicable.

Crown group	Rescaled NMDS equations
1, 2, 3, 4, 8, 9	$$nmds(y)_{RescaledA} = \frac{(nmds(y)_{Original} - nmds(y)_{Min})(crown(y)_{Max} - crown(y)_{Min})}{(nmds(y)_{Max} - nmds(y)_{Min})} + crown(y)_{Min}$$
4	$$nmds(x)_{Rescaled4} = \frac{(nmds(x)_{Original} - nmds(x)_{Min})([0.25(crown(x)_{Max} - crown(x)_{Min}) + crown(x)_{Min}])}{(nmds(x)_{Max} - nmds(x)_{Min})} + [0.5(crown(x)_{Max} - crown(x)_{Min}) + crown(x)_{Min}]$$
5	$$nmds(x)_{Rescaled5} = \frac{(nmds(x)_{Original} - nmds(x)_{Min})([crown(x)_{Min} + 0.64(crown(x)_{Max} - crown(x)_{Min})] - crown(x)_{Min})}{(nmds(x)_{Max} - nmds(x)_{Min})} + crown(x)_{Min}$$
5	$$nmds(y)_{Rescaled5} = \frac{([|nmds(y)_{Original}| + nmds(y)_{Original}] - nmds(y)_{Min})(crown(y)_{Max} - crown(y)_{Min})}{(nmds(y)_{Max} - nmds(y)_{Min})} + crown(y)_{Min}$$

(Continued)

Table 9.4 Table of rescaled NMDS equations for each submorphospace bounded by each crown group from which the new rescaled coordinates are plotted within each crown group submorphospace. The x-direction for crown groups 1, 2 and 3 are colinear so that rescaled NMDS equations are not applicable. (*Continued*)

Crown group	Rescaled NMDS equations
6	$nmds(x)_{Rescaled6} =$ $\dfrac{(nmds(x)_{Original} - nmds(x)_{Min})([crown(x)_{Min}] - [crown(x)_{Min} + 0.875(crown(x)_{Max} - crown(x)_{Min})] - [crown(x)_{Min} + 0.75(crown(x)_{Max} - crown(x)_{Min})])}{(nmds(x)_{Max} - nmds(x)_{Min})}$ $+ [crown(x)_{Min} + 0.75(crown(x)_{Max} - crown(x)_{Min})]$
6	$nmds(y)_{Rescaled6} =$ $\dfrac{\left(([|nmds(y)_{Original}| + nmds(y)_{Original}] - nmds(y)_{Min})(0.5([crown(y)_{Max}] + 0.5(crown(y)_{Max} - crown(y)_{Min})) - crown(y)_{Min})0.5\right)}{(nmds(y)_{Max} - nmds(y)_{Min})}$ $+ crown(y)_{Min} * (0.5([crown(y)_{Max}] + 0.5(crown(y)_{Max} - crown(y)_{Min})) - crown(y)_{Min})$
7	$nmds(x)_{Rescaled7} =$ $\dfrac{(nmds(x)_{Original} - nmds(x)_{Min})([crown(x)_{Min} + 0.5(crown(x)_{Max} - crown(x)_{Min})] - crown(x)_{Min})}{(nmds(x)_{Max} - nmds(x)_{Min})} + crown(x)_{Min}$

(*Continued*)

Table 9.4 Table of rescaled NMDS equations for each submorphospace bounded by each crown group from which the new rescaled coordinates are plotted within each crown group submorphospace. The x-direction for crown groups 1, 2 and 3 are colinear so that rescaled NMDS equations are not applicable. (*Continued*)

Crown group	Rescaled NMDS equations
7	$nmds(y)_{Rescaled7} =$ $$\left(\frac{([|nmds(y)_{Original}| + nmds(y)_{Original}] - nmds(y)_{Min})[crown(y)_{Max} - 0.25(crown(y)_{Max})]0.25}{(nmds(y)_{Max} - nmds(y)_{Min})} + 0.25(crown(y)_{Max})\right)$$ $* (crown(y)_{Max} - 0.25crown(y)_{Max})$
8	$nmds(x)_{Rescaled8} =$ $$\frac{((nmds(x)_{Original} - nmds(x)_{Min})[0.1(crown(x)_{Max} - crown(x)_{Min}) + [0.5(crown(x)_{Max} - crown(x)_{Min})]] - [(crown(x)_{Max} - crown(x)_{Min})])}{(nmds(x)_{Max} - nmds(x)_{Min})}$$ $+ (0.5(crown(x)_{Max} - crown(x)_{Min}) + crown(x)_{Min})$
9	$$nmds(x)_{Rescaled9} = \frac{(nmds(x)_{Original} - nmds(x)_{Min})}{(nmds(x)_{Max} - nmds(x)_{Min})}$$ $$\frac{([[0.2(crown(x)_{Max} - [0.5(crown(x)_{Max} - crown(x)_{Min})] + crown(x)_{Min})]] + (0.5(crown(x)_{Max} - crown(x)_{Min}) + crown(x)_{Min})}{(nmds(x)_{Max} - nmds(x)_{Min})}$$

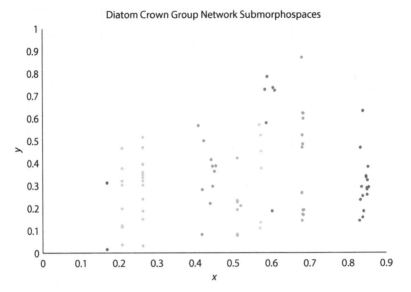

Figure 9.8 Diatom submorphospaces of rescaled NMDS scores for crown groups 1 through 9 positioned in each submorphospace according to Euclidean metric and in the larger morphospace according to NNI metric. See Figure 9.2 legend for color labels.

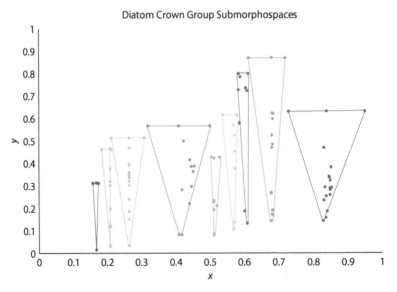

Figure 9.9 Diatom submorphospaces filled with rescaled NMDS scores for each crown group. See Figure 9.2 legend for color labels.

The clustering coefficient calculated for occupied morphospace is similar to $\binom{k}{n} = \dfrac{k!}{n!(k-n)!}$ which is a binomial coefficient so that other possibilities for coefficients may be instituted, depending on the tree and network structures involved or other boundary conditions.

Results from this study indicated that Crown Group 6 from the polar centrics was the largest occupier of morphospace relative to its submorphospace boundaries, while Crown

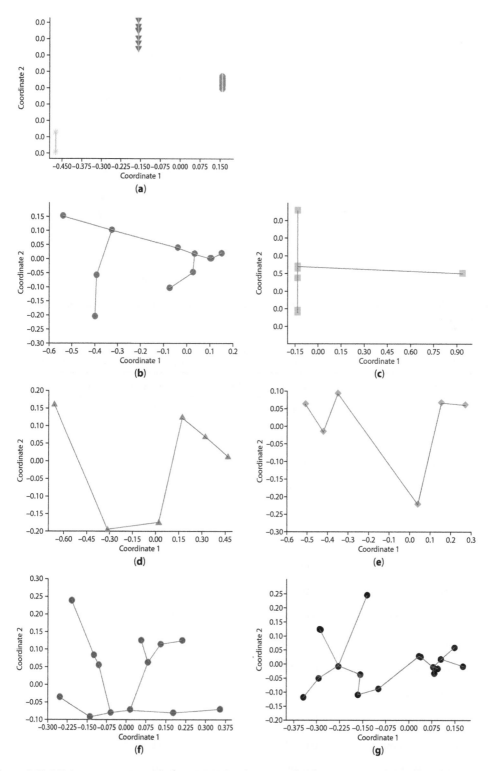

Figure 9.10 Minimum spanning trees from NMDS ordinations of each crown group network submorphospace: a, radial centrics, crown groups 1, 2 and 3 are colinear; polar centrics, crown groups 4 (b), 5 (c) and 6 (d); araphid pennates, crown groups 7 (e) and 8 (f); raphid pennates (g).

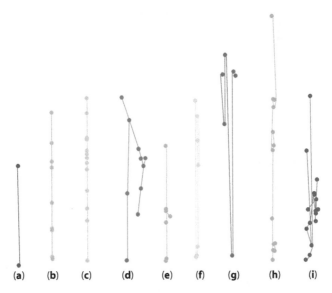

Figure 9.11 Network submorphospaces for crown groups 1 through 9 (a through i) used in calculating clustering coefficients.

Table 9.5 Clustering coefficients as occupied submorphospace for each crown group.

Lineage group	Clustering coefficient	% Occupied morphospace
Crown Group 1	0.00445	0.45
Crown Group 2	0.00398	0.40
Crown Group 3	0.00784	0.78
Crown Group 4	0.00701	0.70
Crown Group 5	0.00288	0.29
Crown Group 6	0.01025	1.03
Crown Group 7	0.00461	0.46
Crown Group 8	0.00781	0.78
Crown Group 9	0.00743	0.74
Radial Centrics	0.01628	1.63
Polar Centrics	0.02014	2.01
Araphid Pennates	0.01242	1.24
Raphid Pennates	0.00743	0.74
Centric Diatoms	0.03642	3.64
Pennate Diatoms	0.01985	1.99
All Genera	0.05627	5.63

Group 5 from the polar centrics was the smallest occupier of morphospace relative to its submorphospace boundaries (Table 9.5). Overall, polar centrics occupied more morphospace than radial centrics, araphid pennates occupied more morphospace than raphid pennates, and centric diatoms occupied more morphospace than raphid diatoms (Table 9.5). From this analysis, the inference is that centric diatoms have a wider range of morphologies than pennate diatoms with respect to lineage boundary conditions (Table 9.5 and Figure 9.9). If directionality is inferred from the x-axis of morphospace, given the position of crown group submorphospaces that match lineage crown groups from Theriot *et al.* [9.13.105.], morphological variation has not evolved in pennate diatoms to the extent that it has in centric diatoms (Figure 9.9). However, caution must be used when trying to make evolutionary inferences using morphospaces (e.g., [9.13.35.]), even though the lineage connections were implicitly embedded via NNI in the resultant lineage bounded morphospace. In general, there is more unoccupied morphospace than there is occupied morphospace (Table 9.5).

9.11.1   Trees, Networks and Morphospaces

Networks have random and non-random properties that are indictive of the distribution of entities as degree of dispersion or clustering [9.13.1.]. Networks are graphs, and randomness in networks is evident via random graphs [9.13.23., 9.13.24., 9.13.38., 9.13.39.]. The properties of random graphs are associated with probability space, and this in turn may induce the representation of unoccupied morphospace via a probability measure. The probability stems from the n-paths (edges) between hubs that may result from a constrained network treatment of the morphospace. For unoccupied network morphospace, a test of randomness may be used to determine whether the probability of an evolutionary basis exists. That is, an evolutionary interpretation of unoccupied morphospace may take the form of directedness in a constrained hierarchical network morphospace. Probabilistic measures of occupied and unoccupied morphospace potentially could be used to represent morphospace dynamics over time.

9.11.2   Probabilistic Distances in Lineage Crown Group Morphospace

From the complete embedding of diatom submorphospaces in lineage crown groups, additional taxa may be added without disruption of the boundaries of the submorphospaces. One result of this is that probabilities of morphology assignment of to each crown group may be determined by setting up each submorphospace as a probability bin. NMDS values rescaled on the interval [0, 1] may enable a probability measure to be obtained. Probability distances (e.g., [9.13.33., 9.13.57.]) may be determined among submorphospaces. Probability determinations may be made with regard to the taxonomic groups radial centrics, polar centrics, araphid pennates, and raphid pennates as pooled values. The total probability of occupied versus unoccupied morphospace may be calculated as well.

9.11.3   Diatom Novelties Versus Repetitive Forms in Occupied Morphospace

Explicit diatom novelties in form should be present in a morphospace, but their identity may be implicit and subject to the definition of what is meant by a novelty form. Novelties displayed as phenotypic external features are morphological outcomes that may need to be accounted for in morphospace analysis. Although such morphological forms have existed for diatoms throughout evolutionary time, they have not counted as explicit entities or subgroups of morphospaces. Having said this, such novelties may be the result of genetic or environmental processes, and the morphospace may account for such a form implicitly. Such forms need not occur outside of occupied morphospace.

Repetitive morphologies exist in diatoms as well, and accounting for repetition would aid in preventing erroneous counts in numbers of morphologies in occupied space. Diatom valves have geometrically repetitive shapes that follow a hypergeometric distribution when calculated via Legendre polynomials (see Chapter 1). These orthogonal polynomials enable the biological interpretation of shape via morphospace analysis when using PCA or DA (e.g., [9.13.87.]). Do repetitive morphologies indicate morphological stability? Questions about morphological stability may be addressed via the intensity of occupation of a diatom morphospace.

9.11.4   Diatom Teratologies and Mutagenicities: Influences on Morphology

Diatom teratologic and mutagenetic forms represent classes of morphologies that should occupy morphospace but would not be typically included in such analyses because they are aberrant forms. However, they are both indicative of actual morphologies that exist while not considered to be typical morphological variants. How should such variants be considered with respect to occupation in morphospaces?

Teratological forms are the result of environmental influences (e.g., [9.13.26.]), while mutagenetic forms are the result of genetic influences (e.g., [9.13.51.]). Both types of aberrant forms are not predictable in the sense that external environmental or subcellular internal stresses may produce a plethora of outcomes, contributing to the notion of all possible morphologies. Often, diatom valve shape outline is abnormal as a result of a teratology [9.13.26.]. To what degree must an aberrant form be considered to be a morphological variant of a given diatom taxon? Slight differences in form may be the result of phenotypic plasticity, or alternatively, the result of environmental stresses as a true teratogenetic form. If a form very different from the normal diatom morphology is evident, identification of the diatom as either an aberrant form or potentially a new species may be difficult to say. Regardless, teratogenetic forms are propagated by developmental miscues and result in actual morphologies that may be accounted for within a diatom morphospace implicitly. Such forms may occupy the same clusters within the morphospace, or they may occupy heretofore unoccupied regions of morphospace. If the latter is true, then part of unoccupied morphospace may have a higher potential to be occupied, say, closer to regions already occupied that those regions farther away in n-dimensional space.

Mutagenicity produces aberrant diatom morphologies via genetic or genomic changes (e.g., [9.13.18., 9.13.41., 9.13.99.]). Such changes may be engineered for medical, industrial, or other commercial applications (e.g., [9.13.116.]). Do bioengineered diatom valve morphologies as mutagenetic forms count as entities that should be included in morphospaces? If bioengineered forms resemble apparent natural forms, should they be counted as part of all possible morphologies that diatoms may produce, and therefore be included in morphospaces? Implicitly, such forms may or may not occupy regions of morphospace.

9.11.5   Understanding Diatom Evolution via Morphospace and Phylogenetic Analyses

Characterizing morphospaces and phylogenies as networks enabled a commonality between different types of data to be combined for an all-inclusive outcome. Networks may provide an avenue for studying processes such as reticulate evolution with regard to hybridization or polyploidy that traditional phylogenetic tree algorithmic approaches do not (e.g., [9.13.6., 9.13.79.]). The inadequacies of traditional phylogenetic trees concerning evolutionary history interpretations are no more evident than when it comes to diatoms. Processes such as hybridization and polyploidy are just starting to be understood for their relevance and importance (e.g., [9.13.4.]) in diatom evolution.

While the study presented here is only a beginning, the results should provide an impetus to advancing diatom morphospace and phylogenetic studies in a more expansive and comprehensive fashion. The concepts presented herein on networks, biologically meaningful distance metrics, and accounting for multiple processes in the evolutionary history of diatoms must be addressed to advance diatom research.

9.12   Summary and Future Research

Morphospace analysis has been changing so that the conventional multivariate statistical analysis and interpretation was augmented by utilizing additional concepts from point-set topology, networks, graph theory, and combinatorics in the presented study. The relation between diatom phylogeny and morphospace must continue to be studied so that a more coherent evolutionary picture of the relation among diatoms may be determined.

Other avenues may be instituted in combination phylogenetic and morphospace methodological studies. Machine learning may be applied to morphospace, tree space (e.g., [9.13.117.]), phylogenetic tree spaces, or a combination of such spaces as networks. Other algorithmic techniques involving networks may be applicable as well. Expanding the methodological application of various mathematical techniques is necessary to enhance morphospace analysis.

9.12.1   What is Morphological Data?

Diatom valve shape is inadequate as a single representation of morphology in morphospace, just as nuclear-encoded small subunit ribosomal RNA is inadequate as a single representation of the diatom genome in phylogenetics. Alternative data to use for morphospace analysis may include 3D surface models for theoretical morphologies (e.g., [9.13.86.]), or images from diatom databases whereby surface morphology is measured via the image Jacobian (see Chapter 2). If the image Jacobian is used, windows on the surface could be used for measurement because measuring the whole valve surface may produce extremely large matrices which induce a negative time cost for analysis. Images could be obtained from any diatom database such as ANSP Diatom New Taxon File [9.13.93.] or the Diatom of North America (DoNA) website [9.13.102.]. In the case of the DoNA website, images of vegetative size reduction series are available for many taxa so that developmental or life cycle morphospaces could be devised. A 3D surface representation may provide a more holistic picture of diatom morphology. Additionally, morphological measurement of micro- and nanostructure may provide an even more inclusive picture of diatom morphology. More work is necessary on what constitutes morphological representation, morphological characters, and morphological character states.

Each set of data could provide potential morphospace outcomes. The nature of the data must be matched to the purposes of the study so that circumscribed results make sense and do not overreach. Evolutionary history depictions are combinatorial in nature, as evidenced by researchers' dependence on phylogenetic trees, so that studies that are statistically-based are necessarily but one of a multitude of possibilities in elucidating ancestor-descendant relations, whether based on morphology, molecular, genetic, genomic, or other data.

9.12.2   Tempo and Mode of Phylomorphogenetic Spaces

Natural selection acts on the phenotype (e.g., [9.13.112.]), and along with epigenetics, phenotypic plasticity is prevalent so we know that morphological evolution, not just genetic evolution, is important. Usually, evolution is said to be directional at large scales, but ultimately randomness may prevail, especially at small scales (e.g., [9.13.40.]), and the frequency of directionality and randomness is referenced with regard to stasis (e.g., [9.13.53.]). However, is evolution deterministically chaotic at all scales as well? Could this be a basis for phylomorphogenetic analyses that better represent the relation among ancestors and descendants in a phylomorphogenetic space? Mathematically and algorithmically, deterministic chaos and network applications may work hand-in-hand and hold some of the answers to these questions.

More work is necessary on applications involving networks in morphospaces as well as phylogenetic tree spaces. Additional work on characterizing morphospaces probabilistically could be an aid in elucidating morphospace dynamics. Using geometric and topological spaces, flexibility in combining morphological and other types of data into a common metric space is one step toward reconciling disparate outcomes for sets of important data that aid in our understanding of evolutionary history. Applying additional concepts, e.g., deterministic chaos theory, may be useful in elucidating the evolutionary history for all organisms, including diatoms.

9.13   References

9.13.1.   Albert, R. and Barabási, A.-L. (2002) Statistical mechanics of complex networks. *Rev. Mod. Phys.* 74(1), 47-97.

9.13.2.   Allman, E.S. and Rhodes, J.A. (2016) Lecture notes: the mathematics of phylogenetics. https://jarhodesuaf.github.io/PhyloBook.pdf

9.13.3.   Angelini, P., Da Lozzo, G., Di Battista, G., Di Donato, V., Kindermann, P., Rote, G., Rutter, I. (2018) Windrose Planarity: embedding graphs with direction-constrained edges. *ACM Transactions on Algorithms* 14(4(54)), 1-24.

9.13.4.   Arnold, M.L. and Fogarty, N.D. (2009) Reticulate evolution and marine organisms: the final frontier? *Int. J. Mol. Sci.* 10, 3836-3860.

9.13.5.   Avena-Koenigsberger, A., Goñi, J., Solé, R., Sporns, O. (2015) Network morphospace. *J. R. Soc. Interface* 12:20140881.

9.13.6.   Baroni, M., Semple, C., Steel, M. (2005) A framework for representing reticulate evolution. *Annals of Combinatorics* 8, 391-408.

9.13.7.   Barrett, M., Donoghue, M., Sober, E. (1991) Against consensus. *Syst. Zool.* 40, 486-493.

9.13.8.   Bentley, J.L. and Friedman, J.H. (1978) Fast algorithms for constructing minimal spanning trees in coordinate spaces. *IEEE Transactions on Computers* C-27(2), 97-105.

9.13.9.   Billera, L.J., Holmes, S.P., Vogtmann, K. (2001) Geometry of the space of phylogenetic trees. *Adv. in Appl. Math.* 27(4), 733-767.

9.13.10. Budd, G.E. (2021) Morphospace. *Current Biology* 31, R1141-R1224.

9.13.11. Budd, G.E. and Mann, R.P. (2019) The dynamics of stem and crown groups. *Science Advances* 6(8), doi: 10.1126/sciadv.aaz1626.

9.13.12. Cayley, A. (1889) A theorem on trees. *Quart. J. Pure Appl. Math.* 23, 376-378.

9.13.13. Clauset, A., Moore, C., Newman, M.E.J. (2008) Hierarchical structure and the prediction of missing links in networks. *Nature* 453, 98-101.

9.13.14. Collienne, L. and Gavryushkin, A. (2021) Computing nearest neighbour interchange distances between ranked phylogenetic trees. *Journal of Mathematical Biology* 82:8.

9.13.15. Collienne, L., Elmes, K., Fischer, M., Bryant, D., Gavryushkin, A. (2019) Geometry of ranked nearest neighbour interchange space of phylogenetic trees. https://www.biorxiv.org/content/10.1101/2019.12.19.883603v3

9.13.16. Corominas-Murtra, B., Rodriguez-Caso, C., Goñi, J., Solé, R. (2011) Measuring the hierarchy of feedforward networks. *Chaos* 016108. https://doi.org/10.1063/1.3562548.

9.13.17. Corominas-Murtra, B., Goñi, J., Solé, R., Rodriguez-Caso, C. (2013) On the origins of hierarchy in complex networks. *PNAS* 110(33), 13316-13321.

9.13.18. Daboussi, F., Leduc, S., Maréchal, A., Dubois, G., Guyot, V., Perez-Michaut, C., Amato, A., Falciatore, A., Juillerat, A., Beurdeley, M., Voytas, D.F., Cavarec, L., Duchateau, P. (2014) Genome engineering empowers the diatom *Phaeodactylum tricornutum* for biotechnology. *Nature Communications* 5, 3831.

9.13.19. DasGupta, B., He, X., Jiang, T., Li, M., Tromp, J., Zhang, L. (2000) On computing the nearest neighbor interchange distance. In: *Discrete Mathematical Problems with Medical Applications: DIMACS Workshop Discrete Mathematical Problems with Medical Applications, December 8–10, 1999, vol 55.* DIMACS Center, American Mathematical Society, p. 19.

9.13.20. deQueiroz, K. (2013) Nodes, branches, and phylogenetic definitions. *Syst. Biol.* 62(4), 625-632.

9.13.21. Devadoss, S.L. and Morava, J. (2015) Navigation in tree spaces. *Advances in Applied Mathematics* 67, 75-95.

9.13.22. Dingle, K., Ghaddar, F., Šulc, P., Louis, A.A. (2021) Phenotype bias determines how natural RNA structures occupy the morphospace of all possible shapes. *Molecular Biology and Evolution* 39(1), msab280.

9.13.23. Erdős, P. and Rényi, A. (1959) On random graphs I. Dedicated to O. Varga, at the occasion of his 50th birthday. *Publ. Math. Debrecen* 6, 290-297.

9.13.24. Erdős, P. and Rényi, A. (1960) On the evolution of random graphs. *Publ. Math. Inst. Hung. Acad. Sci.* 5, 17-61.

9.13.25. Erwin, D.H. (2007) Disparity: morphological pattern and developmental context. *Palaeontology* 50(1), 57-73.

9.13.26. Falasco, E., Bona, F., Badino, G., Hoffmann, L., Ector, L. (2009) Diatom teratological forms and environmental alterations: a review. *Hydrobiologia* 623, 1-35.

9.13.27. Felsenstein, J. (1981) Evolutionary trees from DNA sequences: a maximum likelihood approach. *J. Mol. Evol.* 17, 368-376.

9.13.28. Felsenstein, J. (2008) *Inferring Phylogenies*, Sinauer Associates, Inc., Sunderland, Massachusetts.

9.13.29. Foote, M. (1993) Contributions of individual taxa to overall morphological disparity. *Paleobiology* 19(2), 403–419.

9.13.30. Foote, M. (1997) Evolution of morphological diversity. *Annual Review of Ecology and Systematics* 28, 129–152.

9.13.31. Fráková, M., Poulíčková, A., Neustupa, J., Pichrtová, M., Marvan, P. (2009) Geometric morphometrics – a sensitive method to distinguish diatom morphospecies: a case study on the sympatric populations of Riemeria sinuata and *Gomphonema tergestinum* (Bacillariophyceae) from the River Bečva, Czech Republic. *Nova Hedwigia* 88(1-2), 81-95.

9.13.32. Francis, A., Huber, K.T., Moulton, V., Wu, T. (2018) Bounds for phylogenetic network space metrics. *Journal of Mathematical Biology* 76, 1229-1248.

9.13.33. Garba, M.K., Nye, T.M.W., Boys, R.J. (2018) Probabilistic distances between trees. *Syst. Biol.* 67(2), 320-327.

9.13.34. Garba, M.K., Nye, T.M.W., Lueg, J., Huckermann, S.F. (2021) Information geometry for phylogenetic trees. *Journal of Mathematical Biology* 82(3), 1-39.

9.13.35. Gerber, S. (2017) The geometry of morphospaces: lessions from the classic Raup shell coiling model. *Biol. Rev.* 92(2), 1142-1155.

9.13.36. Gerber, S. (2019) Use and misuse of discrete character data for morphospace and disparity analyses. *Palaeontology, Wiley* 62(2), 305-319.

9.13.37. Ghosh, R.K. and Bhattacharjee, G.P. (1984) Part I Computer Science - A parallel search algorithm for directed acyclic graphs. *BIT Numerical Mathematics* 24, 133-150.

9.13.38. Gilbert, E.N. (1956) Enumeration of labelled graphs. *Canadian Journal of Mathematics* 8, 405-411.

9.13.39. Gilbert, E.N. (1959) Random graphs. *Ann. Math. Statist.* 30(4), 1141-1144.

9.13.40. Gill, M.S., Ho, L.S.T., Baele, G., Lemey, P., Suchard, M.A. (2017) A relaxed directional random walk model for phylogenetic trait evolution. *Syst. Biol.* 66(3), 299-319.

9.13.41. Görlich, S., Pawolski, D., Zlotnikov, I., Kröger, N. (2019) Control of biosilica morphology and mechanical performance by the conserved diatom gene Silicanin-1. *Communications Biology* 2, 245.

9.13.42. Gould, S.J. (1984) Morphological channelling by structural constraint: convergence in styles of dwarfing and gigantism in *Cerion*, with a description of two new fossil species and a report on the discovery of the largest *Cerion*. *Paleobiology* 10, 172-194.

9.13.43. Guillerme, T., Cooper, N., Brusatte, S.L., Davis, K.E., Jackson, A.L., Gerber, S., Goswami, A., Healy, K., Hopkins, M.J., Jones, M.E.H., Lloyd, G.T., O'Reilly, J.E., Pate, A., Puttick, M.N., Rayfield, E.J., Saupe, E.E., Sherratt, E., Slater, G.J., Weisbecker, V., Thomas, G.H., Donoghue, P.C.J. (2020) Disparities in the analysis of morphological disparity. *Biol. Lett.* 16, 20200199.

9.13.44. Hahn, M.W. and Nakhleh, L. (2016) Irrational exuberance for resolved species trees. *Evolution* 70(1), 7-17.

9.13.45. Hammer, Ø., Harper, D.A.T., Ryan, P. D. (2001) PAST: Paleontological Statistics Software Package for Education and Data Analysis. *Palaeontologia Electronica* 4(1): 9pp.

9.13.46. Harrell, R.E. and Karlovitz, L.A. (1975) On tree structures in Banach spaces. *Pacific Journal of Mathematics* 59(1), 85-92.

9.13.47. Hillis, D.M., Heath, T.A., St. John, K. (2005) Analysis and visualization of tree space. *Syst. Biol.* 54(3), 471-482.

9.13.48. Holmes, S. (2003) Statistics for phylogenetic trees. *Theor. Popul. Biol.* 63, 17-32.

9.13.49. Hopkins, M.J. and Gerber, S. (2017) Morphological disparity. In: *Evolutionary Developmental Biology*, Nuño de la Rosa, L. and Müller, G.B. (eds.), Springer International Publishing AG, Cham, Switzerland: DOI 10.1007/978-3-319-33038-9_132-1.

9.13.50. Hohenegger, J. (2014) Species as the basic units in evolution and biodiversity: recognition of species in the Recent and geological past as exemplified by larger foraminifera. *Gondwana Research* 25, 707-728.

9.13.51. Huang, W. and Daboussi, F. (2017) Genetic and metabolic engineering in diatoms. *Phil. Trans. R. Soc. B* 372, 20160411.

9.13.52. Huber, K.T., Linz, S., Moulton, V., Wu, T. (2016) Spaces of phylogenetic networks from generalized nearest-neighbor interchange operations. *Journal of Mathematical Biology* 72, 699-725.

9.13.53. Hunt, G. (2007) The relative importance of directional change, random walks, and stasis in the evolution of fossil lineages. *PNAS* 104(47), 18404-18408.

9.13.54. Huson, D.H. and Scornavacca, C. (2011) A survey of combinatorial methods for phylogenetic networks. *Genome Biology and Evolution* 3, 23-35.

9.13.55. Janssen, R. and Liu, P. (2021) Comparing the topology of phylogenetic network generators. https://doi.org/10.48550/arXiv.2106.06727

9.13.56. Kahn, D.W. (1995) *Topology – An Introduction to the Point-Set and Algebraic Areas.* Dover Publications Inc., New York.

9.13.57. Kaiser, M. and Hilgetag, C.C. (2004) Spatial growth of real-world networks. *Phys. Rev. E* 69, 036103.

9.13.58. Kendall, D. (1984) Shape manifolds: Procrustean metrics and complex projective spaces. *Bulletin of the London Mathematical Society* 16, 81-121.

9.13.59. Kendall, M. and Colijn C. (2015) A tree metric using structure and length to capture distinct phylogenetic signals. https://arxiv.org/pdf/1507.05211.pdf

9.13.60. Kim, J. (2000) Slicing hyperdimensional oranges: the geometry of phylogenetic estimation. *Mol. Phyl. Evol.* 17, 58-75.

9.13.61. Kim, Y.-J., Lee, J., Kim, M.-S., Elber, G. (2011) Efficient convex hull computation for planar freeform curves. *Computers & Graphics* 35, 698-705.

9.13.62. Klimenko, G., Raichel, B., Van Buskirk, G. (2021) Sparse convex hull coverage. *Computational Geometry: Theory and Applications* 98, 101787.

9.13.63. Kotric, B. and Knoll, A.H. (2015) A morphospace of planktonic marine diatoms. I. Two views of disparity through time. *Paleobiology* 41(1), 45-67.

9.13.64. Lewontin, R.C. (1972) The Apportionment of Human Diversity. In: *Evolutionary Biology*, vol. 6, Dobzhansky T., Hecht M.K., and Steere W.C., (eds.), Springer, New York: 381–398.

9.13.65. Li, M., Tromp, J., Zhang, L. (1996) Some notes on the nearest neighborhood interchange distance. In: *Computing and Combinatorics, COCOON 1996, Lecture Notes in Computer Science, vol. 1090*, Springer, Berlin, Heidelberg: https://doi.org/10.1007/3-540-61332-3_168.

9.13.66. Losos, J.B. and Miles, D.B. (1994) Adaptation, Constraint, and the Comparative Method: Phylogenetic Issues and Methods. *Ecological Morphology: Integrative Organismal Biology*, 60.

9.13.67. Lowe, R.L. (1974) *Environmental requirements and pollution tolerance of freshwater diatoms.* EPA-670/4-74-005. United States Environmental Protection Agency, National Environmental Research Center, Office of Research and Development, Cincinnati, Ohio.

9.13.68. MacDonald, J.D. (1869) On the structure of the diatomaceous frustule, and its genetic cycle. *The Annals and Magazine of Natural History* 4, 1.

9.13.69. Mander, L. (2016) A combinatorial approach to angiosperm pollen morphology. *Proceedings: Biological Sciences* 283(1843), 1-10.

9.13.70. Martin, J., Blackburn, D., Wiley, E.O. (2010) Are node-based and stem-based clades equivalent? Insights from graph theory. *PLoS Currents Tree of Life* 2, RRN1196.

9.13.71. McGhee, G.R. (1999) *Theoretical Morphology: The Concept and Its Applications,* Columbia University Press, New York.

9.13.72. Meador, B. (2008) A survey of computer network topology and analysis examples. https://www.cse.wustl.edu/~jain/cse567-08/ftp/topology.pdf

9.13.73. Miller, E., Owen, M., Provan, J.S. (2015) Polyhedral computational geometry for averaging metric phylogenetic trees. *Advances in Applied Mathematics* 68, 51-91.

9.13.74. Mitteroecker, P. and Huttegger, S.M. (2009) The concept of morphospaces in evolutionary and developmental biology: mathematics and metaphors. *Biological Theory* 4(1), 54-67.

9.13.75. Monod, A., Lin, B., Yoshida, R., Kang, Q. (2021) Tropical geometry of phylogenetic tree space: a statistical perspective. https://doi.org/10.48550/arXiv.1805.12400

9.13.76. Moret, B.M.E. and Shapiro, H.D. (1994) An empirical analysis of algorithms for constructing a minimum spanning tree. *DIMACS Series in Discrete Mathematics and Theoretical Computer Science* 15, 99-117.

9.13.77. Morgan, S.E., Achard, S., Termenon, M., Bullmore, E.T., Vértes, P.E. (2018) Low-dimensional morphospace of topological motifs in human fMRI brain networks. *Network Neuroscience* 2(2), 285-302.

9.13.78. Moulton, V. and Steel, M. (2004) Peeling phylogenetic 'oranges.' *Advances in Applied Mathematics* 33, 710-727.

9.13.79. Nakhleh, L., Warnow, T., Linder, C. R., John, K. S. (2005) Reconstructing reticulate evolution in species—theory and practice. *Journal of Computational Biology* 12(6), 796-811.

9.13.80. Nemirko, A.P. and Dulá, J.H. (2021) Machine learning algorithm based on convex hull analysis. *Procedia Computer Science* 186, 381-386.

9.13.81. Nye, T.M.W. (2011) Principal components analysis in the space of phylogenetic trees. *The Annals of Statistics* 39(5), 2716-2739.

9.13.82. Pappas, J.L. (2005) Theoretical morphospace and its relation to freshwater gomphonemoid-cymbelloid diatom (Bacillariophyta) lineages. *Journal of Biological Systems* 13(4), 385-398.

9.13.83. Pappas, J.L. (2008) More on theoretical morphospace and its relation to freshwater gomphonemoid-cymbelloid diatom (Bacillariophyta) lineages. *Journal of Biological Systems* 16(1), 119-137.

9.13.84. Pappas, J.L. (2011) Graph matching a skeletonized theoretical morphospace with a cladogram for gomphonemoid-cymbelloid diatoms (Bacillariophyta) *Journal of Biological Systems* 19(1), 47-70.

9.13.85. Pappas, J.L. (2016) Multivariate complexity analysis of 3D surface form and function of centric diatoms at the Eocene-Oligocene transition. *Marine Micropaleontology* 122, 67-86.

9.13.86. Pappas, J.L. (2021) Quantified ensemble 3D surface features modeled as a window on centric diatom valve morphogenesis. In: *Diatom Morphogenesis [DIMO, Volume in the series: Diatoms: Biology & Applications, series editors: Richard Gordon & Joseph Seckbach].* V. Annenkov, J. Seckbach and R. Gordon, (eds.) Wiley-Scrivener, Beverly, MA, USA: 158-193.

9.13.87. Pappas, J.L., Kociolek, J.P., Stoermer, E.F. (2014) Quantitative morphometric methods in diatom research. In: Nina Strelnikova Festschrift. J.P. Kociolek, M. Kulivoskiy, J. Witkowski, and D.M. Harwood, D.M. (eds.), *Nova Hedwigia, Beihefte* 143, 281-306.

9.13.88. Pfitzer, E. (1869) Über den bau and zellteilung der diatomeen. *Botanische Zeitung* 27, 774-776.

9.13.89. Pfitzer, E. (1871) Untersuchungen über Bau und Entwicklung der Bacillariaceen (Diatomeen). In: *Botanische Abhandlungen aus dem Gebiet der Morphologie und Physiologie.*, J. Hanstein (ed.), 1, 1-189.

9.13.90. Pifer, A. and Bronte, D.L. (1986) Introduction: squaring the pyramid. *Daedalus* 115(1), 1-11.

9.13.91. Pie, M.R. and Weitz, J.S. (2005) A null model of morphospace occupation. *The American Naturalist* 166(1), E1-E13.

9.13.92. Pisier, G. (2016) *Martingales in Banach spaces (Cambridge Studies in Advanced Mathematics*, Cambridge University Press, Cambridge, U.K.

9.13.93. Potapova, M., Veselá, J., Smith, C., Minerovic, A., Aycock, L. (eds.) (2022) Diatom New Taxon File at the Academy of Natural Sciences (DNTF-ANS), Philadelphia. Retrieved on April 1, 2022, from http://dh.ansp.org/dntf

9.13.94. Raup, D.M. and Michelson, A. (1965) Theoretical morphology of the coiled shell. *Science* 147, 1294–1295.

9.13.95. Robinson, D.F. (1971) Comparison of labeled trees with valency three. *J. Comb. Theory Ser. B* 11(2), 105-119.

9.13.96. Schindel, D.E. (1990) Unoccupied morphospace and the coiled geometry of gastropods: architectural constraints or geometric covariation? In: *Causes of Evolution*, Ross, R.A., Allmon, W.D. (eds.), University of Chicago Press, Chicago, Illinois: 270-304.

9.13.97. Schwenk, K. and Wagner, G.P. (2001) Function and the evolution of phenotype stability: connecting pattern to process. *Amer. Zool.* 41, 552-563.

9.13.98. Seoane, L.F. and Solé, R. (2018) The morphospace of language networks. *Scientific Reports* 8:10465.

9.13.99. Serif, M., Dubois, G., Finoux, A.-L., Teste, M.-A., Jallet, D., Daboussi, F. (2018) One-step generation of multiple gene knock-outs in the diatom Phaeodactylum tricornutum by DNA-free genome editing. *Nature Communications* 9, 3924.

9.13.100. Sharov, A.A. and Igamberdiev, A.U. (2014) Inferring directions of evolution from patterns of variation: the legacy of Serfei Meyen. *BioSystems* 123, 67-73.

9.13.101. Siegel, A. (2022) A parallel algorithm for understanding design spaces and performing convex hull computations. *Journal of Computational Mathematics and Data Science* 2, 100021.

9.13.102. Spaulding *et al.* (2022) Diatoms.org: supporting taxonomists, connecting communities. *Diatom Research*, doi: 10.1080/0269249X.2021.2006790.

9.13.103. Stone, J.R. (1997) The spirit of D'Arcy Thompson swells in empirical morphospace. *Mathematical Biosciences* 142(13), 13-30.

9.13.104. Stone, J. (2002) Mapping cladograms into morphospace. *Acta Zoologica* 84, 63-68.

9.13.105. Theriot, E.C., Ashworth, M.P., Nakov, T., Ruck, E., Jansen, R.K. (2015) Dissecting signal and noise in diatom chloroplast protein encoding genes with phylogenetic information profiling. *Molecular Phylogenetics and Evolution* 89, 28-36.

9.13.106. Theriot, E.C., Ashworth, M.P., Nakov, T., Ruck, E., Jansen, R.K. (2016) Data from: Dissecting signal and noise in diatom chloroplast protein encoding genes with phylogenetic information profiling. Dryad, Dataset, https://doi.org/10.5061/dryad.610md

9.13.107. Thomas, R.D.K. and Reif, W.-E. (1993) The skeleton space: a finite set of organic designs. *Evolution* 47, 341-360.

9.13.108. Thompson, D.W. (1917, 1942) *On Growth and Form*, Cambridge University Press, Cambridge, U.K.

9.13.109. Treblico, R., Baum, J.K., Salomon, A.K., Dulvy, N.K. (2013) Ecosystem ecology: size-based constraints on the pyramids of life. *Trends in Ecology & Evolution* 28(7), 423-431.

9.13.110. Wang, J., Qin, Q., Ye, X., Gao, Z. (2014) Hierarchical feature representation of geospatial objects using morphological pyramid exploitation. *2014 IEEE Geoscience and Remote Sensing Symposium* 1789-1792. doi: 10.1109/IGARSS.2014.6946800

9.13.111. Watts, D.J. and Strogatz, S.H. (1998) Collective dynamics of 'small-world' networks. *Nature* 393, 440-442.

9.13.112. Weiss, K.M. and Fullerton, S.M. (2000) Phenogenetic drift and the evolution of genotype-phenotype relationships *Theoretical Population Biology* 57, 187-195.

9.13.113. Weisstein, E.W. (ed.) (2002) *Concise Encyclopedia of Mathematics, 2nd edition*. Chapman and Hall/CRC, New York, USA. https://doi.org/10.1201/9781420035223

9.13.114. Whidden, C. and Matsen, IV, F.A. (2019) Calculating the unrooted subtree prune-and-regraft distance. *IEEE/ACM Transactions on Computational Biology and Bioinformatics* 16(3), 898-911.

9.13.115. Yao, A.C.-C. (1981) A lower bound to finding convex hulls. *J. ACM* 28(4), 780-787.

9.13.116. Yi, Z., Su, Y., Xu, M., Bergmann, A., Ingthorsson, S., Rolfsson, O., Salehi-Ashtiani, K., Brynjolfsson, S., Fu, W. (2018) Chemical mutagenesis and fluorescence-based high-throughput screening for enhanced accumulation of carotenoids in a model marine diatom *Phaeodactylum tricornutum*. *Marine Drugs* 16, 272.

9.13.117. Zairis, S., Khiabanian, H., Blumberg, A.J., Rabadan, R. (2016) Genomic data analysis in tree spaces. https://doi.org/10.48550/arXiv.1607.07503

Part IV

MACROEVOLUTIONARY CHARACTERISTICS OF DIATOMS

10

Diatom Morphological Complexity Over Time as a Measurable Dynamical System

Abstract

Morphological complexity is an important facet of the evolution of organisms. Changes in morphological complexity may be local or global and occur ergodically or chaotically. Diatom morphological complexity was measured using Markov chains with regard to entropy rate and degree of ergodicity and chaoticity. Entropy is a probabilistic measure of complexity, and the normalized compression distance is an approximation measure of Kolmogorov complexity to assess magnitude of change in diatom morphological complexity over time. Cretaceous diatom taxa exhibited more morphological complexity than Cenozoic diatom taxa which was mostly reflected in Kolmogorov complexity assessment. All diatom taxa were determined to exhibit ergodicity with respect to morphological complexity over time, and all diatom morphological complexity and associated chaoticity and instability measured as Lyapunov exponents were determined to increase over time.

Keywords: Cretaceous, Cenozoic, ergodic, chaotic, entropy, Kolmogorov complexity, Lyapunov exponents, instability

10.1 Introduction

Complexity in diatoms is evident when observing their intricate silicate frustules in terms of morphology or functional predation resistance (e.g., [10.14.95.]) or documenting their place in evolutionary history or their life cycle (e.g., [10.14.32.]). Morphological complexity affects functional complexity, and morphology as a record of the phenotype produces the entity of evolution, as evolution acts on the phenotype (e.g., [10.14.97.]). Morphology consists of external and internal structural information as well as shape, pattern, color, and size information. The bauplan has structural characteristics shared via development and has a phylogenetic basis in the evolutionary history of diatoms.

Morphological complexity induces the question, how do we detect morphology? This is a perception and observation problem. Attneave's cat [10.14.7.] illustrates that vertices of maximum curvature, when connected by straight edges, produces the picture of what we perceive to be a sleeping cat. Vertices and edges are used to reconstruct a picture of the morphology of an animal, object or any item [10.14.34.]. Biederman's cup [10.14.10., 10.14.12.] illustrates that contour removal at mid-segment or a vertex reveals the missing parts that we perceive to be a picture of a cup. Unless we perceive curvature or contours (Figure 10.1a) rather than just a series of dots, we have a difficult time perceiving the whole entity as the morphology of an animal, object or any item. For observation, the same holds true.

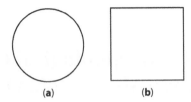

(a)   (b)

Figure 10.1 a, circle; b, square.

If vertices are present, we connect the dots using straight lines as edges to compose the object of our perception (Figure 10.1b). We have a preference for the simplest interpretation over others to infer a structured whole. This is the basis of structural information theory (Table 10.1, column 1) [10.14.65., 10.14.66.].

Curvature or contours induce the perception of a partial entity whereby our observatory powers work to produce a description in our mind of the whole. In this instance, we have a preference for a minimum description to adequately cover what was observed. This is the basis of algorithmic information theory (Table 10.1, column 2) [10.14.26.] from which algorithmic complexity, i.e., Kolmogorov complexity [10.14.26., 10.14.45., 10.14.62., 10.14.63.], emerges.

Perceiving morphology means perceiving in three dimensions (3D). To view the whole, we may observe a scene globally, then select objects to observe within the scene locally [10.14.84.]. We perceive whether a given scene makes sense the longer we observe it. Making sense of a scene involves recognition of the objects and their attributes in it and the relation that those objects have to each other. Morphological attributes include shape, size, pattern, texture, and opaqueness. Perception violations may be recognized when trying to make sense of a scene (e.g., Mezzanotte's scenes in [10.14.12.]). Complexity in a scene involves recognition of the whole or its parts and its attributes. Information gained about complexity is induced by the morphological attributes in part or as a whole.

That means that binocular disparity or object location with depth perception, motion parallax or speed of motion with depth perception, seeing 3D from two-dimensional (2D) structures in motion, shading, texture, and outline contour or edges are important predispositions toward being able to perceive [10.14.56.]. Locally, 3D shape is perceived as the projection of 2D edges or outlines that are invariant over the orientation of the shape [10.14.11.]. Stereoscopic vision and depth perception induces the use of 2D cues to reconstruct a 3D scene [10.14.12., 10.14.56.]. Slant, as the angle between the surface normal and line of sight [10.14.122.], and tilt, as the direction of slant, affect perception and 3D object orientation [10.14.21., 10.14.120.]. While the 2D shape is perceived, the 3D surface

Table 10.1 Characteristics of structural and algorithmic information theories.

Structural information theory	Algorithmic information theory
Occam's Razor	Chatton's Anti-Razor
No more than is necessary	No less than is necessary
Perception to reconstruct a whole	Observation to describe a whole
Least upper bound	Greatest lower bound

has qualities that are perceived as well including smoothness, sharpness and curvedness [10.14.11.]. In terms of symmetry, the 3D surface has compactedness as the degree to which the 2D shape differs from a circle or a 3D shape differs from a sphere [10.14.113.].

Circles and spheres reveal the basis of the geometry of morphology. Symmetry of shape and surface comes into play as well. Symmetry is 2D shape in motion through 3D space and is a special case of shape or surface [10.14.113.]. We perceive recurrent shapes and symmetries, and the interrelation of shape and symmetry go hand-in-hand. Shape and symmetry contribute to perceptions on multiple scales as to whether there are recurrences or not, and recurrences may influence perception of morphological complexity.

Basic geometric shapes may be perceived to be simple or complex when perceived at the same time (Figure 10.2). Viewing such shapes as whole entities provides a test of deciding what degree of complexity is being perceived. A four-sided figure may be simple or complex, depending on the degree of edges or contours present (Figure 10.2a-d). Multi-sided figures may appear to be complex, depending on how concavity contributes to complexity (Figure 10.2e-g). Curvature may induce the perception of complexity, depending on degree of convexity (Figure 10.2h-i). The fact that multiple shapes and symmetries can occur at the same time over multiple scales may contribute to morphological complexity. In turn, shape as an attribute of bauplan and development may be indicative of the adaptive value of a given type of symmetry and complexity present [10.14.121.].

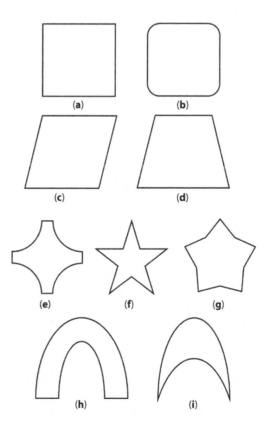

Figure 10.2 Different basic shapes: a, square; b, rounded square; c, rhombus; d, trapezoid; e, cruciform; f, star, deeply cut; g, star, shallowly cut; h, arch; i, moon.

Various procedures may induce changes in complexity. For example, folding may increase complexity via minimal information redundancy or noise (e.g., [10.14.22.]). Folding affects scale, and multiple scales of morphological complexity may induce further complexity via self-similarity (e.g., [10.14.4.]). Folding is an attribute of complexity that may induce a change from a 2D perception to a 3D perception. Stretching (e.g., [10.14.78.]) and squeezing (e.g., [10.14.60.]) are other procedures that may influence complexity. Resultant stability or instability may occur at any time, and instability may be a result of folding or stretching in which changes occur as an exponential rate of divergence (e.g., [10.14.29.]). Squeezing may induce complexity retraction which is also an indication of instability (e.g., [10.14.91.]). Other characteristics of complexity changes include being continuous or discrete as well as random, chaotic or directional at any time, and the magnitude of such changes may be enormous or infinitesimal at any time which may produce an ensemble long term average that is relatively stable or a high rate of exponential divergence inducing instability.

10.1.1 Complexity and Evolution

Evolution may be directed or random, i.e., adaptive or neutral. Evolution as adaptation means that the process of accumulating descendants from ancestors is historically-based (e.g., [10.14.134.]). Neutral evolution means that stochasticity is present, and historicity is not considered (e.g., [10.14.134.]). In any case, complexity occurs.

A phylogenetic viewpoint would characterize complexity as hierarchical with respect to the accumulation of descendants from ancestors over time (e.g., [10.14.134.]). More broadly, from a mutagenetic perspective, algorithmic information theory may be used with respect to genomics and genetic codes in which complexity is a counting process (e.g., [10.14.1., 10.14.2.]) where coevolutionary considerations exist [10.14.58.]. Algorithmic information theory via Kolmogorov complexity may be used as a means of quantifying complexity from an evolutionary viewpoint (e.g., [10.14.106.]). In the same vein, morphological complexity may be quantified via Kolmogorov complexity with respect to morphological evolution.

10.2 Diatom Morphological Complexity

Diatom morphological complexity has been shown to be analyzable from solutions to partial differential equations characterizing 3D surface features of theoretical diatom models [10.14.95.]. Such models have been used to summarize surface morphology (e.g., [10.14.96.]). Diatom morphology consists of ensemble surface features of valves [10.14.96.], and these ensembles are combinations of surface features that change as morphological complexity changes, either increasing, decreasing or staying the same over time [10.14.95.].

Diatom valve surfaces are a record of morphological complexity [10.14.95.]. The history of morphological change may be obtained via measuring image data from which information is extracted and analyzed as long-term evolutionary changes. Using measurement of those surfaces, multiple renditions of morphological change of multiple taxa over time may be obtained.

Valve surface features as morphology may be quantified using digital images (e.g., [10.14.98.]). Compiling such data entails obtaining the information contained within a digital image. Finding a probabilistic basis for comparing diatom valve face images and analyzing the properties of such a compilation may indicate the properties of diatom morphological complexity among n-taxa. Changes in diatom morphological complexity over a long-term time scale would elucidate this aspect of morphological evolution.

Diatom valve face images may be subject to entropy measurement of morphological information [10.14.98.]. Entropy values used to analyze the morphological information relation among n-taxa induces the construction of a Markov chain, providing descriptors of the evolution of morphological change with respect to complexity. Markov chains may be used in time series analysis to provide a probabilistic and predictive picture of the change (e.g., [10.14.132.]), and Markov chains have been used to analyze diatom morphological complexity change over time [10.14.95.]. The time scale used may consist of morphologies from the fossil record as well as present morphologies so that complexity over the long-term will result.

Diatom morphological complexity may be ergodic [10.14.95.]. Diatom morphological complexity may be ergodic and chaotic at the same time. Morphological complexity may be viewed as a dynamical system, and over the long term, different characteristics of complexity changes occur with respect to changes in initial conditions that may not affect the trend (ergodicity), or if they do, changes in stability may affect the trend (chaos). There are local and global components of morphological complexity change over time, and, for example, either ergodic or chaotic characteristics of this process may be present locally at any given time.

10.3   Purposes of this Study

Diatom morphological complexity over time will be treated as a probabilistically-based dynamical system. Input data will consist of entropy values as morphological information. According to image processing techniques (e.g., [10.14.61., 10.14.108.]), the entropy of each diatom image is measured as Shannon entropy [10.14.114.]. A Markov chain will be constructed to model the change in entropy as morphological complexity from one diatom taxon to another over time [10.14.95.], and the time scale will be from the Jurassic though the Cenozoic. Analyses are used to determine whether the long-term trend of diatom morphological complexity depends on the initial state over a given time span. Considering complexity change to be one of instability versus stability, Lyapunov exponents [10.14.74.] will be calculated. If a positive Lyapunov exponent is obtained via entropy values, this means that diatom morphological complexity is unstable, and therefore chaotic. For ergodicity to be present, irreducibility of the Markov chain must be established. Diatom morphological complexity changes for all taxa will be assessed. Taxa from the Cretaceous and Cenozoic will be compared, analyzed and assessed to determine if morphological complexity has changed as well as whether morphological complexity is ergodic or chaotic or both.

10.4   Characterizing Morphological Complexity

Morphological complexity changes over the course of the lifetime of an organism as developmental and evolutionary process occur. Diatom life cycles encompass multiple stages, ranging from size reduction during vegetative reproduction (e.g., [10.14.110.]) to size restoration via sexual reproduction, auxosporulation and initial cell production after reaching the MacDonald-Pfitzer limit [10.14.75., 10.14.101., 10.14.102.]. Resting stages may occur depending on environmental conditions (e.g., [10.14.110.]). On a longer time scale, diatom morphological variation attests to the fact that the variation is "in motion" to the degree that all possible morphological variants are not predictable; rather, such a quantity is combinatorial. Despite this, the redundancy in morphological variants is also indicative of an approximate accounting of morphologies that approach some sort of limit.

Diatom morphological complexity over time may be characterized as a dynamical system. For a single organism, morphological complexity may be linear or non-linear, in equilibrium or non-equilibrium, or stable or unstable at any moment during an organism's lifetime so that multiple states or mixed states of complexity may exist over time. This system's operation may be tracked according to each state. Given its life cycle and multiple morphological states, diatom morphological complexity may be characterized as, e.g., a linear system that is stable or unstable at any moment of equilibrium over a given time span when analyzing each morphological state. Measuring between and at particular morphological states may encompass deterministic, stochastic or chaotic properties.

Morphological complexity between diatom species, genera or other classification or lineage identifiers could be characterized in a similar fashion. The changes in complexity states and complexity trajectories of those changes are geometrically measurable in countable state space (e.g., [10.14.23.]) which is the set of states in phase space on an n-dimensional level. Such a dynamical system measurability may be differentiable, topological or ergodic [10.14.112.], sometimes depending on the time scale used to potentially reach equilibrium or stability states. Short time frames may be measurable via differentiability in which deterministic, stochastic or chaotic properties are used. Topologically, relaxation of geometric measurability occurs and deterministic, stochastic or chaotic properties may be used in measurement over varying time scales. For those cases where equilibrium or stability occurs over the long term, ergodic measurement in association with deterministic, stochastic or chaotic properties may be used. Differentiability, topology or ergodicity may be used in combination, depending on the interplay of morphological complexity measures.

Ergodicity induces the characterization of randomness with respect to determinism and is measurable in a dynamical system [10.14.41.]. The degree of randomness, in turn, induces the characterization of the system based on information or uncertainty. Morphological complexity may be characterizable via ergodic information.

10.5 Information and Morphology

Information theory (e.g., [10.14.16., 10.14.114.]) is based on entropy. Entropy is a measure of the information or uncertainty with regard to an object. As a quantity, entropy is proportionally a logarithmic-based function of all probable states (e.g., [10.14.16., 10.14.114.]). The generalized forms of Shannon entropy are Rényi entropy [10.14.107.] and Tsallis entropy [10.14.127.]. For a dynamical system, Shannon entropy, H_S, of the probability distribution is

$$H_S = -\sum_{k=1}^{n} prob_k \ln\left(\frac{1}{prob_k}\right) \qquad (10.1)$$

from the kth to n-probabilities [10.14.35.]. For a bounded object, entropy may be partitioned so that probabilistically the dynamical system is expressed as

$$H(W_k) = \sum_{k=1}^{n} \mu(W_k) \ln[\mu(W_k)^{-1}] \tag{10.2}$$

where the kth to nth partitions, W, are expressed in the framework of Shannon entropy (e.g., [10.14.112.]). The sum of the partitions is the iteration of the intersection of n-partitions, W, so that

$$h(\mu, \{W_k\}) = \lim_{n \to \infty} \frac{1}{n} H(\{W_k^{(n)}\}) \tag{10.3}$$

and maximizing over μ-countable measurable n-partitions produces

$$h(\mu) = \sup_{W_k} h(\mu, W_k) \tag{10.4}$$

where sup is the supremum (e.g., [10.14.35.]). From this, H and h via maximization may be expressed as the Kolmogorov-Sinai (KS) entropy (e.g., [10.14.117.]) by

$$h_\mu(W, \lambda) := \lim_{n \to \infty} \frac{1}{n} H_\mu\left(\bigvee_{k=0}^{n-1} W^{-k} \lambda \right) \tag{10.5}$$

with λ as positive values for n-states as the average information so that KS entropy is a measure of maximum rate of information at the nth state (e.g., [10.14.112.]). The values of λ will be shown to be a measure of chaotic behavior that is related to KS entropy (e.g., [10.14.39.]).

The bounded object to be used is a diatom valve face, and the partitions represent the changes in information on the surface. In the case of reading digital images of diatom valve faces, entropy is the measure of all probable states [10.14.98.] of an ensemble of surface features and is a measure of surface morphology [10.14.96.]. Morphological entropy over the long term means a change in information that is countable and measurable (e.g., [10.14.15.-10.14.18., 10.14.81., 10.14.114.]). Maximum entropy may be used to measure diatom morphology for comparison among multiple taxa.

10.6 Information and Complexity

Morphology is characterizable in terms of stability of surface features and the amount of information associated with changes in those features over time. Morphological information is evident as the phenotype of organisms. The entropy of morphology over time is one way of reading diatom morphological evolution. Over long periods of time, biological evolution proceeds in which non-equilibrium may occur, then potentially culminating in equilibrium exemplified via a particular morphological state at a particular time (e.g., [10.14.109.]). A state of non-equilibrium means that morphological evolution is at a maximum entropy value more often than not, potentially inducing morphological variability (e.g., [10.14.33.]). The morphological information contained or missing from a biological entity such as a diatom valve face would be measurable as an entropy value that could be related to and compared with other diatom valve faces at particular times throughout the morphological evolution of diatoms.

Regularity is a feature of phenotypes, and therefore morphology (e.g., [10.14.119.]). Recurrent structural features of morphology may occur over time. Time scales to consider for diatoms may be as long as the fossil record, or between periods in which notable environmental, climate (e.g., [10.14.95.]) or extinction events have occurred. Different time scales may produce different pictures of morphology so that probabilistically, stability or instability may be present at any given time. Regularity is a natural characteristic that induces stability of phenotypes (e.g., [10.14.19.]).

From regularity to complexity, entropy may be used to discern diatom morphological complexity over time. Entropy measures are inherently probability measures so that entropy may be used as an indicator of the probability of change in uncertainty in morphological complexity. Mapping a time scale to morphological changes induces a long-term view of complexity. First appearance of diatom taxa on a generic to higher classification level could be used to create a record of morphological complexity over time. This record would lend itself to time scale characterizations with respect to the stepwise changes, direction of changes, and assessing the overall evolution of diatom morphological complexity.

10.7   Markov Chains and their Properties

Maximum entropy is the sum of partitioned surface features (e.g., [10.14.55.]), and the sum of partitioned surface features may be applied to each diatom valve face. Locally, the change in surface features over time, or globally, as the comparison from one taxon to another one may be accounted for as a stretching, folding or squeezing where different states of a taxon or between taxa occur as morphological change. Different states may produce stability or instability, and these states are measurable using the information contained on the valve face surface. Changes on the surface may undergo iterations of exponential divergences that may become fixed, increase or decrease. Such divergences with regard to stability record the degree of chaoticity in the surface. States of equilibrium may be obtained to determine associated stability or instability states.

Entropy values for each diatom taxon valve face form a partially ordered set as a vector, and obtaining the product of this vector and its transpose produces a matrix. When the matrix is transformed to be a stochastic positive square matrix, a transition probability matrix (or transition matrix) results where each row is a conditional stationary distribution (e.g., [10.14.23.]). The transition matrix elements are used to construct a discrete finite Markov chain of a stationary dynamical process of morphological change (e.g., [10.14.95.]) using diatom valve face entropy values as input data. The Markov chain embodies a stochastic process whereby the probabilistic morphological change depends only on the present state rather than past states [10.14.95.].

For a finite discrete Markov chain, the transition matrix is

$$Prob_{Transition}(x,y) = Prob(X_{t+1} = y | X_t = x) \tag{10.6}$$

[10.14.130.]. For a probability distribution, $(prob_{i,j} | i, j \in state\ space)$, of X_t on X_{t+1} for t-time states in which all probability distributions are conditionally independent of one another, a given conditional probability distribution for n-state spaces that are time homogeneous has

$$prob_{ij}(i,j) = Prob(X_{n+1} = j | X_n = i) \tag{10.7}$$

for the ith row and jth column of $Prob^n$ [10.14.23.]. Every row of the transition matrix is $prob_i = Prob(X_n = i)$, and

$$\sum_{i \in state\ space} prob_i = 1 \tag{10.8}$$

[10.14.23., 10.14.52., 10.14.95.].

While a Markov chain has an initial state, X_0, an ergodic Markov chain is independent of this initial state. For a Markov chain to be ergodic for all $x, y \in state\ space$, time $t = t(x, y)$ such that $Prob^t(x, y) > 0$, and $gcd\{t: Prob^t(x, y) > 0\} = 1$ [10.14.130.], where gcd is the greatest common divisor. That is, an ergodic Markov chain has the properties of irreducibility and aperiodicity (e.g., [10.14.23., 10.14.130.]).

An ergodic Markov chain undergoing n-state changes via t-time steps, will ultimately reach a unique stationary distribution, π, given as

$$\pi(y) = \lim_{t \to \infty} Prob^t(x, y) \tag{10.9}$$

that is the limiting distribution, and

$$\pi_j = \lim_{n \to \infty} Prob(X_n = j | X_0 = i) \tag{10.10}$$

for all $i, j \in state\ space$ so that

$$\sum_{j \in state\ space} prob_j = 1 \tag{10.11}$$

and

$$\pi_j = \lim_{n \to \infty} Prob(X_n = j) \tag{10.12}$$

for all $j \in state\ space$ (e.g., [10.14.95.]). For X_n, $\pi = \lim_{n \to \infty} \pi^n = \lim_{n \to \infty} \pi^0 Prob^n$, and for X_{n+1}, $\pi = \lim_{n \to \infty} \pi^{n+1} = \lim_{n \to \infty} [\pi^0 Prob^n] Prob$. If X_n has limiting probability distribution π, then X_{n+1} has limiting probability distribution $\pi Prob$, and

$$\pi_j = \sum_{k \in \text{ state space}} \pi_k Prob_{kj} \tag{10.13}$$

for all $j \in$ *state space* [10.14.95.]. That is, the probability of being in state j at time t is the same probability as being in state j at time $t + 1$.

For a finite discrete ergodic Markov chain, MC,

$$Prob(MC_0^n = mc_0^n) = \pi(mc_0) \prod_{k=0}^{n-1} Prob_{Transition}(mc_k, mc_{k+1}) \tag{10.14}$$

where MC contains mc-state spaces of x and y per row of the transition matrix, $Prob_{Transition}$, with $Prob(MC_0^n = mc_0^n) = Prob(X_0^n, Y_0^n)$ (e.g., [10.14.52., 10.14.53.]). From the logarithmic transformation of the transition matrix,

$$H(X,Y) = -\lim_{n \to \infty} \frac{1}{n} E \log Prob(X_0^n, Y_0^n) = -\lambda(X,Y) \tag{10.15}$$

for entropy, H, where E is the expectation concerning $Prob\ X$ and $Prob\ Y$ [10.14.52., 10.14.67.]. For

$$\lambda(X,Y) = \lim_{n \to \infty} \frac{1}{n} E \log \left[\pi(mc_0) \prod_{k=0}^{n-1} Prob_{Transition}(mc_k, mc_{k+1}) \right] e \tag{10.16}$$

λ is a maximum value that is the Lyapunov exponent, and e is a basis vector with elements equal to one [10.14.52., 10.14.53.]. Calculating the Lyapunov exponent is the same as calculating the expectation of the stationary distribution of the Markov chain [10.14.52., 10.14.67.]. That is, average conditional entropies are $H(X, Y) = -\lambda(X, Y)$ with $H(X) = -\lambda(X)$ and $H(Y) = -\lambda(Y)$ [10.14.52., 10.14.53.].

10.7.1   Markov Chains and Lyapunov Exponents

To obtain an equilibrium state via a finite discrete ergodic Markov chain, the unique stationary distribution and each vector of conditional probabilities from the transition matrix are iterated such that convergence results [10.14.95.]. For large n, the decomposition of the transition matrix in this way produces an eigensystem of eigenvectors, v_k and eigenvalues, λ_k. If the left eigenvector is associated with an eigenvalue equal to one, then the ergodic Markov chain is irreducible (e.g., [10.14.23., 10.14.95.]). The right eigenvectors are associated with the number of transitions needed to reach equilibrium [10.14.95.]. For more details, see Pappas [10.14.95.].

Via the multiplicative ergodic theorem (e.g., [10.14.35., 10.14.94., 10.14.99., 10.14.111.]), the eigenvalues may be expressed with respect to time, t, as

$$\lambda_k = \lim_{t \to \infty} \frac{1}{t} \ln \left[\frac{p_k(t)}{p_k(0)} \right] \tag{10.17}$$

where λ_k is the kth eigenvalue [10.14.129.]. In information theoretic terms, the eigenvalues are

$$\lambda_k \equiv \bar{h}_k = \frac{1}{2n} \ln(H_k) \tag{10.18}$$

for a given time period, and these eigenvalues are Lyapunov exponents (e.g., [10.14.35., 10.14.47., 10.14.112.]) and are equivalent to the average information, $\bar{h}_k$ (e.g., [10.14.43., 10.14.112.]). Lyapunov exponents are also expressible as

$$\lambda_{f(t)} = \lim_{t \to +\infty} \sup_{t \in T} \frac{\ln \| f(t) \|}{t} \tag{10.19}$$

in terms of a time function, $f(t)$, and the norm, $\|\cdot\|$ [10.14.3.]. One point in a chaotic dynamical system may be comparable to another one via n-trajectories as states, and such transitions may be determined to be recurrences [10.14.3.].

The largest Lyapunov exponent signifies the largest component of the dynamical system from the Markov chain (e.g., [10.14.67.]). From the multiplicative ergodic theorem (e.g., [10.14.94.]), the largest Lyapunov exponent will be nearly the same for all states or be dominant so that a single Lyapunov exponent may be used to represent a dynamical system (e.g., [10.14.35.]). Lyapunov exponents form a spectrum and are ordered as $\{\lambda_1, \lambda_2, \cdots, \lambda_n\} \equiv \lambda_{largest} \geq \cdots \geq \lambda_{smallest}$ and represent the degree of state changes (e.g., [10.14.5., 10.14.57.]. The Lyapunov spectrum describes mixing, whether weak or strong (e.g., [10.14.112.]), and degree of mixing may be indicative of degree of stability [10.14.112.] as an entropy rate (e.g., [10.14.133.]). Mixing has both stationary and ergodic properties [10.14.76.], and mixed states are accounted for via the transition probabilities with regard to a Markov chain [10.14.57.].

Degree of chaoticity is expressible via Lyapunov exponents (e.g., [10.14.35.]) in which positive Lyapunov exponents connote instability, negative values connote stability, and a value of zero indicates a combination of stability and instability in dynamical systems (e.g., [10.14.64.]). Positive Lyapunov exponents are indicators of sensitivity to initial conditions (e.g., [10.14.67.]).

10.8   Ergodicity and Chaoticity

Ergodic properties are observable with regard to long term dependent random occurrences that produce non-random regularity (e.g., [10.14.13., 10.14.14., 10.14.16., 10.14.86., 10.14.87., 10.14.103., 10.14.104.]). Ergodicity means that presumably over the long term, every point of complexity is reached over finite time and space (e.g., [10.14.112.]). The implication is that reversals may occur—i.e., complexity may increase or decrease over time—and that eventually, complexity may return to the original state obtained at the start of analysis. That is, ergodicity may be present as forward or backward (e.g., [10.14.128.]). An ergodic state may be observable as a complexity state. Initially, the state may be complex or simple, but eventually over the long term, the complexity state at any time, t, will be independent of the initial state.

For invariant ergodic measures, μ, there are many functions, $f: X \to \mathbb{R}$, and many complexity states, $x_{complexity}(q,r) = x(q_{complexity}, r_{complexity})$, so that the $\lim_{T \to \infty} \dfrac{1}{T} \displaystyle\int_0^T f(T_t(x_{complexity}))dt = \dfrac{1}{\mu(X)} \displaystyle\int f \, d\mu$, for some time, T, results in the ensemble time average being equal to the space average [10.14.13., 10.14.14., 10.14.112.]. That is, over a long time period, changes in the complexity state are concomitant to those changes in time states. In this way, an ergodic state may be viewed to be equivalent to a complexity state.

Conditional complexities may be measured via minimum entropies from one taxon to another one. The cost function is $Cost(y|x) = K(x) - K(y)$ with the transformation of conditional complexities as $Cost(y|x) \pm K(x|y^*) - K(y|x^*)$ where $x^* := \langle x, K(x) \rangle$, and $Cost'(y|x) = K(x\,y) - K(y\,x)$ is the difference between direct conditional complexities [10.14.9.]. Ergodic conditional entropies induce stationary distributions via an ergodic Markov chain, and each stationary distribution has an entropy rate [10.14.76.]. Maximum entropies are measurable via Eqs. (10.3-10.5) as entropy rates, and these rates are functionally Lyapunov functions (e.g., [10.14.88.]) with the maximum Lyapunov exponent as a measure of maximum complexity.

Complexity in terms of stability or instability is measurable via Lyapunov exponents, and degree of stability is an indication of degree of deterministic chaos (e.g., [10.14.98.]). The degree of chaos present in an ergodic system is measurable [10.14.41.]. The presence of chaos is stronger than the presence of ergodicity (e.g., [10.14.92., 10.14.93.]), because the time span dictates whether initial conditions are influencing a given complexity state. Both chaos and ergodicity are characteristics of complexity (e.g., [10.14.57.]).

At the outset, the simplest complexity state change is a Bernoulli shift (i.e., Bernoulli scheme) in which one of only two options are probable over time (e.g., [10.14.36., 10.14.112.]). Mixing, whether weak or strong, is related to ergodicity and chaoticity [10.14.40.] with regard to degree of stability (e.g., [10.14.112.]). From ergodicity to a Bernoulli shift, complexity states may be characterized throughout any time span (Figure 10.3).

10.8.1   Entropy Rates

The entropy rate is approximately the compression rate [10.14.59.]. The compression distance may be use to approximate an entropy rate that is related to a maximum Lyapunov exponent as maximum entropy given as the entropy rate probabilistically as

$$h_\mu = - \sum_{x \in State\ X} \pi_x \sum_{y \in State\ Y} Prob(y|x) \ln Prob(y|x) \tag{10.20}$$

where π_x is a stationary distribution [10.14.57.].

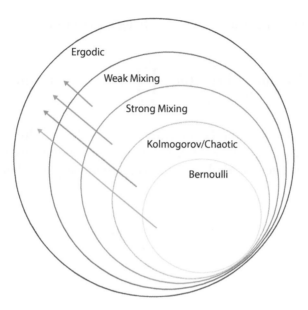

Figure 10.3 Dynamical systems: from Bernoulli to Kolmogorov to strong mixing to weak mixing to ergodicity. Arrows indicate the direction of nesting; the opposite is not true.

Joint entropy may be used to represent complexity changes. The limit of the joint entropy between n-taxa complexity probabilities, X, from a Markov chain [10.14.76.] is given by

$$H(X_{Taxon}) = \lim_{n \to \infty} \frac{1}{n} H(X_1, X_2, \dots, X_n) \tag{10.21}$$

The change in entropy from one taxon to another one is the entropy rate as

$$H(Markov\ chain) = -\sum_{ij}^{n} \mu_i\ prob_{ij} \ln prob_{ij} \tag{10.22}$$

where μ_i is the ith stationary distribution of the Markov chain (e.g., [10.14.35.]), so that

$$h_{Markov\ chain} = -\sum_{x \in State\ X} \pi_x \sum_{y \in State\ Y} Prob(y|x) \ln Prob(y|x) \tag{10.23}$$

is the entropy rate from the transition probabilities and the stationary distribution, π_x [10.14.57.].

The entropy rate is equivalent to the maximum Lyapunov exponent as

$$\lambda_{max} = \lim_{t \to \infty} \lim_{\delta W_0 \to 0} \frac{1}{t} \ln \frac{|\delta W|}{|\delta W_0|} \tag{10.24}$$

[10.14.47.] over time, t for partitions, W. Modifying Eq. (10.16),

$$\lambda(Y|X) = \lim_{n \to \infty} \frac{1}{n} E \log \left[\pi(mc_0) \prod_{k=0}^{n-1} Prob_{Transition}(mc_k, mc_{k+1}) \right] e \qquad (10.25)$$

where the entropy, $H(X, Y)$, from Eq. (10.15) is replaceable with the conditional entropy, $H(Y|X)$ [10.14.52., 10.14.53.]. The weighted averages of n-conditional entropies for n-states comprise the entropy rate of a Markov chain with a stationary distribution [10.14.76.]. In terms of entropy

$$h(\mu) = \lim_{n \to \infty} \frac{1}{n} H(\mu_n) \qquad (10.26)$$

and

$$h_\mu \leq \sum_{\lambda_k > 0} \lambda_k \qquad (10.27)$$

(e.g., [10.14.35.]). In a dynamical system, the entropy rate may be expressed via KS entropy and is given as

$$h_{KS} = \sum_{\lambda_k > 0} \lambda_k \qquad (10.28)$$

[10.14.100.] and is the sum of the positive Lyapunov exponents (e.g., [10.14.47.]). That is, KS entropy is an equivalent measure of chaos via Lyapunov exponents. The sum of the Lyapunov exponents is a measure of the degree of chaoticity among n-states and is an entropy rate.

10.9 Kolmogorov Complexity and Entropy

Kolmogorov complexity is a measure of the complete amount of countable information in an object [10.14.9., 10.14.124.]. Shannon entropy is equivalent to the expected value of Kolmogorov complexity when a recursive probability distribution is established for entropy [10.14.46., 10.14.69., 10.14.124.]. That is, the limit of Shannon entropy is Kolmogorov complexity for the sum of probabilities equal to one (e.g., [10.14.124.]). Kolmogorov complexity is the shortest length that symmetrically reconstructs information between two individual objects [10.14.30., 10.14.46., 10.14.124.].

For conditional entropy between two objects, the limit of the normalized information distance (NID) is

$$\lim_{n \to \infty} NID(x, y) = \frac{\max\{H(X|Y), H(Y|X)\}}{\max\{H(X), H(Y)\}} \tag{10.29}$$

[10.14.59.].

From an ergodic finite-order discrete stationary Markov chain, entropy is relatable to Kolmogorov complexity (e.g., [10.14.9., 10.14.59.]) via an information distance metric. NID expressed in complexity terms is

$$NID(x, y) = \frac{\max\left\{\dfrac{1}{n}K(x|y), \dfrac{1}{n}K(y|x)\right\}}{\max\left\{\dfrac{1}{n}K(x), \dfrac{1}{n}K(y)\right\}} = \frac{\max\{K(x|y^*), K(y|x^*)\}}{\max\{K(x), K(y)\}} \tag{10.30}$$

where $K(x, y)$ is the Kolmogorov complexity, and $K(x|y)$ is the conditional Kolmogorov complexity [10.14.59., 10.14.70.] for objects x and y, representing individual diatom taxa. NID is based on the L^2 formulation of ergodicity that is the distance norm (e.g., [10.14.6., 10.14.48., 10.14.70.] in the form of a vector [10.14.131.]. From an ergodic entropic measure of morphological complexity, this distance metric is a measure of complexity proximity of n-taxa. Morphological complexity represented as Kolmogorov complexity is also algorithmic entropy as an algorithmic information distance metric [10.14.9.].

Kolmogorov complexity as a normalized information distance has been called a universal similarity distance (e.g., [10.14.30., 10.14.70.]). However, the normalized information distance has been termed upper to lower semi-computable (e.g., [10.14.9., 10.14.70.]) at best to non-computable (e.g., [10.14.30., 10.14.59., 10.14.70., 10.14.79., 10.14.80.]). Alternatively, a normalized compression distance (NCD) approximates the normalized information distance, NID, and is given as

$$NCD(x, y) = \frac{C(x\ y) - \min\{C(x), C(y)\}}{\max\{C(x), C(y)\}} \tag{10.31}$$

with C indicating the compression size [10.14.30., 10.14.37., 10.14.70.]. The NCD is a measure that can be used to determine magnitude of complexity as changing ($NCD \approx 1$) or staying the same ($NCD = 0$) among n-states. The NCD is calculated to be a value on the interval [0, 1], and for $C(y) \geq C(x)$,

$$NCD(x, y) = \frac{C(x\ y) - C(x)}{C(y)} \tag{10.32}$$

[10.14.30.].

Although the NCD also has been identified to be a universal similarity distance metric [10.14.30.], the compressor used must be chosen carefully so that the resultant NCD is idempotent, meaning that the results are reproducibly unchanged [10.14.25.]. Many compression algorithms exist that can be used (e.g., [10.14.25.]).

The compressed catenation of the ratio of the x-state with the y-state, the compressed x-state, and the compressed y-state is the compressed distance between n-states where the dominant shared feature induces measurement of similarity among all states that comprises the NCD. That is, the NCD enables the capture of dominant shared features for every pair of n-states that are compared [10.14.30.]. For diatom taxa, the encoded entropy values are a measure of the compressibility of the information contained on their valve surface faces, and these values provide input in calculating NCD.

To illustrate complexity pictorially, the NCD matrix is used to produce a spectral clustering tree (e.g., [10.14.85., 10.14.90., 10.14.135.]). For the spectral clustering tree, dimension reduction is achieved with a Laplacian matrix via the eigenvectors and eigenvalues of the NCD matrix and is used to depict Kolmogorov complexities as relative similarities in a graph structure [10.14.73.]. Complexity comparisons are compiled using the resultant tree.

10.10   Methods

All diatom images and entropy measurements are obtained from the data used for Pappas *et al.* [10.14.98.]. The slant and tilt-corrected entropy value per taxon image is used (hereafter referred to as just "entropy") and is given in Table 10.2. Thirty-eight diatom taxa are used to represent the time line from the Cretaceous to the present, and the entropy values are ordered to reflect this timeline (Table 10.2). For more details on the taxa, images of taxa, and slant and tilt-correction method of images, see Pappas *et al.* [10.14.98.].

The timeline was obtained according to diatom taxonomic, stratigraphic, and other resources and is listed for each taxon in Table 10.2. Each diatom taxon was assigned to a time label, and the entropy values were ordered from oldest taxon from the Jurassic to most recent taxon from the Pliocene. Time labels reflect first appearance as documented in the literature. If more than one time is given, the oldest is used.

To obtain the relation among diatom taxon morphological complexities, an ergodic Markov chain is used. Unlike a previous study that used sums of average coordinate values obtained from solutions to partial derivatives of geometric surface models [10.14.95.], entropy values represent the entire valve face of each diatom image [10.14.98.]. The information on the diatom valve face is a record of the amount of uncertainty associated with that taxon's morphological complexity.

10.10.1   Transition Probability Matrix and Properties of a Markov Chain

A Markov chain is obtained using entropy values. Those values are ordered according to their first appearance in the geologic record from Jurassic to Pliocene. Then, the entropy values are preconditioned via standardization to have a mean of zero and a variance of one. From this, an outer product matrix is formed. The resultant matrix consists of diatom taxa in relation to each other and is a square matrix.

Table 10.2 Diatom taxa with respect to the fossil record* and entropy values.

Taxon	Entropy	Era	Epoch	References
Amphitetras antediluviana	3.88089	Cenozoic	Oligocene	[10.14.83.]*
Asterolampra grevillei	3.42124	Cenozoic	Eocene	[10.14.44., 10.14.125.]
Asterolampra marylandica	3.77453	Cenozoic	Eocene	[10.14.44., 10.14.125.]
Asteromphalus heptactis	3.62449	Cenozoic	Eocene	[10.14.44., 10.14.125.]
Asteromphalus imbricatus	3.49843	Cenozoic	Eocene	[10.14.44., 10.14.125.]
Asteromphalus shadboltianus	3.59599	Cenozoic	Eocene	[10.14.44., 10.14.125.]
Asteromphalus vanheurckii	3.60434	Cenozoic	Eocene	[10.14.44., 10.14.125.]
Cyclotella meneghiniana	3.36278	Cenozoic	Miocene	[10.14.110., 10.14.116.]
Spatangidium arachne	3.78896	Cenozoic	Pliocene	[10.14.38.]
Triceratium bicorne	3.79719	Cenozoic	Eocene	[10.14.110.]
Triceratium castellatum var. *fractum*	3.21872	Cenozoic	Eocene	[10.14.110.]
Triceratium crenulatum	2.36455	Cenozoic	Eocene	[10.14.110.]
Triceratium favus	2.84307	Cenozoic	Eocene	[10.14.110.]
Triceratium favus var. *maxima*	2.71509	Cenozoic	Eocene	[10.14.110.]
Triceratium favus var. *quadrata*	3.51616	Cenozoic	Eocene	[10.14.110.]
Triceratium impressum	3.88434	Cenozoic	Eocene	[10.14.110.]
Triceratium pentacrinus fo. *quadrata*	3.72479	Cenozoic	Eocene	[10.14.110.]
Triceratium sp. [fossil]	3.11817	Cenozoic	Paleocene, Eocene	[10.14.110., 10.14.116.]

(*Continued*)

Table 10.2 Diatom taxa with respect to the fossil record* and entropy values. (*Continued*)

Taxon	Entropy	Era	Epoch	References
Trigonium alternans	3.1035	Cenozoic	Eocene	[10.14.110.]
Trigonium arcticum	3.30671	Cenozoic	Eocene	[10.14.110., 10.14.126.]
Trigonium arcticum var. *kerguelense*	2.50889	Cenozoic	Eocene	[10.14.110.]
Trigonium arcticum var. *quadrata*	3.41985	Cenozoic	Eocene	[10.14.110.]
Trigonium dubium	3.03864	Cenozoic	Eocene	[10.14.110.]
Trigonium formosum fo. *quadrata*	3.64569	Cenozoic	Eocene	[10.14.110.]
Actinoptychus senarius	3.90246	Cretaceous	Upper Cretaceous	[10.14.50.]
Actinoptychus splendens	3.29913	Cretaceous	Upper Cretaceous, Maastrichtian	[10.14.50., 10.14.71.]
Arachnoidiscus ehrenbergii	3.63393	Cretaceous	Upper Cretaceous, Maastrichtian	[10.14.50., 10.14.89.]
Arachnoidiscus ornatus	3.60178	Cretaceous	Upper Cretaceous	[10.14.50.]
Aulacodiscus africanus	3.47433	Cretaceous	Upper Cretaceous	[10.14.50., 10.14.110.]
Aulacodiscus kittonii	3.74032	Cretaceous	Upper Cretaceous	[10.14.50., 10.14.110.]
Aulacodiscus oreganus	3.43433	Cretaceous	Upper Cretaceous	[10.14.50., 10.14.110.]
Aulacodiscus petersii	3.66743	Cretaceous	Upper Cretaceous	[10.14.50., 10.14.110.]
Aulacodiscus rogersii	3.78618	Cretaceous	Upper Cretaceous	[10.14.50., 10.14.110.]
Aulacodiscus scaber	3.89224	Cretaceous	Upper Cretaceous	[10.14.50., 10.14.110.]
Coscinodiscus sp.	3.57415	Cretaceous	Upper Cretaceous, Maastrichtian	[10.14.50., 10.14.89., 10.14.115.]
Eupodiscus radiatus	3.70334	Cretaceous	Upper Cretaceous, Maastrichtian	[10.14.50., 10.14.89.]
Glyphodiscus stellatus	3.92569	Cretaceous	Upper Cretaceous	[10.14.50.]
Lampriscus shadboldtianum	3.136	Jurassic		[10.14.83.]*

* Based on time-calibrated gene-based phylogenetic tree.

Each row of the matrix is linearly rescaled to occur on the interval $[0, 1]$. Normalization of the rescaled matrix is obtained, and as a result, the sum of the elements in each row is one. That is, each row's elements are divided by the sum of each transposed row's elements. Each row now represents a conditional probability distribution for each taxon with respect to every other taxon. The conditional probabilities form the transition matrix which is used to construct the Markov chain, representing the time step transition from one taxon to another one of complexity states [10.14.95.].

From a discrete finite Markov chain, the properties of stationarity and irreducibility must be established in order for the Markov chain to be ergodic. Convergence of the stationary distribution to the limiting distribution, and the limiting distribution being independent of the initial distribution will reveal the ergodic property of the Markov chain [10.14.95.].

To analyze and obtain the properties concerning stability as well as equilibrium or steady-state, eigenvalues and eigenvectors are calculated from the transition matrix [10.14.95.]. A left eigenvector is associated with an eigenvalue equal to one, which is complexity at equilibrium. Right eigenvectors are calculated as the number of transitions necessary for each level of complexity to reach equilibrium. For more details on the transition matrix, Markov chain properties and evaluating stability, see Pappas [10.14.95.]. Additionally, the sum of the eigenvalues as well as the determinant of the eigenvector matrix are calculated to further characterize stability with respect to equilibrium conditions concerning morphological complexity. The determinant of a square matrix with positive elements indicates stretching, while with negative elements indicates squeezing. The relation between the trace of the eigenvalues and the determinant of the eigenvector matrix will be illustrated as a trace-determinant diagram (e.g., [10.14.123.]).

10.10.2 Measuring Morphological Kolmogorov Complexity

In a previous study, the Frobenius norm was used to measure morphological and functional complexity differences among taxa [10.14.95.]. For comparing morphological complexities among diatom taxa over a long time span, the NCD metric is used. NCD has the added advantage that it has been used in phylogenetic applications concerning trees (e.g., [10.14.30.]) and has biologically-meaningful utility.

The NCD induces a compression of one state into another one such that each is describable in terms of their shared similarity with one another [10.14.30.]. Entropy values are compressed using the lossless Lempel-Ziv (LZ) compression algorithm [10.14.31., 10.14.68.], and from this, a distance matrix is calculated and normalized. To measure Kolmogorov complexity, the NCD is applicable via entropy input values (e.g., [10.14.82.]) and are used to assess similarities in Kolmogorov complexity (e.g., [10.14.30.]) among diatom taxa. These similarities are extracted via a spectral clustering tree whereby NCD value clusters are compiled.

10.10.3 Diatom Morphological Complexity over Geologic Time and Comparison of Cretaceous and Cenozoic Taxa

Entropy values representing diatom morphological complexity will be assessed with respect to all taxa. Comparisons between Cretaceous and Cenozoic taxa will be analyzed to determine the differences in morphological complexity. Results from Markov chain analysis will be used to determine ergodicity with respect to morphological complexity.

The sum of the eigenvalues as Lyapunov exponents for all taxa will be calculated as degrees of chaoticity. Change in degree of chaoticity across the Cretaceous-Paleocene boundary will be calculated as well. The determinant of the eigenvector matrix will be calculated and related to the sum of Lyapunov exponents to determine the stability or instability associated with morphological complexity at equilibrium.

NCD matrices will be calculated for Cretaceous and Cenozoic entropy values. Kolmogorov complexity will be assessed from the Cretaceous to the Cenozoic for diatoms. The comparison will be illustrated using spectral clustering trees and diagrams.

10.11   Results

Each diatom taxon was assigned a geologic era according to the scientific literature cited in Table 10.2. The assignment reflects first appearance with respect to the fossil record with two exceptions. *Amphitetras antediluviana* and *Lampriscus shadboldtianum* assignments were obtained from Nakov *et al.* [10.14.83.] that were based on their time-calibrated gene-based phylogenetic tree. As a result, 38 diatom taxa represented the Jurassic, Cretaceous and Cenozoic. Additionally, an epoch was assigned to each taxon. For Cretaceous diatoms, some of the designations included the Upper Cretaceous, specifically the Maastrichtian (Table 10.2). None of the Cretaceous taxa were assigned to be from the Lower Cretaceous. For Cenozoic diatoms, representation included the Paleocene, Eocene, Oligocene, Miocene, and Pliocene (Table 10.2).

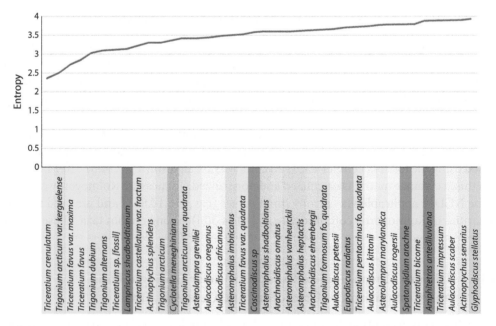

Figure 10.4 Entropy values for all 38 diatom taxa from the Jurassic, Cretaceous and Cenozoic. Genera are color coded.

In all, 38 diatom entropy values as diatom morphological complexities were plotted from lowest to highest values, ranging from around 2.4 to just under 4 (Figure 10.4). A plot of Cretaceous taxa side-by-side with Cenozoic taxa illustrated that the older taxa had a higher complexity range when compared to younger taxa (Figure 10.5). Comparing the plots separately, Cretaceous taxa were more monotonically linear (Figure 10.6), while Cenozoic taxa were slightly logarithmic (Figure 10.7). Both groups of taxa approached a similar approximate asymptotic complexity (Figure 10.5). Cretaceous taxa ranged from *Actinoptychus splendens* at the lowest value to *Glyphodiscus stellatus* at the highest value (Figure 10.6), while Cenozoic taxa ranged from to *Triceratium crenulatum* at the lowest value to *Triceratium impressum* at the highest value (Figure 10.7). Overall, Cretaceous diatom morphological complexity was greater than it was for Cenozoic diatoms (Figure 10.5).

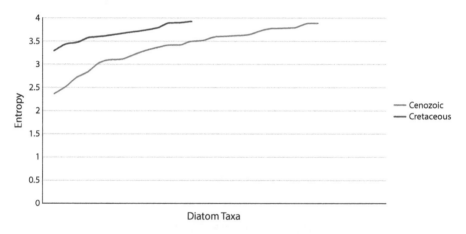

Figure 10.5 Entropy values of morphological complexity comparing diatoms from the Cretaceous with those from the Cenozoic.

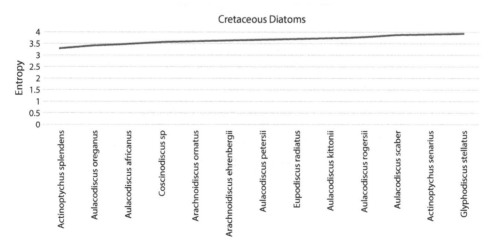

Figure 10.6 Cretaceous diatom taxa ordered from lowest to highest entropies.

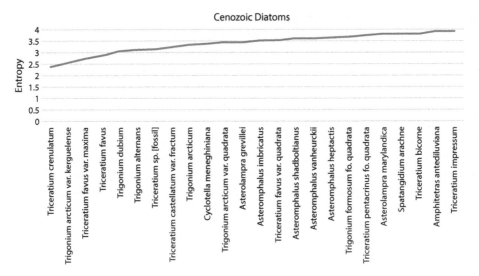

Figure 10.7 Cenozoic diatom taxa ordered from lowest to highest entropies.

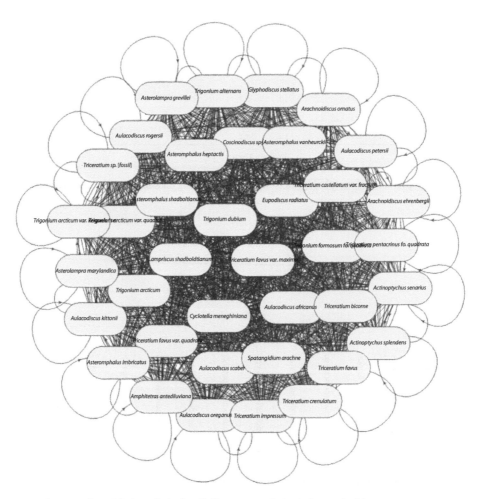

Figure 10.8 Discrete finite Markov chain for all diatom morphological complexities.

Transition probability matrices were calculated from entropy values for all diatom taxa. From this, Markov chains were constructed for all taxa (Figure 10.8), Cretaceous taxa (Figure 10.9) and Cenozoic taxa (Figure 10.10) to illustrate the state space changes in morphological complexity. Properties for each Markov chain were obtained and given in Figure 10.11, and each Markov chain was self-looping and determined to be irreducible and aperiodic, and therefore ergodic. Because of this result, stationary distributions for all taxa, Cretaceous taxa and Cenozoic taxa resulted in unique limiting distributions (Figure 10.12).

Eigenvalues as Lyapunov exponents were calculated for all taxa, Cretaceous taxa and Cenozoic taxa to determine chaoticity. All left eigenvector Lyapunov exponents were equal to one (Table 10.3), signifying that the left eigenvector in each case was related to the stationary distributions. This result confirmed irreducibility for each Markov chain. The Lyapunov exponents for the first right eigenvector were calculated to be the smallest positive value for all taxa, and Cretaceous taxa had a higher positive value than Cenozoic taxa (Table 10.3), indicating degrees of chaoticity. All remaining right eigenvectors had associated Lyapunov exponents approximately equal to zero. Sum of the Lyapunov exponents as

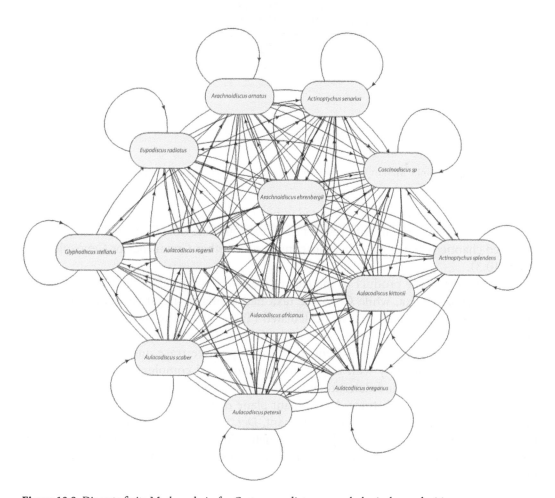

Figure 10.9 Discrete finite Markov chain for Cretaceous diatom morphological complexities.

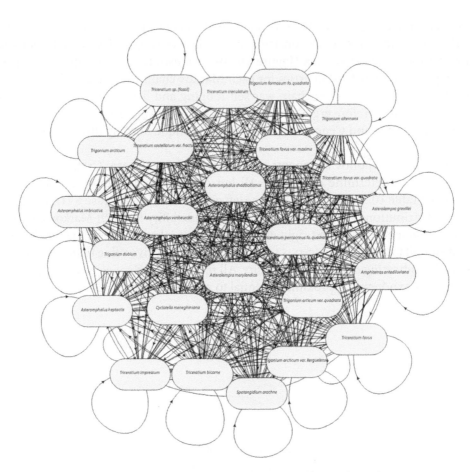

Figure 10.10 Discrete finite Markov chain for Cenozoic diatom morphological complexities.

the trace of the eigenvalues was calculated to be in the same order (Table 10.3), indicating that more chaos was present with respect to Cretaceous taxa than Cenozoic taxa. All taxa indicated chaoticity but at a lower value compared to either Cretaceous or Cenozoic taxa (Table 10.3).

Determinants as the product of the eigenvalues were reported as the largest negative value for Cenozoic taxa, while Cenozoic taxa had a smaller negative value, but for all taxa, a very small positive value was calculated (Table 10.3). The trace of the Lyapunov exponents is related to the determinant and interpretable with respect to stability (e.g., [10.14.64.]). For Cretaceous and Cenozoic taxa, a positive trace and a negative determinant indicated instability with respect to morphological complexity (Table 10.3). For all taxa, a positive trace and determinant indicated instability as well (Table 10.3). These results were illustrated in a trace-determinant plane diagram (Figure 10.13). For a 2 x 2 matrix, $TrDet$, of trace, $Trace$, and determinant, Det, values of Cretaceous and Cenozoic taxa, $Trace^2 = 4Det$ (e.g., [10.14.64.]). A trace-determinant plane diagram illustrated that instability in Cretaceous and Cenozoic diatom complexity occurred as unstable equilibria (Figure 10.13).

Basic Properties	
InitialProbabilities	
TransitionMatrix	
HoldingTimeMean	
HoldingTimeVariance	
Structural Properties	
CommunicatingClasses	{1, ..., 38}
RecurrentClasses	{1, ..., 38}
TransientClasses	None
AbsorbingClasses	None
PeriodicClasses	None
Periods	{}
Irreducible	True
Aperiodic	True
Primitive	True
Limiting Properties	
LimitTransitionMatrix	
Reversible	False

Basic Properties	
InitialProbabilities	
TransitionMatrix	
HoldingTimeMean	
HoldingTimeVariance	
Structural Properties	
CommunicatingClasses	{1, ..., 13}
RecurrentClasses	{1, ..., 13}
TransientClasses	None
AbsorbingClasses	None
PeriodicClasses	None
Periods	{}
Irreducible	True
Aperiodic	True
Primitive	True
Limiting Properties	
LimitTransitionMatrix	
Reversible	False

Basic Properties	
InitialProbabilities	
TransitionMatrix	
HoldingTimeMean	
HoldingTimeVariance	
Structural Properties	
CommunicatingClasses	{1, ..., 24}
RecurrentClasses	{1, ..., 24}
TransientClasses	None
AbsorbingClasses	None
PeriodicClasses	None
Periods	{}
Irreducible	True
Aperiodic	True
Primitive	True
Limiting Properties	
LimitTransitionMatrix	
Reversible	False

Figure 10.11 Markov chain properties for all diatom taxa (first column), Cretaceous taxa (second column) and Cenozoic taxa (third column). Irreducibility is highlighted.

Matrices of NCD values were calculated for all, Cretaceous and Cenozoic taxa. Spectral clustering trees were constructed from each NCD matrix (Figure 10.14). The smaller the NCD value, the more complexity is present. For Cretaceous taxa, the main clustered branches had NCD values of 0.94, 0.89, 0.83, 0.78, 1, and 0, going clockwise from the upper right quadrant to lower left quadrant (Figure 10.14a). Spectral Cretaceous clusters were extracted and depicted that the major cluster membership occurred in the branch at NCD value 0.94 (Figure 10.15a). The most complex Cretaceous taxa mirrored the least complex from *Actinoptychus* to *Arachnoidiscus* to *Aulacodiscus* (Table 10.4).

Cenozoic taxa had main spectral cluster branches at NCD values of 0.89, 0.94, 1, 0.83, and 0.78 going clockwise from upper right to lower left quadrants (Figure 10.14b). Spectral Cenozoic clusters depicted that the major cluster membership occurred in the branch at NCD value 0.89 (Figure 10.15b). *Triceratium* and *Trigonium* were dispersed throughout the complexity range, while *Asterolampra* was the most complex with a mixed outcome for *Asteromphalus* and *Spatangidium* (Table 10.5). Valve shape was not a factor with respect to complexity (Table 10.5).

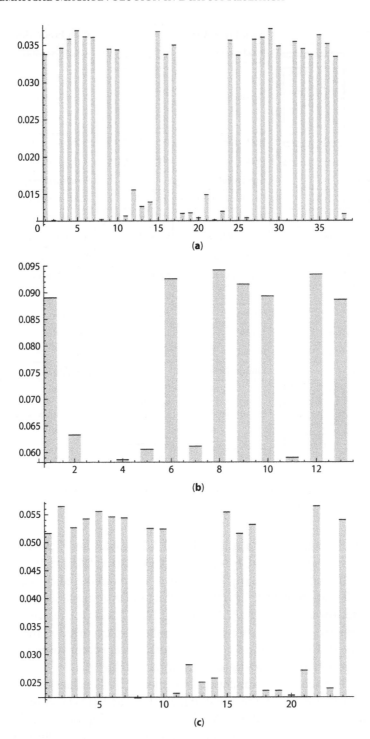

Figure 10.12 Stationary distribution for a, all diatom taxa; b, Cretaceous taxa; c, Cenozoic taxa.

Table 10.3 Lyapunov exponents, sum and determinant.

Taxa	Lyapunov exponents		$\sum \lambda$	$\mathrm{Det} = \prod \lambda$
	λ_1	λ_2		
Cretaceous taxa	1	0.469147	1.46915	-0.0000159788
Cenozoic taxa	1	0.429495	1.4295	-3.66198×10^{-7}
All taxa	1	0.353365	1.35337	1.45635×10^{-19}

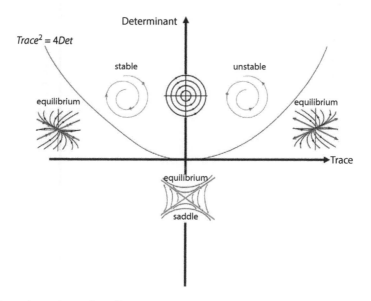

Figure 10.13 Trace-determinant plane diagram.

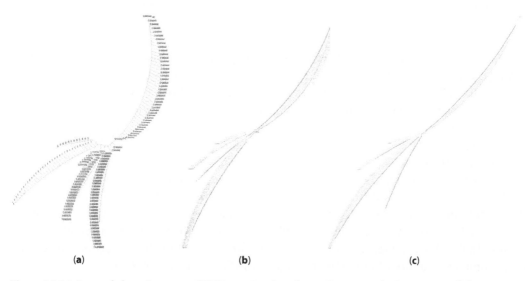

(a)   **(b)**   **(c)**

Figure 10.14 Spectral clustering trees of NCD matrix values for: a, Cretaceous; b, Cenozoic; c, all diatoms.

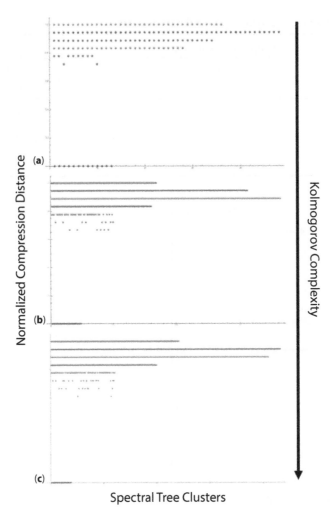

Figure 10.15 Spectral tree clusters based on NCD matrix values: a, Cretaceous; b, Cenozoic; c, all diatoms. The smaller the value, the larger the Kolmogorov complexity.

For all taxa, the main clustered branches were depicted clockwise as NCD values at 0.94 having the longest branch in the upper right quadrant, followed by branches with NCD values of 1, 0.89, 0.83, and 0.78 in the lower left quadrant of the tree (Figure 10.14c). Smaller branches occurred at NCD values between 0.6 and 0.7. Spectral clusters from the tree were diagramed to indicate pattern of cluster membership (Figure 10.15c). Spectral clusters for all taxa depicted that the major cluster membership occurred in the branch at NCD value 0.94 (Figure 10.15c).

For all taxa, Kolmogorov complexity for each taxon relative to all other taxa in the NCD matrix was compiled from spectral clusters and listed in Table 10.6. Taxa from the Asterolampraceae, Aulacodiscaceae and Triceratiaceae were the most complex, indicating that many Cenozoic taxa were more complex than Cretaceous taxa, having smaller NCD values (Table 10.6 and Figures 10.15b and c). However, most of the Cretaceous taxa were more complex than not (Table 10.6 and Figure 10.15a), whereas Cenozoic taxa were more

Table 10.4 For Cretaceous taxa, Kolmogorov complexity relative taxon order from most to least complex diatom morphology based on spectral clustering.

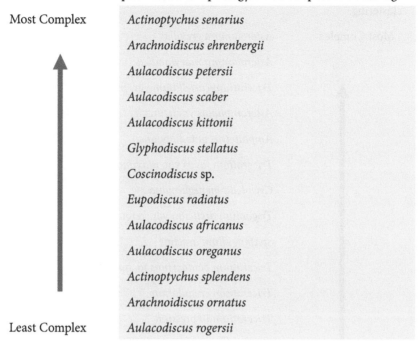

Most Complex	*Actinoptychus senarius*
	Arachnoidiscus ehrenbergii
	Aulacodiscus petersii
	Aulacodiscus scaber
	Aulacodiscus kittonii
	Glyphodiscus stellatus
	Coscinodiscus sp.
	Eupodiscus radiatus
	Aulacodiscus africanus
	Aulacodiscus oreganus
	Actinoptychus splendens
	Arachnoidiscus ornatus
Least Complex	*Aulacodiscus rogersii*

dispersed in terms of degree of complexity, having a wider spread of NCD values (Table 10.6 and Figures 10.15b and c).

10.12 Discussion

Entropy is an approximation for complexity. Diatom morphological complexity was ergodic in the long term, with positive entropy values implicating ergodicity (e.g., [10.14.133.]), as illustrated in Figure 10.3. However, diatom morphological complexity also exhibited instability. It should not be surprising that complexity is unstable, given that morphological variability as a measurable quantity is both probabilistic and NP-hard (e.g., [10.14.24., 10.14.54.]. Multiple diatom valve face feature structures and combinations thereof exemplify this problem. Dynamical complexity is expected to be related to chaotic instability and that the presence of instability means increasing complexity occurs over time [10.14.27.]. Points at which stability occur is a more specific and different problem to solve, requiring different methods. However, it is also evident that stable morphologies exist for diatoms, and complexity changes could be measurable as stable states at finer fixed time scales.

When considering entropy, taxa from the same genus have different morphological complexities (Figure 10.4). For comparison, three-part diatom valve faces occur as the least morphologically complex, having the lowest entropy values including the genera

Table 10.5 For Cenozoic taxa, Kolmogorov complexity relative taxon order from most to least complex diatom morphology based on spectral clustering.

Most Complex	*Asterolampra grevillei*
	Asterolampra marylandica
	Triceratium castellatum var. *fractum*
	Asteromphalus vanheurckii
	Amphitetras antediluviana
	Triceratium favus var. *maxima*
	Cyclotella meneghiniana
	Trigonium arcticum var. *kerguelense*
	Spatangidium arachne
	Triceratium pentacrinus fo. *quadrata*
	Triceratium crenulatum
	Triceratium impressum
	Trigonium arcticum var. *quadrata*
	Triceratium bicorne
	Triceratium favus
	Triceratium sp. [fossil]
	Trigonium formosum fo. *quadrata*
	Asteromphalus imbricatus
	Trigonium alternans
	Asteromphalus shadboltianus
	Asteromphalus heptactis
	Trigonium arcticum
	Triceratium favus var. *quadrata*
Least Complex	*Trigonium dubium*

Triceratium and *Trigonium* (Figure 10.4), but exceptions occur with one *Trigonium* and three *Triceratium* species having higher entropy values. *Asterolampra*, *Asteromphalus* and *Arachnoidiscus* have morphological complexities in the middle range, while *Aulacodiscus*, mostly in the same range, has some higher values (Figure 10.4). With a mixed result, the generic level indicated that similarities in shape and superficial valve face structures are

Table 10.6 For all taxa, Kolmogorov complexity relative taxon order from most to least complex diatom morphology. Jurassic, Cretaceous and Cenozoic taxa are color-coded as light orange, light green and light blue, respectively.

Most Complex

Asterolampra marylandica

Asterolampra grevillei

Trigonium arcticum var. *kerguelense*

Triceratium castellatum var. *fractum*

Aulacodiscus kittonii

Actinoptychus senarius

Spatangidium arachne

Asteromphalus vanheurckii

Triceratium favus var. *maxima*

Aulacodiscus petersii

Cyclotella meneghiniana

Aulacodiscus scaber

Amphitetras antediluviana

Coscinodiscus sp.

Triceratium pentacrinus fo. *quadrata*

Triceratium impressum

Arachnoidiscus ehrenbergii

Aulacodiscus oreganus

Eupodiscus radiatus

Asteromphalus heptactis

Asteromphalus shadboltianus

Triceratium sp. [fossil]

Aulacodiscus africanus

Trigonium formosum fo. *quadrata*

Trigonium arcticum

Triceratium favus var. *quadrata*

Arachnoidiscus ornatus

Triceratium crenulatum

(Continued)

Table 10.6 For all taxa, Kolmogorov complexity relative taxon order from most to least complex diatom morphology. Jurassic, Cretaceous and Cenozoic taxa are color-coded as light orange, light green and light blue, respectively. (*Continued*)

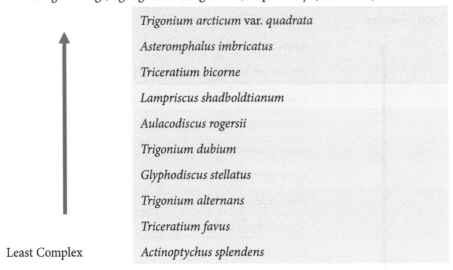

	Trigonium arcticum var. *quadrata*
	Asteromphalus imbricatus
	Triceratium bicorne
	Lampriscus shadboldtianum
	Aulacodiscus rogersii
	Trigonium dubium
	Glyphodiscus stellatus
	Trigonium alternans
	Triceratium favus
Least Complex	*Actinoptychus splendens*

not reliable characteristics to determine morphological complexity (Figures 10.4, 10.6 and 10.7). Possibly, multiple scales of morphological information may be needed to obtain more complete complexity measurement of diatom valve surfaces and subsurfaces.

Sum of the Lyapunov exponents was measured as the degree of chaos. The larger the positive value, the higher the degree of chaos and the higher degree of instability. Cretaceous diatom taxa had more instability associated with morphological complexity than Cenozoic diatom taxa (Table 10.3), and this was reinforced by the trace-determinant plane diagram (Figure 10.13). The change between taxon groups indirectly may be indicative of the change in environments between the Cretaceous and Cenozoic as measured before and after the Cretaceous-Paleocene extinction event. Environmental changes may be recorded in diatom morphology (e.g., [10.14.95.]). Diatom morphological changes are also indicative of predatory pressures (e.g., [10.14.118.]), and all ecological and environmental conditions may contribute to degree of morphological complexity for groups of diatoms from the Cretaceous and Cenozoic. The evolution of morphological complexity and its instability may be tied to other factors (e.g., [10.14.8.]) such as ecological and environmental conditions. The trace-determinant diagram may be extended to a n x n matrix for n-dimensions (e.g., [10.14.20.]) so that multiple time periods covering multiple ecological, environmental and climate situations may be assessed.

Changes in morphological complexity may be expressed as a switching of states. Lyapunov exponents may be separated as constant and intermittent phases, potentially reflecting complexity state trajectories that remain the same and those that change, respectively [10.14.105.]. With this, morphological complexity may be characterized more precisely as either constant or chaotic at a given moment in time [10.14.105.].

NCD and spectral clustering trees were used to illustrate changes in Kolmogorov complexity. NCD may be achieved using any number of compression algorithms (e.g., [10.14.25.]). As with entropy values, shape alone was shown to be an insufficient indicator of complexity

(Tables 10.4-10.6 and Figures 10.14 and 10.15). From NCD analysis, Cretaceous taxa were determined to be more complex than Cenozoic taxa.

While the NCD provides for how similar two entities are with respect to differences in information between two states, the complement of the NCD values provides the similarity in information relative to the compressor used when compressing one state, y, on another one, x, that is already compressed, $C(y) - C(y|x)$, and is given as

$$NCD_{Complement}(x, y) = 1 - \frac{C(y) - C(y|x)}{C(y)} \tag{10.33}$$

[10.14.30.]. The complement NCD may be used as a support vector machine classifier and potentially a general classifier [10.14.30.]. In this way, taxonomic and complexity character-istics of diatoms may be studied concomitantly.

Because spectral clustering trees are based on the Laplacian matrix and eigenvectors and eigenvalues are of importance, this technique could be explored for the relation between the Laplacian matrix and Lyapunov exponents as well as the connection between complex-ity and stability with respect to synchronization (e.g., [10.14.72.]). Applications to diatom life cycle changes, especially with regard to size reduction and synchronization, may be studied in relation to complexity changes. Treating diatom morphology and development as dynamical systems enables the study of complexity with regard to evolution and over different time scales concerning the fossil record as well.

10.13   Summary and Future Research

Additional analyses of diatom morphological complexity may be included in research stud-ies. Lower Cretaceous diatom taxa (e.g., [10.14.42., 10.14.49., 10.14.51.]) could be added to the present study in complexity analysis. Finer timelines may be used with representative diatom morphologies included in complexity analyses. Obtaining a more complete pic-ture from fossil to modern diatom morphological complexity may be obtained by assessing whether morphological complexity is changing substantially over long-term time scales using extinction events as markers with regard to ergodicity and chaoticity. Characterizing complexity analysis as a dynamical system may lead to characterizing other behaviors of changing complexity over time.

Ergodic complexity may be analyzed by other stochastic methods. For example, complex-ity as an optimal evolutionary strategy may be analyzable via a forward exponential perfor-mance criterion with respect to martingale and supermartingale components [10.14.28.]. This criterion may be evaluated using a forward risk entropy measure. An ergodic backward stochastic approach to a forward performance criterion may be used and involves a unique Markov solution [10.14.28.] that may be used. Ergodic limits to evolution may be analyzed with regard to morphological complexity as it has been in genomics (e.g., [10.14.77.]).

Non-ergodic complexity is also another avenue of research. For shorter time periods, non-ergodic morphological complexity of diatoms might be envisaged as a system of dif-ferential or partial differential equations in which solutions may be obtained for various boundary conditions. Such a system would be a record of the history of complexity changes so that the result need not be an ensemble average that is approximately equal to the initial average. Tracking diatom morphological complexity history may indicate that recent complexity is dependent on all previous complexities that have occurred over time.

Delving into the details of using techniques involving NCD could be useful in expanding the analyses of diatom morphological complexity concerning diatoms in relation to their phylogenetically-related groups rather than individual taxa or genera. The relation between morphological complexity and phylogeny may be explored by using NCD with respect to morphological characters and genomic data so that phylogenies may be analyzed as spectra on an evolutionary basis for modern taxa with inferences regarding fossil taxa.

10.13.1 Is Morphological Complexity Related to Morphological Symmetry?

Is there any relation between symmetry and complexity concerning morphology? Symmetry and complexity have been considered in tandem with respect to DNA (e.g., [10.14.106.]). If successive states of instability via symmetry states are found during diatom valve formation or morphogenesis more broadly, how might this impact diatom morphological complexity changes over time? Information-based studies on symmetry in which digital diatom images were used (e.g., [10.14.98.]) may enable a relational connection to complexity via entropy. Algorithmic information theory may be used to tie symmetry to complexity and to determine the role of instability in symmetry and complexity over time. Chaoticity measurable via Lyapunov exponents may be used with regard to symmetry [10.14.98.] as it has been used in this study of complexity. Spectral clustering trees may be illustrative of a connection between symmetry and complexity and applied in such studies as well.

10.14 References

10.14.1.   Adami, C. (2002) What is complexity? *BioEssays* 24, 1085–1094.

10.14.2.   Adami, C., Ofria, C., Collier, T. (2000) Evolution of biological complexity. *Proc. Natl. Acad. Sci.* 97, 4463–4468.

10.14.3.   Alimi, M., Rhif, A., Rebai, A., Vaidyanathan, S., Azar, A.T. (2021) Chapter 13 – Optimal adaptive backstepping control for chaos synchronization of nonlinear dynamical systems. In: *Backstepping Control of Nonlinear Dynamical Systems*, Vaidyanathan, S. and Azar, A.T. (eds.), *Advances in Nonlinear Dynamics and Chaos (ANDC)*, Academic Press, Cambridge, Massachusetts: 291-345.

10.14.4.   Ardakany, A.R., Ay, F., Lonardi, S. (2019) Selfish: discovery of differential chromatin interactions via a self-similarity measure. *Bioinformatics* 35, i1145-i1153.

10.14.5.   Arnold, L. and Crauel, H. (1991) Random dynamical systems. In: *Lyapunov Exponents – Proceeding of a Conference held in Oberwolfach, May 28-June 2, 1990*, Arnold, L., Crauel, H., Eckmann, J.-P. (eds.), Springer-Verlag, Berlin: 1-22.

10.14.6.   Ash, R.B. and Gardner, M.F. (1975) Ergodic theory. In: *Topics in Stochastic Processes*, Academic Press, New York, USA.

10.14.7.   Attneave, F. (1954) Some informational aspects of visual perception. *Psychol. Rev.* 61(3), 183–193.

10.14.8.   Auerbach, J.F. and Bongard, J.C. (2014) Environmental influence on the evolution of morphological complexity in machines. *PLoS Comput. Biol.* 10(1), e1003399.

10.14.9.   Bennett, C.H., Gács, P., Li, M., Vitányi, P.M.B., Zurek, W.H. (1998) Information distance. *IEEE Transactions on Information Theory* 44(4), 1407-1423.

10.14.10.   Biederman, I. (1987) Recognition-by-components: A theory of human image understanding. *Psychol. Rev.* 94(2), 115–147.

10.14.11.   Biederman, I. and Ju, G. (1988) Surface versus edge-based determinants of visual recognition. *Cognitive Psychology* 20, 38-64.

10.14.12.   Biederman, I., Mezzanotte, R.J., Rabinowitz, J.C. (1982) Scene perception: Detecting and judging objects undergoing relational violations. *Cogn. Psychol.* 14(2), 143–177.

10.14.13.   Birkhoff, G. D. (1931) Proof of the ergodic theorem. *Proc. Natl. Acad. Sci.* 17(12), 656–660.

10.14.14.   Birkhoff, G. D. (1942) What is the ergodic theorem? *Amer. Math. Monthly* 49(4), 222–226.

10.14.15.   Boltzmann, L. (1866) Uber die Mechanische Bedeutung des Zweiten Hauptsatzes der Warmetheorie [On the Mechanical Meaning of the Second Law of Heat Theory] [German]. *Wien. Ber.* 53, 195–220.

10.14.16.   Boltzmann, L. (1887) Über die mechanischen Analogien des zweiten Hauptsatzes der Thermodynamik. *J. Mathematik* C(2), 201–212

10.14.17.   Boltzmann, L. (1896) *Vorlesungen uber Gastheorie, vol. I [Lectures on Gas Theory] [German], vol. 1*, J. A. Barth, Leipzig, Germany, 1896.

10.14.18.   Boltzmann, L. (1898) *Vorlesungen uber Gastheorie, [Lectures on Gas Theory] [German], vol. II*, J. A. Barth, Leipzig, Germany.

10.14.19.   Borenstein, E. and Krakauer, D.C. (2008) An end to endless forms: epistasis, phenotype distribution bias, and nonuniform evolution. *PLoS Computational Biology* 4(10): e1000202.

10.14.20.   Brim, L., Demko, M., Pastva, S., Šafránek, D. (2015) High-performance discrete bifurcation analysis for piecewise-affine dynamical systems. In: *Hybrid Systems Biology, HSB 2015, Lecture Notes in Computer Science, vol. 9271*, Abate, A., Šafránek, D. (eds.), Springer Cham, Denmark: 58-74.

10.14.21.   Burge, J., McCann, B.C., Geisler, W.S. (2016) Estimating 3D tilt from local image cues in natural scenes. *J. Vision* 16, 13, 2, 1–25.

10.14.22.   Cardinal, J., Demaine, E.D., Demaine, M.L., Imahori, S., Ito, T., Kiyomi, M., Langerman, S., Uehara, R., Uno, T. (2011) Algorithmic folding complexity. *Graphs and Combinatorics* 27, 341-351.

10.14.23.   Casarotto, C. (2007) Markov chains and the ergodic theorem. http://www.math.uchicago.edu/~may/VIGRE/VIGRE2007/REUPapers/FINALFULL/Casarotto.pdf

10.14.24.   Castañón, C.A.B., Fraga, J.S., Fernandez, S., Gruber, A., Costa, L. da F. (2007) Biological shape characterization for automatic image recognition and diagnosis of protozoan parasites of the genus *Eimeria*. *Pattern Recognition* 40, 1899-1910.

10.14.25.   Cebrián, M., Alfonseca, M, Ortega, A. (2005) Common pitfalls using the normalized compression distance: what to watch out for in a compressor. *Communications in Information and Systems* 5(4), 367-384.

10.14.26.   Chaitin, G.J. (1977) Algorithmic information theory. *Ibm J. Res. Dev.* 21(4), 350–359.

10.14.27.   Chakrabarti, C.G. and Ghosh, K. (2011) Biological evolution: Entropy, complexity and stability. *J. Mod. Phys.* 2(6), 621–626.

10.14.28.   Chong, W.F., Hu, Y., Liang, G., Zariphopoulou, T. (2019) An ergodic BSDE approach to forward entropic risk measures: representation and large-maturity behavior. *Finance Stoch* 23, 239-273.

10.14.29.   Christov, I.C., Lueptow, R.M., Ottino, J.M. (2011) Stretching and folding versus cutting and shuffling: an illustrative perspective on mixing and deformations of continua. https://arxiv.org/pdf/1010.2256.pdf

10.14.30.   Cilibrasi, R. and Vitányi, P. (2005) Clustering by compression. *IEEE Transactions on Information Theory* 51(4), 1523-1545.

10.14.31.   Cormode, G., Paterson, M., Sahinalp, S.C., Vishkin, U. (2000) Communication complexity of document exchange. In: *Proc. 11th ACM-SIAM Symp. On Discrete Algorithms, SODA '00*, Shmoys, D. (chair), Society for Industrial and Applied Mathematics, Philadelphia, Pennsylvania, USA: 197-206.

10.14.32.   Dassow von, P. and Montresor, M. (2011) Unveiling the mysteries of phytoplankton life cycles: Patterns and opportunities behind complexity. *Journal of Plankton Research* 33(1), 3-12.

10.14.33.   Day, T.C., Höhn, S.S., Zamani-Dahaj, S.A., Yanni, D., Burnetti, A., Pentz, J., Honerkamp-Smith, A.R., Wioland, H., Sleath, H.R., Ratcliff, W.C., Goldstein, R.E., Yunker, P.J. (2022) Cellular organization in lab-evolved and extant multicellular species obeys a maximum entropy law. *eLife* 11: e72707.

10.14.34.   De Winter, J. and Wagemans, J. (2008) The awakening of Attneave's sleeping cat: identification of everyday objects on the basis of straight-line versions of outlines. *Perception* 37, 245-270.

10.14.35.   Diamond, P.H. and Thompson, K. (2017) Physics 221A Lecture Notes – Lyapunov exponents and their relation to entropy. https://courses.physics.ucsd.edu/2017/Spring/physics 221a/Phys_221A_Lecture_4-5.pdf

10.14.36.   Donald, A. (2021) Shifts and ergodic theory. https://math.uchicago.edu/~may/REU2021/REUPapers/Donald.pdf

10.14.37.   Emmert-Streib, F. (2010) Statistic complexity: combining Kolmogorov complexity with an ensemble approach. *PloS ONE* 5(8), e12256.

10.14.38.   Fourtanier, E. (1991) 7. Diatom biostratigraphy of equatorial Indian Ocean Site 758. In: *Proceedings of the Ocean Drilling Program, Scientific Results, Vol. 121*, Weissel, J., Taylor, E., Alt, J., *et al.* (eds.), College Station, Texas (Ocean Drilling Program).

10.14.39.   Frigg, R. (2004) In what sense is the Kolmogorov-Sinai entropy a measure for chaotic behaviour?—bridging the gap between dynamical systems theory and communication theory. *British Journal for the Philosophy of Science* 55, 411-434.

10.14.40.   Frigg, R., Berkovitz, J., Kronz, F. (2020) The ergodic hierarchy. Stanford Encyclopedia of Philosophy, Zalta, E.N. (ed.) https://plato.stanford.edu/archives/fall2020/entries/ergodic-hierarchy/

10.14.41.   Galatolo, S., Moyrup, M., Rojas, C. (2009) Effective symbolic dynamics, random points, statistical behavior, complexity and entropy. *Information and Computation* 208(1), 23-41.

10.14.42.   Gersonde, R. and Harwood, D.M. (1990) 25. Lower Cretaceous diatoms from ODP Leg 113 Site 693 (Weddell Sea). Part 1: vegetative cells. In: *Proceedings of the Ocean Drilling Program, Scientific Results Vol. 113*, Barker, P.F., Kennett, J.P., *et al.* (eds.), College Station, Texas (Ocean Drilling Program).

10.14.43.   Gharavi, R. and Anantharam, V. (2005) An upper bound for the largest Lyapunov exponent of a Markovian product of nonnegative matrices. *Theoretical Computer Science* 332, 543-557.

10.14.44.   Gombos, Jr., A.M. (1980) The early history of the diatom family Asterolampraceae. *Bacillaria* 3, 227-272.

10.14.45.   Grünwald, P. and Vitányi, P. (2008) Algorithmic information theory. https://arxiv.org/pdf/0809.2754.pdf

10.14.46.   Grünwald, P. and Vitányi, P. (2010) Shannon information and Kolmogorov complexity. https://arxiv.org/pdf/cs/0410002.pdf

10.14.47.   Guan, K. (2014) Important notes on Lyapunov exponents. https://arxiv.org/ftp/arxiv/papers/1401/1401.3315.pdf

10.14.48.   Hart, A.G., Hook, J.L., Dawes, J.H.P. (2021) Echo state networks trained by Tikhonov least squares are L2(μ) approximators of ergodic dynamical systems. *Physica D* 421, 132882.

10.14.49.   Harwood, D.M. and Gersonde, R. (1990) 26. Lower Cretaceous diatoms from ODP Leg 113 Site 693 (Weddell Sea). Part 2: resting spores, chrysophycean cysts, an endoskeletal dinoflagellate, and notes on the origin of diatoms. In: *Proceedings of the Ocean Drilling Program, Scientific Results Vol. 113*, Barker, P.F., Kennett, J.P., *et al.* (eds.), College Station, Texas (Ocean Drilling Program).

10.14.50.   Harwood, D.M., Nikolaev, V.A. (1995) Cretaceous diatoms: morphology, taxonomy, biostratigraphy. In: *Siliceous Microfossils*, Blome, C.D., Whalen, P.M., Reed, K.M., (Convenors), Paleontological Society Short Courses in Paleontology, The Paleontological Society Papers, No. 8. University of Tennessee, Knoxville, Tennessee: 81–106.

10.14.51.   Harwood, D.M., Nikolaev, V.A., Winter, D.M. (2007) Cretaceous records of diatom evolution, radiation, and expansion. In: *From Pond Scum to Carbon Sink: Geological and Environmental Applications of the Diatoms*, Starratt, S.W. (ed.), Paleontological Society Short Course, The Paleontological Society Papers, No. 13. Paleontological Society, Knoxville, Tennessee: 33–59.

10.14.52.   Holliday, T., Glynn, P., Goldsmith, A. (2003) On entropy and Lyapunov exponents for finite-state channels. *Computer Science, Mathematics* http://citeseerx.ist.psu.edu/viewdoc/download?doi=10.1.1.80.9820&rep=rep1&type=pdf

10.14.53.   Holliday, T., Glynn, P., Goldsmith, A. (2005) Shannon meets Lyapunov: connections between information theory and dynamical systems. *Proccedings of the 44th IEEE Conference on Decision and Control, and the European Control Conference 2005 Seville, Spain, December 12-15*, 2005: 1756-1763.

10.14.54.   Jain, A.K., Duin, R., Mao, J. (2000) Statistical pattern recognition: a review. *IEEE Trans. Pattern Anal. Mach. Intell.* 22, 4-37.

10.14.55.   Jaynes, E.T. (1982) On the rationale of maximum-entropy methods. *Proceedings of the IEEE* 70(9), 939-952.

10.14.56.   Julesz, B. (1964) Binocular depth perception without familiarity cues. *Science* 145(3630), 356-362.

10.14.57.   Jurgens, A.M. and Crutchfield, J.P. (2021) Divergent predictive states: the statistical complexity dimension of stationary, ergodic hidden Markov processes. https://arxiv.org/pdf/2102.10487.pdf

10.14.58.   Kak, S. (2022) The organization of the genetic code, TechRxiv, preprint. https://doi.org/10.36227/techrxiv.19224741.v1

10.14.59.   Kaltchenko, A. (2004) Algorithms for estimating information distance with application to bioinformatics and linguistics. *Canadian Conference on Electrical and Computer Engineering 2004 (IEEE Cat. No.04CH37513)* 4, 2255-2258.

10.14.60.   Keyes, D.E., Ltaief, H., Turkiyyah, G. (2019) Hierarchical algorithms on hierarchical architectures. *Phil. Trans. R. Soc. A* 378: 201990055.

10.14.61.   Khellaf, A., Beghdadi, A., Dupoisot, H. (1991) Entropic contrast enhancement. *IEEE Transactions on Medical Imaging* 110(4), 589-592.

10.14.62.   Kolmogorov, A. (1963) On tables of random numbers. *Sankhyā Ser. A.* 25, 369-375.

10.14.63.   Kolmogorov, A. (1965) Three approaches to the quantitative definition of information. *Probl. Inf. Transm.* 1(1), 1-7.

10.14.64.   Kuznetsov, N.V. and Leonov, G.A. (2005) On stability by the first approximation for discrete systems. *2005 International Conference on Physics and Control, PhysCon 2005, Proceedings Volume*, 2005. 1514053, 596-599.

10.14.65.   Leeuwenberg, E. (1982) The perception of assimilation and brightness contrast as derived from code theory. *Percept. Psychophys.* 32(4), 345-352.

10.14.66. Leeuwenberg, E. and van der Helm, P. A. (2013) *Structural information theory: The simplicity of visual form*, Cambridge University Press, UK.

10.14.67. Leizarowitz, A. 1991 Eigenvalue representation for the Lyapunov exponents of certain Markov processes. In: *Lyapunov Exponents – Proceeding of a Conference held in Oberwolfach*, May 28-June 2, 1990, Arnold, L., Crauel, H., Eckmann, J.-P. (eds.), Springer-Verlag, Berlin: 51-63.

10.14.68. Lempel, A. and Ziv, J. (1976) On the complexity of finite sequences. *IEEE Transactions on Information Theory* IT-22(1), 75-81.

10.14.69. Li, M. and Vitányi, P. (2008) *An Introduction to Kolmogorov Complexity and Its Applications, 3rd ed.*, Springer, New York, NY, USA.

10.14.70. Li, M., Chen, X., Li, X., Ma, B., Vitányi, P. (2004) The similarity matrix. *IEEE Transactions on Information Theory* 50(12), 3250-3264.

10.14.71. Long, J.A., Fuge, D.P., Smith, J. (1946) Diatoms of the Moreno Shale. *Journal of Paleontology* 20(2), 89-188.

10.14.72. Lucas, M., Cencetti, G., Battiston, F. (2020) Multiorder Laplacian for synchronization in higher-order networks. *Physical Review Research* 2, 033410.

10.14.73. Luxburg von, U. (2007) A tutorial on spectral clustering. https://people.csail.mit.edu/dsontag/courses/ml14/notes/Luxburg07_tutorial_spectral_clustering.pdf

10.14.74. Lyapunov, A.M. (1992) The general problem of the stability of motion [Reprinted translation of: Probleme generale de la stabilite du mouvement. *Commun. Soc. Math. Kharkov* 2, 1892, 265–272.

10.14.75. MacDonald, J.D. (1869) On the structure of the diatomaceous frustule, and its genetic cycle. *Annals Magazine Natural History Series* 4(3), 1-8.

10.14.76. Madsen, M.W. (2015) Entropy rates: some definitions, facts, and examples. https://homepages.cwi.nl/~schaffne/courses/inftheory/2015/blackboard/Ergodicity%20-%20Definitions%20and%20Examples.pdf

10.14.77. McLeish, T.C.B. (2015) Are there ergodic limits to evolution? Ergodic exploration of genome space and convergence. *Interface Focus* 5: 20150041.

10.14.78. Meyer, F., Kropfreiter, T., Williams, J.L., Lau, R.A., Hlawatsch, F., Braca, P., Win, M.Z. (2018) Message passing algorithms for scalable multitarget tracking. *Proceedings of the IEEE* 106(2), 221-259.

10.14.79. Mihailović, D.T., Mimić, G., Nikolić-Đorić, E., Arsenić, I. (2015) Novel measures based on the Kolmogorov complexity for use in complex system behavior studies and time series analysis. *Open Phys.* 13, 1-14.

10.14.80. Mihailović, D.T., Nikolić-Đorić, E., Malinović-Milićević, S., Singh, V.P., Mihailović, A., Stošić, T., Stošić, B., Drešković, N. (2019) The choice of an appropriate information dissimilarity measure for hierarchical clustering of river streamflow time series, based on calculated Lyapunov exponent and Kolmogorov measures. *Entropy* 21, 215.

10.14.81. Mihelich, M., Dubrulle, B., Paillard, D., Herbert, C. (2014) Maximum entropy production vs. Kolmogorov-Sinai entropy in constrained ASEP model. *Entropy* 16(2), 1037–1046.

10.14.82. Nagaraj, N., Balasubramanian, K., Dey, S. (2013) A new complexity measure for time series analysis and classification. *Eur. Phys. J. Special Topics* 222, 847-860.

10.14.83. Nakov, T., Beaulieu, J.M., Alverson, A.J. (2018). Accelerated diversification is related to life history and locomotion in a hyperdiverse lineage of microbial eukaryotes (Diatoms, Bacillariophyta) *New Phytologist* 219, 462-473.

10.14.84. Navon, D. (1977) Forest before trees: the precedence of global features in visual perception. *Cognitive Psychology* 9, 353-383.

10.14.85. Ng, A.Y., Jordan, M.I., Weiss, Y. (2001) On spectral clustering: analysis and an algorithm. *Proceedings of the 14th International Conference on Neural Information Processing Systems: Natural and Synthetic, Advances in Neural Information Processing Systems, NIPS January, 2001*, Dietterich, T.G., Becker, S., Ghahramani, Z. (eds.), The MIT Press, Cambridge, Massachusetts, USA: 849-856.

10.14.86. Neumann von, J. (1932a) Proof of the Quasi-ergodic Hypothesis. *Proc. Natl. Acad. Sci.* 18(1), 70–82.

10.14.87. Neumann von, John (1932b) Physical Applications of the Ergodic Hypothesis. *Proc. Natl. Acad. Sci.* 18(3), 263–266.

10.14.88. Nieto-Villar, J.M., Quintana, R., Rieumont, J. (2003) Entropy production rate as a Lyapunov function in chemical systems: proof. *Physica Scripta* 68, 163-165.

10.14.89. Nikolaev, V.A., Kociolek, J.P., Fourtanier, E., Barron, J.A., Harwood, D.M. (2001) Late Cretaceous diatoms (Bacillariophyceae) from the Marca Shale member of the Moreno Formation, California. *Occasional Papers of the California Academy of Sciences* 152, 1-122.

10.14.90. Oikawa, M.A., Dias, Z., de Rezende Rocha, A., Goldenstein, S. (2016) Manifold learning and spectral clustering for image phylogeny forests. *IEEE Transactions on Information Forensics and Security* 11(1), 5-18.

10.14.91. Olfati-Saber, R. (2006) Flocking for multi-agent dynamic systems: algorithms and theory. *IEEE Transactions on Automatic Control* 51(3), 401-420.

10.14.92. Ornstein, D.S. (1989) Ergodic theory, randomness, and "chaos". *Science* 243, 4888, 182–187.

10.14.93. Ornstein, D.S. and Weiss, B. (1991) Statistical properties of chaotic systems. *Bulletin (New Series) of the American Mathematical Society* 24(1), 11-116.

10.14.94. Oseledec, V.I. (1968) A multiplicative ergodic theorem. Lyapunov characteristic numbers for dynamical systems. *Trans. Moscow Math. Soc.* 19, 197–231.

10.14.95. Pappas, J.L. (2016) Multivariate complexity analysis of 3D surface form and function of centric diatoms at the Eocene-Oligocene transition. *Marine Micropaleontology* 122, 67-86.

10.14.96. Pappas, J.L. (2021) Quantified ensemble 3D surface features modeled as a window on centric diatom valve morphogenesis. In: *Diatom Morphogenesis [DIMO, Volume in the series: Diatoms: Biology & Applications, series editors: Richard Gordon & Joseph Seckbach]*. V. Annenkov, J. Seckbach and R. Gordon, (eds.) Wiley-Scrivener, Beverly, MA, USA.

10.14.97. Pappas, J.L. and Miller, D.J. (2013) A generalized approach to the modeling and analysis of 3D surface morphology in organisms. *PLoS ONE* 8(10), e77551

10.14.98. Pappas, J.L., Tiffany, M.A. and Gordon, R. (2021) The uncanny symmetry of some diatoms and not of others: A multi-scale morphological characteristic and a puzzle for morphogenesis. In: *Diatom Morphogenesis [DIMO, Volume in the series: Diatoms: Biology & Applications, series editors: Richard Gordon & Joseph Seckbach]*. V. Annenkov, J. Seckbach and R. Gordon, (eds.) Wiley-Scrivener, Beverly, MA, USA.

10.14.99. Pesin, J.B. (1976) Families of invariant manifolds corresponding to nonzero characteristic exponents. *Math. USSR-Izv.* 10, 1261–1305.

10.14.100. Pesin, Y.B. (1977) Lyapunov characteristic exponents and smooth ergodic theory. *Russian Math. Surveys* 32, (no 4), 55-114.

10.14.101. Pfitzer, E. (1871) Untersuchungen über Bau und Entwicklung der Bacillariaceen (Diatomaceen) [Studies on construction and development of Bacillariaceae (Diatomaceae)] [German]. In: *Botanische Abhandlungen aus dem Gebiete der Morphologie und Physiologie [Botanical Treatises in the Field of Morphology and Physiology]*. J.L.E.R. von Hanstein, (ed.) Adolph Marcus, Bonn, Germany: [i]-vi, 1-189, 186 pls.

10.14.102. Pfitzer, E. (1882) Die Bacillariaceen (Diatomaceae) [Bacillariaceae (Diatomaceae)] [German]. In: *Encyklopaedie der Naturwissenschaften. I. Abteilung. I. Thiel: Handbuch der Botanik*. A. Schenk, (ed.) Verlag von Eduard Trewendt, Breslau. 2: 403-445.

10.14.103. Poincaré, H. (1890a). Sur le problème des trois corps et les équations de la dynamique. *Acta Math*. 13, 1–270. [In French]

10.14.104. Poincaré, H. (1890b) Œuvres VII, 262–490 (theorem 1 section 8), *Mécanique Céleste et Astronomie* (Gauthier-Villars, Paris). [In French]

10.14.105. Prasad, A. and Ramaswamy, R. (1999) Characteristic distributions of finite-time Lyapunov exponents. *Phys. Rev. E* 60(3), 2761–2766.

10.14.106. Pratas, D. and Pinho, A.J. (2017) On the approximation of the Kolmogorov complexity for DNA sequences. In: *Iberian Conference on Pattern Recognition and Image Analysis, Lecture Notes in Computer Science vol. 10255*, Alexandre, L., Salvador Sánchez, J., Rodrigues, J. (eds.), Springer International Publishing AG, New York, USA: 259-266.

10.14.107. Rényi, A. (1961) On measures of entropy and information. In: *Proceedings of the 4th Berkeley Symposium on Mathematics, Statistics and Probability*, 20 June–30 July, 1960, Statistical Laboratory of the University of California, Berkeley, CA, USA: 547–561.

10.14.108. Román, J.C.M., Noguera, J.L.V., Legal-Ayala, H., Pinto-Roa, D.P., Gomez-Guerrero, S., Torres, M.G. (2019) Entropy and contrast enhancement of infrared thermal images using multiscale top-hat transform. *Entropy* 21(3), e21030244.

10.14.109. Rominger, A.J., Fuentes, M.A., Marquet, P.A. (2019) Nonequilibrium evolution of volatility in origination and extinction explains fat-tailed fluctuations in Phanerozoic biodiversity. *Sci. Adv.* 5: eatt0122.

10.14.110. Round, F.E., Crawford, R.M. and Mann, D.G. (1990) *The Diatoms, Biology & Morphology of the Genera*. Cambridge University Press, Cambridge, U.K.

10.14.111. Ruelle, D. (1979) Ergodic theory of differentiable dynamical systems. *Publications mathématiques de l'I.H.S.* 50, 27-58.

10.14.112. Sarig, O. (2020) Lecture notes on ergodic theory. https://www.weizmann.ac.il/math/sarigo/sites/math.sarigo/files/uploads/ergodicnotes.pdf

10.14.113. Sawada, T. (2010) Visual detection of symmetry of 3D shapes. *Journal of Vision* 10(6), 1-22.

10.14.114. Shannon, C.E. (1948) A mathematical theory of communication [corrected version]. *Bell Syst. Tech. J.* 27, 379–423, 623–656.

10.14.115. Sims, P.A. (2011) *Ceratangula* gen. nov. is proposed together with a discussion on the genera *Cerataulus, Eupodiscus, Amphitetras, Amphipentas* and *Diommatetras*. *Diatom Research* 23(2), 435-444.

10.14.116. Sims, P.A., Mann, D.G., Medlin, L.K. (2006) Evolution of the diatoms: insights from fossil, biological and molecular data. *Phycologia* 45(4), 361-402.

10.14.117. Sinai, Y.G. (1959) On the notion of entropy of a dynamical system. *Doklady of Russian Academy of Sciences* 124, 768-771.

10.14.118. Smetacek, V. (2001) A watery arms race. *Nature* 411, 745.

10.14.119. Stanley, K.O. (2006) Exploiting regularity without development. In: *Proceedings of the AAAI Fall Symposium on Developmental Systems*, Technical Report, Menlo Park, CA, USA: 49-56.

10.14.120. Stevens, K.A. (1983) Surface tilt (the direction of slant): A neglected psychophysical variable. *Percept. Psychophys.* 33(3), 241–250.

10.14.121. Su, H., Bouridane, A., Crookes, D. (2006) Scale adaptive complexity measure of 2D shapes. *18th International Conference on Pattern Recognition (ICPR '06)* 2, 134-137.

10.14.122. Sun, C.M. and Si, D.Y. (1997) Skew and slant correction for document images using gradient direction. In: *Proceedings of the Fourth International Conference on Document*

Analysis and Recognition, vol. 1 and 2, IEEE, Computer Society Press, Los Alamitos, California, USA: 142–146.

10.14.123. Sutherland, S. (2011) Linearization, trace and determinant. https://www.math.stonybrook. edu/~scott/mat308.spr11/TraceDet.pdf

10.14.124. Teixeira, A., Matos, A., Souto, A., Antunes, L. (2011) Entropy measures vs. Kolmogorov complexity. *Entropy* 13, 595-611.

10.14.125. Tiffany, M.A. and Hernández-Becerril (2005) Valve development in the diatom family Asterolampraceae H.L. Smith 1872. *Micropaleontology* 51(3), 217-258.

10.14.126. Tiffany, M.A. and Lange, C.B. (2002) Diatoms provide attachment sites for other diatoms: a natural history of epiphytism from southern California. *Phycologia* 41(2), 116-124.

10.14.127. Tsallis, C. (1987) Possible generalization of Boltzmann-Gibbs statistics. *J. Stat. Phys.* 52, 479–487.

10.14.128. Tserunyan, A. and Zomback, J. (2021) A backward ergodic theorem and its forward implications. https://arxiv.org/pdf/2012.10522v2.pdf

10.14.129. Valenza, G., Allegrini, P., Lanatà, A., Scilingo, E.P. (2012) Dominant Lyapunov exponent and approximate entropy in heart rate variability during emotional visual elicitation. *Frontiers in Neuroengineering* 5, 3|1-3|7.

10.14.130. Vigoda, E. (2003) Markov chains, coupling, stationary distributions. https://faculty. cc.gatech.edu/~vigoda/MCMC_Course/MC-basics.pdf

10.14.131. Weisstein, E.W. (ed.) (2002) *CRC Concise Encyclopedia of Mathematics, 2nd ed.*, Chapman & tp:/Hall/CRC, London, United Kingdom.

10.14.132. Wilinski, A. (2019) Time series modeling and forecasting based on a Markov chain with changing transition matrices. *Expert Systems With Applications* 133, 163-172.

10.14.133. Wilkinson, A. (2017) What are Lyapunov exponents, and why are they interesting? *Bulletin (New Series) of the American Mathematical Society* 54(1), 79-105.

10.14.134. Wolf, Y.I., Katsnelson, M.I., Koonin, E.V. (2018) Physical foundations of biological complexity. *PNAS* 115(37), E8678-E8687.

10.14.135. Yan, D., Huang, L., Jordan, M.I. (2009) Fast approximate spectral clustering. Technical Report No. UCP/EECS-2009-45, http://www.eecs.berkeley.edu/Pubs/TechRpts/2009/ EECS-2009-45.html

11

Diatom Surface Symmetry, Symmetry Groups and Symmetry Breaking

Abstract

Symmetry is related to the geometry of organismal form. Specifically, surface symmetry is obtained via geometric representation of diatom surfaces of 3D surface models constructed from systems of parametric equations. While numerical solution to the Jacobian of each system represents the tangent lines and planes of the surface, the inverse Jacobian represents surface symmetry, and the sign of numerical solutions indicates stability if negative and instability if positive. For most diatom surfaces, symmetry is categorizable via symmetry group analysis. Identification of the symmetry groups of cyclic, reflective, dihedral, glide, and scale may be applied to diatom surfaces. Knot symmetry may be applied to diatom surfaces that exhibit twist or writhe and are connected as closed loops, such as those found in *Entomoneis*. Knot surface symmetry is determined by handedness as well. All diatom surface symmetry groups are evaluated in terms of symmetry state, and assessment in terms of symmetry breaking is related to size reduction in diatom vegetative reproduction.

Keywords: Symmetry group, knot symmetry, hyperbolic knot, dihedral, cyclic, glide symmetry, scale symmetry, symmetry breaking, group theory, amphichiral

11.1 Introduction

Two of the most important attributes in describing organismal morphology are geometry and symmetry. To gain a better understanding of morphology as a measurable attribute, three-dimensional (3D) surface models have been used (e.g., [11.9.41.-11.9.43., 11.9.47.]). Using surfaces geometrically entails accounting for shape and surface features. Such models are readily adaptable to very small or very large changes in multiple surface features to study how such features react to morphological changes in a prescribed, stepwise fashion. Measuring changes in this way enables construction of potential scenarios of morphological change during developmental life cycle stages, speciation or other evolutionary events. Because 3D surface models are bounded shapes, symmetry may be assessed with regard to structural surface features. From the geometry of 3D surface models, symmetry may be extracted, analyzed and applied more widely to a given group of organisms, such as diatom genera, to better understand symmetry.

11.1.1 Geometry as a Basis for Form, Surfaces and Symmetry

Characterization of organismal surfaces via systems of parametric 3D equations enables the quantification of those surfaces. Systems of parametric 3D equations are expressible as differentiable functions, and recovering the inclines and declines of a surface are obtained by first partial derivatives. For parametric 3D equations expressed in the form of $x = g(u,v)$, $y = h(u,v)$ and $z = f(u,v)$, the differentials are $dx = \dfrac{\partial x}{\partial u} du + \dfrac{\partial x}{\partial v} dv$, $dy = \dfrac{\partial y}{\partial u} du + \dfrac{\partial y}{\partial v} dv$ and $dz = \dfrac{\partial z}{\partial u} du + \dfrac{\partial z}{\partial v} dv$ [11.9.47.]. The first partial derivatives comprise a 3 x 2 Jacobian matrix (i.e., Jacobian), and the Jacobian represents the tangent lines and planes (e.g., [11.9.40.]) of the surface [11.9.47.]. This matrix represents the surface of an organism as vectors in the x, y, z-directions (rows of the Jacobian), and parameters u, v represent vectors perpendicular to each other in a x-y, y-z or x-z plane (columns of the Jacobian). For a 3D surface, the cross product of vectors u, v is the vector perpendicular to both u and v, and is given as $\|\mathbf{u} \times \mathbf{v}\| = \|\mathbf{u}\| \|\mathbf{v}\| \sin \theta$, where θ is the angle between $\mathbf{u}$ and $\mathbf{v}$ with vector magnitude $\|\mathbf{u} \times \mathbf{v}\|$, and the resultant 3 x 3 Jacobian is

$$J_{3D\,surface} = J_{x,y,z} = \begin{bmatrix} \dfrac{\partial x}{\partial u} & \dfrac{\partial x}{\partial v} & \dfrac{\partial y}{\partial u} \times \dfrac{\partial z}{\partial v} - \dfrac{\partial z}{\partial u} \times \dfrac{\partial y}{\partial v} \\[2ex] \dfrac{\partial y}{\partial u} & \dfrac{\partial y}{\partial v} & \dfrac{\partial z}{\partial u} \times \dfrac{\partial x}{\partial v} - \dfrac{\partial x}{\partial u} \times \dfrac{\partial z}{\partial v} \\[2ex] \dfrac{\partial z}{\partial u} & \dfrac{\partial z}{\partial v} & \dfrac{\partial x}{\partial u} \times \dfrac{\partial y}{\partial v} - \dfrac{\partial y}{\partial u} \times \dfrac{\partial x}{\partial v} \end{bmatrix} \quad (11.1)$$

[11.9.44.].

The generalized 3D surface form of a diatom is a modified capped cylinder [11.9.44., 11.9.45.]. In 2D, a diatom's shape is a modified circle or ellipse. In 2D or 3D, quantified shape or surface, respectively, may be used to inform 2D and 3D symmetry of organisms. The first two columns of the Jacobian are informative of symmetry and whether the symmetry in unstable or stable.

11.1.2 Inverse Functions as a Basis for Symmetry and Stability

Systems of parametric equations represent the 3D organismal form and its 2D shape in a plane for each model. The Jacobian is a summary of the surface in that the differentials are linear functions in which the coefficients are the partial derivatives of functions in the x, y, z directions, and the linear functions are a linear mapping of x, y, z into u, v [11.9.24.]. The 3D surface is mapped onto a 2D plane so that the inverse mapping is u, v into x, y, z where z is a function of x, y. That is, $x = g(u,v)$ and $y = h(u,v)$ with $z = f(g(u,v), h(u,v)) = f(x,y)$ are expressed as implicit functions $g(u,v) - x = 0$ and $h(u,v) - y = 0$ with $f(g(u,v), h(u,v)) - z = 0$, representing the inverse mapping $u = \varphi(x,y)$ and $v = \psi(x,y)$ [11.9.24.]. The Jacobian of the inverse mapping of a tangent vector at a point, p, in $\mathbb{R}^n$ is $J(p)^{-1}_{x,y,z} = J(p)_{u,v}$ that is the inverse Jacobian [11.9.24.].

11.2 Symmetry of 3D Organismal Surfaces

Shape and symmetry have intrinsic, invariant and multivariable properties [11.9.29.]. Symmetry is 2D shape in motion through 3D space and a special case of shape and surface.

Point, shape, surface, and pattern symmetries are interrelated. Symmetry is a continuous, connected and dynamic property of organisms. Shape is one attribute of bauplan and development and is a representation of an interface boundary with the environment. Symmetry changes may be one result and an indicator of stress and associated with developmental instability, depending on evolutionary, ecological and methodological assumptions [11.9.14.].

In development and evolution studies, various types of recurrent symmetries may be applicable including reflective (bilateral, left-right, mirror), cyclic (rotational), scale (dilation, self-similarity), dihedral, and glide. Translation may be viewed as a kind of symmetry, but is more accurately described as motion with respect to repeating symmetries. Dihedral as cyclic plus reflective and glide as translational plus reflective are combination symmetries. Scale symmetry involves folding and multi-scale consideration, indicative of the fractal nature of this type of symmetry. Examples of scale symmetries may be found geometrically in Koch curves, Lindenmayer systems, Sierpinski triangles, Julia sets, and Mandelbrot sets.

Reflective symmetry figures prominently in organismal studies. Developmentally, directional reflective symmetry means that one side of an organism is larger than the other (e.g., [11.9.14.]). Reflective anti-symmetry is indicative of an organism with either side being variable in size [11.9.14.]. Fluctuating reflective asymmetry is instability with respect to either side of an organism in which random deviation from perfect reflective symmetry is evident (e.g., [11.9.37.]) and is indicative of symmetry breaking.

Symmetries are informative in understanding bauplan and morphological evolution in comparative and macroevolutionary studies. For example, symmetry may be adaptive in active organisms with reflective symmetry in contrast to sedentary organisms that have cyclic or dihedral symmetry, although such a dichotomy is not always clear cut (e.g., [11.9.19.]). Organisms with large surface to volume ratio or have repetitive parts as in glide symmetry may be adaptive (e.g., [11.9.20.]). Organisms exhibiting metabolic energy efficiency have scale symmetry which may be adaptive as well (e.g., [11.9.32.]). Surface symmetry considerations may be indicative of stability or instability in organismal development during the life cycle, and such stability or instability may be an evolutionary outcome of the bauplan (e.g., [11.9.19.]).

11.2.1 Shape versus Surface Symmetries

There are differences in assessing symmetry in shapes and surfaces. Shapes, devoid of surface features, are simpler to assess with respect to symmetry. Typically, cyclic and reflective symmetries are assigned to such organismal shapes (Figure 11.1). Cyclic symmetry means that the shape may be rotated n-degrees with respect to a central pivot point so that the same shape is retained. Reflective symmetry means that the shape may be bisected with a line through a central point such that the two shape halves are mirror images of one another. If the reflective symmetric shape is bisected reflectively with a line perpendicular to the original bisection, then the shape may have cyclic symmetry by a 180-degree rotation as well. Surfaces bounded by shape are more complicated in that such surfaces may defy the conventional definitions of 2D shape alone, resulting is a symmetry that may differ from that for the shape. For example, a circular shape may have different surface features, indicating different symmetries (Figure 11.2).

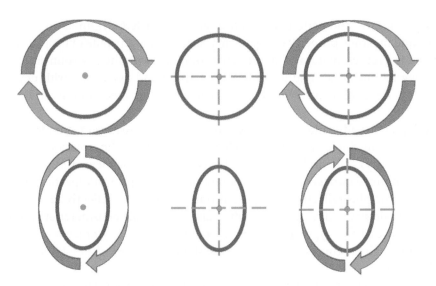

Figure 11.1 Cyclic symmetry (top row) and reflective symmetry (bottom row) of 2D shape.

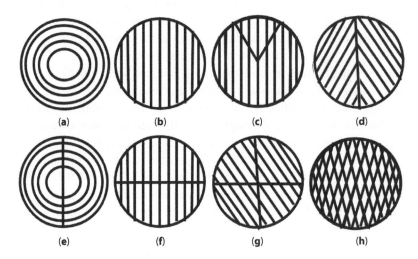

Figure 11.2 Depiction of surface symmetries: a, cyclic, reflective, dihedral, scale; b, reflective, 180-degree cyclic; c, reflective; d, reflective, glide; e, reflective, 180-degree cyclic; f, reflective, 180-degree cyclic, glide; g, reflective, dihedral; h, reflective, glide.

11.2.2 Geometry of Non-Flat 3D Surfaces: Bidirectional Curvature and Its Relation to Twists and Writhes

A graph is a depiction of connected edges between vertices whereby the edges are curves that may exhibit curvature, and bidirectional curvature is torsion where curvature occurs in the x, y and x, z planes simultaneously [11.9.29.]. Torsion is describable as a twist, where a twist is defined as the magnitude of the curvature tracing the motion of the twist about each axis in each plane in space [11.9.29.]. Torsion is also describable as a writhe, where a writhe is the magnitude of each axis being distorted in each plane in space [11.9.1.]. Both twists and writhes are two-way curvatures.

Twists and writhes may be introduced to a line, commencing curvature in one or two directions. When a line becomes connected it forms a closed loop. When twists and writhes become closed loops, they become knots.

11.2.3   Knots: Geometry and Topology of Closed Curved Surfaces

A mathematical knot is a dimensionless closed curve that does not intersect itself and cannot be undone (Figure 11.3; [11.9.1.]). Knot projections in the plane depicted in knot diagrams are used to analyze knots and determine equivalencies. To start, the unknot (or trivial knot), trefoil knot (or the simplest knot) and figure-eight knot are depicted as knot diagrams in Figure 11.3 (top row). Knot projections of Möbius figure-eight knots and Hopf link are also depicted in the same figure (Figure 11.3, bottom row). Each representation of a given knot (or unknot) is equivalent to all others, regardless of the deformation imposed on that given knot if they are ambient isotopic [11.9.1.].

Knots are surfaces that are characterizable as topological manifolds [11.9.1.] which in turn, are topological spaces such that every point in the space has a neighborhood that is homeomorphic to the open set in $\mathbb{R}^n$ [11.9.66.]. Locally, topological spaces are Euclidean with conditions of separability and countability met [11.9.66.]. The topological equivalence between knots and surfaces may be seen in terms of orientability or non-orientability, and examples of such surfaces include Möbius knots and Seifert surfaces [11.9.3.].

Knots have compliments that may be links. A link is a collection of non-intersecting knots or unknots that may be linked or knotted [11.9.1.]. The unlink consists of two unknots, and the simplest link consists of two unknots connected as a Hopf link (Figure 11.3; see [11.9.1.]). There are links that are analogs to knots, and linking may be positive or negative with respect to crossing [11.9.1.].

The type of knot may be useful in determining the properties of an organism's surface curvature. Curvature is also recovered implicitly by determining the crossing number as the minimum number of crossings in a knot diagram. Crossings may be over or under a given part of a knot. Over or under crossings and their direction of a knot are useful descriptors of the orientation of a knot [11.9.1.].

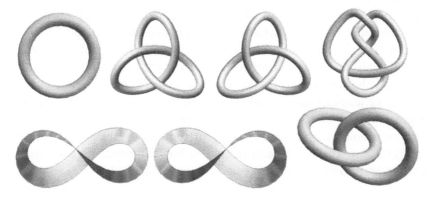

Figure 11.3 Knot projections. First row: unknot; right- and left-handed trefoil knots; figure eight knot. Second row: two Möbius figure eight knots; Hopf link.

Amphichirality is an oriented knot that is equivalent to its mirror image in contrast to a chiral knot [11.9.1., 11.9.66.]. Handedness is related to amphichirality, and handedness and amphichirality are properties that enable the detection of knot symmetry. The classic example of knot symmetry is evident in Perko pair knots [11.9.1.]. Over or under crossings and their direction are indicative of handedness and whether handedness is right or left and if amphichirality exists between a pair of knots. A positive crossing is a right-handed helix, while a negative crossing is a left-handed helix as over and under crossings, respectively. Crossing number indicates the chirality or amphichirality of knots [11.9.1.]. Importantly, for an amphichiral knot, its knot polynomial must be palindromic [11.9.1.].

11.2.4   From Hyperbolic Geometry and Surfaces to Hyperbolic Knots

A hyperbolic paraboloid has the shape of a saddle. Exaggerating the shape by deforming it in multiple directions by including twists or writhes produces an equivalent hyperbolic paraboloid, topologically speaking. If a very thin strand of the surface completes a loop from one edge of the surface to another, a knot may result.

Knots and their relation to hyperbolic geometry was an important advancement in knot theoretical studies [11.9.59., 11.9.60.]. A hyperbolic paraboloid as a saddle has negative curvature [11.9.29.]. Negative curvature with respect to a 3D space induces a hyperbolic metric so that a knot and its complement in a hyperbolic space are measurable via a hyperbolic metric [11.9.1.].

To qualify as a hyperbolic knot, the knot must be a prime knot [11.9.1., 11.9.2., 11.9.21.]. Measurement between a hyperbolic knot and its link compliment as a 3-manifold is accomplished using a hyperbolic geometric distance based on geodesic lines and tetrahedra based on geodesic planes [11.9.1., 11.9.2.]. Identification of hyperbolic knots is determined via their hyperbolic volumes [11.9.1., 11.9.66.] which are almost all invariant [11.9.1.]. Hyperbolic knots are classifiable according to group theory as a finite symmetry group that include cyclic or dihedral groups [11.9.66.]. By extension, reflective or glide symmetry groups may comprise a knot symmetry group.

Hyperbolic structures on manifolds may be decomposed into polygons (e.g., [11.9.61.]). As such, manifolds may undergo Dehn surgery or gluing to create hyperbolic knot structures [11.9.62.]. That is, gluing after polygon decomposition is used in order obtain the homeomorphic structures that qualify as knot complements. A homeomorphism is a one-to-one correspondence of points between two geometric or topological manifolds such that continuous deformation does not change one manifold to be different from the other [11.9.66.]. Orientable surfaces with mirror symmetry are homeomorphic, while Möbius surfaces which may be homeomorphic are non-orientable [11.9.66.]. Self-homeomorphism is characteristic of amphichirality, and hyperbolic knots as a symmetry group exhibit self-homeomorphism, if orientation reversing is isometric [11.9.66.]. Knot symmetry as exhibited by hyperbolic knots may be found in organismal surfaces characterizable as 3-manifolds [11.9.25.]. There is an equivalence of homeomorphisms such that deformations of symmetries mean an equivalence of symmetries [11.9.21.].

11.2.5   Closed Helices, Hyperbolic Knots and Möbius Surfaces

Helices are surfaces that may have twists and writhes, and as closed curved surfaces, helices as knots may exhibit twisting and writhing. As an additional complicating factor, helical knots with twists or writhes may be non-orientable similarly to Möbius surfaces as knots. Other helical knots or Dehn twists may be orientable similarly to hyperbolic knots (e.g., [11.9.31.]). That is, the Möbius surface is a knot and may be a closed helix, while closed helices as Dehn twists may be orientable as are hyperbolic knots.

Organismal surfaces may exhibit hyperbolic geometry and topology, including helical structures. Left- or right-handed helical structures may be evident in knot structure such as a prime knot, and as such, are examples of hyperbolic knots. At the unicellular level, some diatoms exhibit hyperbolic geometry and topology, and may include twists and writhes. The Bacillariales and Surirellales consist of some genera with hyperbolic paraboloid surfaces or frustule twists and writhes. Some of these frustule structures form closed curved surfaces, and as such, are knots. Those genera with such characteristics are few in number and are unlike the majority of diatoms.

11.3   Symmetry Groups

A group is a set of elements in which closure, associativity, identity, and inverse properties are satisfied [11.9.66.]. A symmetry group has all invariant transformations of a geometry object, and is an invertible mapping preserving the form and structure of the object [11.9.66.]. For finite symmetry groups, the number of transformations, inversions, rotations, and reflections is finite, preserving symmetry, and such groups possess multiple symmetries [11.9.13.].

A finite symmetric group of invertible transformations, Γ, of elements f, g in set V have a composition fg, or $f \circ g$, given as $fg(v) = f(g(v))$ with $v \in V$ [11.9.13.]. V has finite dimension in a real vector space, and if the invertible transformations are linear, the symmetry group may also be cyclic, dihedral, cyclic orthogonal, or cyclic-reflection orthogonal [11.9.13.]. For Euclidean groups, rigid transformations are affine as translations, rotations, reflections, and glide reflections.

Cyclic groups contain n-sided polygonal objects having all cyclic symmetries in the plane [11.9.13.]. That is, rotation of a polygonal object may be mapped onto itself by n-rotations, and all n rotations form a cyclic symmetry group. Dihedral groups have n-sided polygonal objects that are cyclic and have reflections at n-angles with respect to lines drawn through the origin [11.9.13.]. That is, for polygonal objects that have cyclic symmetry and also undergo reflection via perpendicular or diagonal bisection have n rotations and n reflections and form a dihedral symmetry group of $2n$ symmetries. Both orthogonal groups given above involve cyclic or cyclic-reflective symmetries with respect to the unit circle [11.9.13.]. Glide symmetry involves reflection perpendicular to the plane and translation parallel to the plane. Repetitive pattern is evident in glide symmetry.

Knot symmetry may be understood as a symmetry group [11.9.21.] and is also representative of cyclic and dihedral symmetry groups via hyperbolic knots [11.9.66.]. Symmetry of a knot is an equivalence relation where a knot is mapped onto itself, and the two knots are the same if one is deformable into the other [11.9.21.]. That is, a knot, K, is homeomorphic to $(\mathbb{R}^3, K)$, which are the spaces that K and $\mathbb{R}^3$ (3D real numbers) occupy. A composition of reversals or invertibility of K or $\mathbb{R}^3$ of a given knot onto itself comprises the knot symmetry group [11.9.21.].

Symmetries contained in a given group depend on the knot characteristics of amphichirality and invertibility. To test for these characteristics, Hoste *et al.* [11.9.21.] devised a table of symmetry types contained in a group based on the orientation of K and $\mathbb{R}^3$ and the combinations of those types that fulfill the requirements of amphichirality and/or invertibility. Knots with a crossing number of eight or less are either amphichiral or invertible [11.9.21.]. There is no knot invariant that can be used to determine amphichirality in all cases [11.9.66.].

Hyperbolic knots as cyclic or dihedral comprise a finite symmetry group [11.9.21.]. With the characteristic of amphichirality or invertibility along with being a hyperbolic knot, a smaller subset of knot symmetries qualifies as a finite symmetry group of knots with eight crossings or fewer. For example, the figure-eight knot is hyperbolic as well as amphichiral and invertible [11.9.63.]. Torus and satellite knots are ineligible members of this group. In Alexander-Briggs notation, the hyperbolic knots that qualify are $4_1, 5_2, 6_1, 6_2, 6_3, 7_2, 7_3, 7_4, 7_5, 7_6, 7_7, 8_1 - 8_{21}$ [11.9.1.].

A knot that involves one half twist or multiples thereof is the Möbius band or strip [11.9.1.]. This surface is bounded and non-orientable, and therefore one-sided. No amount of twisting or untwisting can be done to undo the topological properties of a Möbius strip. The lengthwise cutting of a Möbius strip with one half twist is the unknot, two lengthwise cuttings of a Möbius strip with two half twists produces a Hopf link, but the boundary of a thrice lengthwise cut Möbius strip with three half twists results in the trefoil knot [11.9.1.].

11.3.1   Diatom Surface Symmetry Groups: Cyclic, Reflective, Dihedral, Glide, Scale, and Knot

Diatoms are unicellular pigmented protists that have valves (frustules) generally arranged in a Petri-dish fashion, consisting of amorphous silica. The scaffolded layering of siliceous material is composed of geometric structures and features that are scaled in the nanometer to micrometer range. Patterning of each layer comprises a regularity of pore shapes (such as hexagons or circles), cribra or striae spacings (e.g., [11.9.54.]). Diatom surfaces embody such geometric structures, and these surfaces and their shape determine the symmetry present in diatom valves [11.9.48.].

Diatom surfaces exhibit specific recurrent symmetries when considering their 3D shape. Historically, diatom valve shape has been divided into two main categories: centrics, many with cyclic symmetry; pennates, many with reflective symmetry. Many diatoms are exceptions to the traditional schema, especially those diatoms that are not flat or not lying in a plane. Examples of non-flat diatom 3D surface symmetry may include hyperbolic knots, Möbius surfaces, twisted or writhed closed curved surfaces,

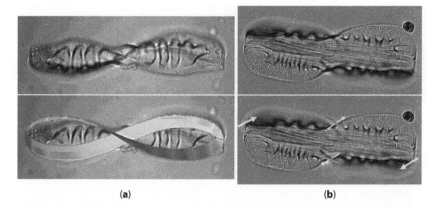

Figure 11.4 *Entomoneis ornata*: a, light micrographs of valve view (top) and the tracing of the boundary shape with a Möbius figure-eight surface (bottom); b, light micrographs of girdle view (top) and the marking of the places where a half twist occurs on the boundary of each valve (bottom). Light micrographs are from [11.9.46.].

or even closed helical surfaces. Twisted and writhed diatoms may include taxa found in the Surirellales and Bacillariales (e.g., [11.9.54.]). In the Surirellales, *Campylodiscus* is a hyperbolic paraboloid surface (Figure 11.5), while *Entomoneis* exhibits a twisted frustule surface with respect to the valve and girdle bands (Figures 11.4 and 11.5). A tracing of the cell boundary is examined to see if a closed curve bounded surface with crossings is present. For *Entomoneis*, a knot diagram reveals that a knot has been formed (Figure 11.4). Amphichirality and invertibility is assigned according to the compiled knot symmetry group classification system based on Hoste *et al.* [11.9.21.] (Table 11.1). From the Bacillariales, *Cylindrotheca* has twists [11.9.53.] and writhes in its surface, with its canalled raphe as well as girdle bands continuously spiralling around the entire frustule (Figure 11.5). However, if *Cylindrotheca* was untwisted, the cell would remain intact and no closed curve boundary is evident, therefore knot symmetry does not apply.

11.3.2   States of Symmetry

In the biological world, perfect symmetry is extremely rare, and symmetry measurements are approximations in the best of circumstances [11.9.13.]. States of symmetry exist, depending on the numerical outcome of symmetry group analysis over time. Often, symmetry measurement is accomplished with respect to a state of equilibrium so that the assessment differentiates between the initial symmetry state and subsequent state after a dynamical change occurs. Symmetry changing patterns may reflect instability as symmetry breaking may be inferred.

Symmetry breaking at equilibrium may be measured such that the equations representing the initial symmetry state do not change, but the solutions to the equations do change. That is, asymmetric perturbations to the initial symmetry state equations results in symmetry breaking. Such symmetry breaking may induce pattern formation [11.9.13.].

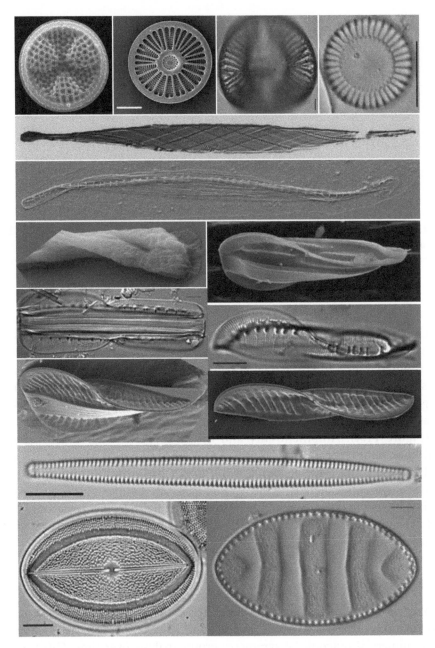

Figure 11.5 Diatom exemplar genera for surface symmetry analysis. Row 1: *Actinoptychus* sp.; *Arachnoidiscus ehrenbergii*; *Campylodiscus hibernicus*; *Cyclotella meneghiniana*. Row 2: *Cylindrotheca* sp. Row 3: *Cylindrotheca* sp. Row 4: Twisted part of *Cylindrotheca* sp.; *Entomoneis alata* (valve). Row 5: *Entomoneis alata* (girdle and valve). Row 6: *Entomoneis grandis* nom. nud. (girdle and valve). Row 7: *Tabularia fasciculata*. Row 8: *Cocconeis grovei*; *Surirella undulata*. *Actinoptychus* SEM from James M. Ehrman Mount Allison University, with permisson. *Arachnoidiscus ehrenbergii* SEM from Mary Ann Tiffany, with permission. *Cylindrotheca* sp.SEMs (Rows 3 and 4) from Shinya Sato, Fukui Prefectural University (SEMS taken at Royal Botanic Garden Edinburgh) with permission. *Entomoneis alata* (Row 4) and *E. grandis* nom. nud. SEMs are from Keigo Osada, The Nippon Dental University School of Life Dentistry at Niigata with permission. Light micrographs of *Campylodiscus* through *Tabularia* are from [11.9.57.].

Table 11.1 Composite compilation of knot symmetry classification by [11.9.21.] and surface identification.

Symmetry group	Surfaces (examples)	K and $\mathbb{R}^3$ orientation	Knot symmetry
Reflective	Polygonal	Both preserved	Chiral and non-invertible
Knot	Möbius band or strip	Both preserved	Chiral and non-invertible
Knot	Trefoil knot; Stevedore knot	Both preserved; or K reversed and $\mathbb{R}^3$ preserved	Chiral and invertible
Reflective	Polygonal	Both preserved	Fully chiral
Reflective	Polygonal	Both preserved; or K preserved and $\mathbb{R}^3$ reversed	Positive amphichiral and non-invertible
Cyclic; reflective; knot	Polygonal; unknot; figure eight knot	Both preserved, or both reversed	Negative amphichiral and non-invertible
Reflective; dihedral	Polygonal	Both preserved or reversed; or K reversed and $\mathbb{R}^3$ preserved; or K preserved and $\mathbb{R}^3$ reversed	Fully amphichiral and invertible
Cyclic or dihedral hyperbolic - knot	Hyperbolic knot	Both preserved or reversed; or K reversed and $\mathbb{R}^3$ preserved; or K preserved and $\mathbb{R}^3$ reversed	Fully amphichiral or invertible

11.4   Purposes of this Study

Using parametric 3D surface models and analyses of those surfaces via the Jacobian to determine 3D surface geometry, surface symmetry analysis many be accomplished. Cyclic, reflective, dihedral, glide, scale, and knot symmetries may be elucidated to describe the range of diatom surface symmetries. Exemplar genera are chosen to represent the diversity of symmetries and determine the relation between symmetry and stability and how changes in stability induce symmetry breaking. The analyses in this study are illustrative of the applicability of group theory to diatom surface symmetries. Each diatom surface symmetry is identifiable as symmetry groups, and group theory may be used to understand diatom surface symmetries.

The diatom developmental life cycle and size reduction may be important components of the adaptive value of surface symmetry. Diatoms undergo size reduction during vegetative reproduction according to the MacDonald-Pfitzer rule [11.9.36., 11.9.49., 11.9.50.].

Analyzing symmetry breaking and determining the role symmetry breaking may have in developmental size reduction during vegetative reproduction is a goal of this study. Spontaneous symmetry breaking is important in pattern formation [11.9.13.], and during vegetative reproduction, many diatom taxa slightly change shape and surface features during size reduction (e.g., [11.9.6.]). Such pattern formation is a dynamic process, and changes in symmetry induced by symmetry breaking may occur during each division during vegetative reproduction.

Gradients in symmetry changes in pattern formation may be established with regard to genera for a variety of diatom surface symmetries. Aside from the importance between symmetry and development, analysis of surface symmetries may be important in understanding aspects of diatom adaptation, including predation avoidance, substrate adherence, frustule strength, and ability to acquire nutrients. Attachment or adhesion capability to particular substrates and deflection of interfering factors to obtain nutrients, light or other elements for living and reproducing is important for diatoms to thrive. Surface symmetries are important for understanding such aspects of diatom evolution.

11.5 Methods

Examining all the possible diatom genera surfaces is a monumental task. However, a very small subset of taxa may be analyzed to illustrate the differences in diatom surface symmetry groups. To this end, *Actinoptychus*, *Arachnoidiscus*, *Campylodiscus* [11.9.33.], *Cocconeis* [11.9.26.], *Cyclotella* [11.9.35.], *Cylindrotheca* [11.9.56.], *Entomoneis* [11.9.9.], *Surirella* [11.9.18.], and *Tabularia* [11.9.27.] are used (Figure 11.5). These taxa were chosen as exemplars to illustrate the methodologies. From a geometrical perspective, systems of parametric 3D equations are used to devise 3D surface models of these exemplar genera for numerical analysis of diatom surface symmetry.

For one genus, surface symmetry is a 3D consideration whether viewing valve or girdle view. *Entomoneis* does not lie flat in either view, and the half-twists in the valves and girdle bands are indicative of its complicated symmetry. *Entomoneis tenera* cultured specimens reveal the sigmoidal shape of the elevated bilobate keel, having apices that are scalpel-like shaped in valve view [11.9.38.]. There are elongated slits at either end of the valves, indicating the half-twists at the apices [11.9.4.]. Light micrographs of *Entomoneis* valves from the Kowie River, South Africa reveal not only the half-twists at the apices, but also the half-twists in the middle where each valve crosses itself [11.9.4.]. The crossing sigmoidal pattern of girdle bands are evident as is the twisted keels as the scalpel-shaped "blades" do not lie in the same plane [11.9.4., 11.9.10.]. These four half-twists indicate the knot symmetry of *Entomoneis* as a figure-eight Möbius surface (Figures 11.4 and 11.5).

Surface symmetry may be understood as a dynamical process with respect to development, and in particular, with respect to size reduction during vegetative reproduction. Of the exemplar genera, some species of *Actinoptychus* [11.9.23.], *Arachnoidiscus* [11.9.55.], *Campylodiscus* [11.9.51.], *Cocconeis* [11.9.67.], *Cyclotella* [11.9.52.], *Cylindrotheca* [11.9.64.], *Entomoneis* [11.9.4., 11.9.10.], *Surirella* [11.9.65.], and *Tabularia* [11.9.11.] have been documented to undergo size reduction. Symmetry is measurable at each step in the size reduction process, and the degree of change is characterizable as degree of stability. The relation between surface symmetry and size reduction during vegetative reproduction is analyzable based on geometrically constructed 3D surface models.

11.5.1 Systems of Parametric 3D Equations for Exemplar Diatom Surface Models

For each exemplar genus, a system of parametric 3D equations represents the geometry that forms the basis from which symmetry groups are extracted. Surface models and systems of parametric 3D equations for each exemplar are given in Table 11.2. All models are constructed on the interval $u,v \in [0,2\pi]$ and illustrate valve view except for *Entomoneis*, which is in girdle view. Because 3D surface models are size invariant, with rotation and reflection involving a switch in parameters and a change in sign [11.9.44.], any view of a diatom frustule qualifies. Comparison is made exclusively on the geometric differences.

11.5.2 Symmetry Groups: Cyclic, Reflective, Dihedral, Glide, Scale, and Knot

Symmetry descriptors for taxa may be determined from an understanding of symmetry groups and applied to diatom surfaces (Figures 11.6-11.10). For 3D surface models of exemplars, symmetry equations may be obtained as the inverse of the parametric 3D equations. As a result, u, v is expressed in terms of x, y, z for each surface model. From group theory, determination of symmetry for each surface model is accomplished via solution to the inverse parametric 3D equations. Cyclic, reflective, dihedral, glide, scale symmetry groups may be determined in this way. Knot symmetry groups are determined with respect to amphichirality which encompasses cyclic, reflective, and dihedral symmetries where applicable to a given 3D surface model (see Table 11.1).

11.5.3 From Partial Derivatives to Ordinary Derivatives to Assess Stability

To determine stability, inverse parametric equations in the 2D u, v-plane are expressed as differential equations $du = \dfrac{\partial u}{\partial x} dx + \dfrac{\partial u}{\partial y} dy$ and $dv = \dfrac{\partial v}{\partial x} dx + \dfrac{\partial v}{\partial y} dy$. A parameterized surface as the height function $z = f(x,y)$ has differential equation $dz = \dfrac{\partial z}{\partial x}\left(\dfrac{\partial x}{\partial u} du + \dfrac{\partial x}{\partial v} dv\right) + \dfrac{\partial z}{\partial y}\left(\dfrac{\partial y}{\partial u} du + \dfrac{\partial y}{\partial v} dv\right)$. From thiserse of the Jacobian yields solutions to du and dv, ordinary derivatives, and when evaluated on the interval $[0,2\pi]$, will produce values that are positive, negative, or equal to zero that are used to determine instability, stability, or a mix of instability and st ordinary derivatives, ability, respectively.

11.5.4 Inverse Jacobian Eigenvalues and Surface Symmetry Analysis

Evaluating the inverse Jacobian at equilibrium and other symmetry states via a positive or negative sign enables the characterization of symmetry in terms of stability or instability. In this way, each symmetry group associated to each genus may be used to assess changes in stability or instability as a means of measuring changes in symmetry.

To identify symmetry groups per 3D surface model of exemplars, symmetry in each equation in a system of parametric 3D equations is used. Symmetry in the solutions to first

Table 11.2 Surface models constructed via systems of parametric 3D equations for exemplar diatom genera showing valve view except where noted.

Exemplar	Surface model	System of parametric 3D equations
Actinoptychus		$x = 18 \cos v (1+ \cos 2u^2)$ $y = 18 \sin v (1+ \cos 2u^2) \sin 2u^2$ $z = \sin 3u - \cos 4v \sin 80u \cos 80u \sin 4v$ $\quad - \cos 3v + \cos(10u - 3) + \cos 30u \sin 60u$
Arachnoidiscus		$x = 80 \cos v (1 + \sin u\, 0.2 \cos u)$ $y = 80 \sin v (1 + \sin u\, 0.2 \cos u)$ $z = 1.4 \sin 22v^3 \sin 0.37u^3\, 0.5 \cos u + \sin 1.9u^2$ $\quad 0.5 \cos 80u^2 + 0.5 \sin(2.9u + 2)^3\, 3 \sin (0.8u$ $\quad + 0.5)^3\, 1.7 \cos(74u^2 + 10) + 2 \sin (u + 1)^3$ $\quad + 0.75 \cos 0.5u^2$
Campylodiscus		$x = \cos u - (0.2 \cos v)^4 (0.2 \sin u)^4$ $y = \cos 1.05v \sin u - (0.2 \sin v)^4 (\cos u)^4$ $z = (0.2v \cos 0.5v)^{0.25} (\sin 1.8v)^{0.25}$ $\quad - ((0.2 \cos 0.5u)^6 (0.2u \sin 1.05u)^6)^{0.25}$ $\quad + (0.1(\sin 2v \sin 16u))^6$ $\quad - (0.5v \sin 0.5v\, u \sin u) (\cos u \cos v)$
Cocconeis		$x = 0.2 \cos v$ $y = \sin v^3 \cos u$ $z = \sin u$ (araphid valve view)
Cyclotella		$x = 80 \cos v (1 + \sin u\, 0.2 \cos u)$ $y = 80 \sin v (1 + \sin u\, 0.2 \cos u)$ $z = 1.4 \sin 7v^4 \sin 0.5u^4\, 0.1 \cos 1.5u \cos u$ $\quad + 0.5 \sin 1.8u^6\, 6 \cos 3.5u^6 \cos 25v \sin 25v$ $\quad + 1.5 \sin 2.9u \sin 0.5u \cos 1.5u \cos 2u$
Cylindrotheca		$x = 0.5 \sin 3v^3\, 0.3 \cos 2u\, 6 \sin u$ $y = 0.2 \tan (u - 5)\, 0.2 \tan (v + 5)$ $z = \sin u - 12 \cos u$

(Continued)

Table 11.2 Surface models constructed via systems of parametric 3D equations for exemplar diatom genera showing valve view except where noted. (*Continued*)

Exemplar	Surface model	System of parametric 3D equations
Entomoneis		$x = 0.1 \cos 12v^3 + 0.1 \cos 16v^3$ $y = 0.6\,(\cos 14.4v^3 \cos 1.5u^2 \sin 18v^3 \sin 1.5u^2$ $\qquad + 0.5 \cos v^3 \cos u^3$ $\qquad + 0.6 \sin u \cos u + 0.5 \cos 3v^{0.5} \sin v^3)$ $z = 0.6 \sin u$ (girdle view)
Surirella		$x = 0.2 \sin 0.3v^6 \cos 0.2u^2$ $y = 2 \cos 0.1v \sin 0.5u$ $z = \sin v^3 \sin u$
Tabularia		$x = 0.1 \cos v$ $y = 0.1 \sin 01.2v^3 \cos u$ $z = \sin u$

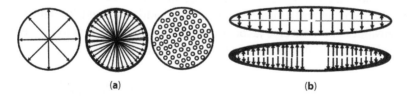

(a)   (b)

Figure 11.6 Depiction of: a, cyclic; b, reflective symmetries in diatom surfaces.

Figure 11.7 Depiction of dihedral symmetry in diatom surfaces.

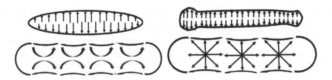

Figure 11.8 Depiction of glide symmetry in diatom surfaces.

Figure 11.9 Depiction of scale symmetry in diatom surfaces.

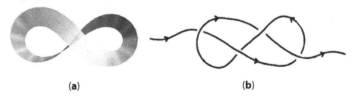

(a)   **(b)**

Figure 11.10 Depiction of: a, Möbius surface figure eight knot symmetry; b, figure eight knot symmetry in diatom surfaces.

partial derivatives of each parametric equation is attained and evaluated. Both the equations and their solutions need not be symmetric [11.9.13.]. If the equations are symmetric and the solutions are not, then symmetry breaking is present [11.9.13.].

11.5.5   Stability and Inverse Jacobian Eigenvalues

To determine the stability of symmetry groups, partial differential equations in x, y, z with parameters u, v are set equal to zero which is equilibrium in each 3D surface direction. At equilibrium, the Jacobian and its inverse is calculated and evaluated. To determine symmetry state, eigenvalues of the inverse Jacobian are calculated according to $J_{Inverse}\mathbf{e} = \lambda\,\mathbf{e}$, where λ is an eigenvalue and $\mathbf{e}$ is an eigenvector with $\mathbf{e} \neq 0$, and

$$J_{Inverse} = J_{u,v} = \begin{bmatrix} \dfrac{\partial u}{\partial x} & \dfrac{\partial v}{\partial x} & \dfrac{\partial u}{\partial y}\times\dfrac{\partial v}{\partial z} - \dfrac{\partial u}{\partial z}\times\dfrac{\partial v}{\partial y} \\[2.5ex] \dfrac{\partial u}{\partial y} & \dfrac{\partial v}{\partial y} & \dfrac{\partial u}{\partial z}\times\dfrac{\partial v}{\partial x} - \dfrac{\partial u}{\partial x}\times\dfrac{\partial v}{\partial z} \\[2.5ex] \dfrac{\partial u}{\partial z} & \dfrac{\partial v}{\partial z} & \dfrac{\partial u}{\partial x}\times\dfrac{\partial v}{\partial y} - \dfrac{\partial u}{\partial y}\times\dfrac{\partial v}{\partial x} \end{bmatrix} \qquad (11.2)$$

From Cramer's rule, the characteristic equation is $\det J_{Inverse} = -\lambda I = 0$, where the left side of the equation in the general form of the characteristic polynomial [11.9.66.] is $f(\lambda) = \lambda^n - \text{Tr}(J_{Inverse})_1\lambda^{n-1} + \cdots + (-1)^n \det(J_{Inverse})$, where Tr is the trace of the inverse Jacobian, and det $(J_{Inverse}) \neq 0$, indicating local invertibility in the neighborhood of a given point. Expanding the characteristic equation results in $\det(J_{Inverse}) - \lambda I = (-1)^n(\lambda - \lambda_1)(\lambda - \lambda_2) \cdots (\lambda - \lambda_n)$. Let $\lambda = 0$, then the characteristic equation becomes $\det J_{Inverse} = \lambda_1\lambda_2 \ldots \lambda_n$, which is the product of the eigenvalues.

For the generalized inverse Jacobian representing a system of homogeneous linear equations as $(J_{Inverse} - \lambda I)\,\mathbf{e} = \mathbf{0}$, with eigenvectors as column vectors $\mathbf{e} = \text{col}(e_1, \ldots, e_n)$, and eigenvalues, λ, the determinantal equation is

$$
\begin{bmatrix}
j_{11} - \lambda & j_{12} & \cdots & j_{1n} \\
j_{21} & j_{22} - \lambda & \cdots & j_{2n} \\
\vdots & \vdots & & \vdots \\
j_{n1} & j_{n2} & \cdots & j_{nn} - \lambda
\end{bmatrix} = 0,
\tag{11.3}
$$

where j_{ik} are the ith, kth elements of the inverse Jacobian, and $i = 1, 2, \ldots n$ for rows and $k = 1, 2, \ldots n$ for columns.

Signs of the eigenvalues are used to determine stability or instability. Complex- or real-valued solutions to the characteristic equation for λ indicate symmetry states and are evaluated with respect to their sign: negative for stability; positive for instability; mixed signs are labelled as instability, although both instability and stability are present. Changing stability into instability or vice versa is obtained from the trace of the inverse Jacobian eigenvalues. The trace is calculated as

$$
\text{Tr}(J_{Inverse}) \equiv \sum_{i=1}^{n} j_{ii}
\tag{11.4}
$$

which is the sum of the diagonal elements [11.9.66.].

11.5.6   Diatom Surface Symmetries and Symmetry Group Assessment

Diatom exemplars used in symmetry analysis are classified according to symmetry groups, and assignment to more than one symmetry group may result as a possibility. Eigenvalues from the inverse Jacobian are symmetry state values on the interval $[0, 2\pi]$ to obtain solutions to du and dv for each exemplar. At each value of u and v used to solve du and dv, instability or stability is assessed based on the sign of the eigenvalues of the inverse Jacobian. A plot of the trace versus the symmetry states is devised for each exemplar, and an nth order polynomial curve is fitted. A coefficient of determination is obtained, and the rate of change in symmetry breaking with respect to symmetry states is calculated as the first derivative and evaluated at a symmetry state of one. Symmetry breaking rate is associated to symmetry group.

11.5.7   Vegetative Size Reduction and Symmetry Breaking

Multiple vegetative size reductions curves are constructed as the change in apical length or diameter over a period of time. Curves range in rates of change from almost no size reduction to fastest size reduction and are patterned after Amato *et al.*'s [11.9.6.] Figure 6. A best-fit is determined for each curve, and first derivatives are calculated as size reduction rates. The relation between symmetry breaking rates and size reduction rates is plotted and assessed.

11.5.8   Relative Stability and Symmetry

Relative stability of exemplar genera is indicative of their surface symmetries with regard to each other. A comparative analysis of this relative stability is compiled as a stacked bar graph and assessed in relation to surface symmetries. Similarities and differences in stability relative to exemplar genera are assessed via symmetry states.

11.5.9   Symmetry Gradients

A stability-instability matrix is devised and based on 0 for instability and 1 for stability for all exemplar genera. Seriation is a combinatorial optimization method [11.9.16.] that is NP-hard in which heuristics are used in achieving an optimal solution. A linear sequence is determined in light of a loss or merit function. Permutation of the matrix occurs in which symmetry groups (columns) and symmetry states (rows) are reordered in unconstrained seriation, while symmetry states (rows) are reordered in constrained seriation [11.9.17.]. Implicitly, exemplar genera are represented by their symmetry group assignments. A symmetry gradient in each case is produced along the diagonal of the reordered matrix as the solution and is illustrated as a visualization by plotting the pattern of each symmetry gradient.

The stability-instability matrix of symmetry states, $Obj = \{obj_1, \ldots, obj_n\}$ consists of dissimilarities, $\mathbf{Dis} = (dis_{ij})$, where $1 \leq i, j \leq n$. Define a permutation function, Ψ_π as $\Psi_\pi = \arg\min_\psi Loss(\Psi_\pi(\mathbf{Dis}))$ for a loss function, $Loss$, or $\Psi_\pi = \arg\max_\psi Merit(\Psi_\pi(\mathbf{Dis}))$ for $Merit$, a merit function [11.9.16.]. Random sampling of dissimilarities ($\mathbf{Dis}$) is used as input for permutation. Monte Carlo simulation is used to produce best-probable solutions with respect to a criterion as a result of permuted outcomes [11.9.17.]. From the permutation function, Ψ_π, $\mathbf{P}_\pi \mathbf{Dis} \mathbf{P}_\pi^T$ results where $\mathbf{P}_\pi = \psi_\pi(\mathbf{I}_\pi)$ is the permutation matrix of permutation π for identity matrix $\mathbf{I}_\pi$ [11.9.15.]. The criterion used for gradient conditions is $dis_{ik} \leq dis_{ij}$ within rows and $dis_{kj} \leq dis_{ij}$ within columns where $1 \leq i < k < j \leq n$ [11.9.15., 11.9.22.]. The generalized combinatorial optimization problem is $Z = \dfrac{dis_{ij} - dis_{Mean}}{dis_{Standard\ Deviation}} = Loss(\Psi_\pi(\mathbf{Dis}))$ minimized with $\psi_\pi \in \Psi$ [11.9.15.] as seriation. Reported outcome is the average of outcomes from Monte Carlo simulation as the optimal unconstrained and constrained symmetry gradients.

11.6   Results

Parametric 3D models were constructed for *Actinoptychus, Arachnoidiscus, Campylodiscus, Cocconeis, Cyclotella, Cylindrotheca, Entomoneis, Surirella,* and *Tabularia* as the surface symmetry groups of dihedral, cyclic-dihedral, scale (hyperbolic), scale (polygonal), cyclic, twist-dihedral, knot, glide-reflective, and reflective-dihedral, respectively. *Entomoneis* was determined to be chiral and cyclic, and dihedral taxa were amphichiral (Table 11.1). Exclusively reflective taxa may be either amphichiral or chiral. The closest any of the exemplar genera fell into this category was *Surirella*, which is chiral (Table 11.1).

Eigenvalues of the inverse Jacobians for each genus were calculated and evaluated with respect to symmetry states selected on the interval $[0, 2\pi]$ and reported in Tables 11.3–11.11. Symmetry states for each genus were assessed in terms of stability or instability. For *Actinoptychus*, four consecutive negative eigenvalue pairs occurred at the higher valued states denoting stability (Table 11.3). Stability extended across states for *Surirella* (Table 11.10) and *Campylodiscus* (Table 11.5). For *Entomoneis*, three stability points extended across all symmetry states (Table 11.9). At two states, *Cylindrotheca* (Table 11.8) exhibited stability, while *Arachnoidiscus, Cocconeis* and *Tabularia*, having only one negative eigenvalue pair exhibited stability at different single lower valued states (Tables 11.4, 11.6 and 11.11). *Cyclotella* exhibited instability at all symmetry states (Table 11.7).

Evaluating symmetry states for diatom surfaces resulted in complex-valued eigenvalues to varying degrees (Tables 11.3–11.11), indicating Hopf bifurcation [11.9.13.]. Such values are expected as a result of the characterization of the geometry of 3D surface models

Table 11.3 *Actinoptychus* values for u and v and Eigenvalues of the inverse Jacobian, indicating stability states. Equilibrium is at symmetry state $u = v = 0$.

Symmetry states: $(u, v) \mapsto (u, v)$ $(u, v) \mapsto (v, u)$ $(u, v) \mapsto (-u, -v)$ $(u, v) \mapsto (-v, -u)$	**Eigenvalue 1 of inverse Jacobian**	**Eigenvalue 2 of inverse Jacobian**	**Stability state**
0	Undefined	Undefined	Unstable
0.02	6.87129+0i	0.226887+0.328612i	Unstable
0.785	0.0434045	-0.0348904	Unstable
1.57	0.000911065+0.019146i	0.000911065-0.019146i	Unstable
2.36	0.026728	0.0132037	Unstable
3.14	-0.04418	-0.00726147	Stable
4.7	-0.00245655+0.0120363i	-0.00245655-0.120363i	Stable
5.5	-0.116803	-0.00487558	Stable
6.08	-0.0488457	-0.00261645	Stable
6.28	-0.558375	0.0260928	Unstable

Table 11.4 *Arachnoidiscus* values for u and v and Eigenvalues of the inverse Jacobian, indicating stability states. Equilibrium is at symmetry state $u = v = 0$.

Symmetry states: $(u, v) \mapsto (u, v)$ $(u, v) \mapsto (v, u)$ $(u, v) \mapsto (-u, -v)$ $(u, v) \mapsto (-v, -u)$	Eigenvalue 1 of inverse Jacobian	Eigenvalue 2 of inverse Jacobian	Stability state
0	0.0625	0.0612379	Unstable
0.02	0.0615606+0.001199i	0.0615606-0.001199i	Unstable
0.785	0.0370945+0i	-0.000469843+0.00192421i	Unstable
1.57	-1.1499+0.06354i	-1.1499-0.06354i	Stable
2.36	0.035365+0i	0.00171146+0.00366533i	Unstable
3.14	0.0624002	-0.0615322	Unstable
4.7	32120.7+0i	-0.0213439+23.6017i	Unstable
5.5	0.211425+0i	0.0000182457+0.000914534i	Unstable
6.08	0.0784042+0i	0.00187204+0.00346599i	Unstable
6.28	0.0626997+0i	0.0000746228+0.000736916i	Unstable

Table 11.5 *Campylodiscus* values for u and v and Eigenvalues of the inverse Jacobian, indicating stability states. Equilibrium is at symmetry state $u = v = 0$.

Symmetry states: $(u, v) \mapsto (u, v)$ $(u, v) \mapsto (v, u)$ $(u, v) \mapsto (-u, -v)$ $(u, v) \mapsto (-v, -u)$	Eigenvalue 1 of inverse Jacobian	Eigenvalue 2 of inverse Jacobian	Stability state
0	Indeterminate	Indeterminate	Unstable
0.02	4.65649+0i	-2.13224+3.68584i	Unstable
0.785	2.0907	-1.87715	Unstable
1.57	-0.985221	-0.934115	Stable
2.36	-0.869229+0.2719i	-0.826152-0.313567i	Stable
3.14	0.276078-0.246465i	-0.064579-0.101075i	Unstable
4.7	-0.969341-0.0746395i	0.965817-0.0000276752i	Unstable
5.5	1.38322+1.45404i	0.766193-0.349867i	Unstable
6.08	-0.700448-0.623335i	-0.752761+0.408853i	Stable
6.28	-11.6979-10.7067i	-8.84359-0.837833i	Stable

Table 11.6 *Cocconeis* values for u and v and Eigenvalues of the inverse Jacobian, indicating stability states. Equilibrium is at symmetry state $u = v = 0$.

Symmetry states: $(u, v) \longmapsto (u, v)$ $(u, v) \longmapsto (v, u)$ $(u, v) \longmapsto (-u, -v)$ $(u, v) \longmapsto (-v, -u)$	Eigenvalue 1 of inverse Jacobian	Eigenvalue 2 of inverse Jacobian	Stability state
0	Undefined	Undefined	Unstable
0.02	77.4741+0i	-4.33494+26.865i	Unstable
0.785	-0.0691868+1.31865i	-0.0691868-1.31865i	Stable
1.57	7.51199+0i	-0.0165071+2.74075i	Unstable
2.36	0.451948	-0.44224	Unstable
3.14	-0.194088	0.194088	Unstable
4.7	26.4158+8.42306i	26.4158-8.42306i	Unstable
5.5	0.176896	-0.176502	Unstable
6.08	0.00031377+0.270195i	0.00031377-0.270195i	Unstable
6.28	0.0984837	-0.0984837	Unstable

Table 11.7 *Cyclotella* values for u and v and Eigenvalues of the inverse Jacobian, indicating stability states. Equilibrium is at symmetry state $u = v = 0$.

Symmetry states: $(u, v) \longmapsto (u, v)$ $(u, v) \longmapsto (v, u)$ $(u, v) \longmapsto (-u, -v)$ $(u, v) \longmapsto (-v, -u)$	Eigenvalue 1 of inverse Jacobian	Eigenvalue 2 of inverse Jacobian	Stability state
0	0.0625	0.0625	Unstable
0.02	0.0627581	0.0623291	Unstable
0.785	0.0534571	-0.0523539	Unstable
1.57	1.28268+0i	-0.857621+0.568279i	Unstable
2.36	0.0495807+0.0456595i	0.0495807-0.0456595i	Unstable
3.14	-0.0624665	0.0624007	Unstable
4.7	82.4915+0i	-0.595967+9.87935i	Unstable
5.5	0.137303+0.0866626i	0.137303-0.0866626i	Unstable
6.08	0.048985+0.00865545i	0.048985-0.00865545i	Unstable
6.28	0.0623024+0i	0.00254994+0.00350022i	Unstable

Table 11.8 *Cylindrotheca* values for u and v and Eigenvalues of the inverse Jacobian, indicating stability states. Equilibrium is at symmetry state $u = v = 0$.

Symmetry states: $(u, v) \mapsto (u, v)$ $(u, v) \mapsto (v, u)$ $(u, v) \mapsto (-u, -v)$ $(u, v) \mapsto (-v, -u)$	Eigenvalue 1 of inverse Jacobian	Eigenvalue 2 of inverse Jacobian	Stability state
0	0+0.771401i	0-0.771401i	Unstable
0.02	0.0000428606+0.639402i	0.0000428606-0.639402i	Unstable
0.785	-10.4737	0.425283	Unstable
1.57	-0.187152+0.0324752i	-0.187152-0.0324752i	Stable
2.36	8.82566+0i	0.11318+0.392586i	Unstable
3.14	0.593299+0i	0.400096+0.413478i	Unstable
4.7	-0.00320836+0.0056039i	-0.00320836-0.0056039i	Stable
5.5	0.414779+0i	-0.113991+0.255723i	Unstable
6.08	0.0744671+0.0885831i	0.0744671-0.0885831i	Unstable
6.28	0.59022+0i	0.407182+0.402409i	Unstable

Table 11.9 *Entomoneis* values for u and v and Eigenvalues of the inverse Jacobian, indicating stability states. Equilibrium is at symmetry state $u = v = 0$.

Symmetry states: $(u, v) \mapsto (u, v)$ $(u, v) \mapsto (v, u)$ $(u, v) \mapsto (-u, -v)$ $(u, v) \mapsto (-v, -u)$	Eigenvalue 1 of inverse Jacobian	Eigenvalue 2 of inverse Jacobian	Stability state
0	Indeterminant	Indeterminant	Unstable
0.02	-0.232356+1.28954i	-0.232356-1.28954i	Stable
0.785	3.87266	-3.75244	Unstable
1.57	34.3423	29.1191	Unstable
2.36	0.245631+2.32679i	0.245631-2.32679i	Unstable
3.14	-0.0263676+0.791777i	-0.0263676-0.791777i	Stable
4.7	-34.6679	1.93554	Unstable
5.5	-0.516999	0.516999	Unstable
6.08	-0.373508+1.06921i	-0.373508-1.06921i	Stable
6.28	0.536846	-0.503104	Unstable

Table 11.10 *Surirella* values for u and v and Eigenvalues of the inverse Jacobian, indicating stability states. Equilibrium is at symmetry state $u = v = 0$.

Symmetry states: $(u, v) \mapsto (u, v)$ $(u, v) \mapsto (v, u)$ $(u, v) \mapsto (-u, -v)$ $(u, v) \mapsto (-v, -u)$	Eigenvalue 1 of inverse Jacobian	Eigenvalue 2 of inverse Jacobian	Stability state
0	Undefined	Undefined	Unstable
0.02	1203+0i	-602.725+1039.27i	Unstable
0.785	-0.642181+0.774496i	-0.642181-0.774496i	Stable
1.57	0.429247+0i	-0.140772+0.365568i	Unstable
2.36	-0.865115	-0.305232	Stable
3.14	-0.344017+0.653215i	-0.344017-0.653215i	Stable
4.7	-0.0949488+0.0539284i	-0.0949488-0.0539284i	Stable
5.5	0.0397434	-0.0309367	Unstable
6.08	-0.00102341+0.0209085i	-0.00102341-0.0209085i	Stable
6.28	0.108279	-0.105889	Unstable

Table 11.11 *Tabularia* values for u and v and Eigenvalues of the inverse Jacobian, indicating stability states. Equilibrium is at symmetry state $u = v = 0$.

Symmetry states: $(u, v) \mapsto (u, v)$ $(u, v) \mapsto (v, u)$ $(u, v) \mapsto (-u, -v)$ $(u, v) \mapsto (-v, -u)$	Eigenvalue 1 of inverse Jacobian	Eigenvalue 2 of inverse Jacobian	Stability state
0	Undefined	Undefined	Unstable
0.02	77.1525+0i	-20.6703+52.897i	Unstable
0.785	6.13159+0i	-0.253688+3.82056i	Unstable
1.57	100.229+0i	-0.00245616+10.0118i	Unstable
2.36	-0.000295393+1.18109i	-0.000295393-1.18109i	Stable
3.14	-0.574539	0.574539	Unstable
4.7	111.015+0i	-2.6681+10.3181i	Unstable
5.5	0.00533392+1.27406i	0.00533392-1.27406i	Unstable
6.08	8.81149E-6+0.299516i	8.81149E-6+0.299516i	Unstable
6.28	0.468086	-0.468086	Unstable

for each genus. Where the complex-valued eigenvalues occurred in relation to real-valued eigenvalues, bifurcation was indicated at a given symmetry state. In some cases, bifurcation was indicated across both eigenvalues, while in other cases, bifurcation was indicated in one or the other eigenvalue. The least number of bifurcation events occurred in *Actinoptychus*, as values increased and stability was evident (Table 11.3). The next fewest bifurcation events were attributed to *Cocconeis* (Table 11.6), then *Cyclotella* (Table 11.7), although instability was highly evident. *Entomoneis* had four bifurcation events, most of which were exemplified by stability (Table 11.9). Most of the bifurcation events occurred with regard to the second eigenvalue for *Arachnoidiscus* (Table 11.4) and *Tabularia* (Table 11.11), and again, instability was the dominant attribution. The majority of eigenvalues were complex-valued for *Surirella* (Table 11.10), *Cylindrotheca* (Table 11.8) and *Campylodiscus* (Table 11.5), which were somewhat aligned with the state of stability in each genus.

The trace of the inverse Jacobian eigenvalues was calculated for each genus and plotted versus the symmetry states of $u = v$ (Figure 11.11). The trace represents total symmetry breaking. Sixth-order polynomial curve fits were obtained for each genus (Table 11.12), and first derivatives were calculated as symmetry breaking rates (Table 11.13). Curve fits for *Arachnoidiscus* and *Cyclotella* were similar, while the curve fit for *Cocconeis* showed similarities except for the first data point (Figure 11.11). *Cylindrotheca* exhibited approximately a rotated and flipped mirror image curve fit when compared to curve fits for *Arachnoidiscus*, *Cyclotella* and *Cocconeis*. Curve fit for *Actinoptychus* was a rotated and flipped mirror image of the curve fit for *Campylodiscus* which was almost a mirror image of the curve fit for *Surirella* (Figure 11.11). The most unique curve fits belong to *Entomoneis* and *Tabularia*, both depicting almost periodic functions ((Figure 11.11).

Size reduction rate curves were constructed (Table 11.14) based on Amato *et al.*'s [11.9.6.] Figure 6 of size versus time. These plots indicated that size reduction occurs as a negative exponential function. The series of size reduction curves illustrated the change in rate from fastest size reduction (Figure 11.12) with the largest negative value to almost no size reduction with a value close to zero (Table 11.15). Size reduction rates were calculated from the first derivative of nth-degree polynomial curve fits, where almost no size reduction and least size reduction exhibited a 1st-order curve fit, lesser size reduction had a 2nd-order curve fit, and modest to fastest size reduction had a 3rd-order curve fit (Figure 11.12).

Symmetry breaking rate versus size reduction rate was plotted (Figure 11.13). The relation between these rates indicated that as size reduction slows, symmetry breaking slows (Figure 11.13). The overall trend followed a negative exponential curve (Figure 11.13).

Stability with respect to symmetry state was plotted as a stacked bar graph depicting the relative stability status of exemplar genera with their symmetry groups (Figure 11.14). Each genus in the Surirellales—*Campylodiscus*, *Entomoneis* and *Surirella*—had a different stability profile with reference to their differences in symmetries (Figure 11.14). Circular genera, typically called centrics—*Actinoptychus*, *Arachnoidiscus* and *Cyclotella*—had different stability profiles, while traditionally identified as pennates, *Cocconeis*, *Cylindrotheca*, and *Tabularia* had different stability profiles as well (Figure 11.14).

Seriation was used to analyze stability status of symmetry for exemplar genera and depicted in plots of surface symmetry gradients (Figure 11.15). Stability status from Tables 11.3 – 11.11 was scored as 0 for unstable and 1 for stable and assembled in a matrix. Preordering the columns of symmetry groups assigned to genera occurred from minimum to maximum number of stabilities (1s). Thirty random matrices were generated via Monte

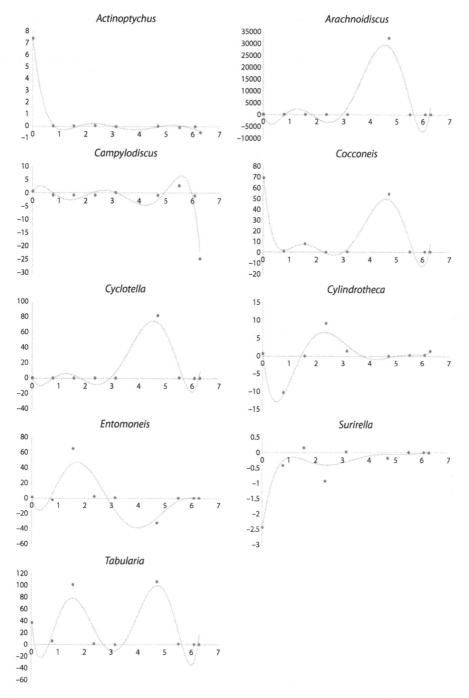

Figure 11.11 Genus plots of total symmetry breaking (y-axes) versus symmetry states (x-axes) from Tables 11.3 – 11.11.

Table 11.12 Best-fit equations for genera of symmetry breaking (*Br*) with respect to symmetry states (*St*).

Genus	Best-fit equation	Coefficient of determination
Actinoptychus	$Br = 0.0097St^6 - 0.2226St^5 + 1.9862St^4 - 8.7248St^3 + 19.512St^2 - 20.56St + 7.7171$	$R^2 = 0.9968$
Arachnoidiscus	$Br = 214.95St^6 - 3751.7St^5 + 23791St^4 - 66780St^3 + 81399St^2 - 33747St + 329.19$	$R^2 = 0.8784$
Campylodiscus	$Br = -0.1207St^6 + 2.0646St^5 - 13.081St^4 + 37.605St^3 - 47.513St^2 + 19.593St - 0.1264$	$R^2 = 0.9119$
Cocconeis	$Br = 0.5222St^6 - 9.7266St^5 + 68.184St^4 - 225.39St^3 + 364.17St^2 - 268.7St + 73.864$	$R^2 = 0.9457$
Cyclotella	$Br = 0.5401St^6 - 9.4231St^5 + 59.719St^4 - 167.49St^3 + 204.08St^2 - 84.873St + 0.9624$	$R^2 = 0.8785$
Cylindrotheca	$Br = 0.0664St^6 - 1.4654St^5 + 12.471St^4 - 50.514St^3 + 95.412St^2 - 64.916St + 1.2832$	$R^2 = 0.9289$
Entomoneis	$Br = 0.2881St^6 - 6.2824St^5 + 51.299St^4 - 189.76St^3 + 298.02St^2 - 136.12St + 2.211$	$R^2 = 0.7161$
Surirella	$Br = -0.0023St^6 + 0.0547St^5 - 0.5193St^4 + 2.4258St^3 - 5.7051St^2 + 6.1738St - 2.5961$	$R^2 = 0.8946$
Tabularia	$Br = 1.4929St^6 - 28.046St^5 + 196.25St^4 - 626.49St^3 + 886.5St^2 - 428.56St + 42.283$	$R^2 = 0.8095$

Carlo simulation, then seriated to achieve the resultant optimal ordering. Unconstrained seriation with a criterion of 0.666667 was used to construct a symmetry gradient (Figure 11.15a). Order of the symmetry gradient was cyclic-dihedral→twist-dihedral→ reflective-dihedral→scale (hyperbolic)→scale (polygonal)→glide-reflective→dihedral→ knot (Figure 11.15a). Constrained seriation with a criterion of 0.461538 produced a second symmetry gradient (Figure 11.15b). Statistics for one permutation cycle of constrained seriation were: mean = 0.396641; standard deviation = 0.0396904; Z-score = -1.6351; p-value = 0.102028. Order of this symmetry gradient was cyclic-dihedral→reflective-dihedral→s-cale (polygonal)→twist-dihedral→knot→scale (hyperbolic)→dihedral→glide-reflective (Figure 11.15b).

Table 11.13 Symmetry group and breaking rates obtained from 6th-order polynomial curve fits for each genus plot of the trace of inverse Jacobian eigenvalues versus symmetry state. Rates calculated as first derivatives with $St = 1$, where St is symmetry state.

Genus	Symmetry breaking rate	Symmetry group
Actinoptychus	-0.8204	Dihedral
Arachnoidiscus	6406.2	Cyclic-dihedral
Campylodiscus	-5.3432	Scale (hyperbolic)
Cocconeis	10.8862	Scale (polygonal)
Cyclotella	15.8181	Cyclic
Cylindrotheca	15.3214	Twist-dihedral
Entomoneis	66.1526	Knot
Surirella	0.2235	Glide-reflective
Tabularia	118.6974	Reflective-dihedral

Table 11.14 Best-fit equations for size reduction curve fits of apical length (l) versus time (t).

Size reduction descriptor	Best-fit equation	Coefficient of determination
Almost no size reduction	$l = -0.0072t + 75.24$	$R^2 = 0.9114$
Least size reduction	$l = -0.0898t + 76.089$	$R^2 = 0.9134$
Lesser size reduction	$l = 0.0012t^2 - 0.4166t + 82.916$	$R^2 = 0.997$
Modest size reduction	$l = -1\mathrm{E} - 05t^3 + 0.0066t^2 - 1.031t + 93.317$	$R^2 = 0.9953$
Faster size reduction	$l = -3\mathrm{E} - 05t^3 + 0.0111t^2 - 1.5164t + 100.53$	$R^2 = 0.9983$
Fastest size reduction	$l = -3\mathrm{E} - 05t^3 + 0.0132t^2 - 1.7758t + 103.96$	$R^2 = 0.9975$

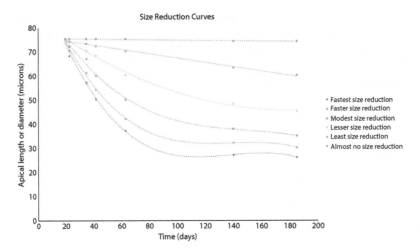

Figure 11.12 Size reduction rate curves constructed as based on [11.9.6.], Figure 6. Rates are illustrated from fastest to almost no size reduction, from bottom to top curve, respectively.

Table 11.15 Size reduction rates from curves plotted of apical length (*l*) versus time (*t*). Rates calculated as first derivatives with *t* = 1.

Size reduction rate curve	Size reduction rate
Fastest size reduction	-1.74955
Faster size reduction	-1.49435
Modest size reduction	-1.01783
Lesser size reduction	-0.4142
Least size reduction	-0.0898
Almost no size reduction	-0.0072
No size reduction	0

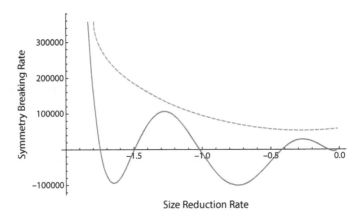

Figure 11.13 Plot of symmetry breaking rate versus size reduction rate. Dashed curve depicts a negative exponential trend.

Relative Stability Status
Stability
4.5
4
3.5
3
2.5
2
1.5
1
0.5
0
0
0.02
0.785
1.57
2.36
3.14
4.7
5.5
6.08
6.28
Symmetry State
Tabularia - reflective-dihedral
Surirella - glide-reflective
Entomoneis - knot
Cylindrotheca - twist-dihedral
Cyclotella - cyclic
Cocconeis - scale (polygonal)
Campylodiscus - scale (hyperbolic)
Arachnoidiscus - cyclic-dihedral
Actinoptychus - dihedral

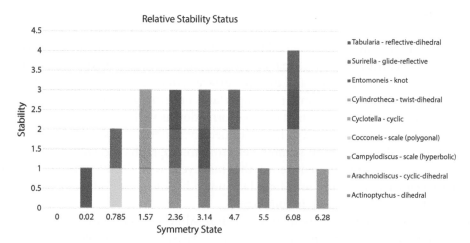

Figure 11.14 Stability versus symmetry state for exemplar genera and their symmetry groups.

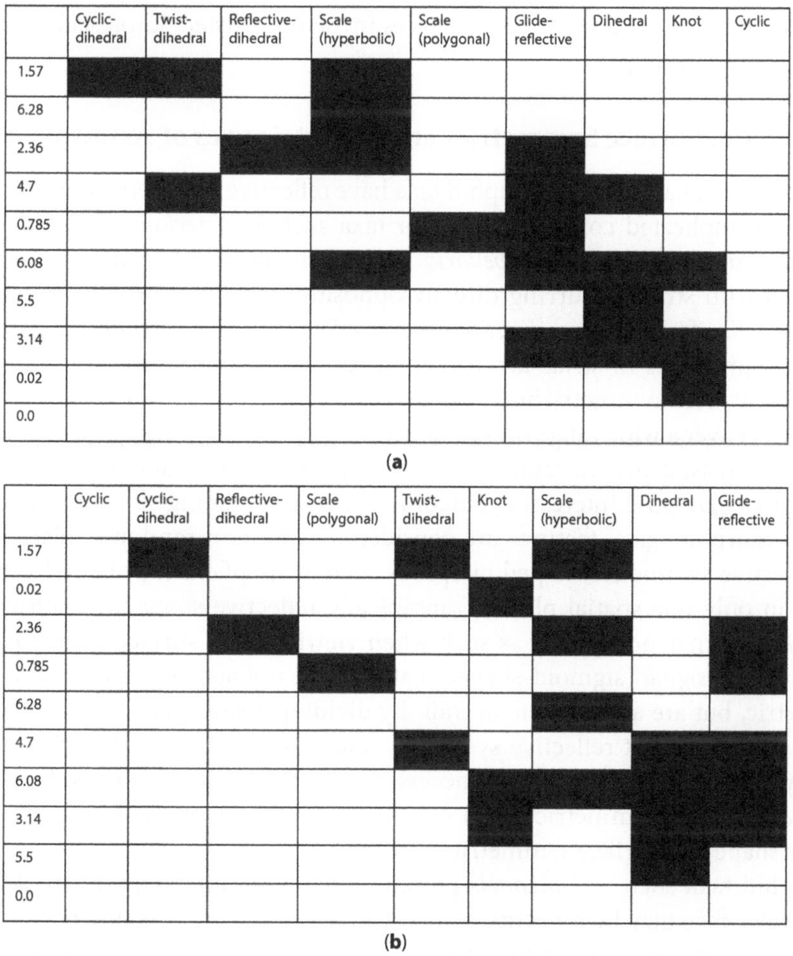

Cyclic-dihedral
Twist-dihedral
Reflective-dihedral
Scale (hyperbolic)
Scale (polygonal)
Glide-reflective
Dihedral
Knot
Cyclic
1.57
6.28
2.36
4.7
0.785
6.08
5.5
3.14
0.02
0.0
(a)
Cyclic
Cyclic-dihedral
Reflective-dihedral
Scale (polygonal)
Twist-dihedral
Knot
Scale (hyperbolic)
Dihedral
Glide-reflective
1.57
0.02
2.36
0.785
6.28
4.7
6.08
3.14
5.5
0.0
(b)

Figure 11.15 Plots of surface symmetry as gradients using a, unconstrained and b. constrained seriation.

11.7   Discussion

Historically, diatoms have been assigned to two types of symmetry, roughly based on valve shape: cyclic for centrics and reflective for pennates. It is known that there are eccentric centrics as well as circular pennates, and diatom surface symmetry analysis reflects this more complicated picture. Surface symmetry is important concerning size reduction during vegetative reproduction and is important in valve formation as well [11.9.48.].

Diatom surface symmetry analysis has identified combinations of symmetries per genus. Perusing a text of diatom light or scanning electron micrographs should invoke a sense of multiple symmetries being present rather that just cyclic or reflective symmetry. Surface symmetries regularly and commonly found include dihedral, glide and scale. In rare cases, knot symmetry is evident as seen in *Entomoneis*.

Diatom surface symmetries are mainly dihedral. Almost every diatom surface involves a combination of cyclic and reflective symmetries. Twist and hyperbolic geometries and knot symmetry typify the 3D nature of diatom symmetry, and dihedral symmetry might be found in a given dimension of similarly projected 3D diatom surfaces. In the constrained symmetry gradient plot (Figure 11.15), genera ranged from dihedral-based symmetry groups to increasingly complicated 3D surfaces involving twisted, knot, scale (hyperbolic), and glide symmetry groups.

11.7.1   Diatom Surface Symmetries and the Intricacies of Assessment

Although the valve shape of many araphid taxa have reflective symmetry, their surfaces may exhibit more complicated configurations. For taxa such as *Asterionella*, *Pseudostaurosira*, *Punctastriata*, *Staurosirella*, and *Tabellaria*, striae are staggered, contrary to other araphid diatoms with striae occurring directly opposite each other on a valve face such as *Fragilaria*, *Fragilariforma*, *Staurosira*, *Strauroforma*, *Synedra*, and *Tabularia* (e.g., [11.9.54.]). Considering differences in striae arrangement, surface symmetry may be informative in assessing stability and symmetry breaking concerning the diatom valve face.

Raphid taxa cover a wide range of symmetries (e.g., [11.9.28.]). Biraphid taxa with asymmetric central areas, deflected raphes, keels, rimoportulae, stigmata, or other asymmetric surface features may pose interesting outcomes when undergoing surface symmetry analysis, if such morphological features are considered to be non-randomly placed. Non-flat apically reflective symmetric shaped biraphids such as *Amphora* may be reflective surface symmetric in only one spatial plane. Transapically reflective shape symmetric biraphids such as *Cymbella* may not qualify as such when viewing their surfaces. Biraphids such as *Actinella* and *Gyrosigma* (sigmoid-shaped diatoms) are not apically or transapically reflective symmetric, but are asymmetric overall. By dividing a valve face in half, *Actinella* and *Gyrosigma* exhibit almost reflective symmetry and translation as partial glide symmetry, respectively, in terms of shape, but not necessarily in their surfaces. Keeled biraphids such as *Nitzschia* are also asymmetric when considering surfaces, if the keel dictates how to assess valve shape and surface symmetries.

Monoraphid taxa such as *Cocconeis* present another symmetry problem. As a case in point, one valve face may be reflective symmetric, while the other is not, or one valve face may have glide symmetry while the other valve face does not. Janus cells such as those found in *Gomphonema* [11.9.7.] exhibit the same problem. Each valve face may belong to different, but potentially overlapping, symmetry groups, for example, even if both valve faces are dihedral.

11.7.2   More on Diatom Surface Symmetries and Handedness

Combination symmetries of helices may be evidence of scale symmetry. Scale symmetry in motion is a feature of conformal symmetries and Möbius transformations and has been found in some *Biddulphia* species ([11.9.48., Figure 3l]). Fractal properties in organisms may be related to adaptive ability or exhibit stability with respect to development or evolution (e.g., [11.9.32.]).

Helices are found in Fibonacci structures, Theodorus spirals or clothoids as Euler spirals [11.9.58.]. Compound or multi-spirals and loop networks may be composed of broken, curved segments as barely fractional helical or spiral parts on a very small scale. Such loop networks as symmetries may be internal phenomena that are not indicative of overall symmetry, but rather as partial surface symmetries. For diatoms, loop networks are evident in the valve faces of taxa such as *Trigonium dubium* and *Triceratium bicorne* ([11.9.48., Figures 3m and n], respectively).

Chirality and handedness are part of helices, and left-right asymmetry is descriptive of chirality as handedness [11.9.30.]. Whether a morphological attribute occurs on the same side or opposite sides of an organism is indicative of chirality, and with diatoms, chirality is evident not only during vegetative reproduction, but also is evident during sexual reproduction (e.g., [11.9.5.]).

Helicity, chirality and handedness are also properties of additional symmetries that may not be easily recognizable. Such asymmetries sometimes require dissection of the whole into its parts. For example, abnormal *Cyclotella meneghiniana* have helicity or spirals in the central area ([11.9.48., Figure 3g]) exhibiting asymmetric surface structures. Morphologically, such apparent asymmetric symmetries may be part of the bauplan, development or some feature on a cellular, molecular, genetic, or genomic scale as well. Helices involve curvature, and along with chirality and handedness, may even be indicative of knots as partial features of an organismal form.

11.7.3   Diatom Vegetative Reproduction and Symmetry Breaking

In the biological world, true symmetry is rare (e.g., [11.9.13.]). Three-dimensional surface models and their geometric properties were used to measure surface symmetry and symmetry breaking in diatom vegetative reproduction. Symmetry breaking may occur during size reduction and may influence reproductive success, phenotypic stability or morphological variation more generally.

Symmetry breaking is more prevalent early in the vegetative reproduction process when size reduction is occurring. Symmetry breaking oscillates as size reduction oscillates sinusoidally, illustrating a damped oscillating curve (Figure 11.13). As the amplitude of size reduction decreases, the overall trend is a negative exponential function (Figure 11.13). As size reduction approaches zero, symmetry breaking approaches zero (Figure 11.13).

If no size reduction occurs, then there should be very slow to no symmetry breaking occurring. This is the most stable situation. If slow size reduction occurs, then a minimal amount of symmetry breaking should occur, and therefore there is a minimum of instability.

In this case, symmetry change is mostly stable. If size reduction occurs rapidly, then a maximum amount of symmetry breaking should occur and is the least stable condition (Figure 11.13). Understanding developmental processes demand a more precise determination of surface symmetry and symmetry breaking regarding size reduction in vegetative reproduction, and morphological variation more broadly, as symmetry in development is instrumental in diatom evolution.

11.7.4   Symmetry Breaking, Vegetative Reproduction, Size Reduction, and Stability

Symmetry breaking was determined to occur most readily and rapidly early in the vegetative reproductive cycle (Figure 11.13). This may be additionally attributable to the change from initial cell to vegetative cell. Symmetry breaking was found to be a damped oscillation as size reduction occurs over the cycle of vegetative reproduction (Figure 11.13). The oscillating nature of size reduction during vegetative reproduction may indicate the expenditure or cost to cell division early in vegetative reproduction, yet there is still a cost to enduring size reduction subsequently in the process, arriving at the MacDonald-Pfitzer limit [11.9.36., 11.9.49., 11.9.50.].

Symmetry gradients indicated that mostly dihedral-based symmetries lead to knot, outright dihedral and glide symmetries in unconstrained seriation (Figure 11.15a), while most dihedral-based symmetries gave way to knot, scale (hyperbolic), outright dihedral, and glide symmetries in constrained seriation (Figure 11.15b). In either case, knot symmetry stands apart from the majority of diatom surface symmetries that involve dihedral symmetry and is one of the more stable symmetries along with outright dihedral and glide symmetries.

11.7.5   Eigenvalues and Variance: Instability and Fluctuating Asymmetry

Fluctuating asymmetry is measured to determine random deviations from reflective symmetry and has been used to infer developmental instabilities, environmental stresses or genetic aberrations inducing a negative outcome during development of a given organism. For a given population, fluctuating asymmetry is a statistical measure where the variance is used to determine the degree of change from one asymmetric state to another during development (e.g., [11.9.37.]). An equivalence is assumed between variance and change in asymmetry [11.9.39.].

Measurement of instability directly from the geometry of an organism is not reliant on statistical characteristics of a single, and potentially non-representative population of a given organism. Across any taxonomic group, instability determined from the inverse Jacobian eigenvalues can be calculated across n-levels for m-organisms without reliance on particular organismal characteristics. The eigenvalues of the inverse Jacobian are also a multidimensional summary of the changes in instability over n-levels, potentially recording the trace (as the sums of squares) of the sums of squares cross product matrix. Because of this, the eigenvalues also may be a record of the percent variance explained as given in various multivariate statistical analyses. While the variance at-large may not necessarily be a measure of fluctuating asymmetry, a narrower slice of that variance, as given by the trace of the inverse Jacobian, may be construed to be a measure of fluctuating asymmetry, providing that an equivalence is established between asymmetry and negative values of the trace of the inverse Jacobian, and that trace records a sums of squares cross product matrix that leads to the covariance matrix.

11.7.6 Symmetry Groups and Evolutionary Dynamics: Symmetry in Diatoms and Adaptation

Environmental conditions may play a role in symmetry. For example, *Craticula cuspidata* has reflective-dihedral symmetry, and with an increase in saline conditions, this taxon's valve surface may become asymmetric which may be an indicator of reproductive success [11.9.34.]. Classification of diatom surface symmetry status may be indicative of environmental conditions in which environmentally-induced surface symmetry changes occur. Whether considering reproduction or environmental stressors, diatom surface symmetries also may have adaptive value via functional morphological considerations.

Surface symmetries may be adaptive at a smaller scale. With dihedral symmetry, patterns of striae or costae may be adaptive in those taxa that exist in a colonial chain formation. In a chain, cells are joined by spines that fit together in an interlocking fashion. In an adaptive sense, there is less stress on spines for cells that fit together than those with disjunct valve faces. The colony with well-fitted cells has more strength to remain as a colony, possibly avoiding breaking apart. When looking at the patterns of striae or costae on diatoms, glide symmetry is prevalent. This "ladder-like" symmetry in concert with various diatom shapes may provide an adaptive value in terms of substrate adherence and frustule strength against the constant movement of water waves [11.9.8.].

In a few diatoms, knot symmetry is present, which may induce expansion of niche tolerance for particular ecological conditions [11.9.4.]. Keeled diatoms have exhibited knotted, twisted or writhed forms which may induce predation avoidance. Such surfaces may be difficult for microcrustaceans to grab hold of while feeding, and knot symmetry may induce deterrence as well as frustule strength as an adaption.

11.8 Summary and Future Research

Understanding surface symmetries may be informative in understanding biological processes such as reproduction. Rather than trying to pigeon hole all organisms into the cyclic or reflective symmetry categories, this study has shown that other forms of symmetry may be determined, quantified and analyzed. Symmetry breaking is evident in size reduction during diatom vegetative reproduction which may be indicative of a potential cost toward success in propagation and survival.

Many questions may be raised concerning diatom surface symmetry. Adaptive radiation and strategies are important with respect to the developmental life cycle. For example, diatom size reduction and knot symmetry occur in *Entomoneis*. Is knot symmetry any less adaptive than other forms of symmetry, and does symmetry breaking affect resultant size-reduced knot symmetric cells? What role might knot symmetry serve at a subcellular level as it relates to the diatom cell?

One item of note in diatom reproduction is that initial cells are often asymmetric. Does this imply the occurrence of symmetry breaking between sexual and asexual reproductive states? What role does asymmetry play in diatom reproduction aside from size reduction? Diatom surface symmetries during vegetative reproduction may have implications for phylogenetic relations. How do surface symmetries match up to phylogenetic relations in contrast to the traditional categories of "centric" and "pennate" identifications?

On a large scale, all diatom genera listed in Fourtanier and Kociolek's [11.9.12.] Catalogue of Diatom Genera could be classified with respect to their surface symmetry groups. What is the distribution of genera with respect to surface symmetry groups? How is this related to diatom valve shape distribution (as determined in Chapter 1)? Surface symmetries are important considerations in multiple areas of diatom macroevolution studies.

11.9 References

11.9.1.   Adams, C.C. (2001) *The Knot Book, An Elementary Introduction to the Mathematical Theory of Knots*, American Mathematical Society, Providence, Rhode Island.

11.9.2.   Adams, C. (2005) Hyperbolic knots. In *Handbook of Knot Theory*, Menasco, W.W. and Thistlethwaite, M.B. (eds.), Elsevier, Amsterdam, The Netherlands: 2-18.

11.9.3.   Adams, C., Bennett, H., Davis, C., Jennings, M., Kloke, J., Perry, N., Schoenfeld, E. (2008) Totally geodesic Seifert surfaces in hyperbolic knot and link complements II. *Journal of Differential Geometry* 79, 1-23.

11.9.4.   Al-Handal, A.Y., Mucko, M., Wulff, A. (2020) *Entomoneis annagodhei* sp. nov., a new marine diatom (Entomoneidaceae, Bacillariophyta) from the west coast of Sweden. *Diatom Research* 35(3), 269-279.

11.9.5.   Amato, A. (2010) Diatom reproductive biology: living in a crystal cage. *The International Journal of Plant Reproductive Biology* 2(1), 1-10.

11.9.6.   Amato, A., Orsini, L., D'Alelio, D., Montresor, M. (2005) Life cycle, size reduction patterns, and ultrastructure of the pennate planktonic diatom *Pseudo-nitzschia delicatissima* (Bacillariophyceae). *Journal of Phycology* 41, 542-556.

11.9.7.   Andrejić, J.Z., Spaulding, S.A., Manoylov, K.M., Edlund, M.B. (2019) Phenotypic plasticity in diatoms: Janus cells in four *Gomphonema* taxa. *Diatom Research* https://doi.org /10.1080/0269249X.2019.1572652.

11.9.8.   Arce, F.T., Avci, R., Beech, I.B., Cooksey, K.E., Wigglesworth-Cooksey, B. (2004) A live bioprobe for studying diatom-surface interactions. *Biophysical Journal* 87, 4284-4297.

11.9.9.   Bahls, L. (2012). *Entomoneis alata*. In Diatoms of North America. Retrieved January 25, 2021, from https://diatoms.org/species/entomoneis_alata; images: https://diatoms.org/ images/9206 and https://diatoms.org/images/9202.

11.9.10. Dalu, T., Taylor, J.C., Richoux, N.B., Froneman, P.W. (2015) A re-examination of the type material of *Entomoneis paludosa* (W Smith) Reimer and its morphology and distribution in African waters. *Fottea, Olomouc* 15(1), 11-23.

11.9.11. Davidovich, N.A., Kaczmarska, I., Ehrman, J.M. (2010) Heterothallic and homothallic sexual reproduction in *Tabularia fasciculata* (Bacillariophyta). *Fottea* 10(2), 251-266.

11.9.12. Fourtanier, E. and Kociolek, J.P. (2011) Catalogue of Diatom Names, California Academy of Sciences, http://research.calacademy.org/research/diatoms/names/index.asp

11.9.13. Golubitsky, M. and Stewart, I. (2015) Symmetry methods in mathematical biology. *São Paulo J. Math. Sci.* 9, 1-36.

11.9.14.   Graham, J.H. (2021) Fluctuating asymmetry and developmental instability, a guide to best practice. *Symmetry* 13(1), 9; https://doi.org/10.3390/sym13010009.

11.9.15.   Hahsler, M. (2017) An experimental comparison of seriation methods for one-mode two-way data. *European Journal of Operational Research* 257, 133-143.

11.9.16.   Hahsler, M., Hornik, K., Buchta, C. (2008) Getting things in order: an introduction to the R package seriation. *Journal of Statistical Software* 25(3), 1-34.

11.9.17.   Hammer, Ø., Harper, D.A.T., Ryan, P.D. (2001) PAST – Palaeontological Statistics. http://www.toyen.uio.no/~ohammer/past.

11.9.18.   Hatcher, K. (2018). *Surirella undulata*. In *Diatoms of North America*. Retrieved January 25, 2021, from https://diatoms.org/species/surirella-undulata; image: https://diatoms.org/images/93943.

11.9.19.   Holló, G. (2015) A new paradigm for animal symmetry. *Interface Focus* 5, 20150032.

11.9.20.   Holló, G. and Novák, M. (2012) The manoeuvrability hypothesis to explain the maintenance of bilateral symmetry in animal evolution. *Biology Direct* 7, 22. doi:10.1186/1745-6150-7-22

11.9.21.   Hoste, J., Thistlethwaite, M., Weeks, J. (1998) The first 1,701,936 knots. *The Mathematical Intelligencer* 20, 33–48.

11.9.22.   Hubert, L., Arabie, P., Meulman, J. (2001) *Combinatorial Data Analysis: Optimization by Dynamic Programming*, Society for Industrial Mathematics, Philadelphia, Pennsylvania.

11.9.23.   Kaczmarska, I., Ehrman, J.M., Bates, S.S. (2000) A review of auxospore structure, ontogeny and diatom phylogeny. In: *Proceedings of the 16th International Diatom Symposium, Athens & Aegean Islands*, Economou-Amilli, A. (ed.), University of Athens, Greece: 153-168.

11.9.24.   Kaplan, W. (2003) *Advanced Calculus, 5th edition*, Addison-Wesley, Reading, Massachusetts, USA.

11.9.25.   Kendall, D.G. (1984) Shape manifolds, procrustean metrics, and complex projective spaces. *Bull Lond Math Soc.* 16(2), 81-121.

11.9.26.   Kimmich, R., Bahls, L. (2020). *Cocconeis grovei*. In Diatoms of North America. Retrieved January 25, 2021, from https://diatoms.org/species/cocconeis-grovei; image: https://diatoms.org/images/96290.

11.9.27.   Kociolek, P. (2011). *Tabularia fasciculata*. In Diatoms of North America. Retrieved January 25, 2021, from https://diatoms.org/species/tabularia_fasciculata; image: https://diatoms.org/images/12964.

11.9.28.   Kociolek, J.P., Spaulding, S.A., Lowe, R.L. (2015) Bacillariophyceae: the raphid diatoms, Chapter 16. In *Freshwater Algae of North America: Ecology and Classification 2nd edition*, Wehr, J.D., Sheath, R.G., Kociolek, J.P. (eds.), Academic Press, San Diego, California: 709-772.

11.9.29.   Koenderink, J.J. (1990) *Solid Shape*, MIT Press, Cambridge, Massachusetts USA.

11.9.30.   Kojić-Prodić, B. and Štefanić, Z. (2010) Symmetry versus asymmetry in the molecules of life: homomeric protein assemblies. *Symmetry* 2, 884-906.

11.9.31.   Kuno, Y., Massuyeau, G., Tsuji, S. (2019) Generalized Dehn twists in low-dimensional topology. Accessed on 26 January 2021, https://arxiv.org/pdf/1909.09496.pdf.

11.9.32.   Kurakin, A. (2011) The self-organizing fractal theory as universal discovery method: the phenomenon of life. *Theoretical Biology and Medical Modelling* 8(4), http://www.tbiomed.com/content/8/1/4.

11.9.33.   Lee, S. (2011). *Campylodiscus hibernicus*. In Diatoms of North America. Retrieved January 25, 2021, from https://diatoms.org/species/campylodiscus_hibernicus; image: https://diatoms.org/images/7850.

11.9.34.   Levkov, Z., Tofilovska, S., Mitić-Kopanja, D. (2016) Species of the diatom genus Craticula Grunow (Bacillariophyceae) from Macedonia. *Contributions, Section of Natural, Mathematical and Biotechnical Sciences, MASA* 37(2), 129-165.

11.9.35.   Lowe, R. and Kheiri, S. (2015). *Cyclotella meneghiniana*. In Diatoms of North America. Retrieved January 25, 2021, from https://diatoms.org/species/cyclotella_meneghiniana; image: https://diatoms.org/images/13283.

11.9.36.   MacDonald, J.D. (1869) On the structure of the diatomaceous frustule, and its genetic cycle. *Annals Magazine Natural History Series* 4(3), 1-8.

11.9.37.   Mather, K. (1953) Genetical control of stability in development. *Heredity* 7, 297-336.

11.9.38.   Mejdandžić, M., Bosak, S., Orlič, S., Udovič, M.G., Štefanič, P.P., Špoljarič, I., Mršič, G., Ljubešič, Z. (2017) *Entomoneis tenera* sp. nov., a new marine planktonic diatom (Entomoneidaceae, Bacillariophyta) from the Adriatic Sea. *Phytotaxa* 292(1), 001-018.

11.9.39.   Palmer, A.R. and Strobeck, C. (1986). Fluctuating asymmetry: measurement, analysis, patterns. *Ann. Rev. Ecol. Syst.* 17, 391-421.

11.9.40.   Panfilov, A.V., ten Tusscher, K.H.S.J., de Boer, R.J. (2020) Matrices, linearization, and the Jacobi matrix, *Theoretical Biology*, Utrecht University, http://tbb.bio.uu.nl/rdb/books/math.pdf.

11.9.41.   Pappas, J.L. (2005a) Geometry and topology of diatom shape and surface morphogenesis for use in applications of nanotechnology. *Journal of Nanoscience and Nanotechnology* 5(1), 120-130.

11.9.42.   Pappas, J.L. (2005b) Theoretical morphospace and its relation to freshwater gomphonemoid-cymbelloid diatom (Bacillariophyta) lineages. *Journal of Biological Systems* 13(4), 385-398.

11.9.43.   Pappas, J.L. (2008) More on theoretical morphospace and its relation to freshwater gomphonemoid-cymbelloid diatom (Bacillariophyta) lineages. *Journal of Biological Systems* 16(1), 119-137.

11.9.44.   Pappas, J.L. (2016) Multivariate complexity analysis of 3D surface form and function of centric diatoms at the Eocene-Oligocene transition. *Marine Micropaleontology* 122, 67-86.

11.9.45.   Pappas, J.L. (2021) Quantified ensemble 3D surface features modeled as a window on centric diatom valve morphogenesis. In: *Diatom Morphogenesis [DIMO, Volume in the series: Diatoms: Biology & Applications, series editors: Richard Gordon & Joseph Seckbach].* V. Annenkov, J. Seckbach and R. Gordon, (eds.) Wiley-Scrivener, Beverly, MA, USA: 158-193.

11.9.46.   Pappas, J.L. and Stoermer, E.F. (1995) Great Lakes Diatoms, (http://umich.edu/~phytolab/GreatLakesDiatomHomePage/top.html) accessed January, 2021.

11.9.47.   Pappas, J.L. and Miller, D.J. (2013) A generalized approach to the modeling and analysis of 3D surface morphology in organisms. *PLoS One* 8(10), #e77551.

11.9.48.   Pappas, J.L., Tiffany, M.A. and Gordon, R. (2021) The uncanny symmetry of some diatoms and not of others: A multi-scale morphological characteristic and a puzzle for morphogenesis [DUNC]. In: *Diatom Morphogenesis [DIMO, Volume in the series: Diatoms: Biology & Applications, series editors: Richard Gordon & Joseph Seckbach].* V. Annenkov, J. Seckbach and R. Gordon, (eds.) Wiley-Scrivener, Beverly, MA, USA: 18-68.

11.9.49.   Pfitzer, E. (1871) Untersuchungen über Bau und Entwicklung der Bacillariaceen (Diatomaceen) [Studies on construction and development of Bacillariaceae (Diatomaceae)] [German]. In: *Botanische Abhandlungen aus dem Gebiete der Morphologie und Physiologie [Botanical Treatises in the Field of Morphology and Physiology].* J.L.E.R. von Hanstein, (ed.) Adolph Marcus, Bonn, Germany: [i]-vi, 1-189, 186 pls.

11.9.50.   Pfitzer, E. (1882) Die Bacillariaceen (Diatomaceae) [Bacillariaceae (Diatomaceae)] [German]. In: *Encyklopaedie der Naturwissenschaften. I. Abteilung. I. Thiel: Handbuch der Botanik.* A. Schenk, (ed.) Verlag von Eduard Trewendt, Breslau. 2: 403-445.

11.9.51.   Poulíčková, A. and Jahn, R. (2007) *Campylodiscus clypeus* (Ehrenberg) Ehrenberg ex Kützing: typification, morphology and distribution. *Diatom Research* 22(1), 135-146.

11.9.52.   Rao, V.N.R. (1978) Studies on *Cyclotella meneghiniana* Kütz. IV. Progressive diminution in cell size. *Proc. Indian Acad. Sci.* 87B(2), 1-15.

11.9.53.   Reimann, F.E.F., Lewin, J.C., Volcani, B.E. (1965) Studies on the biochemistry and fine structure of silica shell formation in diatoms, I. The structure of the cell wall of *Cylindrotheca fusiformis* Reimann and Lewin. *The Journal of Cell Biology* 24, 39-55.

11.9.54.   Round, F.E., Crawford, R.M. and Mann, D.G. (1990) *The Diatoms, Biology & Morphology of the Genera.* Cambridge University Press, Cambridge, U.K.

11.9.55.   Schmid, A.-M.M. and Crawford, R.M. (2001) *Ellerbeckia arenaria* (Bacillariophyceae): formation of auxospores and initial cells. *European Journal of Phycology* 36, 307-320.

11.9.56.   Spaulding, S. and Edlund, M. (2008). *Cylindrotheca.* In Diatoms of North America. Retrieved January 25, 2021, from https://diatoms.org/genera/cylindrotheca and https://diatoms.org/genera/cylindrotheca/guide

11.9.57.   Spaulding, S.A., Bishop, I.W., Edlund, M.B., Lee, S., Furey, P., Jovanovska, E. and Potapova, M. (2019) Diatoms of North America. https://diatoms.org/

11.9.58.   Starostin, E.L., Grant, R.A., Dougill, G., van der Hdijden, G.H.M., Goss, V.G.A. (2020) The Euler spiral of rat whiskers. *Sci. Adv.* 6: eaax5145.

11.9.59.   Thurston, W.P. (2002a) Introduction. In *The Geometry and Topology of Three-Manifolds*, Electronic version 1.1, http://www.msri.org/publications/books/gt3m/.

11.9.60.   Thurston, W.P. (2002b) Chapter 2 Elliptic and hyperbolic geometry. In *The Geometry and Topology of Three-Manifolds*, Electronic version 1.1, http://www.msri.org/publications/books/gt3m/.

11.9.61.   Thurston, W.P. (2002c) Chapter 3 Geometric structures on manifolds. In *The Geometry and Topology of Three-Manifolds*, Electronic version 1.1, http://www.msri.org/publications/books/gt3m/.

11.9.62.   Thurston, W.P. (2002d) Hyperbolic Dehn surgery. In *The Geometry and Topology of Three-Manifolds*, Electronic version 1.1, http://www.msri.org/publications/books/gt3m/.

11.9.63.   Van Buskirk, J.M. (1983) A class of negative-amphicheiral knots and their Alexander polynomials. *Rocky Mountain Journal of Mathematics* 13(3), 413-422.

11.9.64.   Vanormelingen, P., Vanelslander, B., Sato, S., Gillard, J., Trobajo, R., Sabbe, K., Vyverman, W. (2013) Heterothallic sexual reproduction in the model diatom *Cylindrotheca*. *European Journal of Phycology* 48(1), 93-105.

11.9.65.   Veselá, J. and Potapova, M. (2014) *Surirella arctica* comb. et stat. nov. (Bacillariophyta)—a rare arctic diatom. *Phytotaxa* 166(3), 222-234.

11.9.66.   Weisstein, E.W. (ed.) (2002) *CRC Concise Encyclopedia of Mathematics, 2nd ed.*, Chapman & Hall/CRC, London, United Kingdom.

11.9.67.   Wetzel, C.E., Beauger, A., Ector, L. (2019) *Cocconeis rouxii* Héribaud & Brun a forgotten, but common benthic diatom species from the Massif Central, France. *Botany Letters* https://doi.org/10.1080/23818107.2019.1584865.

12

Evolvability of Diatoms as a Function of 3D Surface Phenotype

Abstract

Evolvability is based on outcomes involving phenotype. Three components of evolvability are phenotypic inertia, robustness and stability and are used to test whether diatoms have the capacity to evolve. *Actinoptychus*, *Arachnoidiscus* and *Cyclotella*, representing the time span from the Cretaceous to the Miocene, are used as exemplars via 3D surface models from which phenotypic inertia, robustness and stability are measured using Christoffel symbols, Hessian matrix and Laplacian, respectively, and converted into fitness functions. Evolvability is then computed as an additive fitness function for each genus, and the contribution of phenotypic inertia, robustness and stability to evolvability is calculated as well. The determinant of the differences between Hessian *z* for paired genera is used in a joint entropy measurement for radial plications as phenotypic novelty. Evolvability, calculated as paired fitness decrease between genera and novelty, is illustrated. *Cyclotella* is considered to be evolvable via phenotypic robustness, while *Arachnoidiscus* style radial plications are considered to be a phenotypic novelty.

Keywords: Phenotypic robustness, phenotypic inertia, phenotypic stability, Hessian, Laplacian, Christoffel symbols, evolvability, novelty

12.1 Introduction

Representation of an organism's exterior or interior features may be accomplished in many ways. Morphology of the organism's attributes enables the identification and classification of the organism. The attributes are a result of many biological processes, but the link between genotype and phenotype provides a way to understand the role that inherited characteristics inform the observable features of an organism, including morphology.

Analyzing morphology empirically as the phenotype necessarily involves quantification so that multiple phenotypes may be compared for similarities or differences. Historically, two-dimensional (2D) shape has been used (e.g., [12.7.56.]), and increasingly, three-dimensional (3D) analyses have been applied. The surface of an organism as a 3D entity is quantifiable and analyzable. As such, 3D surface morphology has been used as a proxy for the phenotype, and 3D surface geometric properties are proxies for phenotypic properties [12.7.55.].

Diatoms are unique unicells, having intricate geometric siliceous frustules exhibiting a multitude of patterns often having repetitive geometric features from the nanoscale to the whole unicell itself. They are a diverse phylum of morphologies representing phenotypes that may characterize lineages or be descriptive of genera (e.g., [12.7.64.]). Because diatoms

Janice L. Pappas. *Mathematical Macroevolution in Diatom Research*, (437–474) © 2023 Scrivener Publishing LLC

are geometrically characterizable, quantification of the geometry of their frustules enables quantifying their geometric phenotype.

12.1.1   3D Surface Properties – An Overview

To construct a 3D surface model, parametric equations are used. The 3D surface is defined as a tangent space, and all the points on the surface are used to characterize that surface. This is accomplished using first partial derivatives at each point of this parameterized surface. The first partial derivatives are the elements of the Jacobian matrix (i.e., Jacobian), which represents the slopes at each point on the surface.

The points are unconnected on the 3D surface. Connection of neighboring points in tangent space is accomplished using Christoffel symbols which are related to the Jacobian via the Jacobian determinant [12.7.72.]. Christoffel symbols as connection coefficients [12.7.45.] are the differential linear changes in basis vectors from point to point [12.7.36.] forming the linear connections, giving structure to a tangent space.

The 3D surface has curvature as well at each point, and this is represented by second partial derivatives as elements of the Hessian matrix (i.e., Hessian). Local curvature at points where maxima, minima and saddles occur are recorded by the Hessian, and the sum of change in the surface maxima, minima and saddles are recorded by the sum of diagonal elements from the Hessian which is the Laplacian. The rows of the Jacobian are gradients, and the Laplacian is the divergence of the gradient [12.7.30.], recording the changes in gradients across the surface as a tangent space. For more details, see Pappas [12.7.54.].

12.1.2   From Differential Geometry to the Characterization of 3D Surfaces

Two views are involved in characterizing 3D surface geometry of an organism. One consideration is intrinsic geometry, where geodesic distances between points, Gaussian curvature and connection gradients are expressed. The other consideration is extrinsic geometry. Expressed as curves and surfaces in space, mean curvature and slopes are attributes of the organism as embedded in space using extrinsic geometry. From these views, 3D surfaces of the organism may be analyzed.

To start, the whole organism surface may be characterized as an object in 3D (x, y, z) space. A system of parametric 3D equations is used to model the whole organism surface (e.g., [12.7.49.-12.7.55.), and this enables extracting geometric features from that surface. Variables x, y, z parameterized by u, v enable the characterization of a point anywhere on the surface as well as movement from point to point on the surface along a given curve.

In a moving reference frame, there is a trihedron of vectors at each point on the surface. The Serret-Frenet frame consists of tangent (**t**), unit normal (**n**), and binormal (**b**) vectors for a curve, while a Darboux frame consists of tangent (**t**), unit normal (principal normal) (**u**), and tangent normal vectors (**v**) for a surface. Normal planes at the unit normal are used to define the tangent and tangent normal vectors and define principal directions [12.7.8.].

Curvature, κ, at a point in relation to **n** and **u** is $\kappa n \cdot u$ and the rate of change of a tangent line to a curve [12.7.8.]. Principal curvatures, κ_1 and κ_2, are maximum and minimum curvatures at a point on the surface. Gaussian curvature is $K = \kappa_1 \kappa_2$, and mean curvature is $H = \dfrac{\kappa_1 + \kappa_2}{2}$ [12.7.8.]. For κ_1 and κ_1 as positive or negative, then the surface is locally convex. If κ_1 is positive and κ_2 is negative or vice versa, then the surface is locally a saddle. If κ_1 or κ_1 is equal to 0, then the surface is in between a convexity and saddle. If κ_1 and κ_1 are both equal to 0, then the surface is a plane or a monkey saddle at a flat umbilic [12.7.29.].

Rate of change of a tangent is a first partial derivative with respect to a given curve on a surface with arc length, s. The differential of s is $ds^2 = dx^2 + dy^2 + dz^2$ on the surface, and

$$dx = \frac{\partial x}{\partial u} du + \frac{\partial x}{\partial v} dv, \quad dy = \frac{\partial y}{\partial u} du + \frac{\partial y}{\partial v} dv, \quad \text{and} \quad dz = \frac{\partial z}{\partial u} du + \frac{\partial z}{\partial v} dv \quad (\text{e.g., } [12.7.30.]). \text{ All of}$$

the tangent lines at every point on the surface are summarized in a Jacobian and given as

$$J = \begin{bmatrix} \dfrac{\partial x}{\partial u} & \dfrac{\partial x}{\partial v} \\[2mm] \dfrac{\partial y}{\partial u} & \dfrac{\partial y}{\partial v} \\[2mm] \dfrac{\partial z}{\partial u} & \dfrac{\partial z}{\partial v} \end{bmatrix}$$

as first partial derivatives of x, y, z with respect to u, v. Numerical solution

to the Jacobian characterizes all the tangent lines and planes on the whole organism surface.

Tangent lines define an intrinsic metric via the arc length. This quantity is expressed as $I = Edu^2 + 2Fdudv + Gdv^2$ where E, F, G are coefficients of the first fundamental form (e.g., [12.7.8.]). The second fundamental form defines tangent planes in terms of change in surface shape of the normals with respect to the tangents [12.7.35.] and is expressed as $II = Ldu^2 + 2Mdudv + Ndv^2$, where L, M, M are coefficients of the second fundamental form [12.7.8.].

While consideration of the whole 3D surface is useful, consideration of a given patch on the surface may provide sufficient information about the phenotype. A patch is locally the neighborhood surrounding a point on the 3D surface. Devise a whole 3D surface within a vector space, with continuous vector space norm. Then, the patch is representative of the whole 3D form.

Consider a Monge patch as a tangent plane. Then, only the z-direction (the height function) needs to be analyzed. The parametric equations to be analyzed are $x = g(u, v)$, $y = h(u, v)$ and $z = f(x, y) = f[g(u,v), h(u,v)]$ with differentials $dx = g'(u,v)$

$$du\, dv = \frac{\partial x}{\partial u} du + \frac{\partial x}{\partial v} dv, \quad dy = h'(u,v) \quad du\, dv = \frac{\partial y}{\partial u} du + \frac{\partial y}{\partial v} dv \quad \text{and} \quad dz = f'[g(u,v), h(u,v)]$$

$$dx\, dy = \frac{\partial z}{\partial x}\left(\frac{\partial x}{\partial u} du + \frac{\partial x}{\partial v} dv \right) + \frac{\partial z}{\partial y}\left(\frac{\partial y}{\partial u} du + \frac{\partial y}{\partial v} dv \right), \text{ and this defines the Monge patch [12.7.8.].}$$

At a given point on the parametric surface a patch can be defined and characterized in terms of tangent lines and planes. Tangent vectors are r_u and r_v, and normal vectors are n_u and n_v. Tangent vectors, $r(u, v) = (r_u, r_v) = r_{uv}$, span the tangent plane and serve as a basis in a matrix of coefficients for the differential of the tangent vectors. The matrix expressed in terms of the first and second fundamental forms has a determinant given as $K = \dfrac{LN - M^2}{EG - F^2}$, and half the trace is $H = \dfrac{LG - 2MF + NE}{2(EG - F^2)}$ which are Gaussian and mean curvature, respectively [12.7.8.]. Using K, shape can be determined as a general characterization of the surface [12.7.35.].

The normal to tangent vectors spanning a tangent plane is $n(u, v) = (n_u, n_v) = n_{uv}$, and the unit normal vector is $\dfrac{r_u \times r_v}{\| r_u \times r_v \|}$ (e.g., [12.7.8., 12.7.35.]). The differentiable field of the unit normal vectors on the surface is a Gauss Map of the tangent space of a point [12.7.8.]. The differential of a Gauss Map is expressed as the shape operator or a Weingarten Map and given as $\nabla_{Dir}S_{uv} = (EG - F^2)^{-1}\begin{pmatrix} LG - MF & MG - NF \\ ME - LF & NE - MF \end{pmatrix}$

where $\nabla_{Dir}S_u = \boldsymbol{n}_u = \dfrac{MF - LG}{EG - F^2}\boldsymbol{r}_u + \dfrac{LF - ME}{EG - F^2}\boldsymbol{r}_v$ and $\nabla_{Dir}S_v = \boldsymbol{n}_v = \dfrac{NF - MG}{EG - F^2}\boldsymbol{r}_u + \dfrac{MF - NE}{EG - F^2}\boldsymbol{r}_v$

[12.7.8.]. The shape operator determines all the tangent planes in the neighborhood of a point and is the change in surface normals with tangent vectors [12.7.35.].

The Jacobian determinant of a Gauss Map is Gaussian curvature [12.7.8., 12.7.35.]. Principal curvatures of a Weingarten Map are eigenvalues and are an indicator of how much the surface bends [12.7.8.]. Principal directions are eigenvectors of a Weingarten Map [12.7.8.]. Project the covariant derivative of the Weingarten Map onto the second fundamental form to obtain the connection gradient, ∇. Eigenvalues of the Hessian are measures of Gaussian curvature or the product of principal curvatures, and eigenvectors, as with a Weingarten Map are measures of principal directions [12.7.8., 12.7.35.]. The Laplacian is a measure of mean curvature (e.g., [12.7.8]). Because the surface is characterized in terms of curvature measurements of connected points in a tangent space, each measure of those characterizations—Christoffel symbols, Hessian, Laplacian—represents a quantification of phenotypic surface features that enable analysis of an organism's surface in terms of magnitude and directional changes. For more details, see Pappas [12.7.54.].

12.1.3   The Phenotype Characterized via a 3D Surface and Its Geometric Characteristics

Christoffel symbols, the Hessian and the Laplacian may be used to represent different facets of an organism's phenotypic surface. Christoffel symbols as connection coefficients may be used as representation of that aspect of the phenotypic surface that provides resistance to change. The phenotypic surface is the morphology present that is associated with the organism's identity at any given time. Geometrically, lack of changes in the motion of points, tangent lines and planes on the surface comprise the notion of resistance of that surface to change at any given time. As such, Christoffel symbols are used to characterize and measure the surface as an inertial state of resistance.

The Hessian as the ensemble of height differences of the surface represents the ability of the phenotypic surface to tolerate perturbations. Geometrically, peaks, valleys and saddles on the surface that deviate from flatness induce a degree of robustness that characterizes the surface. As a result, the Hessian is used to characterize persistence of the phenotypic surface in the face of change. That is, the organism's surface phenotype persists despite surface perturbations that have occurred, and the Hessian is a measure of this robustness.

The Laplacian, as the average change in the geometric surface, provides a level of stability of that surface. Stability means that the surface has the capacity to return to a state of original condition. A stable surface is resilient so that this means that the organism's surface phenotype has the capacity to endure, and so the Laplacian is a measure this resilience.

12.1.4   From Geometric Phenotype to Evolvability

Evolvability involves outcomes based on phenotype (e.g., [12.7.17., 12.7.82.]). Phenotypic changes are not merely variation, but a change in the structural or architectural characteristics of the phenotype. Measurement of evolvability is necessarily a geometric quantity in this regard. The geometric status of evolvability of a phenotype may be quantified using minimally the three surface characteristics of inertia, robustness and stability. As specified in the previous section, phenotypic inertia is a measure of surface resistance, phenotypic robustness is a measure of surface persistence, and phenotypic stability is a measure of surface resilience. Measuring these components of phenotype enables understanding the contribution of structural or architectural aspects of phenotype as they relate to evolvability. When considering phenotype, bauplans may be taken into account as a basis for evolvable surface characteristics.

Different time spans dictate the scale of evolvability measured and the kind of expectation of effects. On a microevolutionary scale, heritability is key and occurs on the shortest time scale. That is, short term evolution is considered here on a statistical basis (e.g., [12.7.10.]). Within-population genetic variation and covariation are of importance. The result is that evolvability is a response to natural selection within populations (e.g., [12.7.18., 12.7.38., 12.7.58.]). Heritability for short time spans is defined as $h^2 = R/S$ for R as the population response and S as selective pressure with $CV_a = 100\sqrt{V_a}/X$ for CV_a as the additive coefficient of genetic variation, V_a as the additive genetic variance that responds to selection, and X is the average population character. Lande's equation is $\Delta\bar{z} = G\nabla\ln\bar{W} = G\beta$ where $\Delta\bar{z}$ is response to selection, G is the G-matrix, which is the multivariate additive genetic variance/covariance matrix, and β is a directional selection gradient [12.7.38.]. Lande's equation is structurally related to information theory (e.g., [12.7.10.]).

At an intermediate timescale, evolvability within species not only depends on the genetics, but also genomics and molecular biology as well as developmental biology. Developmental constraints dictate the variability that results with respect to evolvability. As a result, long-term adaptation in terms of directional evolution as well as variability with respect to phenotypic space are the expectations of evolvability [12.7.58.]. At an intermediate timespan, genetic variability induces phenotypic stability in terms of evolvability.

On a macroevolutionary and longest time scale, phenotypic novelty or innovation may occur with respect to evolvability. Capability to transcend genetic or developmental constraints enables the expansion of phenotypic space, and therefore phenotypic evolution. Various phenotypic breakthroughs may result, including not only morphological novelties, but also physiological or behavioral novelties [12.7.57., 12.7.58.].

12.1.5   Evolvability and Phenotypic Novelty

Many definitions exist for evolvability and novelty. Evolvability may involve different aspects of novelty being emphasized, or not. Definitions of novelty are dependent on the emphasis and taxonomic level of a given study as well. No matter what scale evolvability and novelty are being studied, context is most important in clarifying definitions (e.g., [12.7.24., 12.7.48.]).

An origin of bauplans and the evolution of phenotypic novelties is an important facet of evolvability [12.7.59.]. Invoking the notion of the bauplan implicates using phyla or other high level taxonomic levels as the context of study (e.g., [12.7.57.]). One definition of novelty that might be used is that a new trait, new behavior or new combination of existing traits or behaviors enable a new function, in an ecological context, occurring during the evolution of a given lineage [12.7.59.], or origination of a bauplan, a new constructal element, or a new character have evolved such that it is disconnected from the ancestral organism from which it is derived [12.7.57.]. In this context, phenotypic novelty may be a result of genetic innovation [12.7.48.]. When considering structural origination of bauplan, phenotypic novelty entails a homologue that has not heretofore existed (e.g., [12.7.14.]). That is, phenotypic novelty in an organism is exhibited as the very first occurrence of one or more certain characters in contrast to all others in a given lineage [12.7.48.].

Phenotypic novelty from structural origination induces recognition of biological structures as the culmination of biological processes. One such process that plays a large role in the resultant phenotype is development (e.g., [12.7.48.]). As a mechanism, development and its products at each developmental stage may result in distinct phenotypic forms as potential evidence of phenotypic novelty. Exploitation of a new ecological niche or a speciation event may be mechanistic evidence of phenotypic novelty as a key innovation (e.g., [12.7.41.]). Environmental influences during development may play a role in the occurrence of a phenotypic novelty as well [12.7.48.] as a key innovation (e.g., [12.7.23.]).

Because phenotypic novelty may be studied at higher taxonomic levels where no previous homologue exists, it is distinct from morphological variation (e.g., [12.7.15., 12.7.48.]). In a given phylum, introduction of new characters into an existing bauplan is origination of a phenotype that is less defined by genetic inheritance specifically and more so defined by physical and chemical properties at a cellular level [12.7.48.]. Such cellular properties may form the basis of generalized structures of organisms resembling simple geometric structures such as rods (solid cylinders) or tubes (open cylinders) that evolve more complicated structures via natural selection over time [12.7.48.]. Effects from selection may include epigenesis during development. Such changes may affect cell behaviors during development, and in turn, shape, size or other properties may be affected by such modification of developmental processes to produce phenotypic novelty [12.7.48.]. Geometry of structures is one way to delve into the epigenesis of the development of organismal form and the phenotypic novelties that may originate. Discontinuous change in phenotype structure is phenotypic novelty [12.7.57.], and phenotypic change is measurable geometrically.

12.1.6   Evolvability and Diatoms

The modern synthesis [12.7.25., 12.7.69.] has guided evolutionary studies for over a century. As mathematical population genetics was developed, quantitative methods were applied to evolution studies. Advancement of biological studies necessitated the inclusion of findings from developmental biology, ecology and molecular biology to obtain a more complete picture of the interactions and relations of organisms. Advancement also necessitated quantification to find multiple relations among disparate entities that can be assembled in a functional or relational way. Macroevolution of the biological world provides the conceptual means to apply mathematical methods across multiple classification levels and biological disciplines.

Phenotype is an essential component of evolution studies, including diatom evolution (e.g., [12.7.70.]). Applying mathematical tools to phenotype analysis enables studies of any organismal group. Unicells, such as diatoms, have quantifiable geometric phenotypic attributes that are applicable to the study of evolvability. Despite the lack of resolution concerning diatom phylogeny, four structural groups—radial centrics, polar centrics, araphid pennates, and raphid pennates—have been found generally in many molecular phylogenetic analyses, regardless of limitations in using, e.g., small subunit ribosomal RNA genes as input data, sampling schema of taxa, and various optimality criteria [12.7.77.].

As a result, three diatom genera will serve as radial centric structural group exemplars in the study of evolvability: *Actinoptychus*, *Arachnoidiscus* and *Cyclotella*. The phenotype characteristic of interest is different degrees of radial plication or pleating of the valve surface as an ensemble surface feature [12.7.54.]. Radial plications have geometric characteristics (e.g., [12.7.79.]) that are measurable and may act as a functional predation deterrent structure [12.7.71.] or increased surface area for nutrient uptake [12.7.46.]. Phenotype diversity in valve structure in these genera represent a spanning of geologic time, from the Cretaceous to the Miocene (e.g., [12.7.20.]). Geometric phenotype is obtainable from numerical analyses of 3D valve surface models. Evolvability may be quantified via geometric surface analysis quantities that are converted into the phenotypic characteristics of inertia, robustness and stability. These values are also useful in determining whether phenotypic novelty among the genera has occurred and, if so, how novelty relates to evolvability.

12.1.7 Diatom Exemplar Phenotypic and Valve Plication Characteristics

Actinoptychus is an Upper Cretaceous diatom [12.7.20.] currently classified as a member of the Coscinodiscales [12.7.83.]. The valve face has alternating undulations of sectors emanating from the center and is divided into sectors by interior folds or walls [12.7.22., 12.7.39.]. Rimoportulae are curled with an internal slit, and pores on the valve surface have channels [12.7.20.]. For *Actinoptychus senarius*, the areolae are a coarse or reticulate pattern on the valve surface, and there is a marginal ridge from which external projections of the rimoportulae protrude above the valve face [12.7.22.]. *Actinoptychus* are solitary [12.7.20.] and are found in neritic environments [12.7.64.].

Found in the Upper Cretaceous [12.7.20.], *Arachnoidiscus ehrenbergii* may exhibit heterovalvy during auxospore attachment to the epivalve of the parent cell [12.7.34., 12.7.66.]. Different valve patterns may form as a result of differing attachment sites [12.7.28.]. *Arachnoidiscus* has elongated central slits arranged in a ring surrounding a small planar central area on the valve with pores diminishing in view of a planar field [12.7.9.]. Thick ribs emanating from the central area are regularly spaced and may be bisected to look like windows on the pore structured pattern below [12.7.4.]. There are also concentric rings connecting the radial ribs [12.7.64.]. Rimoportulae have internal and external slits and areolae are porous with volae [12.7.20., 12.7.64.]. *Arachnoidiscus* is epiphytic and found in benthic habitats [12.7.20., 12.7.64.].

Cyclotella may be found as solitary or colonial chain cells and date back to the Miocene [12.7.64.]. Valves are undulating with ornamented or plain central areas [12.7.64.]. Areolae have pores with internal cribra [12.7.64.]. *Cyclotella meneghiniana* has distinctive striations as ribs regularly placed at the valve margin covering about half of the valve face [12.7.13., 12.7.67.]. These striations form a regular undulating pattern around the valve periphery giving

Cyclotella meneghiniana its distinct appearance. Marginal rimoportulae are present as well [12.7.13.], and the central area has fultoportulae. The clear central area exhibits a rather uniform surface [12.7.13., 12.7.67.]. *Cyclotella meneghiniana* may exhibit heterovalvy [12.7.64.].

12.1.8   Diatom Architecture and the Geometric Phenotype

Actinoptychus, *Arachnoidiscus* and *Cyclotella* are architecturally similar in that they each exhibit a form of plication or pleating in a radial pattern (Figure 12.1). The broadest pleats that are farthest apart are exhibited by *Actinoptychus*, then next is *Arachnoidiscus* with finer pleats occurring closer together, and finally *Cyclotella* has pleats that are the finest and closest together (Figure 12.1). These plications are connected via a honeycomb pattern of structures, producing a sort of corrugation which increases strength and stiffness [12.7.16.]. Noteworthy is the differences in central areas among the genera, from the smallest central area in *Actinoptychus* to the next largest central area in *Arachnoidiscus* to the largest central area in *Cyclotella* (Figure 12.1).

Plications or pleating on the diatom surface are related to surface area and may be a factor in the strength of the frustule. Plications provide "bumpy" surfaces that may be instrumental in predation deterrence of grazers that crush diatom cells to ingest the contents (e.g., [12.7.44.]). Functionally, plicated diatom valves deter grazing predators [12.7.71.], because the plications provide an increase in depth of the siliceous valve surface, and increasing deterrence is registered by microzooplankton grazers that need to crack diatom frustules to obtain nutrition (e.g., [12.7.74., 12.7.75.]). Radial plications for *Actinoptychus* are low amplitude, rounded swells, for *Arachnoidiscus*, they are alternating ribbed and flat areas, and for *Cyclotella*, they are sharp, close pleated almost sawtooth folds (Figure 12.1).

Another possibility in discerning differences in degree of radial plications and adaptive value is with regard to nutrient uptake. Higher surface area means that there is the potential of higher absorption of nutrients (e.g., [12.7.46.]). However, access to nutrients may be tempered by nanoscale valve features such as pores that enable increasing absorption of nutrients despite increased surface area from radial plications (e.g., [12.7.46.]).

Similarly, adsorption on the diatom surface may be increased as a result of increased surface area via more plications. For example, there may be increasing pollutant adherence to the diatom frustule as a result of increasing plications. Nanoscale features also play a role in increased adsorption along with degree of plications on the diatom surface (e.g., [12.7.12., 12.7.85.]. Diatom valve surface adsorption may be important in diatom-environmental interactions, and possibly, aggregation of particles under various water conditions (e.g., [12.7.73.]).

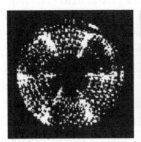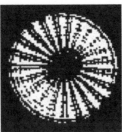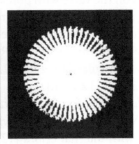

Figure 12.1 Binarized images of *Actinoptychus*, *Arachnoidiscus* and *Cyclotella* are displayed to show radial plications.

The architecture of *Actinoptychus*, *Arachnoidiscus* and *Cyclotella* may be used as exemplars of changing geometric phenotype and its relation to evolvability. Analysis of geometric phenotype may result in not only a determination of evolvability, but also whether phenotypic novelty is present.

12.2 Purposes of this Study

Diatom evolvability of the geometric phenotype is being studied using 3D surface models as a basis. Systems of parametric equations are used to construct 3D surface models of *Actinoptychus*, *Arachnoidiscus* and *Cyclotella* based on scanning electron micrographs (Figure 12.2). From the parametric equations, phenotypic characteristics are quantified as Christoffel symbols, the Hessian, and the Laplacian and used in determining the evolvability components of phenotypic inertia, robustness and stability, respectively. The sign of Christoffel symbols and eigenvalues of the Hessian and Laplacian as well as the Hessian determinant are used to evaluate phenotypic inertia, robustness and stability. Evolvability is determined across genera and across evolvability components as additive fitness functions (e.g., [12.7.5.]). Overall evolvability is characterized by an additive fitness function (e.g., [12.7.47., 12.7.62., 12.7.76.]), while novelty is determined via information theory [12.7.76., 12.7.78.]. Results from evolvability and novelty analysis are compiled graphically to determine the relation between evolvability and novelty.

12.3 Methods

Diatom vegetative cells are represented geometrically by 3D surface models, and the models are constructed via parametric 3D equations, relating 3D surface geometry to the geometric phenotype. For each point on the valve surface, variables x, y, z are parameterized by u, v. Movement on the surface is connected via curves between points. Systems of parametric 3D equations of the general form $x = f(u,v)$, $y = g(u,v)$, $z = h(u,v)$ are evaluated on the interval $[0,2\pi]$ [12.7.50.-12.7.55.]. Parametric 3D equations enable the quantification of points on curves and/or surfaces so that each point in x, y, z Euclidean 3D space can be mapped to a new 2D u, v Euclidean space. Because the surface phenotype resides within a curved boundary (e.g., circle, oval or other polygonal 2D form), the new 2D space contains

Figure 12.2 Scanning electron micrographs of *Actinoptychus*, *Arachnoidiscus* and *Cyclotella*. All micrographs taken by Mary Ann Tiffany used with permission.

information in u, v surface points about the geometric phenotype. A vector emanating from each point on the surface that is perpendicular indicates maximum changes from point to point on the surface. These changes as measured via the solution to a system of parametric 3D equations represent the geometric phenotype. Solutions to systems of parametric 3D equations entails calculating Jacobians, which in turn lead to the calculation of Christoffel symbols, the Hessian and the Laplacian as numerical representation of evolvability characteristics of the diatom valve surface.

12.3.1   Phenotypic Inertia

Phenotypic inertia means connection among phenotype surface features and encompasses a total phenotype response (e.g., [12.7.2.]). That response has adaptive value that results from the phenotype resisting change despite fluctuations in the environment or ecological conditions at various temporal scales (e.g., [12.7.2., 12.7.37.]). Perturbations, including developmental constraints, may restrict phenotype change as well, potentially resulting in stabilizing selection (e.g., [12.7.7.]). Bauplans are evidence of constrained lineages (e.g., [12.7.11., 12.7.60.]), indicating commonality among phenotype characteristics across higher levels of classification despite within population variation. Phenotypic inertia may be a measurement of resistance to change in surfaces across genera within a phylum.

To measure phenotypic inertia, the surface resistance is characterized geometrically. The simplest surface to measure is the one with the least amount of curvature so that the entire surface when viewed in profile is a straight line. A "straight" surface may have many pebbled or ragged surface features at a smaller scale, but overall, the surface projects a straight line (e.g., [12.7.54.]). The more curvature on the surface, the more complicated it is (e.g., [12.7.54.]); however, on the curved, unsmooth surface, there is the potential that at a given point, the surface will more likely fail in its durability in contrast to a flat, unsmooth surface (e.g., [12.7.3.]). That is, the flatter the unsmooth surface, the more likely the surface will resist breakage, or if breakage occurs, the surface will remain mostly intact rather than shattering, at any given point on the surface.

12.3.1.1   *Measurement of Phenotypic Inertia: Christoffel Symbols*

The neighborhood of a point is sufficiently small so that an inertia coordinate system is adequate [12.7.30.]. The Jacobian is the local tangent space in the neighborhood of a point and is transformed into Gauss and Weingarten Maps via the first and second fundamental forms to obtain Gaussian and mean curvatures, respectively. A Gauss Map is definable locally on a patch, and Gaussian curvature is the Jacobian determinant of a Gauss Map (e.g., 12.7.8.). The derivative of a Gauss Map is the shape operator, which is a Weingarten Map [12.7.81.]. Projection of the covariant derivative of the Weingarten Map onto the second fundamental form results in the connection gradient and is the derivative of the trihedron of vectors represented by Christoffel symbols [12.7.8.]. Points connected on the surface are represented by the Christoffel symbols and the moving reference frame at each point on a curve on the surface. Transformation of the Jacobian into Christoffel symbols produces geodesics, i.e., curvature on the surface. The Christoffel symbols measure change in direction of tangent vectors via an inertia coordinate system as a connection gradient (e.g., [12.7.8., 12.7.81.]).

Connection coefficients expressed in terms of the first fundamental form of the connection gradient are $\Gamma_{11}^1 = \dfrac{GE_u - 2FF_u + FE_v}{2(EG - F^2)}$, $\Gamma_{12}^1 = \dfrac{GE_v - FG_u}{2(EG - F^2)}$, $\Gamma_{22}^1 = \dfrac{2EF_v - GG_u - FG_v}{2(EG - F^2)}$, $\Gamma_{11}^2 = \dfrac{2EF_u - EE_v - FE_u}{2(EG - F^2)}$, $\Gamma_{12}^2 = \dfrac{EG_u - FE_v}{2(EG - F^2)}$, and $\Gamma_{22}^2 = \dfrac{EG_v - 2FF_v - FG_u}{2(EG - F^2)}$ as the intrinsic geometry that connects all the points on the surface [12.7.8., 12.7.72.]. For a Monge patch, $z = f(x, y)$, the connection coefficients $r_{ij} = \dfrac{z_{ij}z_k}{1 + z_1^2 + z_2^2}$ in a connection matrix are $\Gamma_{ij}^{k,l} = \begin{bmatrix} \Gamma_{ii}^l & \Gamma_{ij}^l \\ \Gamma_{ij}^k & \Gamma_{jj}^k \end{bmatrix}$ [12.7.81.], where $i, l = 1, \ldots,$ and $j, k = i + 1$. Christoffel symbols of the second kind with contravariant index $k = 3$ are

$$\Gamma^l \equiv \begin{bmatrix} \Gamma_{ii}^l & \Gamma_{ij}^l & \Gamma_{ik}^l \\ \Gamma_{ji}^l & \Gamma_{jj}^l & \Gamma_{jk}^l \\ \Gamma_{ki}^l & \Gamma_{kj}^l & \Gamma_{kk}^l \end{bmatrix} \qquad (12.1)$$

[12.7.81.]. The addition of indices with respect to covariant and contravariant vectors undergoing differentiation will yield covariant and contravariant derivatives such as $\Gamma_{ij}^k = \dfrac{\partial^2 \xi^p}{\partial x^i \partial x^j} \dfrac{\partial x^k}{\partial \xi^p} = \Gamma_{ji}^k$ where ξ^p are standard coordinates from a fixed coordinated system [12.7.30.].

Evaluation of the Christoffel symbols in the connection matrix is used to determine phenotypic inertia. The differences among geometric phenotypes for each genus are recorded in the Christoffel symbols. Those differences measure how difficult it is for the phenotype to be changed in curvature in one direction or curvature in two directions (torsion) at a given point on the surface. Overcoming the inertia that is required to "straighten" the surface with regard to curvature or torsion is the sum of the Christoffel symbols (e.g., [12.7.65.]). For each genus, the differences in geometric phenotypes are measured as

$$\textit{Phenotypic Inertia}_{genus} = \sum_l^n \Gamma_{i,j,k}^l \leq \Gamma_{ii}^l + \Gamma_{ij}^l + \Gamma_{ik}^l + \Gamma_{jj}^l + \Gamma_{jk}^l + \Gamma_{kk}^l \qquad (12.2)$$

where $l = 1, 2, 3$ for the contravariant lth Christoffel symbol, Γ, with covariant indices i, j, k defining n-Christoffel symbols that have different values when comparing all genera. The maximum number of Christoffel symbols included in the sum is the upper diagonal values of Γ^l in Eq. (12.1).

12.3.2   Phenotypic Robustness

Phenotypic robustness means maximum extent of phenotype surface features expressed with respect to the magnitude and direction of those features. Canalized traits, developmental constraints, fitness, and epistasis are implicitly involved (e.g., [12.7.5., 12.7.63.]). The persistence of the phenotype surface is a hallmark of robustness.

To measure phenotypic robustness, the maximum displacement between the heights and troughs on the surface are used, regardless of the amount of curvature in the surface overall. Magnitude of the phenotype is reflected in the peaks and valleys on the surface as well as the inflection points as saddles. The more peaks, valleys and saddles on the surface, the higher the magnitude of the phenotype and the more robust is the surface. High surface area is determined by increasing peaks, valleys and saddles, so that the higher the surface area, the higher the phenotypic robustness of the surface.

12.3.2.1 Measurement of Phenotypic Robustness: the Hessian

Critical points or extrema on the surface are phenotypic constraints measured as the Hessian. The Hessian is a summary of phenotypic surface conditions with respect to allelic interactions (e.g., [12.7.63.]). Eigenvalues of the Hessian are measures of Gaussian curvature, and eigenvectors of the Hessian are measures of principal directions.

For a surface patch, $z = f(x, y)$, z is the height function. From the Jacobian, the phenotype gradient is $\nabla z = \begin{bmatrix} \dfrac{\partial z}{\partial u} \\ \dfrac{\partial z}{\partial v} \end{bmatrix}$, and the magnitude of ∇z is $\| \nabla z \| = \sqrt{\left(\dfrac{\partial z}{\partial u} \right)^2 + \left(\dfrac{\partial z}{\partial v} \right)^2}$ which is the magnitude of maximum slopes.

Allelic interactions are represented by $A_{Hessian} = \begin{bmatrix} \dfrac{\partial^2 z}{\partial u^2} & \dfrac{\partial^2 z}{\partial u \partial v} \\ \dfrac{\partial^2 z}{\partial u \partial v} & \dfrac{\partial^2 z}{\partial v^2} \end{bmatrix}$, which is the dominance-epistasis matrix [12.7.63.]. The product of the phenotype gradient and the allelic matrix is equivalent to the product of the maximum heights and maximum slopes and is given as

$$Phenotypic\ Robustness_{genus} = \nabla z \cdot A_{Hessian}$$

$$= \begin{bmatrix} \dfrac{\partial z}{\partial u} \\ \dfrac{\partial z}{\partial v} \end{bmatrix} \begin{bmatrix} \dfrac{\partial^2 z}{\partial u^2} & \dfrac{\partial^2 z}{\partial u \partial v} \\ \dfrac{\partial^2 z}{\partial u \partial v} & \dfrac{\partial^2 z}{\partial v^2} \end{bmatrix} = \begin{bmatrix} \left(\dfrac{\partial z}{\partial u} \right)\left(\dfrac{\partial^2 z}{\partial u^2} \right) + \left(\dfrac{\partial z}{\partial v} \right)\left(\dfrac{\partial^2 z}{\partial u \partial v} \right) \\ \left(\dfrac{\partial z}{\partial u} \right)\left(\dfrac{\partial^2 z}{\partial u \partial v} \right) + \left(\dfrac{\partial z}{\partial v} \right)\left(\dfrac{\partial^2 z}{\partial v^2} \right) \end{bmatrix} \quad (12.3)$$

and this is the measure of phenotypic robustness. This measure may be normalized by $\| \nabla z \|$.

Evaluation of the eigenvalues, ψ, and the Hessian determinant is used to determine degree of robustness with respect to Eq. (12.3). Hessian eigenvalues are calculated from the characteristic equation, $A_{Hessian} - \psi I = 0$, which is the general form of the characteristic polynomial [12.7.81.]. The eigenvalues are the roots of the characteristic polynomial and are obtained from $\det A_{Hessian} - \psi I = \prod_i (\psi - \psi_i)$, and the Hessian determinant is $A_{HessDet} = \begin{vmatrix} \dfrac{\partial^2 z}{\partial u^2} & \dfrac{\partial^2 z}{\partial u \partial v} \\ \dfrac{\partial^2 z}{\partial u \partial v} & \dfrac{\partial^2 z}{\partial v^2} \end{vmatrix}$ [12.7.81.].

12.3.3   Phenotypic Stability

Phenotypic stability means phenotype surface features are counted that contribute to variation. This is a dynamic measure of phenotype and is implicitly represented as developmental noise, mutational influences and environmental influences. As a dynamic occurrence of the surface, resilience of the phenotype is evidenced by its longevity via the capability to withstand deleterious influences. A summation of the surface attributes that account for resilience involve perturbations that affect stability. Surfaces that have the fewest features have the fewest phenotypic attributes where deleterious changes could occur, so that the surface that is the most uniform, regardless of the amount of overall curvature, is the more resilient surface and will be the most stable.

12.3.3.1   Measurement of Phenotypic Stability: the Laplacian

Stability involves the divergence of the gradient on the surface. The Laplacian, or more generally, the Laplace-Beltrami operator, is a summary of the dynamics of phenotype surface changes as small perturbations. The Laplacian is a measure of mean curvature (e.g., [12.7.8.]). The trace of the Hessian is the Laplacian and is given as $\text{Tr}(A_{Hessian}) \equiv \sum_{i=1}^{n} a_{ii}$, where a_{ii} are the diagonal elements of $A_{Hessian}$. The Laplacian is the average value of the phenotype.

In terms of phenotypic stability, we want to measure surface equilibrium as

$$Phenotypic\ Stability(Equilibrium)_{genus} = div\ (grad\ z) = \nabla^2 z = \frac{\partial^2 z}{\partial u^2} + \frac{\partial^2 z}{\partial v^2} = 0$$

(12.4)

and at non-equilibrium

$$Phenotypic\ Stability(Non-equilibrium)_{genus} = div\ (grad\ z) = \nabla^2 z = \frac{\partial^2 z}{\partial u^2} + \frac{\partial^2 z}{\partial v^2} \lessgtr 0.$$

(12.5)

Solution is obtained by separation of variables where $\nabla^2 z$ is an initial value problem with a homogeneous boundary condition (e.g., [12.7.80.]). For $z = f(x,y)$; $x = g(u,v)$, $y = h(u,v)$, quadratic expansion of $\nabla^2 z$ is

$$\frac{\partial^2 z}{\partial u^2} = \frac{\partial z}{\partial x}\frac{\partial^2 x}{\partial u^2} + \left(\frac{\partial^2 z}{\partial x^2}\right)\left(\frac{\partial x}{\partial u}\right)^2 + 2\frac{\partial^2 z}{\partial x \partial y}\frac{\partial x}{\partial u}\frac{\partial y}{\partial u} + \left(\frac{\partial^2 z}{\partial y^2}\right)\left(\frac{\partial y}{\partial u}\right)^2 + \frac{\partial z}{\partial y}\frac{\partial^2 y}{\partial u^2}$$

(12.6)

and

$$\frac{\partial^2 z}{\partial v^2} = \frac{\partial z}{\partial x}\frac{\partial^2 x}{\partial v^2} + \left(\frac{\partial^2 z}{\partial x^2}\right)\left(\frac{\partial x}{\partial v}\right)^2 + 2\frac{\partial^2 z}{\partial x \partial y}\frac{\partial x}{\partial v}\frac{\partial y}{\partial v} + \left(\frac{\partial^2 z}{\partial y^2}\right)\left(\frac{\partial y}{\partial v}\right)^2 + \frac{\partial z}{\partial y}\frac{\partial^2 y}{\partial v^2}.$$

(12.7)

Second partial derivatives are discretized with difference quotients (finite differences) as determined by Taylor expansion of the Laplacian [12.7.80.] as

$$\nabla^2 z_{i,j} \approx \frac{z_{i+1,j} - 2z_{i,j} + z_{i-1,j}}{\Delta u^2} + \frac{z_{i,j+1} - 2z_{i,j} + z_{i,j-1}}{\Delta v^2} + \cdots \tag{12.8}$$

Each term is approximately equal to the eigenvectors of $\nabla^2 z$ in u and v so that

$\dfrac{\partial^2 z}{\partial u^2} = \left(\dfrac{d^2}{du^2} \otimes I_v \right) z(u,v)$ and $\dfrac{\partial^2 z}{\partial v^2} = \left(I_u \otimes \dfrac{d^2}{dv^2} \right) z(u,v)$, where each term of the Laplacian

is a tensor product [12.7.80.]. For $\nabla^2 z_{i,j}$, the eigenvalue problem is $\dfrac{d^2 z}{du^2} = \lambda_u z(u)$ and $\dfrac{d^2 z}{dv^2} = \lambda_v z(v)$ so that

$$\nabla^2 z = \left[\left(\frac{d^2}{du^2} \otimes I_v \right) + \left(I_u \otimes \frac{d^2}{dv^2} \right) \right] z(x,y) \tag{12.9}$$

The eigenvectors, $eig_{i,j}(u,v) = eig_u \otimes eig_v$, for eigenvalues, $\lambda_{i,j}(u,v) = \lambda_{u_i} + \lambda_{v_j}$, are $eig_{u_i}(u) = \sqrt{\dfrac{2}{L}} sin\left(\dfrac{i\pi u}{L} \right)$ and $eig_{v_j}(v) = \sqrt{\dfrac{2}{L}} sin\left(\dfrac{j\pi v}{L} \right)$ for Dirichlet boundary ($\partial\Omega = 0$) [12.7.80.].

Evaluation of the eigenvalues, $\lambda = \lambda_u + \lambda_v$, is used to determine degree of stability. From the eigenvalues of the Hessian z matrix, the sum of the diagonal elements in u and v are the Laplacian eigenvalues.

12.3.4   Evolvability of each Diatom Genus: *Actinoptychus*, *Arachnoidiscus* and *Cyclotella*

Actinoptychus, *Arachnoidiscus* and *Cyclotella* are compared in terms of phenotypic inertia, robustness and stability. The comparisons are accomplished via looking at the phenotypic change from a defined original basal surface to that surface subsequently. This approach induces fitness functions with respect to phenotypic inertia, robustness and stability as resistance, persistence and resilience components of evolvabililty [12.7.76.], and a composite fitness function of phenotypic components is equivalently a measure of evolvability [12.7.62.].

The change in surface phenotype is a change in fitness and is measured geometrically from 3D surface models as fitness at a basal value of 0 and change in fitness at $2\pi = 6.28$. Phenotypic inertia, robustness and stability are calculated as summation values for each genus as

$$Actinoptychus_{Fitness} = \begin{bmatrix} \sum_{Actinoptychus} Phenotypic\ Inertia \\ \sum_{Actinoptychus} Phenotypic\ Robustness \\ \sum_{Actinoptychus} Phenotypic\ Stability \end{bmatrix}, \qquad (12.10)$$

$$Arachnoidiscus_{Fitness} = \begin{bmatrix} \sum_{Arachnoidiscus} Phenotypic\ Inertia \\ \sum_{Arachnoidiscus} Phenotypic\ Robustness \\ \sum_{Arachnoidiscus} Phenotypic\ Stability \end{bmatrix} \qquad (12.11)$$

and

$$Cyclotella_{Fitness} = \begin{bmatrix} \sum_{Cyclotella} Phenotypic\ Inertia \\ \sum_{Cyclotella} Phenotypic\ Robustness \\ \sum_{Cyclotella} Phenotypic\ Stability \end{bmatrix}. \qquad (12.12)$$

where $Phenotypic\ Inertia = \sum_i^n \Gamma_{jl}^i$, $Phenotypic\ Robustness = \sum_i \begin{bmatrix} Robust_u \\ Robust_v \end{bmatrix}_i$, and $Phenotypic\ Stability = \sum_i Stab_i$ for each genus.

To assess evolvability in *Actinoptychus*, *Arachnoidiscus* and *Cyclotella*, an additive fitness function is devised for each genus. The generalized fitness function is

$$Evolvability_{genus\ fitness}$$

$$= \left(\sum_l^n \Gamma_{i,j,k}^l \right)_{\substack{Phenotypic \\ Inertia}} + (\nabla z \cdot A_{Hessian})_{\substack{Phenotypic \\ Robustness}} + (\nabla^2 z)_{\substack{Phenotypic \\ Stability}}$$

$$= \sum_l^n \left(\frac{\partial^2 \xi^p}{\partial x^i \partial x^j} \frac{\partial x^k}{\partial \xi^p} \right)_l + \begin{bmatrix} \left(\frac{\partial z}{\partial u} \right)\left(\frac{\partial^2 z}{\partial u^2} \right) + \left(\frac{\partial z}{\partial v} \right)\left(\frac{\partial^2 z}{\partial u \partial v} \right) \\ \left(\frac{\partial z}{\partial u} \right)\left(\frac{\partial^2 z}{\partial u \partial v} \right) + \left(\frac{\partial z}{\partial v} \right)\left(\frac{\partial^2 z}{\partial v^2} \right) \end{bmatrix} + \begin{bmatrix} \frac{\partial^2 z}{\partial u^2} + \frac{\partial^2 z}{\partial v^2} \end{bmatrix} \qquad (12.13)$$

for any coordinate system, ξ, over contravariant and covariant indices l and i, j, k, respectively, and parameters u, v for z, where the second term results in a vector quantity.

12.3.5   Evolvability Among Diatom Genera

The sum of fitness functions characterizes each genus as evolvability that may have resulted in adaptability in spite of developmental constraints or other influences. The decrease in fitness from one genus to another is indicative of the change in evolvability as a comparative value. Comparative evolvability as fitness decrease is calculated as

$$Evolvability_{fitness1, fitness2} = fitness_{decrease} = \frac{(fitness_{Genus1} - fitness_{Genus1+Genus2})}{fitness_{Genus1+Genus2}}. \quad (12.14)$$

Resultant quantities are evaluated relative to zero [12.7.76.]. If $fitness_{decrease} \approx 0$, then the outcome is neutral, and evolvability may or may not have occurred. If $fitness_{decrease} > 0$, then the outcome is favorable, and evolvability may have occurred. If $fitness_{decrease} < 0$, then no evolvability has occurred as a result of fitness decrease.

12.3.6   Contribution of Phenotypic Inertia, Robustness and Stability to Evolvability

For all genera, phenotypic inertia, robustness and stability contribute to evolvability. To determine what the proportional contributions are, discrete vectors are compiled across all genera from equilibrium at $u, v = 0$ and non-equilibrium at $u, v = 6.28$.

Let the phenotypic inertia for all genera be compiled as discrete vectors,

$$PF = Phenotypic\ Inertia_{u,v=0} = \begin{bmatrix} PF_{Actinoptycus} \\ PF_{Arachnoidiscus} \\ PF_{Cyclotella} \end{bmatrix}$$

for $u, v = 0$ and

$$PG = Phenotypic\ Inertia_{u,v=6.28} = \begin{bmatrix} PG_{Actinoptycus} \\ PG_{Arachnoidiscus} \\ PG_{Cyclotella} \end{bmatrix}$$

for $u, v = 6.28$. The change in surfaces from PF to PG is obtained as the sum of the differences in vector elements given as

$$\sum_{\substack{i,j \\ u,v=0 \\ u,v=6.28}} (PG_i - PF_j) \quad (12.15)$$

Sum of the vector elements results in phenotypic inertia at non-equilibrium.

For phenotypic robustness, HN and HM may be obtained via results from Eq. (12.3). Let the phenotypic robustness for all genera be compiled as the matrices,

$$HN = Phenotypic\ Robusness_{u,v=0} = \begin{bmatrix} hn_{Actinoptychus1} & hn_{Actinoptychus2} \\ hn_{Arachnoidiscus1} & hn_{Arachnoidiscus2} \\ hn_{Cyclotella1} & hn_{Cyclotella2} \end{bmatrix} \quad (12.16)$$

at equilibrium and

$$HM = Phenotypic\ Robusness_{u,v=6.28} = \begin{bmatrix} hm_{Actinoptychus1} & hm_{Actinoptychus2} \\ hm_{Arachnoidiscus1} & hm_{Arachnoidiscus2} \\ hm_{Cyclotella1} & hm_{Cyclotella2} \end{bmatrix} \quad (12.17)$$

at non-equilibrium. The sum of the differences in the vector elements results in phenotypic robustness.

Phenotypic stability, expressed as LD at equilibrium and LE at non-equilibrium, are

$$LD = Phenotypic\ Stability_{u,v=0} = \begin{bmatrix} ld_{Actinoptychus} \\ ld_{Arachnoidiscus} \\ ld_{Cyclotella} \end{bmatrix} and \quad (12.18)$$

$$LE = Phenotypic\ Stability_{u,v=6.28} = \begin{bmatrix} le_{Actinoptychus} \\ le_{Arachnoidiscus} \\ le_{Cyclotella} \end{bmatrix}. \quad (12.19)$$

Difference in the sum of the vector elements at equilibrium and non-equilibrium is

$$\sum_{\substack{i,j \\ u,v=0 \\ u,v=6.28}} (LE_i - LD_j) \quad (12.20)$$

and results in phenotypic stability.

Contribution of phenotypic inertia, robustness and stability to evolvability will be illustrated as a stacked percentage plot. Each contribution will be parsed according to each genus to identify the component contribution to evolvability.

12.3.7   Phenotypic Novelty Measurement

Fitness measurement does not necessarily result in assessing evolvability of phenotypic novelty (e.g., [12.7.76.]), but discontinuity in phenotype structure is useful as a measure of novelty [12.7.57.]. To account for the discontinuous phenotype of radial plication geometry from *Actinoptychus* to *Arachnoidiscus* to *Cyclotella*, the portion of the geometric phenotype that characterizes radial plications is used. From systems of parametric equations, genera were identical in x and y, and this 2D geometry is a solid circle (Figure 12.3). In the z-direction, the profiles of the radial plications exhibit maxima, minima and saddle points and are recovered quantitatively by the Hessian in z determinant of the differences between paired comparisons of genera.

To calculate novelty, information theory is used (e.g., [12.7.76., 12.7.78.]). Entropy for a genus is

$$Entropy_{Genus_i} = -\sum_{Genus_i} Probability(Genus_i)\log_2 Probability(Genus_i) \quad (12.21)$$

where $i = 1, 2$ *or* 3. Comparative information between genera is given as

$$Joint\ Entropy_{Genus_i, Genus_j} =$$
$$-\sum_{Genus_i}\sum_{Genus_j} Probability(Genus_i, Genus_j)\log_2 Probability(Genus_i, Genus_j)$$

$$(12.22)$$

where $i, j = 1, 2$ *or* 3 with $i \neq j$. Resultant values are used as information that novelty has been identified in one of the genera.

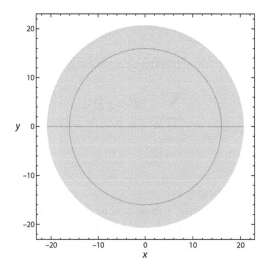

Figure 12.3 Plot of x, y parametric equations for *Actinoptychus*, *Arachnoidiscus* and *Cyclotella*.

To determine novelty, the efficiency of change from one genus to another is used and calculated as the inverse normalized joint information

$$Novelty_{Genus1,Genus2} = Joint\ Information_{Phenotype}$$

$$= 1 - \frac{(Probability_{Genus1} + Probability_{Genus2}) - Probability_{Genus1,Genus2}}{\max(Probability_{Genus1}, Probability_{Genus2})}$$

$$(12.23)$$

Because the comparative measurement of novelty is based on a distribution, the scale comprising a threshold is 0 to 1. If $Joint\ Information_{Phenotype} > 0.5$ [12.7.76.], then the phenotypic characteristic is novel.

12.3.8 Evolvability and Phenotypic Novelty

To illustrate the relation between evolvability and phenotypic novelty, a stacked percentage plot of evolvability as the fitness decrease between pairs of genera and novelty as the joint entropy of pairs of genera was used. From this, identification of the genus (or genera) that exhibited evolvability and/or have radial plications that are a phenotypic novelty will be determined.

12.4 Results

From parametric equations, 3D surface models of *Actinoptychus*, *Arachnoidiscus* and *Cyclotella* were constructed (Figure 12.4). For all models, $x = 16 \cos u \cos v (1 + \sin u)$ and $y = 16 \cos u \sin v (1 + \sin u)$, respectively. Because the x- and y-equations were the same for all 3D models, each model is treated as a Monge patch, and the z-equations are given in Table 12.1. Phenotypic surface features, mathematical operators and general characterization of the numerical results are given in Table 12.2.

Christoffel symbols of the second kind were calculated [12.7.19.] with contravariant index $k = 3$ and are reported in Table 12.3. For contravariant indices $k = 1, 2, 3$, covariant indices i, j are symmetric so that $\Gamma_{12}^k = \Gamma_{21}^k$, $\Gamma_{13}^k = \Gamma_{31}^k$, and $\Gamma_{23}^k = \Gamma_{32}^k$.

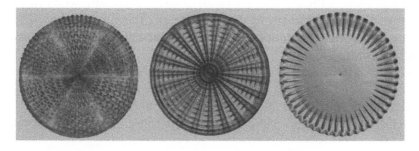

Figure 12.4 3D surface models of *Actinoptychus*, *Arachnoidiscus* and *Cyclotella*.

Table 12.1 The z-equations for *Actinoptychus*, *Arachnoidiscus* and *Cyclotella*. For all parameters, $u, v \in [0, 2\pi]$.

Taxon	z-equation[†]
Actinoptychus senarius	$-\alpha u \cos 3u \cos 3v + \beta \sin 3u - \gamma \sin(6u)^5$ $-\zeta u \cos(5 - 4u) \sin(3 - 4v)^4 + \mu \sin 120u \sin 6v$ $-\varphi \cos(0.5u)^2 \cos(40u)^4 \cos 3v \sin 40u \sin(40v)^3$ $-\psi u \sin 40u \sin 160v$
Arachnoidiscus ehrenbergii	$v \cos(0.5u^2) + \rho \cos(80u)^2 \sin(1.9u)^2 + \Xi \sin(1 + u)^3$ $+ \tau \cos(10 + 74u^2) \sin(0.5 + 0.8u)^3 \sin(2 + 2.9u)^3$ $+ \vartheta \cos u \sin(0.37u)^3 \sin(22v)^3$
Cyclotella meneghiniana	$\cos 1.5u[\xi \cos 2u \sin 0.5u \sin 2.9u$ $+ 0.14 \cos u \sin(0.5u)^4 \sin(7v)^4]$ $+ \eta \cos(3.5u)^6 \cos 25v \sin 1.8u^6 \sin 25v$

[†] Coefficients: $0.0008 \leq \alpha \leq 0.64$, $0.01 \leq \beta \leq 0.1$, $0.001 \leq \gamma \leq 1$, $0.003 \leq \zeta \leq 0.09$, $0.0025 \leq \mu \leq 0.125$, $0.2 \leq \varphi \leq 10$, $0.0025 \leq \psi \leq 0.025$, $0.004 \leq \vartheta \leq 0.7$, $0.05 \leq v \leq 0.75$, $0.025 \leq \rho \leq 0.5$, $0.051 \leq \tau \leq 2.55$, $2 \leq \Xi \leq 3$, $0.01 \leq \xi \leq 0.5$, $0.18 \leq \eta \leq 3$.

Table 12.2 Phenotypic surface features, mathematical operator and numerical results.

Surface feature	Mathematical operator	Analytical/numerical results
Peaks	Hessian	Maxima - Positive
Valleys	Hessian	Minima - Negative
Saddles	Hessian	Zero
Convex or Concave Smoothness	Laplacian	Positive or Negative
Smoothness	Laplacian	Zero
Curvature	Christoffel symbols	Positive or Negative
Flatness	Christoffel symbols	Zero

At equilibrium where $u, v = 0$, *Actinoptychus* and *Cyclotella* had identical values for the Christoffel symbols, while *Arachnoidiscus* differed only in Christoffel symbols Γ^1_{33}, Γ^3_{31} and Γ^3_{32} (Table 12.3). Most of the entries were equal to zero, which signifies an almost flat surface and defines the starting point. At non-equilibrium, there were identical values at or around zero as well with the exception of Christoffel symbol Γ^2_{22} that was an identical value but a negative number (Table 12.3).

At non-equilibrium where $u, v = 6.28$, Christoffel symbols for the genera were identical except for Γ^1_{33}, Γ^2_{33}, Γ^3_{31}, and Γ^3_{32} (Table 12.3). For Γ^1_{33} and Γ^2_{33}, the smallest to largest values occurred from *Arachnoidiscus* to *Actinoptychus* to *Cyclotella*, and for Γ^3_{31} and Γ^3_{32}, *Cyclotella*, *Actinoptychus* and *Arachnoidiscus* occurred from smallest to largest values (Table 12.3).

Table 12.3 Results from calculation of Christoffel symbols for *Actinoptychus*, *Arachnoidiscus* and *Cyclotella* at equilibrium and non-equilibrium.

	Actinoptychus		*Arachnoidiscus*		*Cyclotella*	
	$u, v = 0$	$u, v = 6.28$	$u, v = 0$	$u, v = 6.28$	$u, v = 0$	$u, v = 6.28$
$\Gamma111$	0.5	0.503188	0.5	0.503188	0.5	0.503188
$\Gamma121$	0	0.00159266	0	0.00159266	0	0.00159266
$\Gamma122$	0	0.00160281	0	0.00160281	0	0.00160281
$\Gamma131$	0	0	0	0	0	0
$\Gamma132$	0	0	0	0	0	0
$\Gamma133$	Indeterminate	-0.0957139	-0.0694528	-10.5865	0	-0.0037189
$\Gamma211$	0.5	0.5	0.5	0.5	0.5	0.5
$\Gamma221$	0.5	0.503188	0.5	0.503188	0.5	0.503188
$\Gamma222$	Complex Infinity	-156.97	Complex Infinity	-156.97	Complex Infinity	-156.97
$\Gamma231$	0	0	0	0	0	0
$\Gamma232$	0	0	0	0	0	0
$\Gamma233$	Indeterminate	46.7439	Indeterminate	0.864298	Indeterminate	545.13
$\Gamma311$	0	0	0	0	0	0
$\Gamma321$	0	0	0	0	0	0
$\Gamma322$	0	0	0	0	0	0
$\Gamma331$	Indeterminate	-0.378678	0.629821	148.014	Indeterminate	-0.336028
$\Gamma332$	Indeterminate	-0.589079	0	0.0384916	Indeterminate	-156.895
$\Gamma333$	0	0	0	0	0	0

For the Hessians calculated for the genera, mixed second partial derivatives in terms of uv are symmetrical; that is, $uv = vu$. Hessian values are reported in columns 2, 3 and 4 for u, uv and v, respectively, in Table 12.4. For all genera, Hessian values were identical in x and y at $u, v = 0$ as were Hessian values at $u, v = 6.28$ (Table 12.4).

All differences in the Hessian were found in the z-values. At $u, v = 0$, Hessian z-values were zero for all genera except for those at u and uv for *Actinoptychus* and at u for *Arachnoidiscus*

Table 12.4 Results from calculation of the Hessian and the Laplacian for *Actinoptychus*, *Arachnoidiscus* and *Cyclotella* at equilibrium and non-equilibrium.

Actinoptychus	at $u, v = 0$				
Hessian x	-16	0	-16	Laplacian x	-32
Hessian y	0	16	0	Laplacian y	0
Hessian z	0.000274	89.9989	0	Laplacian z	0.000274
Actinoptychus	at $u, v = 6.28$				
Hessian x	-15.79598	0.051126	-15.948873	Laplacian x	-31.7449
Hessian y	0.0503152	16.05056	0.0508022	Laplacian y	0.101117
Hessian z	91.551754	-1475.584	-1179.0407	Laplacian z	-1087.49
Arachnoidiscus	at $u, v = 0$				
Hessian x	-16	0	-16	Laplacian x	-32
Hessian y	0	16	0	Laplacian y	0
Hessian z	26.8525	0	0	Laplacian z	26.8525
Arachnoidiscus	at $u, v = 6.28$				
Hessian x	-15.79598	0.051126	-15.948873	Laplacian x	-31.7449
Hessian y	0.0503152	16.05056	0.0508022	Laplacian y	0.101117
Hessian z	595118.2615	-0.09104	-54.91261	Laplacian z	595133.3
Cyclotella	at $u, v = 0$				
Hessian x	-16	0	-16	Laplacian x	-32
Hessian y	0	16	0	Laplacian y	0
Hessian z	4.35	0	0	Laplacian z	4.35
Cyclotella	at $u, v = 6.28$				
Hessian x	-15.79598	0.051126	-15.948873	Laplacian x	-31.7449
Hessian y	0.0503152	16.05056	0.0508022	Laplacian y	0.101117
Hessian z	18.516375	-177.6099	444.83417	Laplacian z	463.351

and *Cyclotella* (Table 12.4). At $u, v = 6.28$, differences in the Hessian were evident at u, uv and v for all genera. For *Actinoptychus* and *Arachnoidiscus*, the u-value in z was positive and the uv- and v-values in z were negative (Table 12.4). The v-value in z was positive for *Cyclotella* (Table 12.4). The largest u-value in z was positive and occurred for *Arachnoidiscus*. while the largest negative value occurred for *Actinoptychus* in the uv-value in z (Table 12.4).

Eigenvalues were calculated for the Hessian (Table 12.5, column 1). *Arachnoidiscus* for x and y and *Cyclotella* for y and z have complex-valued eigenvalues at $u, v = 0$, and at u,

Table 12.5 Hessian and Laplacian x, y, z eigenvalues for *Actinoptychus*, *Arachnoidiscus* and *Cyclotella* at $u, v = 0$ and $u, v = 6.28$.

Actinoptychus	at $u, v = 0$		
Hessian eigenvalue x	16	Laplacian eigenvalue x, y, z	-2.79×10^{16}
Hessian eigenvalue y	-15.9997		
Hessian eigenvalue z	-0.000274		
Actinoptychus	at $u, v = 6.28$		
Hessian eigenvalue x	-1177.72	Laplacian eigenvalue x, y, z	-1178.79
Hessian eigenvalue y	-17.086		
Hessian eigenvalue z	16.0202		
Arachnoidiscus	at $u, v = 0$		
Hessian eigenvalue x	$-8 + 19.1217i$	Laplacian eigenvalue x, y, z	0
Hessian eigenvalue y	$-8 - 19.1217i$		
Hessian eigenvalue z	16		
Arachnoidiscus	at $u, v = 6.28$		
Hessian eigenvalue x	$-35.3544 + 3080i$	Laplacian eigenvalue x, y, z	-54.658
Hessian eigenvalue y	$-35.3544 - 3080i$		
Hessian eigenvalue z	16.0507		
Cyclotella	at $u, v = 0$		
Hessian eigenvalue x	16	Laplacian eigenvalue x, y, z	-8.88×10^{-16}
Hessian eigenvalue y	$8 + 2.36643i$		
Hessian eigenvalue z	$8 - 2.36643i$		
Cyclotella	at $u, v = 6.28$		
Hessian eigenvalue x	444.172	Laplacian eigenvalue x, y, z	445.089
Hessian eigenvalue y	16.0615		
Hessian eigenvalue z	-15.1445		

$v = 6.28$, *Arachnoidiscus* for x and y has complex-valued eigenvalues. *Actinoptychus* and *Arachnoidiscus* have eigenvalues of 16 for z at u, $v = 6.28$, and *Actinoptychus* and *Cyclotella* have the same value for x at u, $v = 0$. The largest negative eigenvalue occurred at x for *Actinoptychus*, while the largest positive eigenvalue occurred at x for *Cyclotella* (Table 12.5).

The Laplacian was calculated as the trace of the diagonal elements of the Hessian, and for each genus, values were reported in column 6 in Table 12.4. Laplacian values in x were approximately equal at u, $v = 0$ and u, $v = 6.28$, values in y were zero at u, $v = 0$ and the same value at u, $v = 6.28$, and differed among genera only in z at u, $v = 0$ and u, $v = 6.28$ (Table 12.4). The Laplacian in z at u, $v = 0$ ranged from near zero to a very small positive value to a small positive value for *Actinoptychus*, *Cyclotella* and *Arachnoidiscus*, respectively. At u, $v = 6.28$, the same order for the genera occurred, ranging from a large negative value to a positive value to a very large positive value (Table 12.4). Eigenvalues for the Laplacian in x, y, z were around zero at u, $v = 0$ and ranged from a large negative value for *Actinoptychus* to a smaller negative value for *Arachnoidiscus* to a positive value for *Cyclotella* at u, $v = 6.28$ (Table 12.5).

12.4.1   Phenotypic Inertia, Robustness and Stability

To calculate phenotypic inertia from Christoffel symbols, those values that differed among genera were Γ^1_{33}, Γ^2_{33}, Γ^3_{31}, and Γ^3_{32} at non-equilibrium, were used (Table 12.3). The connection among these Christoffel symbols on the surface is the sum and ranged as smallest to largest values from *Actinoptychus* to *Arachnoidiscus* to *Cyclotella* (Table 12.6).

For phenotypic robustness calculated according to Eq. (12.3), the phenotype gradient is necessary, and values for this quantity with respect to the Hessian in z were calculated at u, $v = 0$ and u, $v = 6.28$ (Table 12.7). At equilibrium, all values for the phenotype gradient were near zero, and at non-equilibrium, values were positive and smallest for *Actinoptychus*, with the first value being large for *Arachnoidiscus* and very small for *Cyclotella* (Table 12.7). The dot product of the phenotype gradient and Hessian in z resulted in phenotypic robustness values (Table 12.8). Except for the first value for *Arachnoidiscus*, all other values were negative or zero at u, $v = 0$, while at u, $v = 6.28$, a very large first positive value occurred for *Arachnoidiscus*, while large negative first values occurred for *Actinoptychus* and *Cyclotella* (Table 12.8). Second values for phenotypic robustness occurred as a large negative value for

Table 12.6 Sum of the Christoffel symbols that are different among genera at u, $v = 6.28$.

Christoffel symbol	Actinoptychus	Arachnoidiscus	Cyclotella
Γ133	-0.09571	-10.5865	-0.00372
Γ233	46.7439	0.864298	545.13
Γ331	-0.37868	148.014	-0.33603
Γ332	-0.58908	0.038492	-156.895
Sum	45.68043	138.3303	387.8953

Table 12.7 Values of the phenotype gradient in the z-value from the Hessian at equilibrium and non-equilibrium.

	$u, v = 0$	$u, v = 6.28$
Actinoptychus	-0.34001, 0	3.05306, 4.74939
Arachnoidiscus	2.22249, 0	337.685, 0.087817
Cyclotella	0, 0	0.118626, 55.3876

Table 12.8 Phenotypic robustness as the dot product of the phenotype gradient and Hessian in z at equilibrium and non-equilibrium.

	$u, v = 0$	$u, v = 6.28$
Actinoptychus	-0.0000931, -30.6035	-6728.62, -10104.8
Arachnoidiscus	59.6793, 0	2.0099×10^8, 0.087817
Cyclotella	0, 0	-9835.2, 24617.2

Actinoptychus, a large positive value for *Cyclotella*, and a value near zero for *Arachnoidiscus* (Table 12.8).

To evaluate phenotypic robustness, Hessian z eigenvalues (Table 12.9) and the Hessian determinant were calculated (Table 12.10). At $u, v = 0$ and $u, v = 6.28$, large negative Hessian determinantal values occurred for all genera except for zero values for *Arachnoidiscus* and *Cyclotella* at $u, v = 0$ (Table 12.10). Hessian z first eigenvalues indicated that there was a large positive value for *Actinoptychus*, a smaller positive value for *Arachnoidiscus*, and a very small positive value for *Cyclotella* at equilibrium (Table 12.9). At non-equilibrium, the eigenvalues were a very large positive value for *Arachnoidiscus*, a much smaller positive value for *Cyclotella*, and a large negative value for *Actinoptychus* (Table 12.9). *Actinoptychus* had a second Hessian z eigenvalue that was negative, while the value was zero for *Arachnoidiscus* and *Cyclotella* at equilibrium (Table 12.9). Comparable negative second eigenvalues occurred for *Arachnoidiscus* and *Cyclotella*, while *Actinoptychus* had a large positive value (Table 12.9).

Comparison of the first and second Hessian z eigenvalues has been summarized as a pair of doughnut plots of equilibrium and non-equilibrium (Figure 12.5). *Arachnoidiscus* is the most phenotypically robust, followed by *Actinoptychus* for the first Hessian z eigenvalue, while *Actinoptychus* is the most phenotypically robust, followed by *Arachnoidiscus* for the second Hessian z eigenvalue. From equilibrium to non-equilibrium, phenotypic robustness is ordered as *Arachnoidiscus* to *Actinoptychus* to *Cyclotella* for the first and second Hessian z eigenvalues. From the first to the second Hessian z eigenvalues, there is an increase in phenotypic robustness for *Actinoptychus*, a decrease for *Arachnoidiscus*, and almost equivalent or slight increase for *Cyclotella* from equilibrium to non-equilibrium (Figure 12.5).

Table 12.9 Hessian and Laplacian z eigenvalues for *Actinoptychus*, *Arachnoidiscus* and *Cyclotella* at equilibrium and non-equilibrium.

Actinoptychus	at $u, v = 0$
Hessian eigenvalues $z(u, v)$	89.999, -89.9987
Laplacian eigenvalue $z(u, v)$	0.000274
Actinoptychus	at $u, v = 6.28$
Hessian eigenvalues $z(u, v)$	-2150.28, 1062.79
Laplacian eigenvalue $z(u, v)$	-1087.49
Arachnoidiscus	at $u, v = 0$
Hessian eigenvalues $z(u, v)$	26.8525, 0
Laplacian eigenvalue $z(u, v)$	26.8525
Arachnoidiscus	at $u, v = 6.28$
Hessian eigenvalues $z(u, v)$	595188, -54.9126
Laplacian eigenvalue $z(u, v)$	595133
Cyclotella	at $u, v = 0$
Hessian eigenvalues $z(u, v)$	4.35, 0
Laplacian eigenvalue $z(u, v)$	4.35
Cyclotella	at $u, v = 6.28$
Hessian eigenvalues $z(u, v)$	509.132, -45.781
Laplacian eigenvalue $z(u, v)$	463.351

Table 12.10 The Hessian determinants from the z-values.

	at $u, v = 0$	at $u, v = 6.28$
Actinoptychus	-8099.8	-2.29×10^6
Arachnoidiscus	0	-3.27×10^7
Cyclotella	0	-23308.6

The Laplacian z eigenvalue as the trace of the Hessian z eigenvalues was calculated (Table 12.9). The Laplacian z eigenvalue mirrored the first value for Hessian z eigenvalues for *Arachnoidiscus* and *Cyclotella* at equilibrium, but the value for *Actinoptychus* was near zero (Table 12.9). At non-equilibrium the Laplacian eigenvalue was close to the first values for the Hessian z for all genera (Table 12.9).

Hessian *z* 1st Eigenvalues

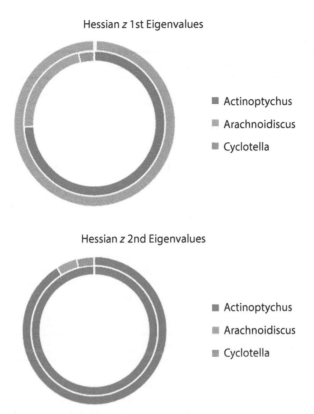

- Actinoptychus
- Arachnoidiscus
- Cyclotella

Hessian *z* 2nd Eigenvalues

- Actinoptychus
- Arachnoidiscus
- Cyclotella

Figure 12.5 First and second eigenvalues of the Hessian *z* comparing *Actinoptychus*, *Arachnoidiscus* and *Cyclotella* at *u*, *v* = 0 and *u*, *v* = 6.28.

12.4.2   Evolvability across Genera

Phenotypic inertia as fitness function values for each genus resulted in all positive sums (Table 12.11). Values ranged from the smallest to largest from *Actinoptychus* to *Arachnoidiscus* to *Cyclotella*. For phenotypic robustness, *Arachnoidiscus* had equilibrium and non-equilibrium values that were positive at 2.0×10^8 and 0.088, *Actinoptychus* had values that were negative at -6728.6 and -10135.4, and *Cyclotella* had negative and positive values at -9835.2 and 24617.2. Phenotypic stability values ranged from *Arachnoidiscus* at 595160.2 to *Cyclotella* at 467.7 to *Actinoptychus* at -1087.5 (Table 12.11).

Evolvability as an additive fitness function for each genus produced the largest positive value for *Arachnoidiscus*, with *Cyclotella* having a much smaller positive number, and *Actinoptychus* having a smaller negative number (Table 12.11). Evolvability of each genus was depicted in a bar plot showing the increase or decrease in evolvability for each genus (Figure 12.6). *Arachnoidiscus* had the largest increase, with *Cyclotella* showing a relative very slight increase, and *Actinoptychus* showing a very small decrease (Figure 12.6). In terms of fitness decrease, evolvability resulted in approximately -1 for *Actinoptychus-Arachnoidiscus* to a very small negative number near zero for *Arachnoidiscus-Cyclotella* to almost seven for *Actinoptychus-Cyclotella* (Table 12.12).

Table 12.11 Values from additive fitness functions for all genera and contribution of each evolvability component.

Phenotypic inertia at u,v = 6.28; values at u,v = 0 are 0.

	Actinoptychus	Arachnoidiscus	Cyclotella	SUM
Γ133	-0.09571	-10.5865	-0.00372	-10.6859
Γ233	46.7439	0.864298	545.13	592.7382
Γ331	-0.37868	148.014	-0.33603	147.2993
Γ332	-0.58908	0.038492	-156.895	-157.446
SUM	45.68043	138.3303	387.8953	571.906

Phenotypic robustness

	Actinoptychus		Arachnoidiscus		Cyclotella		SUM
u,v = 0	0	-30.6	59.68	0	0	0	29.08
u,v = 6.28	-6728.6	-10104.8	2×10^8	0.088	-9835.2	24617.2	2×10^8
SUM	-6728.6	-10135.4	2×10^8	0.088	-9835.2	24617.2	2×10^8

Phenotypic stability

	Actinoptychus	Arachnoidiscus	Cyclotella	SUM
u,v = 0	0.00027	26.85	4.35	31.20027
u,v = 6.28	-1087.49	595133.3	463.351	594509.2
SUM	-1087.49	595160.2	467.701	594540.4
Evolvability	-17905.8	2.01×10^8	15637.6	

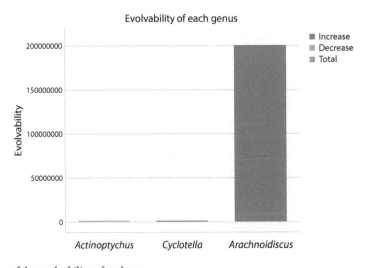

Figure 12.6 Plot of the evolvability of each genus.

12.4.3 Evolvability and the Fitness Function Components of Phenotypic Inertia, Robustness and Stability

For phenotypic inertia, summed values for each Christoffel symbol that contributed to fitness were ordered from $\Gamma 332$ to $\Gamma 133$ as negative values to $\Gamma 331$ to $\Gamma 233$ as positive values at non-equilibrium (Table 12.11). For phenotypic robustness, values ranged from mostly zero at equilibrium to large negative to positive values at non-equilibrium, and these quantities were mirrored as summed values of the Hessian z (Table 12.11). Vectors for the Laplacian in z as

$$\text{phenotypic stability were} \left(\begin{bmatrix} 0.00027 \\ 26.85 \\ 4.35 \end{bmatrix}, \begin{bmatrix} -1087.49 \\ 595133.3 \\ 463.35 \end{bmatrix} \right) \text{ for equilibrium and non-equilibrium,}$$

respectively. Sums at u, $v = 0$ and u, $v = 6.28$ were 31.20027 and 594509.2, respectively (Table 12.11).

Contribution of each evolvability component was depicted in a stacked percentage plot (Figure 12.7). Phenotypic inertia provided the largest contribution for *Cyclotella*, and to a much lesser extent, *Actinoptychus*, while phenotypic robustness and stability provided the largest contribution for *Arachnoidiscus* (Figure 12.7). Comparing all components, phenotypic robustness was shown to be the largest overall contributor (Figure 12.7).

12.4.4 Phenotypic Novelty and Comparison to Evolvability

Novelty of radial plications in the geometric phenotype was determined via information theory. Radial plications were obtained via the z-equation from systems of parametric equations for each genus that were used to construct 3D surface models (Figure 12.8). Joint entropy values were a large negative number for *Arachnoidiscus-Cyclotella* to close to zero for *Actinoptychus-Cyclotella* to almost two for *Actinoptychus-Arachoidiscus* (Table 12.12).

Comparison of evolvability to novelty was accomplished with a stacked percentage plot (Figure 12.9). *Arachnoidiscus* had the largest positive percentage for novelty, while *Cyclotella*

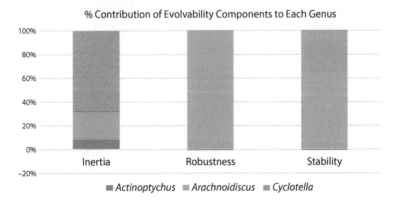

Figure 12.7 Stacked percentage plot of contribution of phenotypic inertia, robustness and stability to evolvability of *Actinoptychus*, *Arachnoidiscus* and *Cyclotella*.

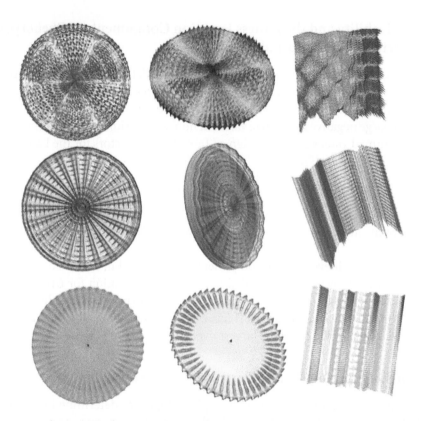

Figure 12.8 3D surface models of *Actinoptychus*, *Arachnoidiscus* and *Cyclotella* with close-ups of z-direction plications. Hessian in z was calculated from the z-equation from systems of parametric equations for each genus model.

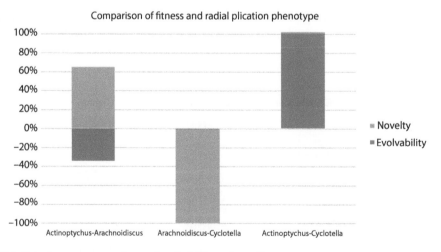

Figure 12.9 Paired genera comparisons of evolvability and novelty.

Table 12.12 Values for comparative evolvability based on fitness decrease and novelty based on joint information between phenotypes.

	Evolvability	Novelty
Actinoptychus-Arachnoidiscus	-1.000089271	1.851305775
Arachnoidiscus-Cyclotella	-7.79498E-05	-216.3311699
Actinoptychus-Cyclotella	6.894236089	0.005577418

had the largest positive percentage for evolvability (Figure 12.9) when compared to the other genera.

12.5 Discussion

In this study, evolvability is about the whole geometric phenotype. Evolvability is about being a fitness function with components of phenotypic inertia, robustness and stability. Novelty is about partial geometric phenotype with regard to a particular architectural or structural characteristic, and information on the surface as the probability of a given architectural or structural characteristic being novel. Evolvability and novelty are comparative measurements which define the context of a given analysis in a global or local sense. Assessment of evolvability and novelty depended on thresholds defined on a scale commensurate with the results that is based on a distribution, norm or some other generalizable metric. This study embodies all of these characteristics.

For *Actinoptychus*, *Arachnoidiscus* and *Cyclotella*, evolvability of the geometric phenotype is the evolvability of architecture or structure that may be applicable in terms of a functional adaptive response. Upper Cretaceous genera such as *Actinoptychus* and *Arachnoidiscus* had frustules that were more heavily silicified than later Cenozoic genera such as *Cyclotella* (e.g., [12.7.61.]). With silica limiting conditions existing since at least the Eocene-Oligocene transition (e.g., [12.7.6.]), *Cyclotella* may be an example of finer radial plications used for predation deterrence to compensate for its thinner siliceous frustules [12.7.42.]. This structural adaptation may be evidence of evolvability as selection acts on the phenotype (e.g., [12.7.32., 12.7.55.]) to induce an adaptive response. Alternatively, finer radial plications of the phenotype may be a non-selective response, induced by developmental constraints (e.g., [12.7.60.]) with respect to less silica availability.

Phenotypic inertia represents resistance to change of the surface phenotype. The highest negative and positive phenotypic inertia values belong to *Cyclotella* (Table 12.3). These values may be construed to indicate large changes in negative and positive curvature, meaning that phenotypic inertia has been overcome, so that *Cyclotella* may have the capacity of evolvability, while *Actinoptychus* and *Arachnoidiscus* have net positive values that are less than *Cyclotella* (Table 12.3). Phenotypic inertia values reflect the amount of resistance to change in curvature with respect to the z-axis via a sweeping over the x, y plane of the valve surface. The sweep occurs from annulus to edge, whereby deflection of the path is either negative (down) or positive (up). The net positive value for *Cyclotella* indicates that the net upward deflection of the geometric phenotype is greater than it is for *Actinoptychus* and *Arachnoidiscus*. Phenotypic inertia for *Cyclotella* is greater and may be an indicator of evolvability (Table 12.3; Figure 12.7). As a measurement of the geometrical path dynamics, Christoffel symbols represent the motion of angular surface changes as torque (e.g., [12.7.65.]) and may be indicative of evolvability for *Cyclotella*.

Phenotypic robustness as persistence (e.g., [12.7.43.]) of the geometric phenotype is indicated by the "thickness" of phenotypic surface features. The building up over time of surface features may be viewed as a response to predation pressure. Minimum slope, or the Hessian z that is negative with respect to movement along a curve in measuring z, is the canalization of the phenotype (e.g., [12.7.63.]) and an increase in robustness. At maximum slope, or when the Hessian z is positive, decanalization occurs (e.g., [12.7.63.]) indicating phenotypic plasticity (e.g., [12.7.84.]), and a decrease in robustness results. If the Hessian z has large positive values as indicative of decanalization and a decrease in robustness, potentially, evolvability may have occurred. However, robustness and evolvability are not necessarily proportional (e.g., [12.7.1., 1.7 .81.].

The summed and net positive highest value in z occurred for *Arachnoidiscus* (Table 12.8). Negative values occurred for *Actinoptychus*, and *Cyclotella* had a mix of negative and positive values (Table 12.8). For the Hessian z, the largest positive eigenvalue at u, $v = 0$ occurred for *Actinoptychus*, while the largest positive first eigenvalue at u, $v = 6.28$ occurred for *Arachnoidiscus* (Table 12.9). For the z-Hessian determinant at equilibrium, negative to zero values occurred, while at non-equilibrium, the largest negative value occurred for *Arachnoidiscus* (Table 12.10). The disparate results between *Arachnoidiscus* and the other genera indicate the large influence that phenotypic robustness has on evolvability. Phenotypic robustness may increase with increasing evolvability by establishment of fitness at the current phenotype with the potential to evolve (e.g., [12.7.26.]) or via mutational robustness at the level of the genotype (e.g., [12.7.33.]).

The geometric phenotype is interpretable in terms of valve surface resilience as an indicator of longevity in the environment. Such phenotypic stability may occur despite changes in nutrient availability, temperature shifts and microzooplankton predation risks. From the eigenvalues for the Laplacian in z, the smallest net value (non-equilibrium – equilibrium) for phenotypic stability occurred for *Cyclotella*, and therefore exhibits an approximate lack of evolvability as compared to *Actinoptychus* and *Arachnoidiscus* (Table 12.9). However, *Cyclotella*'s phenotypic stability measurement came closest to exhibiting a harmonic function on average with respect to its radial plications (Table 12.5). *Actinoptychus* with a net large negative eigenvalue, indicated reduced evolvability in terms of negative curvature on average, while *Arachnoidiscus* with a net large positive eigenvalue indicated expanding evolvability (Figure 12.7) in terms of positive curvature (Table 12.9). However, the geometric phenotype may be maintained despite large changes in positive curvature, and therefore *Arachnoidiscus* may not necessarily have the capacity of evolvability. For surface architecture with a narrowly constrained geometric phenotype in x, y, z, *Arachnoidiscus*'s phenotypic stability net eigenvalue is closer to zero (Table 12.5), and therefore evolvability may not be present. Narrowly defined phenotypic constraints may be indicative of lack of evolvability [12.7.68.]. However, *Arachnoidiscus* may exhibit evolvability with respect to phenotypic stability from the results of this study.

Overall, when considering all the results, the genus that may best exhibit evolvability is *Arachnoidiscus* (Figure 12.6) via phenotypic robustness and stability (Figure 12.7) in contrast to *Cyclotella* exhibiting evolvability via phenotypic inertia (Table 12.3; Figure 12.7). While *Arachnoidiscus* first appeared in the Cretaceous, modern representatives exist (e.g., [12.7.83.], accessed on 3 March 2021). This is not the case for *Actinoptychus*, and *Cyclotella* has representation only in the Cenozoic. However, on a comparative basis for proportional fitness decrease, *Cyclotella* exemplified evolvability when compared with *Actinoptychus*, and *Arachnoidiscus* (Table 12.12; Figure 12.9).

To determine if valve radial plication is a novelty, the fraction of this feature with respect to the total geometric phenotype was measured. By quantifying radial plications geometrically, comparative measurement indicated that with a specified threshold, this phenotypic characteristic could be identified as novel in *Arachnoidiscus* when compared to the other genera (Table 12.12; Figure 12.9). Plications as alternating ribs and flat areas have been a unique characteristic for *Arachnoidiscus* that does not appear to be evolvable (Table 12.12; Figure 12.9). The incongruence of evolvability and novelty measurement is dependent on evolvability being based on separate measures and novelty being based on a paired comparative measure.

Degree of radial plications as part of the geometric phenotype were measured and compared with respect to evolvability. This phenotypic feature may be effective in predation deterrence as differences in kind rather than number of plications. Oceanic silica limiting conditions have not prevented diatom diversification (e.g., [12.7.21., 12.7.27.]) and the potential for evolvability as well as phenotypic novelty. Coevolution of diatoms and microzooplankton grazing predators occurred during the Cenozoic [12.7.31.], and with such grazers coevolving as species-specific predators with respect to diatoms [12.7.74., 12.7.75.], the potential for novel structural changes on diatom valves exists. Coevolution may induce an increase in morphological and functional complexity [12.7.53.] and may be a sign of evolvability as well.

12.6   Summary and Future Research

This study was an illustration of how to measure the geometric phenotype in terms of evolvability characteristics. While only three exemplars were used, the focus was on the phenotypic characteristic of radial plications in diatoms for illustrative purposes. Computational time is intensive which must be taken into consideration when calculating evolvability and its components. Rather than using 3D surface models, evolvability characteristics may be calculated from the genotypic phenotype via the image Jacobian (Chapter 2) of the diatom valve surfaces from digital images. Application of the methodology would be the same whether using models or images. To improve on the results of this study, many genera or lineages may be used in the schema presented to measure multiple taxa and their evolvability status. Again, computational time must be considered when designing a study involving the measurement of evolvability from the geometric phenotype.

Application of algorithmic techniques may prove fruitful in studying evolvability as an optimization problem. Such analyses would require high computational time, and possibly be NP-hard. However, at the very least, computation of geometric phenotypes representing diatom lineages may prove to be insightful with respect to quantifying and analyzing evolvability characteristics and novelty assessments.

Information content regarding novelty may be assessed further using the Kullback-Leibler divergence as $\mathcal{D}(q'_m \| q) = \sum q'_m \log\left(\dfrac{q'_m}{q_m}\right)$ with the change in phenotype information content as $\Delta_m novelty = \mathcal{D}(q' \| q) + \mathcal{D}(q \| q')$, which is the Jeffreys divergence as the sum of differences between genera [12.7.10.]. Additionally, novelty differences could be measured via one of a number of distance metrics, especially concerning lineages. For all assessments of novelty, just as with evolvability, a threshold must be defined within the context of the analysis. For a more formal application, a potential threshold may be obtained by using a Pareto front while treating evolvability or novelty analysis as an optimization problem (e.g., [12.7.23.]).

Novelty may lead to an increase in complexity [12.7.40.], and an increase in complexity may be a sign that evolvability has occurred. For some diatoms, complexity has increased over geologic time [12.7.53.], and this may lead to the evolution of novel whole phenotypes or specific phenotypic surface structures. Evolvability and novelty studies may provide avenues of additional studies with applications of complexity or other macroevolutionary characteristics regarding the diatom phenotype.

12.7 References

12.7.1.   Bauer, C.R., Li, S., Siegal, M. (2015) Essential gene disruptions reveal complex relationships between phenotypic robustness, pleiotropy, and fitness. *Mol. Syst. Biol.* 11, 773.

12.7.2.   Blomberg, S.P. and Garland, Jr., T. (2002) Tempo and mode in evolution: phylogenetic inertia, adaptation and comparative methods. *J. Evol. Biol.* 15, 899-910.

12.7.3.   Bradt, R.C. (2011) The fractography and crack patterns of broken glass. *J. Fail. Anal. and Preven.* 11, 79-96.

12.7.4.   Brown, N.E. (1933) *Arachnoidiscus*, W. Watson and Sons, Ltd., London, United Kingdom.

12.7.5.   Carter, A.J.R., Hermisson, J., Hansen, T.F. (2005) The role of epistatic gene interactions in the response to selection and the evolution of evolvability. *Theoretical Population Biology* 68, 179-196.

12.7.6.   Cermeño, P., Falkowski, P.G., Romero, O.E., Schaller, M.F., Vallina, S.M. (2015) Continental erosion and the Cenozoic rise of marine diatoms. *PNAS* 112(14), 4239-4244.

12.7.7.   Charlesworth, B., Lande, R., Slatkin, M. (1982) A neo-Darwinian commentary on macroevolution. *Evolution* 36(3), 474-498.

12.7.8.   do Carmo, M.P. (1976) *Differential Geometry of Curves and Surfaces*, Prentice-Hall, Englewood Cliffs, New Jersey, USA.

12.7.9.   Ferrara, M.A., Dardano, P., De Stefano, L., Rea, I., Coppola, G., Redina, I., Congestri, R., Atonucci, A., De Stefano, M., De Tommasi, E. (2014) Optical properties of diatom nanostructured biosilica in Arachnoidiscus sp.: micro-optics from mother nature. *PLoS One* 9(7), e103750.

12.7.10. Frank, S.A. (2012) Natural selection. V. How to read the fundamental equations of evolutionary change in terms of information theory. *Journal of Evolutionary Biology* 25, 2377-2396.

12.7.11.   Gould, S.J. and Lewontin, R. (1979) The spandrels of San Marco and the Panglossian paradigm: a critique of the adaptionist programme. *Proc. R. Soc. Lond. B* 205, 581-598.

12.7.12.   Gutiérrez Moreno, J.J., Pan, K., Wang, Y., Li, W. (2020) A computational study of APTES surface functionalization of diatom-like amorphous SiO_2 surfaces for heavy metal adsorption. *ChemRxiv* doi.org/10.26434/chemrxiv.11473080.v1.

12.7.13.   Håkansson, H. and Chepurnov, V. (1999) A study of variation in valve morphology of the diatom Cyclotella meneghiniana in monoclonal cultures: effect of auxospore formation and different salinity conditions. *Diatom Research* 14(2), 251-272.

12.7.14.   Hall, B.K. and Kerney, R. (2011) Levels of biological organization and the origin of novelty. *Journal of Experimental Zoology (Mol. Dev. Evol.)* 314B, 1-10.

12.7.15.   Hallgrímsson, B., Jamniczky, H.A., Young, N.M., Rolian, C., Schmidt-Ott, U., Marcucio, R.S. (2012) *J. Exp. Zool. (Mol. Dev. Evol.)*, 3188, 501-517.

12.7.16.   Hamm, C. (2005) The evolution of advanced mechanical defenses and potential technological applications of diatom shells. *Journal of Nanoscience and Nanotechnology* 5(1), 108-119.

12.7.17.   Hansen, T.F. (2006) The evolution of genetic architecture. *Annu. Rev. Ecol. Syst.* 37, 123-157.

12.7.18.   Hansen, T.F. and Houle, D. (2004) Evolvability, stabilizing selection, and the problem of stasis. In: *Phenotypic Integration: Studying the Ecology and Evolution of Complex Phenotypes*, Pigliucci, M., Preston, K. (eds.), Oxford University Press, Oxford, UK: 130-150.

12.7.19.   Hartle, J.B. http://web.physics.ucsb.edu/~gravitybook/mathematica.html, Accessed on 31 Aug 2018.

12.7.20.   Harwood, D.M. and Nikolaev, V.A. (1998) Cretaceous diatoms: morphology, taxonomy, biostratigraphy. In: *Siliceous Microfossils*, C.D. Blome, P.M. Whalen, K.M. Reed (convenors), Paleontological Society Short Courses in Paleontology, Number 8, University of Tennessee Publication, Knoxville, Tennessee: 81-106.

12.7.21.   Harwood, D.M., Nikolaev, V.A., Winter, D.M. (2007) Cretaceous records of diatom evolution, radiation, and expansion. In: *From Pond Scum to Carbon Sink: Geological and Environmental Applications of the Diatoms, Paleontological Society Short Course*, The Paleontological Society Papers, No. 13, Starratt, SW. (ed.), Paleontological Society, Knoxville, Tennessee: 33-59.

12.7.22.   Hasle, G.R. and Sims, P.A. (1986) The diatom genera *Stellarima* and *Symbolophora* with comments on the genus *Actinoptychus*. *British Phycological Journal* 21(1), 97-114.

12.7.23.   Hu, X.-B., Wang, M., Di Paolo, E. (2013) Calculating complete and exact Pareto front for multiobjective optimization: a new deterministic approach for discrete problems. *IEEE Transactions on Cybernetics* 43(3), 1088-1101.

12.7.24.   Hunter, J.P. (1998) Key innovations and the ecology of macroevolution. *TREE* 13(1), 31-36.

12.7.25.   Huxley, J. (1942) *Evolution: The Modern Synthesis*, George Allen & Unwin Ltd, London UK.

12.7.26.   Jiang, P., Kreitman, M., Reinitz, J. (2018) The relationship between robustness and evolution. https://www.biorxiv.org/content/10.1101/268862v1.

12.7.27.   Jordan, R.W. and Stickley, C.E. (2010) Diatoms as indicators of paleoceanographic events. In: *The Diatoms: Applications for the Environmental and Earth Sciences, 2nd edition*, Smol, J.P. and Stoermer, E.F. (eds.), Cambridge University Press, Cambridge, UK: 424-453.

12.7.28.   Kaczmarska, I., Ehrman, J.M., Bates, S.S. (2001) A review of auxospore structure, ontogeny and diatom phylogeny. In: *Proceedings of the 16th International Diatom Symposium*,

Athens & Agegean Islands, Economou-Amilli, A. (ed.), University of Athens, Greece: 153-168.

12.7.29.   Kaplan, W. (1999) *Maxima and Minima with Applications – Practical Optimization and Duality*, John Wiley & Sons, Inc., New York, USA.

12.7.30.   Kaplan, W. (2003) *Advanced Calculus, 5th edition*, Addison-Wesley, Reading, Massachusetts.

12.7.31.   Katz, M.E., Finkel, Z.V., Grzebyk, D., Knoll, A.H., Falkowski, P.G. (2004) Evolutionary trajectories and biogeochemical impacts of marine eukaryotic phytoplankton. *Annu. Rev. Ecol. Evol. Syst.* 35, 523-556.

12.7.32.   Kirschner, M. (2013) Beyond Darwin: evolvability and the generation of novelty. *BMC Biology* 11:110.

12.7.33.   Klug, A., Park, S.-C., Krug, J. (2019) Recombination and mutational robustness in neutral fitness landscapes. *PLoS Comput. Biol.* 15(8):e1006884.

12.7.34.   Kobayashi A., Osada K., Nagumo T., and Tanaka J. (2001) An auxospore of *Arachnoidiscus ornatus* Ehrenberg. In: *Proceedings of the 16th International Diatom Symposium*, A. Economou-Amilli (ed.), Amvrosiou Press, University of Athens, Athens, Greece: 197–204.

12.7.35.   Koenderink, J.J. (1990) *Solid Shape*, MIT Press, Cambridge, Massachusetts USA.

12.7.36.   Koop, R. (1993) *Global Gravity Field Modelling Using Satellite Gravity Gradiometry*, W.D. Meinema B.V., Delft, The Netherlands.

12.7.37.   Kuzawa, C.W. (2005) Fetal origins of developmental plasticity: are fetal cues reliable predictors of future environments? *American Journal of Human Biology* 17, 5-21.

12.7.38.   Lande, R. (1979) Quantitative genetic analysis of multivariate evolution, applied to brain: body size allometry. *Evolution* 33(1) Part 2, 402-416.

12.7.39.   Lee, J.H. and Chang, M. (1996) Morphological variations of the marine diatom genus *Actinoptychus* in the coastal waters of Korea. *Algae* 11(4), 365-374.

12.7.40.   Lehman, J. and Stanley, K.O. (2011) Abandoning objectives: evolution through the search for novelty alone. *Evolutionary Computation* 19(2), 189-223.

12.7.41.   Liem, K.F. (1990) Key evolutionary innovations, differential diversity, and symecomorphosis. In: *Evolutionary Innovations*, Nitecki, M.H. (ed.), University of Chicago, Chicago, Illinois: 147-170.

12.7.42.   Liu, H., Chen, M., Zhu, F., Harrison, P.J. (2016) Effect of diatom silica content on copepod grasin, growth and reproduction. *Frontiers in Marine Science* 3:89.

12.7.43.   Masel, J. and Siegal, M.L. (2009) Robustness: mechanisms and consequences. *Trends Genet.* 25(9), 395-403.

12.7.44.   Michels, J., Vogt, J., Gorb, S.N. (2012) Tools for crushing diatoms—opal teeth in copepods feature a rubber-like bearing composed of resilin. *Scientific Reports* 2, 465-470.

12.7.45.   Misner, C.W, Thorne, K.S., Wheeler, J.A. (1973) *Gravitation*, W.H. Freeman and Company, San Francisco, California, USA.

12.7.46.   Mitchell, J.G., Seuront, L., Doubell, M.J., Losic, D., Voelcker, N.H., Seymour, J., Lai, R. (2013) The role of diatom nanostructures in biasing diffusion to improve uptake in a patchy nutrient environment. *PLoS ONE* 8(5), e59548.

12.7.47.   Mouret, J.-B. and Doncieux, S. (2012) Encouraging behavioral diversity in evolutionary robotics: an empirical study. *Evolutionary Computation* 20(1), 91-133.

12.7.48.   Müller, G.B. and Newman, S.A. (2005) The innovation triad: an evodevo agenda. *Journal of Experimental Zoology (Mol. Dev. Evol.)* 304B, 487-503.

12.7.49.   Pappas, J.L. (2005a) Geometry and topology of diatom shape and surface morphogenesis for use in applications of nanotechnology. *Journal of Nanoscience and Nanotechnology* 5(1), 120-130.

12.7.50.   Pappas, J.L. (2005b) Theoretical morphospace and its relation to freshwater gomphon-emoid-cymbelloid diatom (Bacillariophyta) lineages. *Journal of Biological Systems* 13(4), 385-398.

12.7.51.   Pappas, J.L. (2008) More on theoretical morphospace and its relation to freshwater gom-phonemoid-cymbelloid diatom (Bacillariophyta) lineages. *Journal of Biological Systems* 16(1), 119-137.

12.7.52.   Pappas, J.L. (2011) Graph matching a skeletonized theoretical morphospace with a clado-gram for gomphonemoid-cymbelloid diatoms (Bacillariophyta) *Journal of Biological Systems* 19(1), 47-70.

12.7.53.   Pappas, J.L. (2016) Multivariate complexity analysis of 3D surface form and function of centric diatoms at the Eocene-Oligocene transition. *Marine Micropaleontology* 122, 67-86.

12.7.54.   Pappas, J.L. (2021) Quantified ensemble 3D surface features modeled as a window on centric diatom valve morphogenesis. In: *Diatom Morphogenesis [DIMO, Volume in the series: Diatoms: Biology & Applications, series editors: Richard Gordon & Joseph Seckbach].* V. Annenkov, J. Seckbach and R. Gordon, (eds.) Wiley-Scrivener, Beverly, MA, USA: 158-193.

12.7.55.   Pappas, J.L. and Miller, D.J. (2013) A generalized approach to the modeling and analysis of 3D surface morphology in organisms. *PLoS ONE* 8(10), e77551.

12.7.56.   Pappas, J.L., Kociolek, J.P., Stoermer, E.F. (2014) Quantitative morphometric meth-ods in diatom research. In: Nina Strelnikova Festschrift. J.P. Kociolek, M. Kulivoskiy, J. Witkowski, and D.M. Harwood, D.M. (eds.), *Nova Hedwigia, Beihefte* 143, 281-306.

12.7.57.   Peterson, T. and Müller, G.B. (2016) Phenotypic novelty in evodevo: the distinction between continuous and discontinuous variation and its importance in evolutionary theory. *Evolutionary Biology* 43, 314-335.

12.7.58.   Pigliucci, M. (2008) Is evolvability evolvable? *Nature Reviews|Genetics* 9, 75-82.

12.7.59.   Pigliucci, M. (2008) What, if anything, is an evolutionary novelty? *Philosophy of Science* 75, 887-898.

12.7.60.   Pigliucci, M. and Kaplan, J. (2000) The fall and rise of Dr Pangloss: adaptationism and the Spandrels paper 20 years later. *TREE* 15(7), 66-70.

12.7.61.   Raven, J.A. and Waite, A.M. (2004) The evolution of silicification in diatoms: inescapable sinking and sinking as escape? *New Phytologist* 162, 45-61.

12.7.62.   Reisinger, J., Stanley, K.O., Miikkulainen, R. (2005) Towards an empirical measure of evolvability. *GECCO '05: Proceedings of the 7th Annual Workshop on Genetic and Evolutionary Computation*, ACM, 257-264, doi.org/10.1145/1102256.1102315.

12.7.63.   Rice, S.H. (1998) The evolution of canalization and the breaking of Von Baer's laws: modeling the evolution of development with epistasis. *Evolution* 52(3), 647-656.

12.7.64.   Round, F.E., Crawford, R.M., Mann, D.G. (1990) *The Diatoms – Biology and Morphology of the Genera*, Cambridge University Press, Cambridge, United Kingdom.

12.7.65.   Safeea, M. Neto, P., Bearee, R. (2019) Robot dynamics: a recursive algorithm for efficient calculation of Christoffel Symbols. *Mechanism and Machine Theory* 142, 103589.

12.7.66.   Sato, S., Nagumo, T., Tanaka, J. (2004) Auxospore formation and the morphology of the initial cell of the marine araphid diatom *Gephyria media* (Bacillariophyceae). *Journal of Phycology* 40, 684-691.

12.7.67.   Schmid, A.-M.M. and Volcani, B.E. (1983) Wall morphogenesis in Coscinodiscus wail-esii I. valve morphology and development of its architecture. *Journal of Phycology* 19(4), 387-402.

12.7.68.   Schwenk, K. and Wagner, G.P. (2001) Function and evolution of phenotypic stability: connecting pattern to process. *Amer. Zool.* 41, 552-563.

12.7.69. Simpson, G.G. (1944) *Tempo and Mode in Evolution*. Columbia University Press, New York, USA.

12.7.70. Sims, P.A., Mann, D.G., Medlin, L.K. (2006) Evolution of the diatoms: insights from fossil, biological and molecular data. *Phycologia* 45(4), 361-402.

12.7.71. Smetacek, V. (2001) A watery arms race. *Nature* 411, 745.

12.7.72. Sochi, T. (2016) *Principles of Differential Geometry* arXiv:1609.02868v1 [math.HO] 9 Sep 2016.

12.7.73. Sternberg, E., Tang, D., Ho, T.-Y., Jeandel, C., Morel, F.M.M. (2005) Barium uptake and adsorption in diatoms. *Geochimica et Cosmochimica Acta* 69(11), 2745-2752.

12.7.74. Tall, L., Cloutier, L., Cattaneo, A. (2006a) Grazer-diatom size relationships in an epiphytic community. *Limnol. Oceanogr.* 51, 1211–1216.

12.7.75. Tall, L., Cattaneo, A., Cloutier, L., Dray, S., Legendre, P. (2006b) Resource partitioning in a grazer guild feeding on a multilayer diatom mat. *J. N. Am. Benthol. Soc.* 25, 800–810.

12.7.76. Tarapore, D. and Mouret, J.-B. (2015) Evolvability signatures of generative encodings: beyond standard performance benchmarks. *Information Sciences* 313(20), 43-61.

12.7.77. Theriot, E.C., Ashworth, M.P., Nakov, T., Ruck, E., Jansen, R.K. (2015) Dissecting signal and noise in diatom chloroplast protein encoding genes with phylogenetic information profiling. *Molecular Phylogenetics and Evolution* 89, 28-36.

12.7.78. Wagner, A. (2017) Information theory, evolutionary innovations and evolvability. *Phil. Trans. R. Soc. B* 372, 20160416.

12.7.79. Wakeman, R.J., Hanspal, N.S., Waghode, A.N., Nassehi, V. (2005) Analysis of pleat crowding and medium compression in pleated cartridge filters. *Chemical Engineering Research and Design* 83(A10), 1246-1255.

12.7.80. Weinberger, H.F. (1965) *A First Course in Partial Differential Equations with Complex Variables and Transform Methods*. Dover Publications, Inc., New York, New York, USA.

12.7.81. Weisstein, E.W. (ed.) (2002) *CRC Concise Encyclopedia of Mathematics, 2nd ed.*, Chapman & tp:/Hall/CRC, London, United Kingdom.

12.7.82. Whitacre, J.M. (2010) Degeneracy: a link between evolvability, robustness and complexity in biological systems. *Theoretical Biology and Medical Modelling* 7:6. http://www.tbiomed.com/content/7/1/6.

12.7.83. WoRMS Editorial Board (2021). World Register of Marine Species. Available from http://www.marinespecies.org at VLIZ. Accessed 2021-03-03. doi:10.14284/170; http://marinespecies.org/aphia.php?p=browser&id=616462, accessed on 3 March 2021.

12.7.84. Yadav, A., Dhole, K., Sinha, H. (2016) Genetic regulation of phenotypic plasticity and canalization in yeast growth. *PLoS ONE* 11(9), e0162326.

12.7.85. Zhang, D., Wang, Y., Zhang, W., Pan, J., Cai, J. (2011) Enlargement of diatom frustules pores by hydrofluoric acid etching at room temperature. *Journal of Material Science* 46, 5665-5671.

Epilogue – Findings and the Future

The chapters of this book represent a compendium of efforts to demonstrate that a mathematical approach to analyzing biological processes and attributes may be used to further our understanding of the macroevolution of diatoms. Each chapter elucidates as well as proposes the need for further studies along the lines of the matters espoused within or as brand-new research. What do each of these chapters contribute to our understanding of diatom macroevolution? The following is a synopsis of the findings and some additional questions and ideas concerning potential future research projects.

Valve shape distribution was demonstrated to be hypergeometric in Chapter 1 based on Legendre polynomials that were used to obtain shape coefficients. Because Legendre polynomials were used, the results from Chapter 1 extend to any of the Jacobi polynomials as well as to Fourier functions. Now, this statement would need to be proven through formal mathematics, even though this chapter takes us through this journey algebraically. Diatom valve shape evolution could be traced from at least the Cretaceous to the Recent with respect to a hypergeometric distribution. How do the valve shapes actually vary over geologic time? What are the recurrent shape patterns over geologic time, if any? Because diatoms are heavily invested in valve shape, such questions may give important insight into diatom morphological evolution.

By selecting morphological features on the valve face, diatom digital images were matched to each other with respect to their current identity, according to the scientific literature. Matching surface models to digital images enabled the degree to which we can ascribe identification with morphological geometric characteristics. In Chapter 2, the idea that a surface model can be distinguished from a digital image and be assigned to one of two species is a surprising outcome. Can it be that artificially assigning points or curves to diatom outlines has gone by the wayside? If we can use digital images to distinguish one species from the next morphologically, then we have no reason to force diatoms into geometric 2D schema that are not representative of the 3D valve face surface. Much more research is obviously needed here.

The overarching theme of Chapter 3 was that machine learning could be used for taxonomic chores. Many, many more digital images on the order of thousands would be needed to have a proper assessment of *Navicula* taxonomy and naviculoid taxa that were the diatoms used for study in Chapter 3. Possibly, there may be a need to establish working groups of, say, this genus or any other genus with many species to circumscribe such taxonomic studies. The goal is to vanquish "junk" taxonomic bins no matter which genus is under consideration. Cooperative digital image databases could be set up so that higher percentages of machine learned genera are obtained. This would be a large, but potentially, rewarding task, if everyone in diatom research were amenable to participating. Diatom experts would have the last word on the results. A combination of machine learning coupled with long standing expertise could be fruitful on the taxonomic front.

Diatom adaptive radiation in the Antarctic is a huge and important topic. The myriad of ways to study adaptive radiation is astounding. The approach given in Chapter 4 showed that stochastic, deterministic and chaotic characteristics of adaptive radiation can be included in a single model. In this case, short term evolution was the outcome being studied, but that does not preclude looking at long term evolution. In the Antarctic, comparison of diatom lineages elicited the result that Bacillariales were favored over Chaetocerotales in terms of adaptive radiation. How do the Bacillariales compare to lineages other than Chaetocerotales? More research is definitely needed here. This chapter along with Chapter 6 on diatom food web dynamics in the Arctic may provide an interesting avenue for research projects and programs concerning two areas of the world important in climate matters and the role that diatoms play.

Diatom diversity is a marvel of the Cenozoic attested to by the importance of diatoms' presence via evolution. In Chapter 5, Cenozoic diatom origination was found to have a larger influence on diversity than extinction, although accumulated switching from origination to extinction produced a larger influence for extinction rather than origination. Origination and extinction were found to be influenced by chaos rather than stochasticity in some cases. The influence of chaos on diatom origination and extinction early (PETM) and later (MMCO) in the Cenozoic contrasts with chaotic influences on diatom diversity at approximately 34 Ma (EOT). Because such events can be parsed in such terms, exploration of the relation among diatom origination, extinction and diversity can be taken in a multitude of directions.

Returning to Chapter 6, diatom food web dynamics in the Arctic was determined to be related to the hydrodynamical regimes concerning ice formation as well as nutrient and bloom dynamics. The upshot of simulation results enabled the association of diatoms to Arctic Ocean conditions that may be useful in climate assessments. Such simulation results are indicative of the dynamical system of the Arctic in which diatoms may be used as indicators of a shifting climate. Studies on diatoms as indicators of the amount of ice present seasonally in the Arctic is of the utmost importance concerning climate. Methodologically, this chapter illustrated the interrelationship among diatoms, their food web dynamics and hydrodynamics which could be applied to other aquatic regimes.

The extent and travels of diatoms were on full display in terms of their biogeography in Chapter 7. Using a combinatorial approach, freshwater diatom clades were optimized as shortest cyclical worldwide routes. As a result, diatoms were shown to be indicators of environmental status in terms of phenotypic integration or related to climate on a worldwide basis. In this chapter, endemism was readily distinguishable from other modes of diatom distribution. This avenue of research could be extended to, e.g., diatom distribution based on phenotypic plasticity. Assessment of endemism may be more thoroughly studied using biogeographic and climatologic methods.

The life cycle illustrates the complicated nature of diatoms in their capacity to exist, propagate and thrive. By modeling the life cycle, as given in Chapter 8, as a series of stages and switches that exhibited similar oscillatory properties, the timing and mode was revealed in a generalized schema. Alterations in initial conditions for each stage may be used to fit particularities to individual diatom taxa. The differences among genera or at the lineage level could be specified and introduced into the modeling schema presented. This may enable direct comparisons of multiple taxa at each stage to gain a better understanding of the life cycle differences.

Usually, morphospaces are depicted as affine or statistically-based spaces. By using networks as graphs and embedding such graphs in a metric space, the connection between morphospaces and phylogenies was accomplished in Chapter 9. The results from this chapter induces the call for more comprehensive studies between the morphospace and phylogenetic pillars of diatom evolution studies. Diatom morphospace occupation was calculated with respect to lineage-based diatom morphospace. Using this approach in diatom morphospace analysis illustrated the necessity of combining morphological and phylogenetic results that may provide further utility in diatom evolution studies.

Diatom morphological complexity in Chapter 10 was tackled and shown to be amenable to analyses based on information theory. Even more telling, complexity as entropy induced the assessment that diatom morphological complexity was ergodic, and more than this, was probabilistically unstable and chaotic in the long run. Cretaceous taxa were determined to be more morphologically complex than Cenozoic taxa. The inference is that there is a ceiling to the maximum complexity that would occur in diatom morphology over time so that chaotic outcomes do not occur. As a measure of diatom morphology, what is the relation between complexity and symmetry? What is the relation between diatom morphological complexity and phylogenetic inference concerning evolution? What is a morphological character in terms of complexity and relation to phylogeny over geologic time? Diatom complexity analysis enables addressing such questions.

Leading up to Chapter 11, diatom symmetry was determined to be a 3D problem to be solved. One has only to look at accounts of 2D diatom symmetry as only "bilateral" or "radial" to see that many diatoms unnecessarily are excluded from being categorized or analyzed with regard to symmetry. The 3D approach enabled the inclusion of multiple surface symmetry categories, including the knot symmetry of *Entomoneis*. The measurement of 3D surface symmetry enabled symmetry groupings and measurement of symmetry breaking in all diatoms. A potential research project could be the role of symmetry in diatom reproduction or the potential relation between complexity, symmetry and the topic of Chapter 12, evolvability.

What is the connection between phenotypic novelty and the evolvability of diatoms? In Chapter 12, phenotypic inertia, robustness and stability were the evolvability characteristics as resistance, persistence and resilience measured from 3D diatom surface models. The phenotypic trait of radial plications was evaluated as phenotypic novelty. In contrast to *Actinoptychus*, *Arachnoidiscus* best exhibited evolvability in terms of phenotypic robustness and stability, while *Cyclotella* exhibited evolvability with regard to phenotypic inertia. Phenotypic novelty as partial evolvability was found to exhibit an increase in complexity, potentially returning to the theme of Chapter 10. Other questions to explore include: What is the relation among evolvability, complexity and symmetry? What is the relation among evolvability, phenotypic novelty and endemism? Many research projects could be devised to address such questions.

There are recurrent themes that emerge from the studies presented in each chapter. Image processing techniques are invaluable in measuring diatom surfaces. Stability, or alternatively, instability as chaos, 3D surfaces, dynamical systems, or optimization were some of the modes of study that recurred throughout. Dynamical systems induce the study of stability, and optimization is a goal for such outcomes. Except for valve shape distribution, there was a reliance on 3D surface models or digital images to obtain data for analysis. In all of the studies, geometric, algebraic, combinatorial, and statistical approaches were used singly or in combination. With a number of studies expressing chaotic characteristics of outcomes or thresholds, the natural progression to self-similarity and fractal studies is evident. The application of knot symmetry is but one surprising hint of this. The limits to life cycle stages as chaotic is another outcome inducing the need for more work on self-similarity in diatom research. With ever more methodological, technological, and algorithmic advances along with access to large data sets, mathematics will increasingly be needed to describe and analyze complicated or multifaceted biological problems such as those posed in this book on diatom morphological macroevolution.

On a final note, while the mathematical approach to diatom macroevolution studies has been used in this book, the same methodologies may be applied to other organisms, combination of organisms, biological systems, or biological processes, including diatoms or not. The generalized *modus operandi* of finding a relation among defined variables and parameters, then describing those relations via geometric and algebraic expressions, combinatorial algorithms, statistical assessments, or some combination enables a plethora of possibilities in analyzing organisms, their interrelations, and the processes that govern evolution across multiple phylogenies and classification levels as well as temporal and spatial scales. The idea is to have analytical and mathematical tools that are not dependent on the particular biological characteristics of the organisms. A goal of this book is to illustrate that a mathematical approach will enable more widely applied methodologies in research that could, in turn, be the impetus for new studies in macroevolution. At least, this is the promise which awaits us.

Index

Also of Interest

Check out these published and forthcoming titles in "Diatoms: Biology and Applications" series from Scrivener Publishing

www.scrivenerpublishing.com

Printed in the USA/Agawam, MA
September 13, 2023

851434.001